D1825143

J. BRENNAN

Protein Kinase C Protocols

METHODS IN MOLECULAR BIOLOGY™

John M. Walker, SERIES EDITOR

METHODS IN MOLECULAR BIOLOGY™

Protein Kinase C Protocols

Edited by

Alexandra C. Newton

Department of Pharmacology, University of California at San Diego, La Jolla, CA

Humana Press ✳ Totowa, New Jersey

© 2003 Humana Press Inc.
999 Riverview Drive, Suite 208
Totowa, New Jersey 07512
www.humanapress.com

This publication is printed on acid-free paper. ∞
ANSI Z39.48-1984 (American Standards Institute) Permanence of Paper for Printed Library Materials.

Production Editor: Robin B. Weisberg.

Cover Illustration: Representation of protein kinase C in its active, membrane-bound conformation, superimposed on image of COS cell expressing green fluorescent (GFP)-tagged protein kinase C. The membrane targeting C1 and C2 domains of PKC are shown engaged on the membrane and bound to the second messengers diacylglycerol and calcium. Small circles represent the three processing phosphorylation sites. The upper left oval represents the exposed pseudosubstrate sequence. Picture is courtesy of Jon Violin.

Cover design by Patricia F. Cleary.

For additional copies, pricing for bulk purchases, and/or information about other Humana titles, contact Humana at the above address or at any of the following numbers: Tel: 973-256-1699; Fax: 973-256-8341; E-mail: humana@humanapr.com, or visit our Website at www.humanapress.com

Printed in the United States of America. 10 9 8 7 6 5 4 3 2 1

Library of Congress Cataloging in Publication Data

Protein kinase C protocols / edited by Alexandra C. Newton.
      p. cm. — (Methods in molecular biology ; v. 233)
   Includes bibliographical references and index.
   ISBN 1-58829-068-9 (alk. paper) 1-59259-397-6 (e-ISBN)
   ISSN 1064-3745
      1. Protein kinase C—Laboratory manuals. I. Newton, Alexandra C. II. Methods in molecular biology (Totowa, NJ) ; 233.

QP606.P76P7353 2003
572'.792—dc21
                                                            2003042346

This book is dedicated to my father, Brian Newton, who introduced me to the science of data collection one idyllic summer on the island of Corfu when I helped him tabulate data for one of his own books on the Greek language, and to my mother, Niki Lardner, who, among everything else, always knew I would get the job done in time.

Preface

Since the discovery that protein kinase C (PKC) transduces the abundance of signals that result in phospholipid hydrolysis, this enzyme has been at the forefront of research in signal transduction. *Protein Kinase C Protocols* covers fundamental methods for studying the structure, function, regulation, subcellular localization, and macromolecular interactions of PKC.

Protein Kinase C Protocols is divided into 11 sections representing the major aspects of PKC regulation and function. Part I contains an introduction and a historical perspective on the discovery of PKC by Drs. Yasutomi Nishizuka and Ushio Kikkawa. Part II describes methods to purify PKC. Part III describes the standard methods for measuring PKC activity: its enzymatic activity and its stimulus-dependent translocation from the cytosol to the membrane. Part IV describes methods for measuring the membrane interaction of PKC in vivo and in vitro. Part V provides methodologies and techniques for measuring the phosphorylation state of PKC, including a protocol for measuring the activity of PKC's upstream kinase, PDK-1. Novel methods for identifying substrates are described in Part VI. Part VII presents protocols for expressing and analyzing the membrane targeting domains of PKC. Part VIII provides a comprehensive compilation of methods used to identify binding partners for PKC. Part IX describes pharmacological probes used to study PKC. The book ends with a presentation of genetic approaches to study PKC (Part X) and a discussion of approaches used to study PKC in disease (Part XI). Each section begins with an introduction placing the protocols in the context of PKC function or regulation.

Protein Kinase C Protocols is targeted to biochemists, cell biologists, and molecular biologists whose research has brought them to PKC, as well as to PKC researchers who may need more information in new and developing methodologies.

Alexandra C. Newton

Contents

Contributors

BHARATH ANANTHANARAYANAN • *Department of Chemistry, University of Illinois at Chicago, Chicago, IL*

JACQUELYN K. BEALS • *Department of Anesthesiology, University of Virginia, Charlottesville, VA*

PETER M. BLUMBERG • *Laboratory of Cellular Carcinogenesis and Tumor Promotion, National Cancer Institute, National Institutes of Health, Bethesda, MD*

RACHEL BRIGHT • *Department of Molecular Pharmacology, Stanford University School of Medicine, Stanford, CA*

CHRISTINE A. CARTWRIGHT • *Department of Medicine, Stanford University School of Medicine, Stanford, CA*

BETTY Y. CHANG • *Department of Medicine, Stanford University School of Medicine, Stanford, CA*

WONHWA CHO • *Department of Chemistry, University of Illinois at Chicago, Chicago, IL*

DOO-SUP CHOI • *Department of Neurology, Ernest Gallo Clinic and Research Center, University of California, San Francisco, Emeryville, CA*

FENG CHU • *Department of Cancer Biology, University of Texas M.D. Anderson Cancer Center, Houston, TX*

MICHELLE DIGMAN • *Department of Chemistry, University of Illinois at Chicago, Chicago, IL*

GERDA ENDEMANN • *Department of Molecular Pharmacology, Stanford University School of Medicine, Stanford, CA*

KAREN ENGLAND • *Biochemistry Department, University College Cork, Cork, Ireland*

ALAN P. FIELDS • *The Mayo Clinic Cancer Center, Jacksonville, FL*

JENNIFER GIORGIONE • *Department of Pharmacology, University of California at San Diego, La Jolla, CA*

MICHAEL GSCHWENDT • *German Cancer Research Center, Heidelberg, Germany*

W. CLAY GUSTAFSON • *The Sealy Center for Cancer Cell Biology, University of Texas Medical Branch, Galveston, TX*

JAMES H. HURLEY • *Laboratory of Molecular Biology, National Institute of Diabetes and Digestive and Kidney Diseases, National Institutes of Health, Bethesda, MD*

ISA M. HUSSAINI • *Department of Pathology, University of Virginia, Charlottesville, VA*

JUDITH R. C. JACOBS • *Joslin Diabetes Center, Harvard Medical School, Boston, MA*

SUSAN JAKEN • *Lilly Research Laboratories, Eli Lilly and Company, Indianapolis, IN*

MARCELO G. KAZANIETZ • *Center for Experimental Therapeutics and Department of Pharmacology, University of Pennsylvania School of Medicine, Philadelphia, PA*

USHIO KIKKAWA • *Biosignal Research Center, Kobe University, Kobe, Japan*

GEORGE L. KING • *Joslin Diabetes Center, Harvard Medical School, Boston, MA*

NANCY E. LEWIN • *Laboratory of Cellular Carcinogenesis and Tumor Promotion, National Cancer Institute, National Institutes of Health, Bethesda, MD*

PATRICIA S. LORENZO • *Cancer Research Center of Hawaii, University of Hawaii, Honolulu, HI*

HIDENORI MATSUZAKI • *Biosignal Research Center, Kobe University, Kobe, Japan*

ROBERT O. MESSING • *Department of Neurology, Ernest Gallo Clinic and Research Center, University of California, San Francisco, Emeryville, CA*

DARIA MOCHLY-ROSEN • *Department of Molecular Pharmacology, Stanford University School of Medicine, Stanford, CA*

HIDEYUKI MUKAI • *Biosignal Research Center, Kobe University, Kobe, Japan*

NICOLE R. MURRAY • *The Mayo Clinic Cancer Center, Jacksonville, FL*

CHRISTOPHER MURRIEL • *Department of Molecular Pharmacology, Stanford University School of Medicine, Stanford, CA*

ERIC A. NALEFSKI • *US Genomics, Woburn, MA*

ALEXANDRA C. NEWTON • *Department of Pharmacology, University of California at San Diego, La Jolla, CA*

YASUTOMI NISHIZUKA • *Biosignal Research Center, Kobe University, Kobe, Japan*

CATHERINE A. O'BRIAN • *Department of Cancer Biology, University of Texas M.D. Anderson Cancer Center, Houston, TX*

M. FOSTER OLIVE • *Department of Neurology, Ernest Gallo Clinic and Research Center, University of California, San Francisco, Emeryville, CA*

YOSHITAKA ONO • *Biosignal Research Center, Kobe University, Kobe, Japan*

AMADEO M. PARISSENTI • *Department of Chemistry and Biochemistry, Laurentian University, Sudbury, Ontario, Canada*

PETER J. PARKER • *Cancer Research UK, London Research Institute, London, UK*

JASON M. PASS • *Departments of Physiology and Medicine, University of California at Los Angeles, Los Angeles, CA*

ANTONIO M. PEPIO • *Elan Pharmaceuticals, San Diego, CA*

PEIPEI PING • *Departments of Physiology and Medicine, University of California at Los Angeles, Los Angeles, CA*

HEIMO RIEDEL • *Division of Cell, Development, and Neurobiology, Department of Biological Sciences, Wayne State University, Detroit, MI*

DORIT RON • *Department of Neurology, Ernest Gallo Clinic and Research Center, University of California, San Francisco, Emeryville, CA*

MARTIN RUMSBY • *Department of Biology, University of York, York, UK*

NAOAKI SAITO • *Biosignal Research Center, Kobe University, Kobe, Japan*

JULIANNE J. SANDO • *Department of Anesthesiology, University of Virginia, Charlottesville, VA*

DEBORAH SCHECHTMAN • *Department of Molecular Pharmacology, Stanford University School of Medicine, Stanford, CA*

KAVITA SHAH • *Genomics Institute of the Novartis Research Foundation, San Diego, CA*

GARRY X. SHEN • *Department of Medicine, University of Manitoba, Winnipeg, Manitoba, Canada*

KEVAN M. SHOKAT • *Department of Cellular and Molecular Pharmacology, University of California, San Francisco, San Francisco, CA, and Department of Chemistry, University of California, Berkeley, Berkeley, CA*

WAYNE S. SOSSIN • *Department of Neurology and Neurosurgery, McGill University, Montreal, Quebec, Canada*

ROBERT V. STAHELIN • *Department of Chemistry, University of Illinois at Chicago, Chicago, IL*

LUISE STEMPKA • *German Cancer Research Center, Heidelberg, Germany*

JUBILEE R. STEWART • *Department of Cancer Biology, University of Texas M.D. Anderson Cancer Center, Houston, TX*

MIKIKO TAKAHASHI • *Biosignal Research Center, Kobe University, Kobe, Japan*

HISAAKI TANIGUCHI • *Institute for Enzyme Research, University of Tokushima, Tokushima and Harima Institute at SPring-8, RIKEN, Sayo, Hyogo, Japan*

ALEX TOKER • *Department of Pathology, Beth Israel Deaconess Medical School, Harvard Medical School, Boston, MA*

THOMAS M. VONDRISKA • *Departments of Physiology and Medicine, University of California at Los Angeles, Los Angeles, CA*

NANCY E. WARD • *Department of Cancer Biology, University of Texas M.D. Anderson Cancer Center, Houston, TX*

KERRIE J. WAY • *Joslin Diabetes Center, Harvard Medical School, Boston, MA*

MICHAEL B. YAFFE • *Center for Cancer Research, Massachusetts Institute of Technology, Cambridge, MA*

TOSHIYOSHI YAMAMOTO • *Biosignal Research Center, Kobe University, Kobe, Japan*

EMIKO YAMAUCHI • *Institute for Enzyme Research, University of Tokushima, Tokushima and Harima Institute at SPring-8, RIKEN, Sayo, Hyogo, Japan*

GONGYI ZHANG • *Laboratory of Molecular Biology, National Institute of Diabetes and Digestive and Kidney Diseases, National Institutes of Health, Bethesda, MD*

JUN ZHANG • *Departments of Physiology and Medicine, University of California at Los Angeles, Los Angeles, CA*

I

INTRODUCTION

1

The Ins and Outs of Protein Kinase C

Alexandra C. Newton

1. Protein Kinase C

The seminal discovery of protein kinase C (PKC) by Nishizuka and co-workers *(1)* in the late 1970s provided the first chapter in the story of one of the most studied enzymes in biology. The subsequent finding that PKC transduces signals that cause lipid hydrolysis, followed shortly thereafter by the discovery that PKC was the long sought-after receptor for the potent tumor-promoting phorbol esters, catapulted PKC to the forefront of research on signal transduction. More than 35,000 research articles have been published on PKC. An abundance of reviews describe the structure, regulation, and biological function of PKC and the interested reader is referred to these for a more detailed overview than the brief one provided below *(1–11)*. The goal of this volume is to present a lab manual on classic and novel techniques that are currently used for studying PKC.

2. PKC Primer

PKC comprises a family of serine/threonine kinases whose members play a pivotal role in cell signaling. Specifically, PKC isozymes transduce the myriad of signals that produce lipid second messengers. Examples of physiological stimuli that activate PKC are mitogens, which operate through tyrosine kinase receptors, or catecholamines, which operate through G protein-coupled receptors. PKC can also be hyperactivated by treating cells with phorbol esters (*see* Chapter 34). These molecules bind PKC over two orders of magnitude more tightly than diacylglycerol, and coupled with their resistance to metabolism, cause essentially constitutive activation of PKC.

From: *Methods in Molecular Biology, vol. 233: Protein Kinase C Protocols*
Edited by: A. C. Newton © Humana Press Inc., Totowa, NJ

There are three subclasses of PKCs, which have in common an amino-terminal regulatory domain linked to a carboxyl-terminal kinase domain (**Fig. 1**). The regulatory domain comprises at least one membrane-targeting module that directs PKC to the membrane in response to generation of lipid second messengers (*see* Chapter 8). Engaging these modules on the membrane activates PKC by providing the energy to release an autoinhibitory pseudosubstrate sequence from the substrate-binding cavity in the kinase domain, thus activating the enzyme (*see* Chapter 5). Conventional (α, βI, βII, γ) and novel (δ, ϵ, η/L, θ) PKCs are allosterically regulated by diacylglycerol, which binds to the C1 domain (this domain is a tandem repeat in most isozymes; *see* **Fig. 1**). Conventional isozymes are under additional control by Ca^{2+}, which binds to the C2 domain and promotes its interaction with anionic phospholipids. Although novel PKCs contain this domain, the Ca^{2+} binding pocket lacks essential aspartates involved in coordinating Ca^{2+} and thus does not bind Ca^{2+}. Atypical PKCs (ζ, $\iota\lambda$) contain a single membrane-targeting module, the C1 domain, but the ligand-binding pocket is compromised so that it is unable to bind diacylglycerol. The function of this domain in atypical PKCs is not known. In addition to second messenger regulation, the activity of all isozymes of PKC is stimulated by phosphatidylserine: this lipid participates in anchoring the C1 domain to the membrane. The structures of the isolated C1 and C2 domains have been solved; that of the kinase domain has been modeled based on the crystal structure of the related protein kinase A (*see* Chapter 23).

Before PKC is competent to respond to second messengers, it must first be phosphorylated at three conserved positions in the kinase core (*see* Chapter 13). The first phosphorylation is catalyzed by the recently discovered phosphoinositide-dependent kinase, PDK-1 (*12*). For conventional and novel PKCs, this phosphorylation is constitutive and part of the maturation process of the enzyme. It serves to correctly position residues in the active site for catalysis without directly activating the enzyme. Activation requires removal of the pseudosubstrate from the active site, which depends on the second messenger-mediated membrane translocation. In contrast, the phosphorylation of atypical PKCs is under moderate regulation by 3' phosphoinositides. These isozymes do not appear to be allosterically regulated by second messengers, and it may be that phosphorylation by PDK-1 serves as the direct on/off switch for enzymatic activity.

The function of PKC is exquisitely sensitive to its subcellular location (*see* Chapter 26). This location is not only dictated by the protein:lipid interactions described above but also by protein:protein interactions. A variety of anchoring proteins for PKC have been described. These proteins tether both inactive and active PKC at specific intracellular locations. A striking example of the importance of such scaffold proteins is in the *Drosophila* phototransductive

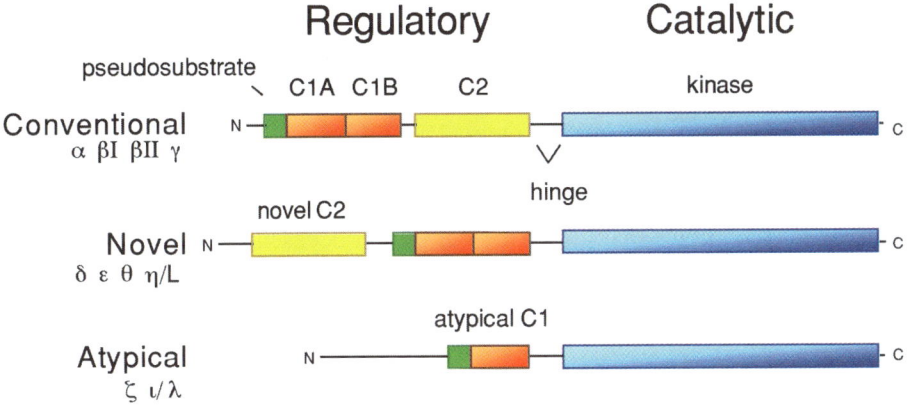

Fig. 1. Schematic showing the domain structure of the conventional, novel, and atypical subclasses of PKC. Indicated are the pseudosubstrate, C1 and C2 domains in the regulatory moiety, and the carboxyl-terminal kinase domain. Note that the kinase domain is sometimes subdivided into C3 and C4 domains representing the ATP-binding and substrate-binding lobes of the kinase (e.g., *see* **Fig. 3** in Chapter 32). Adapted from **ref. 9**.

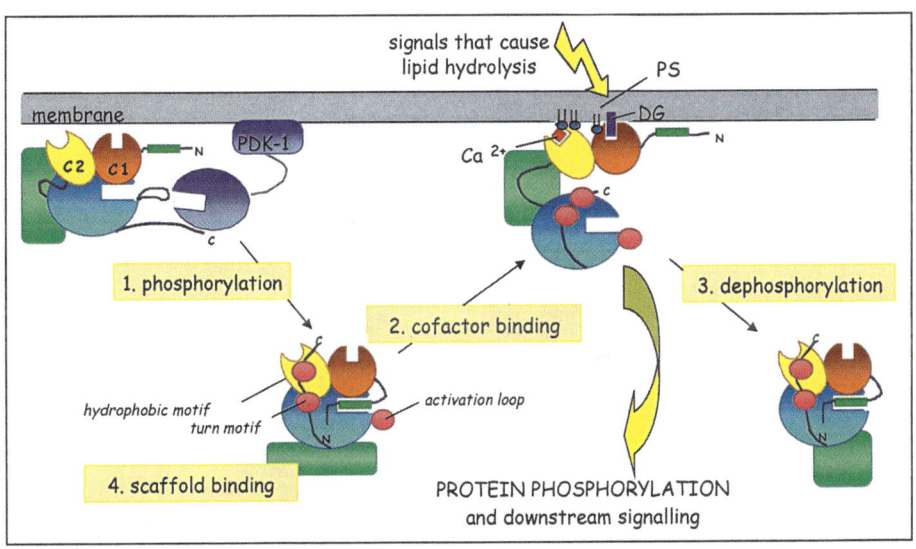

Fig. 2. Model illustrating how the function and subcellular location of protein kinase C is under the coordinated regulation of 1. phosphorylation mechanisms, 2. cofactor binding, 3. activation-dependent dephosphorylation mechanisms, and 4. binding to scaffold proteins. Pink circles represent phosphorylation sites. The upstream kinase, PDK-1, is shown in purple. Shaded green boxes represent scaffold proteins for the various activation states of protein kinase C. Note that engaging the C2 (yellow) and C1 (orange) domains on the membrane locks PKC in the active conformation. In this conformation, the pseudosubstrate (green rectangle) is removed from the substrate-binding cavity in the kinase domain (blue circle) allowing substrate phosphorylation and down-stream signaling.

cascade. In this system, components of the signaling cascade are coordinated on a single scaffold, the inaD protein. Disruption of the interaction of PKC with the scaffold abolishes signaling through this pathway.

Once activated, PKC phosphorylates an abundance of downstream substrate proteins that regulate many distinct cellular processes (*see* Chapter 20). It should be noted, however, that despite two decades of research on PKC, a unifying signaling mechanism centered on PKC has remained elusive. Nonetheless, some themes are emerging. For example, various PKC family members, in particular PKCε and PKCζ have been shown to regulate cell growth and gene transcription by activating the mitogen-activated protein kinase pathway. Similarly, PKC has been implicated in cellular differentiation, such as neurite extension in the rat pheochromocytoma cell line PC12. Activated PKC is very sensitive to dephosphorylation and prolonged activation results in dephosphorylation and eventual proteolysis, a process referred to as downregulation.

The study of PKC has been greatly helped by the use of pharmacological probes, most notably phorbol esters (*see* Chapter 32). Genetic approaches have also provided much insight into the function of PKC (*see* Chapter 36).

References

1. Nishizuka, Y. (1986) Studies and perspectives of protein kinase C. *Science* **233,** 305–312.
2. Blumberg, P. M., Acs, G., Areces, L. B., Kazanietz, M. G., Lewin, N. E., Szallasi. Z. (1994) Protein kinase C in signal transduction and carcinogenesis. *Prog. Clin. Biol. Res.* **387,** 3–19.
3. Nishizuka, Y. (1995) Protein kinase C and lipid signaling for sustained cellular responses. *FASEB J.* **9,** 484–496.
4. Mellor, H. and Parker, P. J. (1998) The extended protein kinase C superfamily. *Biochem. J.* **332,** 281–292.
5. Toker, A. (1998) Signaling through protein kinase C. *Front. Biosci.* **3,** D1134–D1147.
6. Jaken, S. and Parker P. J. (2000) Protein kinase C binding partners. *Bioessays* **22,** 245–254.
7. Parekh, D. B., Ziegler, W., and Parker P. J. (2000) Multiple pathways control protein kinase C phosphorylation. *EMBO J.* **19,** 496–503.
8. Mochly-Rosen, D. and Gordon A. S. (1998) Anchoring proteins for protein kinase C: a means for isozyme selectivity. *FASEB J.* **12,** 35–42.
9. Newton, A.C. and Johnson J. E. (1998) Protein kinase C: a paradigm for regulation of protein function by two membrane-targeting modules. *Biochim. Biophys. Acta* **1376,** 155–172.
10. Ron, D. and Kazanietz M. G. (1999) New insights into the regulation of protein kinase C and novel phorbol ester receptors. *FASEB J.* **13,** 1658–1676.

11. Newton, A. C. (2001) Protein kinase C: structural and spatial regulation by phosphorylation, cofactors, and macromolecular interactions. *Chem. Rev.* **101,** 2353–2364.
12. Toker, A. and Newton A. (2000) Cellular signalling: pivoting around PDK-1. *Cell* **103,** 185–188.

2

Early Studies of Protein Kinase C

A Historical Perspective

Yasutomi Nishizuka and Ushio Kikkawa

1. Introduction

The covalent attachment of phosphate to either seryl or threonyl residues of proteins was identified first by F. Lipmann and P. A. Levene at the Rockefeller Institute for Medical Research (Rockefeller University, New York) in 1932; these researchers were interested in the chemical nature of acidic macromolecules present in the cell nucleus (paranucleic acid was the term used by Levene). This nuclear material was presumably a mixture of what we now call transcription factors. The enzyme responsible for this protein modification, casein kinase (phosvitin kinase), was subsequently found by M. Rabinowitz and Lipmann, but no obvious function was assigned for this enzyme. Another line of study focusing on glycogen metabolism initiated by C. F. Cori and G. T. Cori at St. Louis in the early 1940s, and by eminent investigators, such as E. W. Sutherland, T. W. Rall, E. G. Krebs, E. Fischer, and J. Larner, clarified the role of reversible phosphorylation in controlling the breakdown and resynthesis of glycogen.

In the mid-1960s, Y. Nishizuka spent 1 year as an NIH International Postdoctoral Research Fellow in the laboratory of Lipmann to work on the elongation factors of protein synthesis in *Escherichia coli*. Discussions there about a possible relationship between nuclear phosphoproteins and bacterial adaptive enzymes induced by cyclic AMP sparked Nishizuka's lifelong interest in protein kinases in hormone actions. At the end of the 1960s, when Nishizuka moved to Kobe, Krebs and his colleagues announced that cyclic AMP activates glycogen phosphorylase kinase kinase, known as protein kinase A (PKA) today

From: *Methods in Molecular Biology, vol. 233: Protein Kinase C Protocols*
Edited by: A. C. Newton © Humana Press Inc., Totowa, NJ

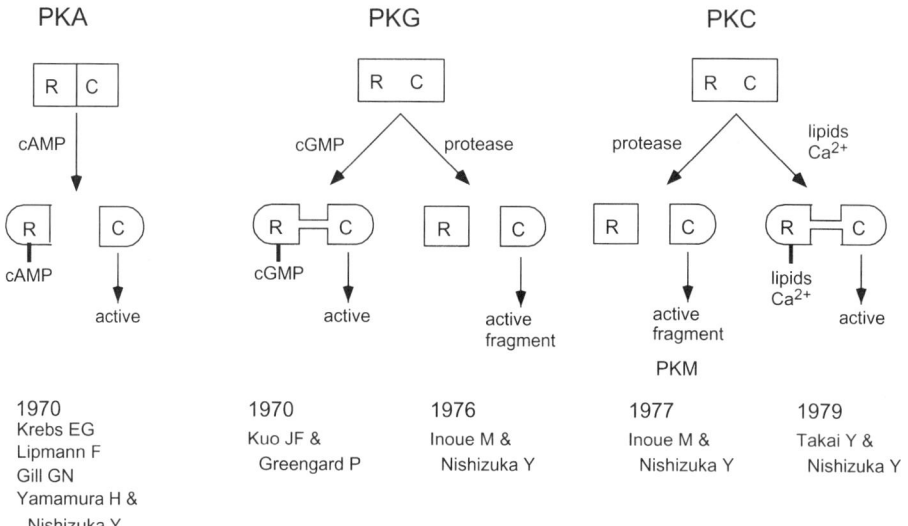

Fig. 1. Mode of activation of three protein kinases.

(1). H. Yamamura and Nishizuka at that time in Kobe isolated a functionally unidentified kinase from rat liver with histone as phosphate acceptor and confirmed that cyclic AMP greatly stimulated its catalytic activity. Soon, in 1970, four laboratories (Krebs, Lipmann, G. N. Gill, and Yamamura and Nishizuka) concurrently reported that PKA consists of catalytic and regulatory subunits and that cyclic AMP activates the enzyme by dissociating these subunits (**Fig. 1**).

The 1970s marked the initiation of several important studies of protein kinases (**Table 1**). In the case of cyclic GMP-dependent protein kinase (PKG), discovered by J. F. Kuo and P. Greengard in the brain in 1970, M. Inoue in the Kobe group found that this enzyme, unlike PKA, is a single polypeptide chain and is activated by cyclic GMP, which binds simply to its regulatory region to promote catalytic activity. They found that a constitutively active enzyme fragment insensitive to the cyclic nucleotide could be generated by limited proteolysis *(2)*. This enabled us to find a new enzyme, Ca^{2+}-activated, phospholipid-dependent protein kinase that is protein kinase C (PKC).

2. PKC and Link to Receptor

Higher levels of an active fragment named protein kinase M (PKM; M for its only known requirement, Mg^{2+}), which was assumed to be derived from PKG, were found in previously frozen rat brain compared with freshly obtained brain, where not much fragment was detected. Freezing and thawing resulted in

Table 1
Discovery of Protein Kinases (1955–1980)

Years	Researchers	Studies
1955	E. Fischer and E. G. Krebs	Glycogen phosphorylase kinase
1960	M. Rabinowitz and F. Lipmann	Casein kinase (phosvitin kinase)
1963	J. Larner	Glycogen synthetase kinase
1964	T. Langan and F. Lipmann	Histone kinase
1968	D. A. Walsh, J. P. Parkins, and E. G. Krebs	Protein kinase A
1970	J. F. Kuo and P. Greengard	Protein kinase G
1977	P. Greengard	Calmodulin-dependent protein kinase
	K. Yagi; D. G. Hartshorne	Myosin light chain kinase
	P. Cohen	Calmodulin in phosphorylase kinase
	Y. Nishizuka and colleagues	Protein kinase C
1978	R. L. Erikson	Src protein kinase
1979	H. Hunter	Protein tyrosine kinase
1980	S. Cohen	Receptor tyrosine kinase (epidermal growth factor receptor)

the appearance of the active fragment, suggesting the presence of a proenzyme susceptible to limited proteolysis, perhaps by a Ca^{2+}-dependent protease, later called calpain *(3)*. We soon noticed the existence of such a putative proenzyme in the brain; curiously, it was not sensitive to any cyclic nucleotide, but produced PKM upon limited proteolysis. Attempts to identify the protease responsible for this proteolysis led us to uncover large quantities of an activating substance associated with the membrane. It was not a protease, but simply anionic phospholipids, particularly phosphatidylserine. Even more curiously, crude phospholipids extracted from brain membranes could support activation of the enzyme in the absence of added Ca^{2+}, whereas pure phospholipids obtained from erythrocyte membranes could not produce any enzyme activation unless a higher concentration of Ca^{2+} was added to the reaction mixture (**Fig. 2**).

Analysis of the lipid impurities on a silicic acid column led us to conclude that diacylglycerol was an essential activator *(4)*. This observation suggested a critical link between protein phosphorylation and the signal-induced hydrolysis of inositol phospholipids that was described by M. R. Hokin and L. E. Hokin with acetylcholine-stimulated pancreatic acinar cells in the early 1950s *(5)*. A ubiquitous distribution of PKC in mammalian and other animal tissues was immediately confirmed by J. F. Kuo in Atlanta. Today, we know that the enzyme is distributed more widely, and it is also extensively studied for its role in yeast, Nematoda, and the fly.

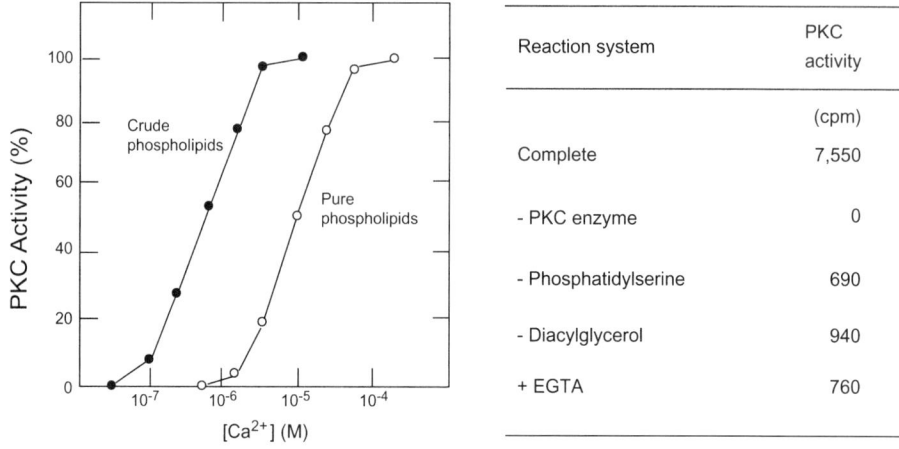

Fig. 2. Activation of PKC by phospholipids, diacylglycerol, and Ca^{2+}.

To obtain evidence that diacylglycerol is the intracellular mediator of hormone actions, we needed a method to activate PKC within intact cells. Diacylglycerols that have two long chain fatty acids could not be readily intercalated into the cell membrane. If, however, one of the fatty acids is replaced with a short chain, acetyl group, then the resulting diacylglycerol, such as 1-oleoyl-2-acetyl-glycerol, obtains some detergent-like properties and could be dispersed into the membrane lipid bilayer and could thus activate PKC directly. Incidentally, two observations reported in the literature attracted our attention. The first was the observation by S. Rittenhouse-Simmons in Boston that diacylglycerol accumulated transiently in thrombin-stimulated platelets, possibly as a result of inositol phospholipid hydrolysis. Second were the articles by P. W. Majerus in St. Louis and by R. J. Haslam in Hamilton both describing that upon stimulation of platelets with thrombin two endogenous proteins with 20- and 47-kDa molecular size became heavily phosphorylated. It was already known that the 20-kDa protein is myosin light chain and is phosphorylated by a specific calmodulin-dependent kinase. Using the fingerprint technique, we discovered that the 47-kDa protein, known as pleckstrin today, was a substrate specific to PKC both in vitro and in vivo. Thus, these two proteins served as excellent markers for the increase of Ca^{2+} and diacylglycerol-dependent activation of PKC, respectively. In the spring of 1980, we were able to show that both Ca^{2+} increase and PKC activation were essential and acted synergistically for the full activation of platelets to release serotonin *(6)*. Similarly, it was possible to show unequivocally that PKC activation is indispensable for neutrophil release reaction and T-cell activation *(7)*, establishing a link between PKC activation and receptor stimulation *(8)*. At that time, we used Ca^{2+} ionophore to increase intracellular concentrations of this cation and did not know where Ca^{2+} may come from, although R. Michell had proposed that inositol phospholipid breakdown may open Ca^{2+} gates *(9)*. In September 1983, at a meeting in Zeist, M. Berridge and his colleagues in Cambridge first described the important inositol 1,4,5-trisphosphate (IP_3) story *(10)*.

3. Phorbol Ester and Exploration

On July 25, 1980 in Brussels, at a garden party in Prof. H. de Wulf's home on the occasion of the Fourth International Conference on Cyclic Nucleotides, Nishizuka had an exciting time discussing with M. Castagna in Villejuif a potential connection between PKC activation and phorbol ester that mimics a variety of hormone actions. Castagna had spent the previous summer in the laboratory of P. Blumberg, who was then at Harvard in Boston, where the phorbol ester receptor was characterized. In August 1981, Castagna joined us in Kobe to test whether PKC had any connection to phorbol ester action. In contrast to usual carcinogens that bind to DNA to produce mutations, phorbol ester appeared to bind to a membrane-associated receptor, eventually leading to gene expression, differentiation, and proliferation *(11,12)*. We had already established experimental systems that were needed for testing whether phorbol ester could cause inositol phospholipid hydrolysis to activate PKC. However, it was extremely disappointing to find that in platelets phorbol ester did not show any evidence of producing diacylglycerol. Instead, this tumor promoter induced remarkable phosphorylation of the endogenous 47-kDa protein. We interpreted this to mean that our already published idea that diacylglycerol is the mediator for PKC activation was not correct.

During a sleepless night, reading the review article written by Blumberg *(11,12)*, an idea occurred: what if phorbol ester could activate PKC directly because it contains a diacylglycerol-like structure very similar to the membrane-permeant lipid molecule that we had used (**Fig. 3**)? This revelation occurred at the end of August. A series of subsequent experiments performed that fall were able to show that phorbol ester mimicked diacylglycerol action by increasing the affinity of PKC for Ca^{2+} and phosphatidylserine, thereby activating the enzyme directly, eventually leading to cellular responses *(13)*. In March 1982, these results were presented at the UCLA-NCI Symposium, Evolution of Hormone Receptor System, organized by R. A. Bradshaw and G. N. Gill in Squaw Valley. This talk under the chair of G. Todaro from NCI obviously attracted great attention, and the results were confirmed immediately. In the following year, several groups of investigators showed that PKC and the phorbol ester receptor can be copurified (J. E. Niedel, P. Blumberg, J. J. Sando, and C. L. Ashendel), and U. Kikkawa described the stoichiometric binding of

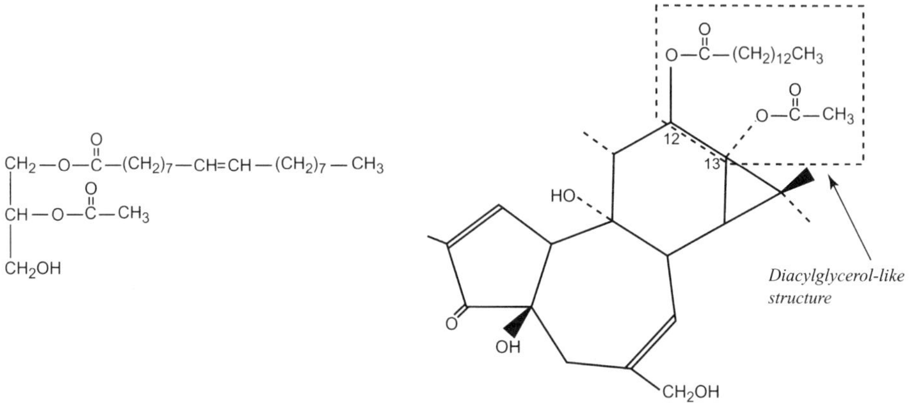

CH$_2$—O—C(=O)—(CH$_2$)$_7$—CH=CH—(CH$_2$)$_7$—CH$_3$
CH—O—C(=O)—CH$_3$
CH$_2$OH

1-Oleoyl-2-acetyl-glycerol
Membrane permeant Diacylglycerol

12-*O*-Tetradecanoylphorbol-13-acetate
Phorbol ester

Diacylglycerol-like structure

Fig. 3. The structure of synthetic diacylglycerol and phorbol ester. The chemical structure of the tumor promoter was previously identified by E. Hecker and B. L. Van Duuren in the late 1960s.

phorbol ester to PKC using the enzyme in a pure form *(14)*. Before long it was shown that phorbol ester could cause translocation of PKC from the cytosol to the membrane (W. B. Anderson). As a result, the traditional concept of tumor promotion proposed originally by I. Berenblum from Oxford as early as 1941 had been replaced by an explicit biochemical explanation that centered on understanding the role of PKC. Phorbol esters and membrane-permeant diacylglycerols, including dioctanoylglycerol, later developed by R. M. Bell, have since been used as crucial tools for the manipulation of PKC in intact cells and have allowed the wide range of cellular processes regulated by this enzyme to be determined *(15)*. It was realized much later, however, that phorbol ester can bind to other cellular proteins, such as chimaerin *(16)* and RasGRP3 *(17)*, and potentially affect cell functions through additional targets.

4. Structural and Functional Diversity

A more detailed molecular understanding of PKC came after the cloning and sequencing of the enzyme in the mid-1980s. On the occasion of the J. Folch's Memorial Colloquium on Inositol Phospholipids, organized by J. N. Hawthorne at Nottingham in September 1981, P. Cohen introduced P. Parker in his laboratory at Dundee to Nishizuka. Since then, their paths crossed frequently and, in October 1985, after a seminar on PKC, at the laboratory of M. Waterfield in the Ludwig Institute London, Parker and Nishizuka both agreed that PKC may not be a single entity because the enzyme often shows double, sometimes triple bands upon gel electrophoresis. Complete sequences of several isoforms thus appeared from both laboratories in the next year *(18,19)*. We now know that the PKC family consists of at least 10 isoforms encoded by nine genes. The structural feature and several functional domains of each isoform are well investigated as repeatedly documented in excellent reviews *(20–22)*. In addition, protein kinases that share kinase domains closely related to the PKC family have been isolated and characterized *(23)*.

It has also become clear in the last decade from numerous investigations in many places, especially those led by A. Newton in San Diego and by Parker in London, that the mechanism of activation of the PKC family enzymes is far more complicated than we initially thought. Newton has shown elegantly that the newly synthesized PKC appears inert and is maturated by phosphorylation by itself and also by other kinases, including phosphoinositide-dependent kinase 1 *(22)*. Thus a cross-talk has emerged between the signal pathways of inositol phospholipid hydrolysis and phosphatidylinositol 3-kinase activation, which was described first by L. Cantley in Boston 10 years ago *(24)*. Another cross-talk for PKC activation through tyrosine phosphorylation of the enzyme molecule is becoming clearer *(25)*.

The catalytic and functional competence of the isoforms appears also to depend on the specific intracellular localization of each, namely their translocation or targeting to particular compartments, such as plasma membrane, Golgi complex, and cell nucleus, as directed by lipid mediators. In other words, targeting appears to represent the activation and isoform-specific functions at the site of destination. Curiously, such targeting is sometimes oscillating. The dynamic behavior of the PKC isoforms was visualized first by N. Saito in our Kobe group with the enzymes fused to green fluorescent protein *(26)*. The structures of some of these targeting domains in the enzyme molecule have been clarified *(27)*.

Indeed, it has been a while since we proposed phospholipase A_2 as an additional player in signal transduction because free fatty acids, especially arachidonic acid, and lysophospholipids are frequently synergistic with diacylglycerol to activate PKC both in vitro as well as in vivo systems *(28)*. It is plausible then that, in addition to diacylglycerol, many of the lipid products produced transiently in membranes by the action of phospholipases A_2 and D, sphingomyelinase, and phosphatidylinositol 3-kinase also play key roles in the translocation and targeting of PKC family members and related protein kinases. These lipid products could direct PKC to distinct intracellular compartments to perform their specific biological functions, for example by producing so-called lipid rafts through lipid–protein and/or protein–protein interactions.

Such a fascinating idea was first proposed by D. Mochly-Rosen *(29)* and S. Jaken *(30)*.

5. Coda

For many years, the cell membrane was thought to be a biologically inert entity that splits the exterior and interior cellular compartments. Inositol was recognized as a constituent of plant tissues in the 19th century, but was found in the mammalian brain by D. W. Woolley in 1941. The following year, J. Folch and Woolley at Rockefeller University in New York described the chemical structure of inositol phospholipid. In the 60 years since then, our knowledge as to the biological role of membrane lipids as well as of protein kinases in cell-to-cell communication has expanded enormously, as briefly described above. More developments in the PKC story will certainly be unveiled in the future, especially for medical and therapeutic usage; for example, for treatment of cancer (A. Fields) and diabetes (G. King).

Acknowledgments

We acknowledge all our colleagues who participated in the early studies on the discovery of PKC. We wish to dedicate this chapter to our friend, Alexandra Newton, the editor of this book; we have been greatly admiring her beautiful work on PKC for many years.

References

1. Walsh, D. A., Perkins, J. P., and Krebs, E. G. (1968) An adenosine 3′,5′-monophosphate-dependent protein kinase from rabbit skeletal muscle. *J. Biol. Chem.* **243,** 3763–3765.
2. Inoue, M., Kishimoto, A., Takai, Y., and Nishizuka, Y. (1976) Guanosine 3′:5′-monophosphate-dependent protein kinase from silkworm, properties of a catalytic fragment obtained by limited proteolysis. *J. Biol. Chem.* **251,** 4476–4478.
3. Inoue, M., Kishimoto, A., Takai, Y., and Nishizuka, Y. (1977) Studies on a cyclic nucleotide-independent protein kinase and its proenzyme in mammalian tissues. II. Proenzyme and its activation by calcium-dependent protease from rat brain. *J. Biol. Chem.* **252,** 7610–7616.
4. Takai, Y., Kishimoto, A., Kikkawa, U., Mori, T., and Nishizuka, Y. (1979) Unsaturated diacylglycerol as a possible messenger for the activation of calcium-activated, phospholipid-dependent protein kinase system. *Biochem. Biophys. Res. Commun.* **91,** 1218–1224.
5. Hokin, M. R. and Hokin, L. E. (1953) Enzyme secretion and the incorporation of ^{32}P into phospholipids of pancreatic slices. *J. Biol. Chem.* **203,** 967–977.
6. Kawahara, Y., Takai, Y., Minakuchi, R., Sano, K., and Nishizuka, Y. (1980) Phospholipid turnover as a possible transmembrane signal for protein phosphoryla-

tion during human platelet activation by thrombin. *Biochem. Biophys. Res. Commun.* **97,** 309–317.

7. Kaibuchi, K., Takai, Y., and Nishizuka, Y. (1985) Protein kinase C and calcium ion in mitogenic response of macrophage-depleted human peripheral lymphocytes. *J. Biol. Chem.* **260,** 1366–1369.

8. Nishizuka, Y. (1984) The role of protein kinase C in cell surface signal transduction and tumour promotion. *Nature* **308,** 693–697.

9. Michell, R. H. (1975) Inositol phospholipids and cell surface receptor function. *Biochim. Biophys. Acta* **415,** 81–147.

10. Streb, H., Irvine, R. F., Berridge, M. J., and Schulz, I. (1983) Release of Ca^{2+} from a nonmitochondrial intracellular store in pancreatic acinar cells by inositol-1,4,5-trisphosphate. *Nature* **306,** 67–69.

11. Blumberg, P. M. (1980) In vitro studies on the mode of action of the phorbol esters, potent tumor promoters: part 1. *Crit. Rev. Toxicol.* **8,** 153–197.

12. Blumberg, P. M. (1981) In vitro studies on the mode of action of the phorbol esters, potent tumor promoters, part 2. *Crit. Rev. Toxicol.* **8,** 199–234.

13. Castagna, M., Takai, Y., Kaibuchi, K., Sano, K., Kikkawa, U., and Nishizuka, Y. (1982) Direct activation of calcium-activated, phospholipid-dependent protein kinase by tumor-promoting phorbol esters. *J. Biol. Chem.* **257,** 7847–7851.

14. Kikkawa, U., Takai, Y., Tanaka, Y., Miyake, R., and Nishizuka, Y. (1983) Protein kinase C as a possible receptor protein of tumor-promoting phorbol esters. *J. Biol. Chem.* **258,** 11442–11445.

15. Kikkawa, U. and Nishizuka, Y. (1986) The role of protein kinase C in transmembrane signalling. *Annu. Rev. Cell Biol.* **2,** 149–178.

16. Caloca, M. J., Fernandez, N., Lewin, N. E., Ching, D., Modali, R., Blumberg, P. M., et al. (1997) β2-Chimaerin is a high affinity receptor for the phorbol ester tumor promoters. *J. Biol. Chem.* **272,** 26488–26496.

17. Lorenzo, P. S., Kung, J. W., Bottorff, D. A., Garfield, S. H., Stone, J. C., and Blumberg, P. M. (2001) Phorbol esters modulate the Ras exchange factor RasGRP3. *Cancer Res.* **61,** 943–949.

18. Coussens, L., Parker, P. J., Rhee, L., Yang-Feng, T. L., Chen, E., Waterfield, M. D., et al. (1986) Multiple, distinct forms of bovine and human protein kinase C suggest diversity in cellular signaling pathways. *Science* **233,** 859–866.

19. Ono, Y., Kurokawa, T., Fujii, T., Kawahara, K., Igarashi, K., Kikkawa, U., et al. (1986) Two types of complementary DNAs of rat brain protein kinase C. Heterogeneity determined by alternative splicing. *FEBS Lett.* **206,** 347–352.

20. Toker, A. (1998) Signaling through protein kinase C. *Front. Biosci.* **3,** D1134–D1147.

21. Parekh, D. B., Ziegler, W., and Parker, P. J. (2000) Multiple pathways control protein kinase C phosphorylation. *EMBO J.* **19,** 496–503.

22. Newton, A. C. (2001) Protein kinase C: structural and spatial regulation by phosphorylation, cofactors, and macromolecular interactions. *Chem. Rev.* **101,** 2353–2364.

23. Mellor, H. and Parker, P. J. (1998) The extended protein kinase C superfamily. *Biochem. J.* **332,** 281–292.

24. Toker, A. and Cantley, L. C. (1997) Signalling through the lipid products of phosphoinositide-3-OH kinase. *Nature* **387,** 673–676.

25. Kikkawa, U., Matsuzaki, H., and Yamamoto, T. (2002) Activation mechanisms and functions of protein kinase Cδ. *J. Biochem.* **132,** 831–839.

26. Sakai, N., Sasaki, K., Ikegaki, N., Shirai, Y., Ono, Y., and Saito, N. (1997) Direct visualization of the translocation of the γ-subspecies of protein kinase C in living cells using fusion proteins with green fluorescent protein. *J. Cell Biol.* **139,** 1465–1476.

27. Hurley, J. H. and Misra, S. (2000) Signaling and subcellular targeting by membrane-binding domains. *Annu. Rev. Biophys. Biomol. Struct.* **29,** 49–79.

28. Nishizuka, Y. (1995) Protein kinase C and lipid signaling for sustained cellular responses. *FASEB J.* **9,** 484–496.

29. Mochly-Rosen, D. and Gordon, A. S. (1998) Anchoring proteins for protein kinase C: a means for isozyme selectivity. *FASEB J.* **12,** 35–42.

30. Jaken, S. (1996) Protein kinase C isozymes and substrates. *Curr. Opin. Cell Biol.* **8,** 168–173.

II

Expression and Purification

3

Expression and Purification
of Protein Kinase C from Insect Cells

Hideyuki Mukai and Yoshitaka Ono

1. Introduction

Isolation of protein kinase C (PKC) cDNAs has led to the identification of 10 related gene products. Some of these isoforms have been reported to have been purified to homogeneity by standard chromatographic techniques from native tissues, but the purity of each isoform of PKC preparation must be limited due to the potential contamination of known or unknown isoforms and related kinases. We have expressed each recombinant isoform of PKC and its related protein kinase PKN by using the baculovirus expression system in eukaryotic insect cells. The advantages of this system are the potential for high yields and the production of defined, isotype pure PKCs and PKNs. We have also combined this system with a specific affinity chromatography system, which helps isolate the highly purified active recombinant enzyme easily and quickly.

The baculovirus *Autographa californica* nuclear polyhedrosis virus (AcNPV) has become a popular vehicle for the cloning and expression of recombinant proteins in insect cells. The transcriptional activity in the very late phase of insect cells infected with AcNPV is dedicated to the polyhedrin and p10 promoters, which makes them ideal for use in driving the high-level expression of introduced foreign genes that replace these viral genes. Uninfected insect cells does not contain any measurable PKC activity or phorbol ester binding activity *(1,2)*, and the posttranslational modifications of the gene products of these viruses, such as myristoylation, phosphorylation, amidation, and fatty acid acylation, occur in virus-infected insect cells as well as in mammalian cells. Thus, this expression system seems to be very useful for obtaining

From: *Methods in Molecular Biology, vol. 233: Protein Kinase C Protocols*
Edited by: A. C. Newton © Humana Press Inc., Totowa, NJ

sufficient amounts of functional PKCs and related kinases, especially in cases where activity is known to be regulated by phosphorylation.

AcNPV has a large (130 kb) circular double-stranded DNA genome with multiple recognition sites for many restriction endonucleases and does not permit easy manipulation for the insertion of foreign genes. Therefore, the strategy usually used depends on homologous recombination between viral DNA and an appropriate bacterial plasmid, so-called transfer vector when cotransfected in insect cells. The transfer vector consists of the recombinant gene inserted downstream of the polyhedrin promoter flanked by the same sequences that flank the polyhedrin gene in the intact virus. In vivo recombination between the homologous flanking sequences on the virus and transfer vector results in the replacement of the polyhedrin gene with the foreign gene. In the early years, 0.1 to 1% of the resulting progeny were recombinant, with the heterologous gene inserted into the genome of the parent virus by homologous recombination in vivo. To improve the frequency of recombination, a number of modifications to the virus genome were performed. Linearization of the baculovirus genome at the unique restriction enzyme site inserted at the polyhedrin locus drastically reduced the infectivity of the virus, and the recombination frequency was improved to nearly 30% by using a parent virus that was linearized at the unique site located near the target site for insertion of the foreign gene into the baculovirus genome *(3)*. Further modification was to disrupt an ORF 1629, an essential gene for virus replication in cell culture. Infectious virus can, therefore, only be produced by recombination with a transfer vector containing the intact ORF 1629, along with the foreign gene under the control of the polyhedrin promoter. Using this method, recombination frequencies reach 80–100% *(4)*. Nowadays, this kind of pretreated virus DNA and various transfer vectors that allow insertion of a foreign gene into a multiple-cloning site downstream of the polyhedrin promoter are commercially available (Clontech, Novagen, Pharmingen, etc.). Most of these transfer vectors contain an intact ORF 1629 and can, therefore, be used in cotransfection with the pretreated linearized virus DNA.

Some methods have been proposed to permit the cloning of foreign genes into the virus genome outside of the insect cells and the introduction of only modified recombinant virus DNA into insect cell culture. There are at least four types of technology: 1) recombination in *Saccharomyces cerevisiae (5)*; 2) enzymatic in vitro recombination using Cre-lox *(6)*; 3) direct cloning at a unique restriction site *(7)* or specific vector-compatible overhangs *(8)*; and 4) site-specific transposition in *Escherichia coli (9)*. (Types 3 and 4 are now commercially available; Novagen's pBac-LIC and Invitrogen's Bac-to-Bac, respectively.) These technologies offer greater flexibility and reduce the time it takes for further isolation of a single recombinant virus by plaque assay because recombinant viral DNA can be checked before its introduction into insect cells. These methods are effective and have technical superiority to recombination. However, we usually use the recombination system in Sf9 cells for expressing various proteins, including isoforms of PKC and PKN, because many kinds of transfer vectors that can be easily manipulated and various virus DNA optimized for expression of foreign gene are now commercially available.

Purification of recombinant proteins from culture cells has been facilitated by the development of techniques that allow specific affinity purification. Glutathione S-transferase (GST) fusion proteins can be purified to >90% in a single chromatographic step under mild elution conditions (10 mM glutathione), which preserve functionality of proteins *(10)*. By applying this GST purification system to isolate recombinant PKC and PKN in insect cells, we can easily prepare abundant active enzymes quickly. Further purity of these enzymes can be obtained by addition of 6x His tag/Ni-NTA system *(11)* as another affinity chromatography step.

2. Materials

2.1. Maintenance of Sf9 Cells

1. EX-CELL 420 medium (JRH Bioscience).
2. Fetal bovine serum (Gibco).
3. Antibiotic-antimycotic (×100, Gibco)
4. 28°C incubator.
5. 60-mm, 100-mm culture dishes.
6. Erlenmeyer flask (500 mL).
7. Orbital incubator shaker (Taitek).

2.2. Virus DNA and Transfer Vectors

1. BacVector 3000 triple cut virus DNA (Novagen).
2. pBlueBacHis (Invitrogen).
3. pGEX4T-1 (Pharmacia).
4. pAcGHLT (Pharmingen).
5. cDNA clones for isoforms of PKC and PKN.
6. Reagents for plasmid isolation from *E. coli* (CsCl, NaAc, etc.).

2.3. Screening of Recombinant Viruses

1. Pasteur pipette.
2. Polystyrene tube (6 mL; Falcon 2058).
3. BacPlaque agarose (Novagen).
4. Eufectin reagent (Novagen).
5. Neutral Red solution (0.33% in phosphate-buffered saline; Sigma).
6. Anti-GST antibody (Santa Cruz), anti-His tag antibody (QIAGEN).

2.4. Purification of the Recombinant Proteins

1. Buffer A: 50 m*M* Tris-HCl pH 7.5, 1 µg/mL of leupeptin, 1 m*M* ethylenebis (oxyethylenenitrilo) tetraacetic acid, 1 m*M* ethylenediamine tetraacetic acid, 3 m*M* MgCl$_2$, 1 m*M* dithiothreitol.
2. Buffer B: 50 m*M* Tris-HCl, pH 8.0, 1 µg/mL of leupeptin, 1 m*M* ethylenediamine tetraacetic acid, 3 m*M* MgCl$_2$, 1 m*M* dithiothreitol, 10 m*M* glutathione.
3. Buffer C: 50 m*M* Tris-HCl, pH 8.0, 1 µg/mL of leupeptin, 1 m*M* MgCl$_2$, 0.3 *M* NaCl, 10 m*M* β-ME.

3. Methods

The methods described here outline 1) the construction of baculovirus for expression of His$_6$ and GST-fused isoforms of PKC and PKN; 2) the expression and purification of these protein; and 3) the characterization of isoforms.

3.1. Maintenance of Sf9 Cells (see Note 1)

Sf9 cells have been adapted to grow in 5% FCS/EX-CELL420 (JRH Biosciences) medium. Fungi can easily grow in the medium, and thus, we usually use antimycotic drug such as amphotericin in addition to Penicillin G and streptomycin in the medium. We routinely maintain these cells at 5×10^5 -2×10^6 cells/mL in a 100-mL culture scale in a 500-mL shake flask operating at 150 rpm at 28°C in atmospheric air. When we infect a virus and express target proteins in Sf9 cells, we spread an aliquot of shake culture to 10-cm dish (1×10^7/dish) and add virus at the indicated titer (*see* **Note 2**).

3.2. Virus DNA

We used BacVector-3000 triple cut virus DNA (Novagen) for construction of baculovirus for expression of the isoforms of PKC and PKN. This virus DNA lacks the polyhedrin gene and seven additional nonessential genes, including cysteine protease and a chitinase, to eliminate proteins that could compete with target genes for cellular resources and thus be deleterious to the expression of some gene products. This vector is compatible with many commercially available transfer plasmids used for gene insertion at the polyhedrin locus.

3.3. Transfer Vectors

3.3.1. Backbone Vectors

The following vectors were used for the production of GST alone and His$_6$-GST alone, as well as in conjunction with polypeptide coding for the full length and the catalytic domain of isoforms of PKC [α*(12)*, βI, βII*(13,14)*, γ*(12)*, δ, ε, ζ *(15,16)*], and PKN [α*(17)*, β*(18)*, γ/PRK2*(19)*; *see* **Note 3**]. High amounts of each vector can be recovered by conventional plasmid isolation techniques from bacterial culture. Successful recombination efficiency seems to depend on the purity of the transfer vectors. Plasmid prepared from standard preparation techniques, such as PEG precipitation or QIAGEN prep kit, usually work. If positive plaques are not obtained in a first trial, however, we recommend purifying transfer DNA by CsCl gradient preparation method.

3.3.1.1. PBLUEBACHIS

pBlueBacHis vector is derived from pJVETL-Z *(20)* and contains twin promoters derived from the early transcript large (ETL) and polyhedrin genes of AcNPV. The ETL promoter directs the synthesis of β-galactosidase whereas the polyhedrin promoter controls the synthesis of foreign gene products. The ATG start codon is followed by an open reading frame coding for a polyhistidine (His$_6$) metal binding domain and enterokinase cleavage site. Next is a multiple cloning site.

3.3.1.2. PBLUEBACHISGST (PR538)

pBlueBacHisGST (pR538; **Fig. 1A**) was made by subcloning the coding region for GST and a thrombin recognition site in frame to the multiple cloning site of pBlueBacHis-B. Thus, the pBlueBacHisGST (pR538) multiple cloning site contains unique *Bam*HI, *Sma*I, and *Hin*dIII, and the reading frame is the same as pGEX4T-1 (expression vector for GST fusion protein in *E. coli*). If desired, the His$_6$ tag can be cleaved by enterokinase, yielding GST fusion protein. The His$_6$ tag and GST tag can be cleaved by thrombin, yielding essentially native recombinant protein.

3.3.1.3. PACGHLT

The pAcGHLT vectors are derivatives of the pAcCL29 vector *(21)*. This vector contains a GST tag and a *Bam*HI cloning site upstream of a His$_6$ tag, protein kinase A (PKA) site, thrombin cleavage site, and the multiple-cloning site following to them. Incorporation of a PKA site allows the purified proteins to be phosphorylated with PKA, and in general purpose it might be useful for the radiolabeling of the resultant protein. However, both PKCs and PKNs can phosphorylate this PKA site efficiently. In some cases, we eliminated the PKA site from this vector, and in other cases we inserted the coding sequences of these kinases to the *Bam*HI site of this vector to avoid expressing the PKA site, yielding GST fusion protein without His$_6$ tag and thrombin cleavage site.

3.4. Recombinant Virus Screening

The procedures described here are based on the Novagen's Direct Plaquing Method. For the detailed protocols, refer to the manufacturer's instruction manual. By using this method the average time from transfection to finished high titre stocks of virus was approx 2 wk.

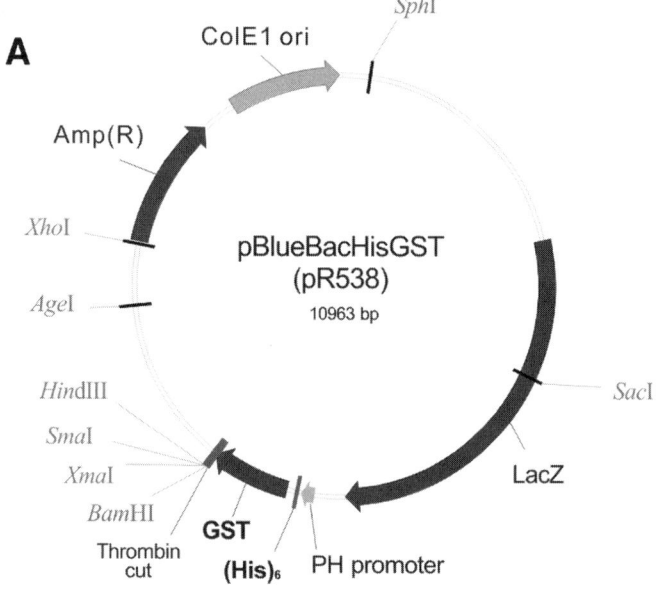

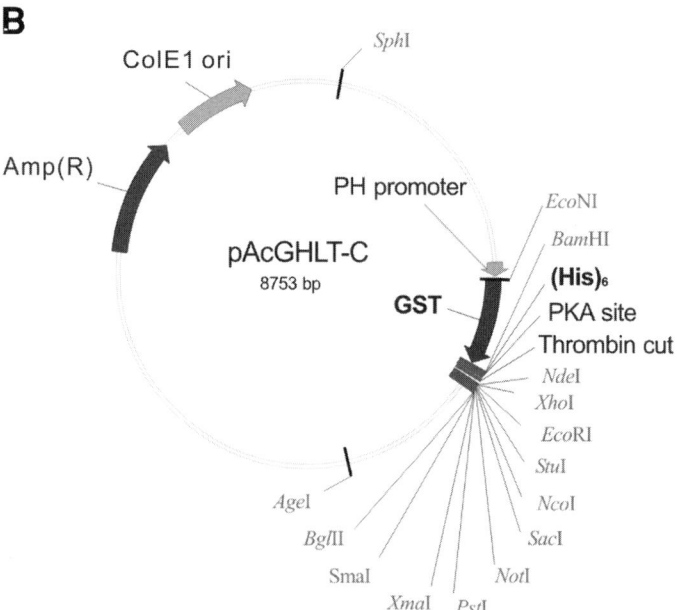

Fig. 1. Transfer vectors. The unique restriction sites are indicated. (**A**) Schematic drawing of pBlueBacHisGST (pR 538) transfer plasmid. (**B**) Schematic drawing of pAcGHLT-C transfer plasmid.

1. Prepare 4- × 60-mm culture dishes, each seeded with a total of 2.5×10^6 Sf9 cells. Label them 1/10, 1/50, 1/250, and 1/1250.
2. Allow the cells to attach to the dishes (about 20 min at 28°)
3. Mix the following components in a sterile 6-mL polystyrene tube:
   - 15 µL of medium (no antibiotics and no serum)
   - 5 µL of BacVector-3000 virus DNA (20 ng/µL).
   - 5 µL of recombinant transfer plasmid (50 ng/µL).
4. Prepare the following reagent mix in a separate sterile 6 mL polystyrene tube labeled with 1/10 and immediately add the DNA mixture **(step 3)** prepared above, then incubate at room temperature for 15 min:
   - 20 µL of D.W.
   - 5 µL of Eufectin transfection reagent.
5. During the incubation, wash the cells in the dishes twice with medium (no antibiotics and no serum). To avoid drying, the second wash should be removed just prior to adding the transfection mixture **(step 7)**.
6. Add 0.45 mL of medium (no antibiotics and no serum) to the mixture **(step 4)**. Set up three polystyrene tubes, each containing 0.4 mL of medium (no antibiotics and no serum) labeled with 1/50, 1/250, and 1/1250. Serially transfer 0.1 mL from the mixture to the next tube and so forth, to produce the indicated dilutions.
7. Transfer 0.1 mL of each dilution to the labeled, freshly drained dishes.
8. Incubate the dishes at room temperature for 1 h.
9. While the transfection mixtures are incubating, prepare the melted 1% agarose solution in a complete medium at 37°C.
10. Add 6 mL of the agarose solution **(step 9)** each dish by pipetting slowly.
11. Incubate at room temperature until the agarose has solidified (about 20 min).
12. Add 2 mL of complete medium to the agar layer.
13. Incubate at 28°C for 3 d.
14. Remove the liquid overlay from the plates, add 2 mL of the freshly diluted Neutral Red solution [1:13 with phosphate-buffered saline(–)] onto the center of each plate. Incubate the plates at 28°C for 2 h.
15. Remove the stain and store the plates at room temperature overnight (*see* **Note 4**).
16. Pick up agarose plug at each well-isolated plaque, and transfer to 1 mL medium in an Eppendorf tube. Incubate the tube at room temperature for 2 h to allow the virus to elute from the agarose (virus stock 1).
17. Seed a 60-mm dish with 2.5×10^6 Sf9 cells. After the cells have attached to the bottom of the dishes, aspirate medium and add 200 µL of each virus suspension **(step 16)** to each dish. Incubate at 28°C for 1 h.
18. Add 5 mL of medium to the dish and incubate at 28°C for 3 d.
19. After aspiration of the medium (virus stock 2), add sodium dodecyl sulphate (SDS) sample buffer to the cell pellet, and subject to Western blotting with anti-tag or anti-target protein antibody. Select virus clone for higher expression of target proteins, and amplify the virus again by the procedure in **steps 17** and **18** (*see* **Note 5**).

3.5. Expression of PKC and PKN in Sf9 Cells for Purification

We usually seed a 10-cm culture dish with 1×10^7 cells and incubate at 28°C for 1 h to allow the cells to attach to the dish. After removing the medium, we add recombinant baculovirus at high multiplicity (5 to 10 PFU/cell) to cells. After incubating virus/cells for 1 h at room temperature, we add 10 mL of medium to the dish and incubate at 28° for 2–3 d until the cytopathic effect is observed (*see* **Note 6**). The cells are harvested by scraping them from the dish with a rubber policeman. The cell pellets are collected by centrifugation at 1000*g* for 5 min at 4°C and stored at –80°C.

3.6. Affinity Purification of Recombinant Protein

3.6.1. Single-Affinity Purification of GST-Tagged PKC and PKN (Starting from a 5 × 10-cm Dish)

1. Thaw the cell pellet.
2. Resuspend cell pellet with 4 mL of buffer A with 1 mM phenylmethylsulfonyl fluoride. and incubate on ice for 10 min.
3. Homogenize using Dounce homogenizer with 30 strokes
4. Add Triton X-100 (final concentration = 1%; *see* **Note 7**).
5. Incubate at 4°C for 10 min
6. Centrifuge at 100,000*g* for 30 min and collect supernatant.
7. Add 400 µL of 50% slurry of glutathione Sepharose 4B (Pharmacia) equilibrated with buffer A to the lysate, and rotate at 4°C for 30 min.
8. Load sample onto an empty column and wash with 30 column volumes or more of buffer A.
9. Elute the recombinant protein four times with 100 µL of buffer B in each tube on ice.
10. Dialyze the eluate to remove glutathione when necessary (*see* **Note 8**).
11. Conventionally purified PKC is able to be preserved at –80°C in 0.05% Triton X-100/10% glycerol for at least 1 yr *(22)*. We stored the enzymes in 50% glycerol without detergent at –80°C. Without addition of glycerol PKN can be preserved for about 1 mo at –80°C without significant loss of enzyme activity. Avoid freeze/thaw cycle (*see* **Note 9**).

3.6.2. Double-Affinity Purification of His$_6$- and GST-Tagged PKC and PKN (see **Note 10**)

1. Thaw the cell pellet.
2. Resuspend cell pellet with 4 mL of buffer C containing 1 mM phenylmethylsulfonyl fluoride and 10 mM imidazole and incubate on ice for 10 min.
3. Homogenize using Dounce homogenizer with 30 strokes.
4. Add Triton X-100 (final concentration = 1%).
5. Incubate at 4° for 10 min.
6. Centrifuge at 100,000*g* for 30 min and collect supernatant.

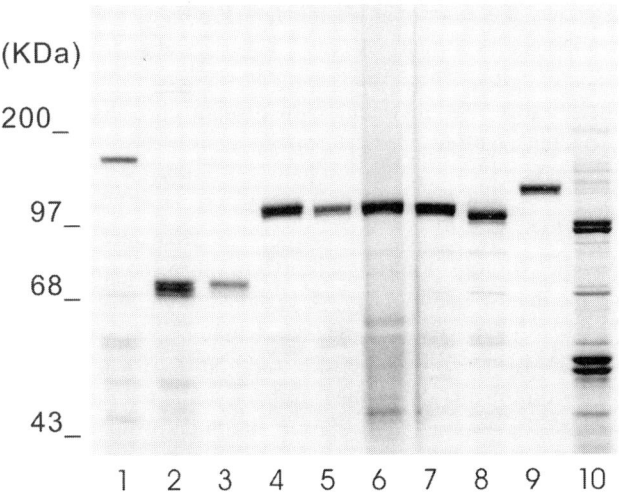

Fig. 2. Silver staining of the purified recombinant enzymes fused to GST. 1, full length of PKNα; 2, catalytic domain of PKNα; 3, catalytic domain of PKNα (K644E mutant); 4, PKCα; 5, PKCβ1; 6, PKCβ2; 7, PKCγ; 8, PKCδ; 9, PKCε; 10, PKCζ.

7. Add 400 µL of 50% Ni-NTA agarose equilibrated with buffer C to the lysate and rotate at 4°C for 1 h.
8. Load the lysate-resin mixture into an empty column, and wash with 30 column volumes or more of buffer C containing 10 mM imidazole.
9. Elute the recombinant protein with 800 µL of buffer C containing 250 mM imidazole.
10. Add 200 µL of 50% glutathione Sepharose equilibrated with buffer A to the eluate and rotate at 4°C for 1 h.
11. Load sample onto an empty column and wash with 30 column volumes or more of buffer A.
12. Elute the recombinant protein four times with 100 µL of buffer B in each tube (*see* **Note 11**).
13. Dialyze the eluate to remove glutathione when necessary, and store the enzyme as described above.

3.7. Analysis of the Purified Protein

3.7.1. SDS-Polyacrylamide Gel Electrophoresis (PAGE)

1. Generally, the estimated total protein content in insect cells is approx 20 mg per 10^7 cells with recombinant protein expression levels ranging between 0.05 and 50%, and the theoretical maximum protein yield is 10 µg to 10 mg per 10^7 cells. For PKCs and PKNs, we usually obtained several µg to 100 µg of the purified enzyme from 10^7 cells.

2. **Figure 2** shows the examples of the SDS-PAGE of purified enzymes. Other minor protein species sometimes could be observed at around the major protein band in each lane. It was reported that the larger protein band could be converted to the smaller form by treatment with protein phosphatases, suggesting the different phosphorylation state of the enzyme in some isoform cases (*see* **Note 12** and **refs.** *1,23,24*).

3.7.2. Protein Kinase Assay

1. To assess the protein kinase activity of the enzymes, purified enzyme (~10 ng) is incubated for 5 min at 30°C in a reaction mixture (final volume 25 µL) containing 20 mM Tris-HCl, at pH 7.5, 4 mM MgCl$_2$, 40 mM ATP, 18.5 kBq of [γ-^{32}P]ATP, 40 µM δPKC peptide (*see* **Note 13**) as phosphate acceptor, 0.1 mg/mL recombinant GST as the stabilizer, with or without 40 µM arachidonic acid. Reactions are terminated by spotting them onto Whatman P81 phosphocellulose papers, submersing them in 75 mM phosphate, and then washing three times for 10 min. The incorporation of ^{32}P phosphate into the δPKC peptide is assessed by liquid scintillation counting.
2. Burns et al. *(2)* reported that recombinant PKCα, βII, and γ isoforms obtained from Sf9 cells behaved exactly as the rat brain enzyme, reflecting similar processing event occurs in Sf9 cells. The wild-type enzyme expressed as a fusion protein with GST in Sf9 cells had the same properties as native PKN purified from rat testis with regard to substrate specificity and response to effectors (*see* **Note 14** and **ref.** *24*). The catalytic domains of PKCs are structurally highly similar to those of PKNs; however, the substrate specificity is different between these recombinant PKCs and PKNs *(25)*. Differences of substrate specificity among recombinant isoforms of PKC from Sf9 cells are described in **refs.** *26* and *27*.

4. Notes

1. We obtained the two types of Sf9 cells from different sources. Even when we use the same lot of recombinant baculovirus, recovery of the recombinant enzyme in the soluble fraction was clearly different between these two Sf9 cells. It might also depend on the medium of Sf9 cells. If the resultant enzyme cannot be found in the soluble fraction, Sf9 cells should be changed.
2. Sf9 cells stick to the culture dish at least in this medium, so passage from culture dish is relatively hard. Recovery of cells from culture dish is low, and sometimes mechanical stimulation to cells during passaging can potentially damage cells. Some researchers use mild detergent with the medium. Shaking the flask at more than 100 rpm prevents Sf9 cells from attaching themselves to the wall of the flask. We usually seed the dish with the cells from the shaken culture flask before we infect virus.
3. When cDNA of each isoform of PKC and PKN was inserted at the cloning site of these vectors, we took care not to fuse the first Met to the Ser at the *Bam*HI site (GGA-$_{Gly}$:TTC-$_{Ser}$). Because PKC and PKN can efficiently phosphorylates Ser of this Gly-Ser-Met sequence.

4. Novagen's BacVector triple cut viral DNA contains a *lacZ* gene that is replaced after rescue by the plasmid containing the foreign gene, all recombinants will produce colorless plaques when stained with X-gal. You can increase the recombination efficiency further by using this selection system when a pAcGHLT transfer vector is co-transfected with this viral DNA. However, pBlueBacHis or pBlueBacHisGST(pR538) transfer vector contains 5′ lacZ fragment, so even the recombinant plaques produced are blue. In both cases, staining by X-gal is usually not necessary because of the high recombination efficiency using the triple cut viral DNA.

5. It is not recommended that the virus stocks are consecutively passaged many times as this will lead to the occurrence and accumulation of virus mutations. It is recommended to go back to the master stock to prepare further quantities of high volume virus stock. We usually take care not to passage more than five times.

6. To increase the recovery of these recombinant enzymes in the supernatant fraction, multiplicity of infection and the incubation time after infection with virus should be determined carefully. Addition of high concentration of virus and long incubation times results in the decrease of the recovery of the enzyme in the supernatant fraction.

7. Triton X-100 is not necessary to recover most of these recombinant enzymes. However, the recovery of novel PKC and atypical PKC isozymes is relatively low in the absence of Triton X-100. Detergent including non-ionic Triton X-100 affects the kinase activity of PKN in a biphasic manner *(28)*, so the final eluate from the affinity column should not include these detergents.

8. Ward et al. *(29)*described the preincubation of purified PKC from a rat brain with 10 mM glutathione at 30°C for 5 min resulted in 90% inactivation of the Ca^{2+}- and PS-dependent Histone kinase activity of the enzyme *(29)*. They also reported that no inhibition was observed if the preincubation step was omitted, and glutathione was added directly to reaction mixtures at the corresponding final (30-fold diluted) concentrations and incubated at 30°C for 10 min, providing evidence for an irreversible inactivation mechanism *(29)*. We could get enough enzyme activity of every isoform of PKC by glutathione Sepharose purification method as described above, then glutathione seems not to irreversibly inhibit the enzyme activity at low temperatures or at concentrations lower than 4 mM.

9. Glycerol does not inhibit the kinase activity of PKN; rather, it mildly increases it.

10. In case of sequential affinity chromatography, Ni-NTA affinity chromatography should not be the last step if the dialysis step is omitted because imidazole in the elution buffer is inhibitory to the kinase activity of PKC and PKN. Dialysis can remove imidazole, but an additional incubation at 4°C is not recommended.

11. As described in **Subheading 3.3.**, the tag (His_6 and GST) of these recombinant proteins can be cleaved by thrombin. However, cleavage by thrombin needs incubation for a few hours at near room temperature, and the activity of the purified enzyme may decrease especially if the concentration of the purified enzyme is low. From our preliminary work, the substrate specificity in vitro and fatty acid sensitivity in vitro seem to be similar between tagged and non-tagged PKC.

The GST and His_6 tag might induce steric hindrance or conformational changes of the enzyme itself and affect substrate recognition due to dimer formation of GST *(30)*. Of course removal of the tags is necessary for rigorous analyses, but these tags might have already affected maturation processes, such as modification by phosphorylation before cleavage of the tags.

12. The reason has not been clarified yet, but smaller fragments of PKCζ were frequently observed. Increasing the quantity and variety of protease inhibitors might help reduce the presence of additional protein bands.

13. δPKC peptide was synthesized based on a peudosubstrate site of δPKC (corresponds to the amino acid 137–153, substituting Ser for Ala; AMFPTMNRRG SIKQAKI).

14. However, we should keep in mind that the purified recombinant enzyme from insect cells still might be composed of heterogeneous molecules with different modifications and that it is "recombinant" and "from insect cells."

References

1. Patel, G. and Stabel, S. (1989) Expression of a functional protein kinase C-gamma using a baculovirus vector: purification and characterisation of a single protein kinase C iso-enzyme. *Cell Signal.* **1,** 227–240.
2. Burns, D. J., Bloomenthal, J., Lee, M. H., and Bell, R. M. (1990) Expression of the alpha, beta II, and gamma protein kinase C isozymes in the baculovirus-insect cell expression system. Purification and characterization of the individual isoforms. *J. Biol. Chem.* **265,** 12044–12051.
3. Kitts, P. A., Ayres, M. D., and Possee, R. D. (1990) Linearization of baculovirus DNA enhances the recovery of recombinant virus expression vectors. *Nucleic Acids Res.* **18,** 5667–5672.
4. Kitts, P. A. and Possee, R. D. (1993) A method for producing recombinant baculovirus expression vectors at high frequency. *BioTechniques* **14,** 810–817.
5. Patel, G., Nasmyth, K., and Jones, N. (1992) A new method for the isolation of recombinant baculovirus. *Nucleic Acids Res.* **20,** 97–104.
6. Peakman, T. C., Harris, R. A., and Gewert, D. R. (1992) Highly efficient generation of recombinant baculoviruses by enzymatically medicated site-specific in vitro recombination. *Nucleic Acids Res.* **20,** 495–500.
7. Ernst, W. J., Grabherr, R. M., and Katinger, H. W. (1994) Direct cloning into the *Autographa californica nuclear polyhedrosis virus for generation of recombinant baculoviruses. Nucleic Acids Res.* **22,** 2855–2856.
8. Bishop, D. H. L., Novy. R., and Mierendorf, R. (1995) The BacVector system: simplified cloning and protein expression using novel baculovirus vectors. *Innovations* **4,** 1–6.
9. Luckow, V. A., Lee, S. C., Barry, G. F., and Olins, P. O. (1993) Efficient generation of infectious recombinant baculoviruses by site-specific transposon-mediated insertion of foreign genes into a baculovirus genome propagated in *Escherichia coli. J. Virol.* **67,** 4566–4579.

10. Smith, D. B. and Johnson, K. S. (1988) Single-step purification of polypeptides expressed in *Escherichia coli as fusions with glutathione S-transferase. Gene* **67,** 31–40.
11. Janknecht, R., de Martynoff, G., Lou, J., Hipskind, R. A., Nordheim, A., and Stunnenberg, H. G. (1991) Rapid and efficient purification of native histidine-tagged protein expressed by recombinant vaccinia virus. *Proc. Natl. Acad. Sci. USA* **88,** 8972–8976.
12. Ono, Y., Fujii, T., Igarashi, K., Kikkawa, U., Ogita, K., and Nishizuka, Y. (1988) Nucleotide sequences of cDNAs for alpha and gamma subspecies of rat brain protein kinase C. *Nucleic Acids Res.* **16,** 5199–5200.
13. Ono, Y., Kurokawa, T., Fujii, T., Kawahara, K., Igarashi, K., Kikkawa, U., Ogita, K., and Nishizuka, Y. (1986) Two types of complementary DNAs of rat brain protein kinase C. Heterogeneity determined by alternative splicing. *FEBS Lett.* **206,** 347–352.
14. Ono, Y., Kikkawa, U., Ogita, K., Fujii, T., Kurokawa, T., Asaoka, Y., Sekiguchi, K., Ase, K., Igarashi, K., and Nishizuka, Y. (1987) Expression and properties of two types of protein kinase C: alternative splicing from a single gene. *Science* **236,** 1116–1120.
15. Ono, Y., Fujii, T., Ogita, K., Kikkawa, U., Igarashi, K., and Nishizuka, Y. (1988) The structure, expression, and properties of additional members of the protein kinase C family. *J. Biol. Chem.* **263,** 6927–6932.
16. Ono, Y., Fujii, T., Ogita, K., Kikkawa, U., Igarashi, K., and Nishizuka, Y. (1989) Protein kinase C zeta subspecies from rat brain: its structure, expression, and properties. *Proc. Natl. Acad. Sci. USA* **86,** 3099–3103.
17. Mukai, H. and Ono, Y. (1994) A novel protein kinase with leucine zipper-like sequences: its catalytic domain is highly homologous to that of protein kinase C. *Biochem. Biophys. Res. Commun.* **199,** 897–904.
18. Oishi, K., Mukai, H., Shibata, H., Takahashi, M., and Ona, Y. (1999) Identification and characterization of PKNbeta, a novel isoform of protein kinase PKN: expression and arachidonic acid dependency are different from those of PKNalpha. *Biochem. Biophys. Res. Commun.* **261,** 808–814.
19. Palmer, R. H., Ridden, J., and Parker, P. J. (1995) Cloning and expression patterns of two members of a novel protein-kinase-C-related kinase family. *Eur. J. Biochem.* **227,** 344–351.
20. Vialard, J., Lalumiere, M., Vernet, T., Briedis, D., Alkhatib, G., Henning, D., et al. (1990) Synthesis of the membrane fusion and hemagglutinin proteins of measles virus, using a novel baculovirus vector containing the beta-galactosidase gene. *J. Virol.* **64,** 37–50.
21. Livingstone, C. & Jones, I. (1989) Baculovirus expression vectors with single strand capability. *Nucleic Acids Res.* **17,** 2366.
22. Kikkawa, U., Go, M., Koumoto, J., and Nishizuka, Y. (1986) Rapid purification of protein kinase C by high performance liquid chromatography. *Biochem. Biophys. Res. Commun.* **135,** 636–643.

23. Takahashi, M., Mukai, H., Oishi, K., Isagawa, T., and Ono, Y. (2000) Association of immature hypophosphorylated protein kinase C epsilon with an anchoring protein CG-NAP. *J. Biol. Chem.* **275,** 34592–34596.

24. Yoshinaga, C., Mukai, H., Toshimori, M., Miyamoto, M., and Ono, Y. (1999) Mutational analysis of the regulatory mechanism of PKN: the regulatory region of PKN contains an arachidonic acid-sensitive autoinhibitory domain. *J. Biochem. (Tokyo)* **126,** 475–484.

25. Taniguchi, T., Kawamata, T., Mukai, H., Hasegawa, H., Isagawa, T., Yasuda, M., Hashimoto, T., Terashima, A., Nakai, M., Mori, H., Ono, Y., and Tanaka, C. (2001) Phosphorylation of tau is regulated by PKN. *J. Biol. Chem.* **276,** 10025–10031.

26. Kazanietz, M. G., Areces, L. B., Bahador, A., Mischak, H., Goodnight, J., Mushinski, J. F., and Blumberg, P. M. (1993) Characterization of ligand and substrate specificity for the calcium-dependent and calcium-independent protein kinase C isozymes. *Mol. Pharmacol.* **44,** 298–307.

27. Liyanage, M., Frith, D., Livneh, E., and Stabel, S. (1992) Protein kinase C group B members PKC-delta, -epsilon, -zeta and PKC-L(eta). Comparison of properties of recombinant proteins in vitro and in vivo. *Biochem. J.* **283(Pt 3),** 781–787.

28. Kitagawa, M., Mukai, H., Shibata, H., and Ono, Y. (1995) Purification and characterization of a fatty acid-activated protein kinase (PKN) from rat testis. *Biochem. J.* **310(Pt 2),** 657–664.

29. Ward, N. E., Pierce, D. S., Chung, S. E., Gravitt, K. R., and O'Brian, C. A. (1998) Irreversible inactivation of protein kinase C by glutathione. *J. Biol. Chem.* **273,** 12558–12566.

30. Kaplan, W., Husler, P., Klump, H., Erhardt, J., Sluis-Cremer, N., and Dirr, H. (1997) Conformational stability of pGEX-expressed Schistosoma japonicum glutathione S-transferase: a detoxification enzyme and fusion-protein affinity tag. *Protein Sci.* **6,** 399–406.

4

Expression and Purification of Protein Kinase Cδ from Bacteria

Michael Gschwendt and Luise Stempka

1. Introduction

Protein kinase C (PKC) covers a family of 12 isoenzymes known so far to possess phospholipid-dependent serine and threonine kinase activity *(1,2)*. These kinases play a key role in signal transduction and are involved in the regulation of numerous cellular processes. PKCδ is a member of the so-called novel PKC subgroup consisting of Ca^{2+}-unresponsive diacylglycerol- and 12-*O*-tetradecanoylphorbol-13-acetate (TPA)-activated isoenzymes. It is ubiquitously expressed and exhibits some unique properties *(3)*. For example, contrary to other PKC isoforms, such as PKCα *(4)* and β_{II} *(5)*, phosphorylation of the activation loop (threonine 505 in PKCα) is not a prerequisite for enzymatic activity *(6,7)*. Thus, whereas other PKC isozymes cannot be expressed in active form in bacteria because they require phosphorylation by the upstream kinase phosphoinositide-dependent kinase 1, active PKCδ can be obtained from bacteria *(6)*.

2. Materials

1. Expression vector pET28 (Novagen, Madison, WI).
2. *Escherichia coli* (*E. coli*) strain BL21(DE3)pLysS (AGS, Heidelberg, Germany).
3. Bacto-Trypton, yeast extract, Bacto-Agar (Difco, Detroit, MI).
4. LB medium: 10 g of NaCl, 10 g of Bacto-Trypton, and 5 g of yeast extract in 1 L of H_2O.
5. Buffer 1: 30 m*M* K-acetate, pH 5.8, 100 m*M* RbCl, 50 m*M* $MnCl_2$, 10 m*M* $CaCl_2$, 15% glycerol.

From: *Methods in Molecular Biology, vol. 233: Protein Kinase C Protocols*
Edited by: A. C. Newton © Humana Press Inc., Totowa, NJ

6. Buffer 2: 10 m*M* 3-(*N*-morpholino) propane sulfonic acid, pH 6.8, 10 m*M* RbCl, 75 m*M* CaCl$_2$, 15% glycerol.
7. SOC-medium: a solution of 20 g of Bacto-Tryptone, 5 g of yeast extract, 0.5 g of NaCl, 10 mL of 250 m*M* KCl in 970 mL of H$_2$O is autoclaved and then 10 mL of sterile 1 *M* MgCl$_2$ and 20 mL of sterile 1 *M* glucose is added.
8. Isopropyl-1-thio-β-D-galactopyranoside (Sigma, Munich, Germany).
9. Buffer N (BN): 50 m*M* sodium phosphate, pH 8.0, 150 m*M* NaCl, 1% Triton X-100, 10% glycerol, 1 m*M* phenylmethylsulfonyl fluoride, 10 μg/mL each of leupeptin, aprotinin, and pepstatin, 5 m*M* imidazole. BN 20, 50, 100, 200, and 500 contain 50, 100, 200, and 500 m*M* imidazole, respectively.
10. Mono-Q-Sepharose (Amersham Pharmacia, Freiburg, Germany).
11. Buffer Q (BQ): 10 m*M* Tris-HCl, pH 7.5, 1 m*M* phenylmethylsulfonyl fluoride, 10 μg/mL each of leupeptin, aprotinin, and pepstatin. BQ 200, 300, 400, and 500 contain 200, 300, 400, and 500 m*M* NaCl, respectively.
12. Buffer P: 20 m*M* Tris-HCl, pH 7.5, 10 m*M* β-mercaptoethanol.
13. Phosphatidyl serine liposomes: The commercially available solution of phosphatidyl serine (Sigma, Munich, Germany) is evaporated, buffer P is added (final concentration 1 mg/ml), and the suspension is sonicated on ice with a Branson sonifier (3 min pulse at setting 6).
14. Liquid scintillation cocktail, for example, Ready Safe (Beckman, Palo Alto, CA).
15. The pseudosubstrate-related peptide δ: MNRRGSIKQAKI.
16. Protein dye reagent concentrate (Bio-Rad, Munich, Germany).

3. Methods

3.1. Expression of His-Tagged PKCδ in Bacteria

3.1.1. Polymerase Chain Reaction Amplification of PKCδ cDNA and Cloning into the Vector pET28

For the construction of a PKCδ full-length cDNA with an *Nde*I restriction site at the initiation signal ATG and an *Eco*RI restriction site behind the stop codon TGA, the following oligonucleotide primers are used: 5′-AAA GGA TCC CAT ATG GCA CCG TTC CTG CGC-3′ as the 5′-primer and 5′-TCT GGG AAT TCA CTA CTA TTC CAG GAA TTG CTC-3′ as the 3′-primer. For polymerase chain reaction amplification (cycle profile: 94°C/5 min; 10 × 94°C/15 sec, 56°C/30 sec, 72°C/2 min; 15 × 94°C/15 sec, 56°C/30 sec, 72°C/2 min, plus cycle elongation of 20 sec for each cycle; 72°C/7 min), a rat PKCδ full-length cDNA clone is used as template. The resulting cDNA of 2048 bp is cut with *Nde*I and *Eco*RI and cloned into the *Nde*I-*Eco*RI-cut expression vector pET28. The resulting plasmid coding for PKCδ with an N-terminal His-tag (six histidine residues) and termed pET28δ is used for transformation of the *E. coli* cells BL21(DE3)pLysS.

3.1.2. Preparation of Chemically Competent E. coli Cells

1. The *E. coli* strain BL21 (DE3)pLysS is grown on 1.5% Bacto-Agar in LB-medium for 16 h at 37°C.
2. An *E. coli* colony is transfered from the agar to 2 mL of LB-medium and incubated for another 16 h at 37°C (all incubations are performed under shaking).
3. Upon a 1:50 dilution with LB-medium the incubation is continued until the O.D. at 600 nm reaches 0.5.
4. All further procedures are performed at 0–4°C. The bacteria are sedimented from the suspension by centrifugation at 3000*g* for 10 min, resuspended in 30 mL of sterile buffer 1 and incubated at 0°C for 30 min.
5. Upon centrifugation as above, the bacteria are resuspended in 10 mL of sterile buffer 2 and incubated at 0°C for 16 h.
6. Aliquots (500 μL each) of the competent bacteria are stored at –70°C.

3.1.3. Transformation of E. coli Cells with the Plasmid pET28δ and Extraction of Cells

1. Competent bacteria stored at –70°C are thawed on ice for about 15 min.
2. Cell suspension (100 L) is mixed with 50 ng of the plasmid pET28δ and incubated on ice for 1 h.
3. Upon a heatshock at 42°C for 45 sec and cooling down on ice for 2 min, the bacteria are mixed with 400 μL of SOC medium and incubated at 37°C for 1 h.
4. Transformed bacteria (200 μL) are plated on 1.5% Bacto-Agar in LB medium in the presence of 50 μg/mL kanamycin and 30 μg/mL chloramphenicol.
5. The plates are dried for 10 min and incubated at 37°C for 12 h.
6. Bacterial colonies are transfered to 50 mL of LB medium containing the same antibiotics as above and incubated at 37°C for another 12 h.
7. Upon dilution with 1 L of LB medium plus the antibiotics incubation is continued at 24°C until the absorbance at 600 nm reaches 0.5–0.7.
8. Induction is now performed with 0.5 m*M* (final concentration) isopropyl-1-thio-δ-D- galactopyranoside at 24°C for 12 h.
9. Thereafter the bacteria are sedimented at 5000*g* for 10 min (4°C), washed once with ice cold phosphate-buffered saline and frozen at –20°C. Frozen cells are thawed at 4°C, lysed with 25 mL of BN, sonicated on ice with a Branson sonifier (three times for 1 min each at setting 6 and 50% intensity), and centrifuged at 80,000*g* for 45 min at 4°C. The supernatant, termed bacterial extract, is used for the purification of His-tagged PKCδ (His-PKCδ).

3.2. Partial Purification of His-PKCδ

3.2.1. Chromatography on Nickel-Nitrilo-Triacetic Acid (Ni-NTA)

As the result of an interaction of histidine residues with nickel ions His-tagged PKCδ can be purified by affinity chromatography on Ni-NTA. Elution

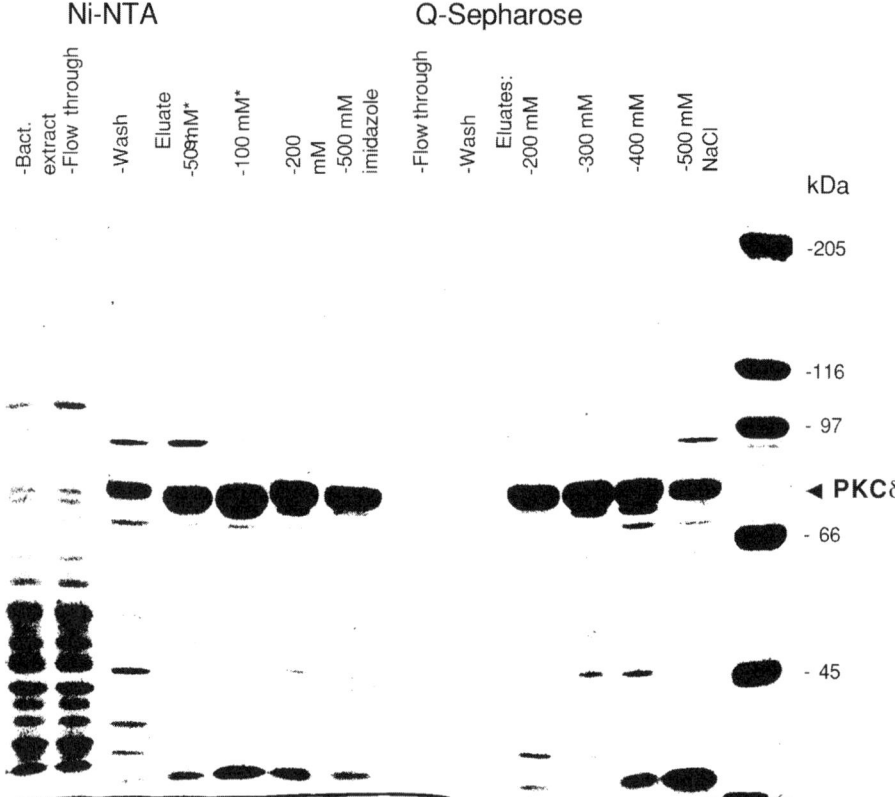

Fig. 1. Purification of recombinant His-PKCδ from bacteria. The purification was performed as described in the **Subheadings 3.2.1.** and **3.2.2.** An aliquot of each fraction containing 20 μg of protein was applied to sodium dodecyl sulfate-polyacrylamide gel electrophoresis (7.5%) and proteins were stained with coomassie blue. Molecular weight markers (kDa) are shown on the right side of the gel.

of His-PKCδ is performed with imidazole, which competes with the histidine residues for the binding to nickel ions.

1. A Ni-NTA column (0.5 mL) is washed with 5 mL of H_2O and equilibrated with 5 mL of BN. No pressure is applied to the column.
2. The bacterial extract resulting from 1 L of cultured bacteria (see above) is then loaded onto the column.
3. The column is washed with 5 mL each of BN and BN20.
4. Then the kinase is eluted stepwise with 1 mL each of BN50, BN100, BN200, and BN500.

Table 1
Partial Purification of Recombinant PKCδ

	Protein conentration [mg/mL]	Total protein (Σ) [mg]	Spec. Activity [nmol/ min/mg]	Yield [nmol/min/ Σ Prot.]	Purification -fold
Bacterial extract	27	540	3.0	1650 (100%)	
Ni-NTA					
Run through	26.5	530	1.2	636	
Wash	0.7	2.1	25	53	
Eluate* (50 m*M*)	1.6	1.0	127	127 (8%)	
Eluate* (100 m*M*)	3.7	2.2	144	317 (19%)	48
Eluate (200 m*M*)	1.5	0.9	72	64	
Eluate (500 m*M* imidazole)	1.0	0.6	20	12	
Mono-Q-Sepharose					
Run through	<0.05	<0.27	0	0	
Wash	<0.05	<0.35	0	0	
Eluate (200 m*M*)	0.7	0.73	258	188 (11%)	86
Eluate (300 m*M*)	1.0	1.1	119	131 (8%)	
Eluate (400 m*M*)	0.86	0.95	64	61	
Eluate (500 m*M* NaCl)	0.46	0.5	21	11	

The Ni-NTA and Mono-Q-Sepharose chromatographies were performed as described in the **Subheadings 3.2.1.** and **3.2.2.** The eluates of the Ni-NTA marked with an asterix were combined and applied to the Mono-Q-Sepharose. The kinase activity of His-PKCδ was measured in 5-μL aliquots of each fraction by the kinase assay (*see* **Subheading 3.3.**). Protein concentration was determined with the protein dye reagent concentrate using bovine serum albumin as standard.

His-PKCδ is monitored upon sodium dodecyl sulfate-polyacrylamide gel electrophoresis of the fractions by either staining with coomassie blue or immunoblotting with a PKCδ specific antibody *(8)*. Moreover, the kinase activity in each fraction is determined by the kinase assay (*see* **Subheading 3.3.**). All four eluates contain His-PKCδ, as shown in **Fig. 1**. The BN50 and BN100 fractions with the highest specific activity (*see* **Table 1**) are combined and chromatographed on the strong anion exchanger Mono-Q-Sepharose.

3.2.2. Chromatography on Mono-Q-Sepharose

1. A Mono-Q-Sepharose column (0.7 mL) is equilibrated with 10 mL of BQ.
2. The combined fractions BN50 and BN100 of the Ni-NTA chromatography are diluted 1:5 with BQ and loaded onto the column.

3. The column is washed with 10 mL of BQ.
4. Then the kinase is eluted with 1 mL each of BQ200, 300, 400, and 500. 10% glycerol and 10 mM β-mercaptoethanol are added to the eluates immediately.

His-PKCδ is detected in the fractions and its kinase activity is measured as described above for the Ni-NTA chromatography. All the eluates contain His-PKCδ as shown in **Fig. 1**. The fraction BQ200 with the highest specific activity is purified 86-fold and is at least 80% pure (*see* **Table 1** and **Fig. 1**). Aliquots (50 mL) of this fraction are stored at −70° and used for kinase assays. The purity of fraction BQ300 is comparable to that of BQ200. However, its specific activity is somewhat lower (*see* **Fig. 1** and **Table 1**). Taking the BQ200 and BQ300 fractions together (1.8 mg protein), around 80% pure enzymatically active His-PKCδ is obtained with a yield of 19%. From these data it can be calculated that 1 L of the *E. coli* strain produces around 7.5 mg of soluble His-PKCδ (1.4% of total protein in the bacterial extract).

3.3. Kinase Assay

Phosphorylation reactions are performed in a total volume of 100 μL containing buffer P, 4 mM MgCl$_2$, 10 μg of phosphatidyl serine (10 μL of liposomes), 100 nM TPA, 5 μg of pseudosubstrate-related peptide δ as substrate (*6*), 1 μg (0.26 units) of purified His-PKCδ (fraction BQ200) or 5 μL of each fraction of the purification procedure (*see* **Table 1**), and 37 μM ATP containing 1 μCi [γ-^{32}P] ATP. After incubation for 7 min at 30°C the phosphorylation reaction is terminated by transferring 50 μL of the assay mixture onto a 20-mm square piece of phosphocellulose paper (Whatman p81). After washing the paper three times in deionized water and twice in acetone, radioactivity is determined by liquid scintillation counting, using the liquid scintillation cocktail Ready Safe from Beckman. For the features of the purified kinase, *see* **Notes 1–4**.

4. Notes

1. The specific activity of the partially purified recombinant His-PKCδ (258 units/mg protein, BQ200 fraction) is comparable to that of native PKCδ purified from porcine spleen (304 units/mg protein, *see* **ref. 8**).
2. The kinase activity of His-PKCδ, like that of the native enzyme, is absolutely dependent on phosphatidyl serine and TPA, as shown in **Fig. 2**.
3. The K$_m$ and V$_{max}$ values with the pseudosubstrate-related peptide δ as substrate are 5 μM and 400 nmol/min/mg, respectively.
4. Also other properties of the recombinant enzyme are very similar to that of the native enzyme (*see* **ref. 6**).

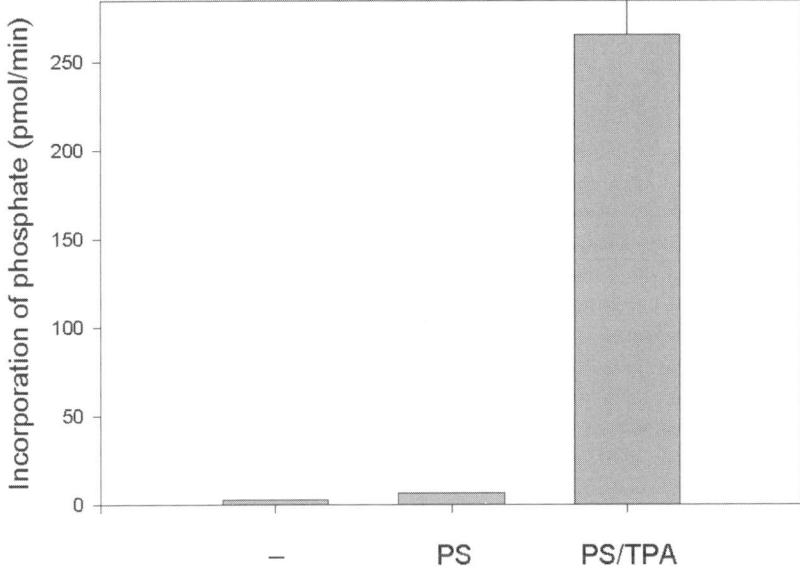

Fig. 2. Phosphorylation of the pseudosubstrate-related peptide δ (*see* **ref. 6**) by His-PKCδ purified from a bacterial extract. His-PKCδ (BQ200 fraction) was used for the phosphorylation of peptide δ in the absence or presence of phosphatidyl serine (PS) and PS/TPA, as described in **Subheading 3.3.** The values are the mean of two determinations.

Taken together, the expression of PKCδ in bacteria is a convenient and suitable method for the preparation of large amounts of an enzymatically active kinase, that behaves like the native enzyme.

Acknowledgment

This work was supported by the Wilhelm Sander-Stiftung, Grant 97.090.19.

References

1. Marks, F. and Gschwendt, M. (1996) Protein kinase C, in *Protein Phosphorylation* (Marks, F., ed.), Verlag Chemie, Weinheim, Germany, pp. 81–116.
2. Blobe, G. C., Stribbing, S., Obeid, L. M., and Hannun, Y. A. (1996) Protein kinase C isoenzymes. Regulation and function. *Cancer Surveys* **27,** 213–248.
3. Gschwendt, M. (1999) Portein kinase Cδ. *Eur. J. Biochem.* **259,** 555–564.
4. Cazaubon, S., Bornancin, F., and Parker, P. J. (1994) Threonine-497 is a critical site for permissive activiation of protein kinse Cα. *Biochem. J.* **301,** 443–448.

5. Orr, J. W. and Newton, A. C. (1994) Requirement for negative charge on "activation loop" of protein kinase C. *J. Biol. Chem.* **269,** 27715–27718.

6. Stempka, L., Girod, A., Müller, H.-J., Rincke, G., Marks, F., Gschwendt, M., et al. (1997) Phosphorylation of protein kinase Cδ (PKCδ) at threonine 505 is not a prerequisite for enzymatic activity. *J. Biol. Chem.* **272,** 6805–6811.

7. Stempka, L., Schnölzer, M., Radke, S., Rincke, G., Marks, F., and Gschwendt, M. (1999) Requirements of protein kinase Cδ for catalytic function. Role of glutamic acid 500 and phosphorylation on serine 643. J Biol. Chem. **274,** 8886–8892.

8. Leibersperger, H., Gschwendt, M., and Marks, F. (1990) Purification and characterization of a calcium-unresponsive, phorbol ester/phospholipid-activated protein kinase from porcine spleen. *J. Biol. Chem.* **265,** 16108–16115.

III

MEASURING THE ACTIVITY OF PROTEIN KINASE C

5

Complexities in Protein Kinase C Activity Assays

An Introduction

Julianne J. Sando

1. A Definition of Protein Kinase C (PKC) Activity and Differences Among Isozymes

PKC activity is defined, operationally, as phospholipid-dependent kinase activity. Nishizuka's group first identified PKC as a precursor for a cofactor-independent protein kinase that was generated by calcium-dependent proteolysis *(1,2)*. However, they then observed that intact PKC could be activated reversibly by association with membrane lipids *(3)*, and they went on to identify phosphatidylserine (PS), diacylglycerol (DAG), and calcium as the major activators *(3,4)*. Castagna et al. *(5)* later showed that phorbol ester tumor promoters could replace DAGs in activating PKC.

Numerous laboratories were involved in identifying the isozymes that make up the PKC family (reviewed in **refs. *6–9***). The isozymes differ in their requirements for calcium or DAG. The classic or cPKCs (α, βI, βII, and γ) have homologous constant domains C1, C2, C3, and C4. C1 contains two cysteine-rich subdomains, C1a and C1b, at least one of which interacts with DAGs or phorbol esters; C2 confers calcium dependence; C3 has a typical ATP-binding consensus; and C4 contains a substrate-binding site. The novel or nPKCs (δ, ε, η, θ) are calcium-independent and have a noncalcium-binding C2-like domain N-terminal to the C1 domain. The atypical or aPKCs (ζ, λ/ι) have only one of the cysteine-rich C1 subdomains and are not activated by phorbol esters. These isozyme differences are illustrated in **Fig. 1**. All of the isozymes require acidic phospholipids, particularly PS. Thus, whereas other

From: *Methods in Molecular Biology, vol. 233: Protein Kinase C Protocols*
Edited by: A. C. Newton © Humana Press Inc., Totowa, NJ

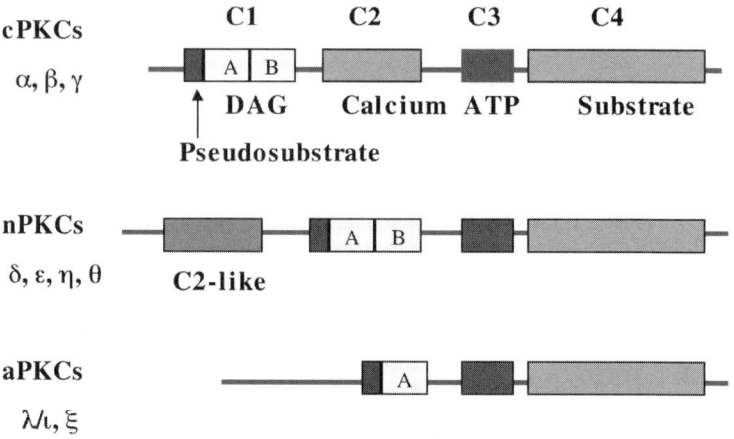

Fig. 1. Illustration of the differences in PKC isozymes.

lipid-dependent kinases (e.g., phosphatidylinositide-dependent kinase-1) have been identified subsequently, PS-dependent kinase activity has remained the operational definition of PKC activity.

2. Problems with in Vitro Assays for PKC Activity

Despite more than 25 years of studies on PKC, assays for its activity remain problematic. The lipid concentration, composition, and physical properties all affect PKC activity assays in vitro (reviewed in **refs. *10–13***). The cofactor requirements are interdependent, as was apparent even in the early experiments from the Nishizuka lab *(3–5)*. Thus, the concentrations of calcium or diacylglycerol required, for those isozymes that require them, depend on the lipid composition and concentration. The concentrations of cofactors required for maximal activity depend also on the concentrations of enzyme and substrate used in the reaction. Further complicating this situation is the lability of the enzyme and its propensity to stick to all surfaces, thus reducing the concentration present in solution as a function of time.

A basic, inexpensive method for in vitro assay of PKC activity is described in the Chapter 6. This assay is quite suitable for following elution of activity from purification columns where the goal is to identify PKC-containing fractions for collection and pooling. The more difficult task is to quantitate the PKC activity of a given preparation for comparison between various experimental conditions. It is essential that the investigator optimize assays for the conditions used. It is reasonably straightforward to optimize concentrations of water-soluble components, such as ATP. Optimization of lipid and lipid-binding constituents is more difficult and will be discussed here in more detail.

2.1. Selection and Optimization of Lipids Used to Activate PKC

2.1.1. Specificity for PS and DAG Headgroups

Although other acidic phospholipids can support partial activation of PKC *(3,14)*, PS provides much greater activity. Lee and Bell *(15)* conducted extensive structure–activity analysis of the PS head group and demonstrated stereospecificity for L-serine. Newton and Keranin *(16)* showed that other acidic lipids could contribute to PKC membrane binding, but that the activation by DAG requires PS.

The classic and novel subgroups of PKC isozymes are activated by DAGs or phorbol esters. Stereospecific activation by 1,2-*sn-diolein (17)*, competition by DAGs for phorbol ester binding *(18)*, and deletion analysis of PKC *(19)* argued for specific binding of DAG or phorbol ester to a site(s) in the C1 domain of the enzyme. Crystallization of the C1b domain of PKCδ bound to phorbol ester *(20)* confirmed this hypothesis. Slater et al. *(21)* used fluorescent phorbol ester derivatives to argue for two phorbol ester sites of greatly differing affinity. More recently, Conesa-Zamora et al. *(22)* and Ochoa et al. *(23)* have proposed that the C1 subdomain that binds phorbol esters in the intact enzyme differs in cPKCs versus nPKCs because of the opposite order of the C1 and C2 or C2-like domains. They proposed that C1a of cPKCs or C1b of nPKCs interacts with phorbol esters or DAGs in the membrane, whereas the C1 subdomain that is adjacent to the C2 or C2-like domain is oriented away from the membrane. These models are consistent with data from Medkova and Cho *(24)* and Hurley et al. *(25)* on high-affinity phorbol ester binding to C1a in cPKCs or to C1b in nPKCs.

2.1.2. Effects of Lipid Acyl Chains

Even if one uses the simplest system of PS and DAG to activate PKC, one must select the type of PS and DAG. The cheapest sources of PS are often extracts from brain lipids (e.g., bovine brain PS from Sigma Chemical Co.). These extracts contain an often uncharacterized mixture of acyl chains. Phospholipids with defined acyl chains can be purchased at greater cost (e.g., from Avanti Polar Lipids, Inc.). The acyl chain composition of the lipids has a dramatic effect on PKC activity. If all of the acyl chains are saturated, minimal or no PKC activity can be observed at low concentrations of DAG (1–5 mol% with respect to total lipid) *(26)*. The activity of PKC correlates dramatically with the mol% of unsaturated acyl chains in the system. However, even fully saturated phospholipids can be rendered effective PKC activators at very high molar ratios of DAG *(26)*.

These observations suggested that PKC also requires certain physical properties of the lipid, not just the chemical structure of the head group. In further support of that hypothesis, we observed that the number of double bonds per acyl chain and their position in the chain are of less importance *(26)*. The more critical issue is that at least one *cis* double bond is present in the phospholipid; *trans* double bonds provide lesser activation. Branched chain phospholipids similarly enhance PKC activation *(27)*.

Several physical properties of the membrane have been proposed to contribute to PKC activation. Saturated phospholipids with a cylindrical shape, such as phosphatidylcholine (PC), pack tightly into lamellar structures with strong hydrogen bonding between the acyl chains. Phospholipids with a small ratio of head group to acyl chain volume, such as phosphatidylethanolamine (PE) and branched chain phospholipids, have a more conical shape and tend to pack into inverted hexagonal structures or Hex_{II} phases, which do not activate PKC. However, inclusion of some Hex_{II}-forming lipids in a bilayer structure increases PKC activity *(28)*. Properties of such bilayers that may correlate with PKC activation include increased lipid head-group spacing *(26,29)* and the *tendency* to form nonlamellarphases with associated curvature strain *(26–29)*. Giorgione et al. *(30)* have shown increased activation of PKC with cubic phase lipid as well.

DAGs, too, can destabilize the lamellar structure of the membrane *(31–33)*, in addition to their ability to bind specifically to PKC. They can affect PKC membrane binding, can alter the requirement for additional cofactors such as calcium and PS, and can affect maximal activity. The acyl chain composition of the DAGs contributes to these effects. An illustration of how various lipid head groups and acyl chains contribute to PKC activation is shown in **Fig. 2**.

2.1.3. Form of Lipid Vesicles Provided

The next issue to consider is the form in which the lipids are provided to the enzyme. Naturally occurring long chain lipids are not soluble in aqueous solution and must be provided in some sort of vesicle or micelle form. Generally, all of the lipid components in organic solvents are combined in a glass tube and the solvent is evaporated with an inert gas such as N_2 or Ar. For the best preparations, it is ideal to redissolve the lipids in benzene/methanol (19:1, v/v) and to lyophilize them overnight. All unsaturated lipids should be kept under inert gas in the dark to protect the double bonds from oxidation.

The lipids are then hydrated in an appropriate buffer to make some form of vesicle. If the lipids are sonicated, small unilamellar vesicles (SUVs) of approximately 20 nm in diameter are formed. These consist of a single lipid bilayer enclosing some of the buffer. Vortexing instead of sonicating the lipids, at temperatures above their phase transition temperatures, generates multilamellar vesicles (MLVs), concentric bilayers somewhat like onionskin.

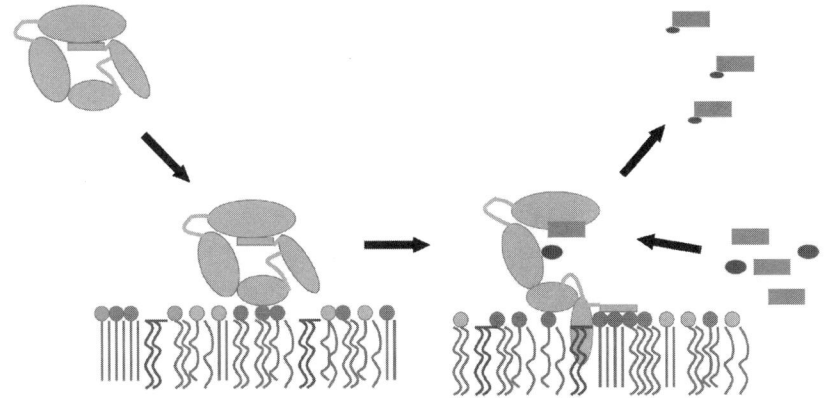

Fig. 2. A model for PKC activation. PKC is shown to bind to regions of lipid membranes rich in acidic headgroups (darker circles attached to phospholipid acyl chains) and binding contributes further to clustering of acidic lipids. Physical features of the lipid affected by acyl chain unsaturation (wavy lines) and DAG (lipid lacking headgroup) contribute to insertion of the PKC C1 domain into the membrane with release of the pseudosubstrate from the active site. Binding of DAG to the C1 domain and binding of the pseudosubstrate to acidic lipids help to stabilize the active form of the enzyme. Substrate proteins (small rectangles) and ATP (dark ovals) can now enter the active site and undergo the phosphorylation reaction.

These larger structures present a more planar lipid surface than do the small, highly curved SUVs; however, only a small percentage of the lipid is exposed on the surface for access to PKC. Thus, if one is investigating the stoichiometry of some lipid constituent required for PKC activation, SUVs may provide a better system. MLVs may be preferable for other experiments, such as examining PKC-membrane binding, because they are heavier and it is possible to pellet them at high centrifugal forces.

It also is possible to generate large unilamellar vesicles (LUVs) from MLVs by passing them repeatedly through a polycarbonate filter of defined pore size (e.g., 100 nm from Costar Scientific Corp, Cambridge, MA) using a hand-held extruder (Avestin). Certain lipid mixtures are not amenable to LUV generation because of the clogging of the extruder membrane *(26)*, which is sometimes caused by phase separation *(34)*. It is appropriate to verify generation of appropriately sized LUVs. This can be done with a particle sizer using quasi-elastic light scattering (e.g., Nicomp Model 370) *(35)* or by electron microscopy. The LUVs can be prepared in buffer solutions containing sucrose to make heavier, sucrose-loaded large unilamellar vesicles *(36)*. After their preparation, the suspension can be diluted such that the heavier sucrose-loaded large unilamellar vesicles can be pelleted easily by centrifugation. It should be noted, however, that the pelleting of lipid vesicles is seldom quantitative. When quantitation is required, the vesicles can be spiked with a radiolabeled lipid to assess the effectiveness of centrifugation *(36)*.

The lipid can be provided to PKC at temperatures below (more ordered gel phase) or above (more fluid liquid crystalline phase) their phase transition temperatures. Boni and Rando *(37)* and Senisterra and Epand *(28)* showed that gel and fluid phase lipids both could activate PKC. PKC activation does not correlate with fluidity of the lipid *(26)*. Hinderliter et al. *(38)* observed increased PKC activity when a specific lipid mixture was held at temperatures above versus below its phase transition; however, analysis of Arrhenius plots suggested that this effect may be attributed simply to increased intrinsic activity of the enzyme at higher temperature.

2.1.4. Biphasic Activation/Inhibition of PKC with Lipid Micelles

In an attempt to activate PKC with lipid monomers in aqueous solution instead of with vesicles, Walker et al. *(39)* synthesized some short chain PSs that were water-soluble up to high concentrations. These short chain PSs did not activate PKC at monomeric concentrations but did activate the enzyme at concentrations near their critical micelle concentration (cmc) *(39)*. In fact, the short-chain phosphatidylcholines, from which the short chain PSs were synthesized, also activated PKC in the absence of PS *(40)*. A peak in activity was observed that correlated with the cmc of each short chain PC. This result, very surprising at the time, argues further for the importance of physical properties of the lipid and provided the first demonstration that the PS requirement could be bypassed under very special conditions.

Detergents also can both stimulate and inhibit PKC activity depending on their concentration (unpublished observations), although with very low maximal activation unless PS and DAG are included. Mixed micelles containing PS and DAG with detergents activate PKC quite well *(41)*; however, these structures are less representative of cellular membranes than are vesicles.

The decrease in PKC activity at short-chain lipid concentrations above the cmc *(40)* may relate to interactions of PKC with lipid, substrate, or other PKC molecules (see below). The PC micelles, although a nonphysiological artificial system, have proven useful for activation of PKC in the absence of significant metal ion chelation *(42)* or substrate binding *(43)* or when subsequent removal from lipid is desired, because dilution of short chain PC below the cmc does not support activation.

2.1.5. Use of Lipid Mixtures for PKC Activation

Mixtures of phospholipids have certain advantages in PKC assays. First, pure PS vesicles are difficult to characterize physically because they form huge aggregates in the presence of calcium, even though the aggregates do provide PKC activation. Inclusion of at least 25% PS in a PC vesicle allows for PKC-membrane binding and activation with lipid vesicles that are more stable in the presence of calcium *(37,44)*. Replacement of the PC with PE provides a mixture more similar to that in the inner leaflet of cell membranes. PE, with its greater propensity to generate nonlamellar phases, also affords greater PKC activation than do comparable molar ratios of PC *(28,45)*.

2.1.6. Biphasic Activation/Inactivation of PKC in Lipid Vesicles

We now know that biphasic activation of PKC is not unique to micelles. Virtually any lipid-soluble or lipid-binding molecule can both activate and inhibit PKC activity as a function of its concentration. Such modulators include lysolipids *(46)*, alcohols *(35)* and steroids, such as tamoxifen *(47)* and bile acids *(48)*. Even the originally identified activators DAG and PS can be present in excess such that PKC activation is diminished *(49)*.

There are several reasons for the maxima in PKC activation as a function of mol% of any particular lipid. First, it is not the aqueous concentration of a lipid constituent that is relevant to PKC activation, but its molar ratio in the total lipid. As the molar ratio of one lipid is increased, the molar ratio of others necessarily decreases. Second, lipid membranes and vesicles are not static structures; the individual lipids in the mixtures can move in the plane of the bilayer and they tend to cluster into domains of varying composition and size that exist for varying, usually unknown, periods of time. Differential scanning calorimetry has been used to demonstrate that 1:1 mixtures of PC:PS do not mix uniformly *(34)*. Addition of DAG to these mixtures causes more pronounced lateral domain segregation as determined by differential scanning calorimetry *(34)* and nuclear magnetic resonance *(33)*. Dibble et al. *(34)* used Fourier-transformed infrared spectroscopy to argue for preferential clustering of the DAG with the PS in these mixtures *(34)*. A cartoon illustrating this effect is shown in **Fig. 3.**

Gel phase lipids require lower molar ratios of DAG for maximal activation than do liquid-crystalline phase lipids, possibly because the DAG-rich domains would be expected to have a longer half-life in gel phase lipids *(34,38)*.

After reaching a maximum at a certain molar ratio of a lipid, PKC activity often decreases with further increase in that lipid. Such biphasic effects argue against interaction of the lipid at a single site on PKC. A substance cannot

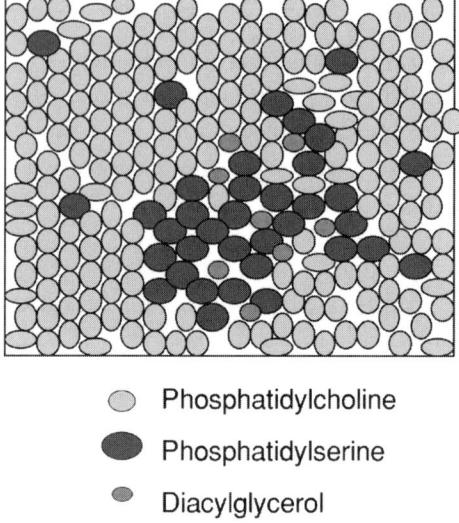

○ Phosphatidylcholine
● Phosphatidylserine
◓ Diacylglycerol

Fig. 3. Illustration of preferential clustering of DAG with PS vs PC.

exhibit agonist and antagonist effects through the same site. The following possibilities exist for the biphasic effects: 1) activation of PKC may require a specific lipid domain composition, which is lost as more of one constituent is added; 2) increase in the molar ratio of one lipid may dilute out another necessary component in that lipid, for example, dilution of DAG in PS (illustrated in **Fig. 4**); 3) Increase in the molar ratio of one lipid may dilute any constituents that bind to that lipid (e.g., substrates; see below); and 4) PKC activation may be facilitated by the interface between domains, which would diminish as one lipid component increases in molar ratio. It is important to note that virtually any lipid component will exhibit preferential partitioning into certain lipid domains and thus will affect the domain formation, therefore PKC activation, even if the lipid does not interact directly with PKC.

2.2. Optimizing PKC and Substrate Concentrations

The PKC recognition motif in substrate proteins consists of serine or threonine residues surrounded by various basic amino acids. (See Nishikawa et al., **ref. 50**, for optimal linear recognition motifs for individual isozymes.) Like the PKC pseudosubstrate motifs *(51)*, which contribute to membrane binding (*see* Fig. 2; **refs. 52** and *53*), the basic PKC substrates often bind to the acidic lipids. In fact, this colocalization of PKC with substrates probably contributes to substrate specificity of these otherwise promiscuous enzymes. Binding of substrates or PKC to acidic lipid domains can increase domain formation and

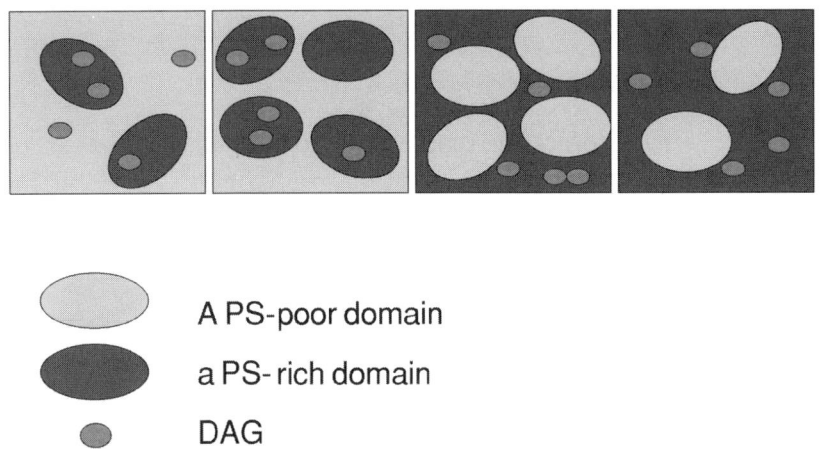

A PS-poor domain

a PS- rich domain

DAG

Fig. 4. Illustration of the concentration of DAG in PS-rich domains and dilution of DAG in those domains as the molar ratio of PS increases.

longevity, alter domain composition, and, at high substrate concentrations, compete with PKC for binding to the lipid domains *(54,55)*. Thus, plots of PKC activity as a function of substrate concentration often fail to plateau, but decrease after a maximum is achieved (e.g., **ref.** *56*). The concentration of substrate required for maximal PKC activity depends on the nature of the substrate and the lipid mixtures used. Furthermore, in the case of some substrates, such as neuromodulin, binding to acidic lipids may stabilize a secondary structure of the substrate, such as an α-helix, that may be required for recognition by the enzyme *(43)*. For all of these reasons, the concentration of substrate used must be optimized for the lipid and enzyme conditions used.

The enzyme concentration also must be optimized for particular lipid and substrate conditions. The concentration of an activating lipid at which PKC activity begins to decrease can be pushed higher if enzyme concentration is increased *(49)*. This observation suggests that enzyme–substrate or enzyme–enzyme complexes are diluted in the plane of activating lipid domains when such domains are in excess. Even when PKC autophosphorylation is monitored rather than phosphorylation of an exogenous substrate, the decrease in activity with excess lipid activator can be moved to higher lipid concentrations when PKC concentration is increased *(49)*. This result argues either for an intermolecular autophosphorylation mechanism or for the possibility that PKC-PKC dimers (or higher order aggregates) are more active than monomers. (*see* **Fig. 5** for an illustration of these possibilities.) Huang et al. *(57)* have provided evidence for lipid-dependent dimerization of PKC.

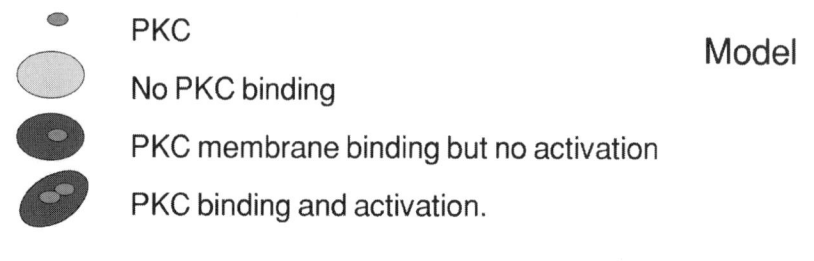

Model

(increasing generation of activating domains) ⟶

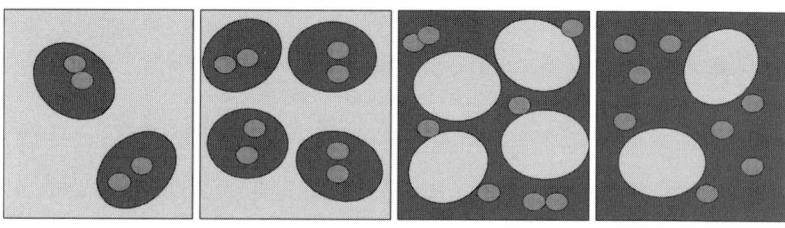

◄─── (increasing competition for (removal of) active domains)

Fig. 5. Illustration of how an increase in activating domains (dark gray) may dilute out PKC–PKC or PKC–substrate complexes that bind to those domains. An insufficient number of activating domains or competition for domain binding also may limit the PKC activation.

3. Problems with Assay of PKC Activity in Whole Cells

A much bigger problem than measurement of PKC activity in vitro is the assessment of PKC activation in cells. To implicate PKC in a response to an agonist, one typically wants to show 1) that the agonist activates PKC; 2) that specific PKC activators replicate the effect; and 3) that specific PKC inhibitors block it. Downregulation or knockout of PKC isozymes followed by regain of function with add-back of the isozyme(s) strengthens the argument. Problems exist with each of these approaches.

3.1. Limited Specificity of PKC Activators and Inhibitors

Originally, any cellular response to addition of phorbol ester tumor promoters was taken as evidence of PKC involvement. However, we now know that several signaling molecules in addition to PKC, such as N-chimaerin *(58)*, a Rac-GTPase activating protein, also bind to and are activated by phorbol esters (reviewed in **refs. 9** and **59**). Nor does lack of a cellular response to phorbol esters rule out involvement of PKC because some isozymes (ζ and λ/ι) do not bind to or become activated by these agents (reviewed in **refs. 6–9**). Similar problems in specificity arise with PKC inhibitors (*see* **part IX** of this volume).

3.2. Difficulties in Assessing the Activation State of PKCs in Cells

A specific assay for PKC activation in a cell is especially problematic because a specific PKC or PKC isozyme substrate been identified in very few cells. PKCs are notably promiscuous and phosphorylate numerous substrates that also are phosphorylated by other kinases. To a large extent, the substrate specificity is dictated by the subcellular location of enzyme activation and by the existence of particular substrates in the vicinity.

PKC activity can be examined in crude cell fractions with endogenous or exogenous substrates. However, addition of lipids to cytosolic enzyme can show only that the enzyme is activatable, not that is was active in the unbroken cell. Extraction of the enzyme from membranes removes it from any lipids that may (nor may not) have activated it in the cell. Furthermore, the detergents used in the membrane extraction may activate or inhibit PKC depending on their concentration.

The operational definition of PKC activity as PS dependent allows for some estimation of PKC activity in the presence of the numerous other kinases present in crude cell extracts. It is essential, in such assays, to examine activity in the presence vs the absence of PS, not PS and calcium, so as not to include the many kinases dependent on calcium only. Calcium can be included in tubes with and those without the PS so that the discriminator is the presence of PS only. Assay of the activity of individual PKC isozymes is much more difficult, however, and requires their separation from other PKC isozymes. This can be accomplished by standard purification methods (ion exchange, hydrophobic, and affinity chromatography; *see* **part II** of this volume). However, standard purification of enzyme from multiple cell treatment conditions is prohibitive in terms of the cost of starting material for each sample and the time required for purification; and it is not quantitative due to variable (and often poor) yields at each step.

3.3. Problems with Immune Complex Kinase Assays

Immunoprecipitation with isozyme-specific antibodies is much faster but still suffers from both quantitative and qualitative limitations. Because PKC is so hydrophobic, significant amounts can be precipitated with unrelated antibodies (e.g., antibodies reactive with other PKC isozymes or even the secondary antibody used in the immunoprecipitation). Thus, some copelleting of other isozymes, in addition to co-precipitation of any PKC-binding proteins, is possible and must be assessed by Western blotting of immunoprecipitates.

However, many antibodies that are effective in Western blotting are poor at precipitating native protein and quantitative immunoprecipitation is rare.

When immunoprecipitation is used in conjunction with kinase assays to measure activation of enzymes in cells, several problems must be recognized. First, because immunoprecipitation buffers often contain detergents, the enzymes usually have been removed from any activating lipid. Thus add-back of lipid may be able to demonstrate activatable enzyme, but would not be expected to reflect the activation state of the isozyme in the cell before extraction. In addition, the detergent must be diluted out to non-activating as well as non-inhibiting concentrations. However, many have used this sort of assay to assess activation of PKC in a cell after some appropriate stimulus and agonist-induced changes in PKC have been reported (e.g., **refs. 60** and **61**).

For an immunoprecipitation assay to reflect a change in activity of PKC in the cell, the activation would have to cause some sort of stable modification of the enzyme that survives the immunoprecipitation/removal of activating lipid. It is possible that some phosphorylation of PKC accompanying its activation accounts for the agonist-increased PKC activity some have observed in immune complex kinase assays. Auto- or heterologous PKC phosphorylation also may affect the stability of the enzyme or its susceptibility to proteolysis (e.g., **refs. 62** and **63**) with consequent alteration in its lipid requirements. Konishi et al. (**64**) have identified phosphorylation of Y-311 in the hinge region of PKCδ to be responsible for H_2O_2-induced constitutive activation of PKCδ without lipid. (For a review on effects of PKC phosphorylation, *see* **ref. 65** and **part V** of this volume.)

Other possibilities for observation of altered PKC activity in immune complex assays include activation or inhibition of PKC by the antibody (**66**), agonist-induced inclusion of additional substrates in the form of antibody or co-precipitated proteins, and noncovalent forms of enzyme activation that might survive immunoprecipitation. Examples of the latter may include co-precipitation of activating or inhibitory lipids or proteins. Polyanionic substrates such as protamine sulfate can bypass the acidic lipid requirement for PKC activation (**2,56**), and nonsubstrate polyanions can activate PKC as well (**67**). Numerous proteins have been identified that interact with PKC at sites other than the substrate site and many stabilize an activated or inhibited form of the enzyme. These include receptors for activated C-kinase, which selectively co-immunoprecipitate with specific isozymes, and substrates that interact with C-kinase (for reviews, *see* **refs. 9, 68, 69**, and **part VIII** of this volume). Included among the protein-dependent mechanisms for PKC activation is the possibility that PKC-PKC complexes may exhibit greater activity than do monomers (**49,57**).

When immune complex assays do reveal a difference in activity of PKC isozymes before and after cell stimulation, an explanation for the change seldom is provided. Because of their unknown complexity and the difficulties in their interpretation, immune complex PKC assays will not be described specifically. It is possible to adapt the typical PKC kinase assay described in the next chapter for use with immunoprecipitated enzymes. However, the performer of such assays should go to some effort to characterize the conditions achieved in the assay (amount and phosphorylation state of PKC; identity and amount of co-precipitated proteins). And conclusions should be qualified appropriately.

3.4. Use of Subcellular Translocation as a Measure of Activation

Because many of the PKC isozymes move from one subcellular compartment to another upon activation (often from cytosol to plasma membrane), this translocation often reflects activation and sometimes is used instead of direct assays as a measure PKC activation within cells. Translocation can be both isozyme- and cell type-specific and time course experiments may be required to catch a transient translocation. Thus, lack of translocation in response to an agonist does not serve to rule out PKC activation. A basic protocol for translocation experiments is provided in the chapter following in vitro assays. More detail on the nature of the PKC–lipid interaction and more sophisticated kinetic analysis of PKC translocation can be found in **part IV** of this volume.

Acknowledgments

This work was supported by USPHS grant GM31184 (to J. J. Sando). I thank Dr. Jacquelyn Beals for her helpful comments on the manuscript.

References

1. Takai, Y., Kishimoto, A., Inoue, M., and Nishizuka, Y. (1977) Studies on a cyclic nucleotide-independent protein kinase and its proenzyme in mammalian tissues. I. Purification and characterization of an active enzyme from bovine cerebellum. *J. Biol. Chem.* **252,** 7603–7609.
2. Inoue, M., Kishimoto, A., Takai, Y., and Nishizuka, Y. (1977) Studies on a cyclic nucleotide-independent protein kinase and its proenzyme in mammalian tissues. II. Proenzyme and its activation by calcium-dependent protease from rat brain. *J. Biol. Chem.* **252,** 7610–7616.
3. Takai, Y., Kishimoto, A. Iwasa, Y., Kawahara, Y., Mori, T., and Nishizuka, Y. (1979) Calcium-dependent activation of a multifunctional protein kinase by membrane phospholipids. *J. Biol. Chem.* **254,** 3692–3695.
4. Kishimoto, A., Takai, Y., Mori, T., Kikkawa, U., and Nishizuka, Y. (1980) Activation of calcium and phospholipid-dependent protein kinase by diacylglycerol, its possible relation to phosphatidylinositol turnover. *J. Biol. Chem.* **255,** 2273–2276.

5. Castagna, M., Takai, Y., Kaibuchi, K., Sano, K., Kikkawa, U., and Nishizuka, Y. (1982) Direct activation of calcium-activated, phospholipid-dependent protein kinase by tumor-promoting phorbol esters. *J. Biol. Chem.* **257**, 13,193–13,196.
6. Stabel, S. and Parker, P. (1991) Protein kinase C. *Pharmacol. Ther.* **51**, 71–95.
7. Newton, A. C. (1995) Protein kinase C: structure, function and regulation. *J. Biol. Chem.* **270**, 28,495–28,498.
8. Mellor, H. and Parker, P. J. (1998) The extended protein kinase C superfamily. *Biochem. J.* **332**, 281–292.
9. Ron, D. and Kazanietz, M. G. (1999) New insights into the regulation of protein kinase C and novel phorbol ester receptors. *FASEB J.* **13**, 1658–1676.
10. Sando, J. J., Maurer, M. C., Bolen, E. J., and Grisham, C. M. (1992) Role of cofactors in protein kinase C activation. *Cell. Signal.* **4**, 595–609.
11. Zidovetski, R. and Lester, D.S. (1992) The mechanism of activation of protein kinase C: a biophysical perspective. *Biochem. Biophys. Acta* **1134**, 261–272.
12. Epand, R. M. (1994) In vitro assays of protein kinase C activity. *Anal. Biochem.* **218**, 241–247.
13. Stubbs, C. D. and Slater, S. J. (1996) The effects of nonlamellarforming lipids on membrane protein-lipid interactions. *Chem. Physics Lipids* **81**, 185–196.
14. Lee, M.-H. and Bell, R. M. (1992) Supplementation of the phosphatidyl-L-serine requirement of protein kinase C with nonactivating phospholipids. *Biochemistry* **31**, 1576–1582.
15. Lee, M.-H. and Bell, R. M (1989) Phospholipid functional groups involved in protein kinase C activation, phorbol ester binding, and binding to mixed micelles. *J. Biol. Chem.* **264**, 14,797–14,805.
16. Newton, A. C. and Keranen, L. M. (1994) Phosphatidyl-L-serine is necessary for protein kinase C's high affinity interaction with diacylglycerol-containing membranes. *Biochemistry* **33**, 6651–6658.
17. Young, N. and Rando, R. R. (1984) The stereospecific activation of protein kinase C. *Biochem. Biophys. Res. Comm.* **122**, 818–823.
18. Sharkey, N. A., Leach, K. L. and Blumberg, P. M. (1984) Competitive inhibition by diacylglycerol of specific phorbol ester binding. *Proc. Natl. Acad. Sci. USA* **81**, 607–610.
19. Burns, D. J. and Bell, R. M. (1991) Protein kinase C has two phorbol ester binding domains. *J. Biol. Chem.* **266**, 18,330–18,338.
20. Zhang, G., Kazanietz, M. G., Blumberg, P. K., and Hurley, J. H. (1995) Crystal structure of the Cys2 activator-binding domain of Protein Kinase Cδ in complex with phorbol ester. *Cell* **81**, 917–924.
21. Slater, S. J., Ho, C., Kelly, M. B., Larkin, J. D., Taddeo, F. J., Yeager, M. D., et al. (1996) Protein kinase Cα contains two activator binding sites that bind phorbol esters and diacylglycerols with opposite affinities. *J. Biol. Chem.* **271**, 4627–4631.
22. Conesa-Zamora, P., Gomez-Fernandez, J. C., and Corbalan-Garcia, S. (2000) The C2 domain of protein kinase Cα is directly involved in the diacylglycerol-dependent binding of the C1 domain to the membrane. *Biochim. Biophys. Acta* **1487**, 246–254.

23. Ochoa, W. F., Garcia-Garcia, J., Fita, I., Corbalan-Garcia, S., Verdaguer, N., and Gomez-Fernandez, J. C. (2001) Structure of the C2 domain from novel protein kinase C epsilon. A membrane binding model for Ca^{2+}-independent C2 domains. *J. Mol. Biol.* **311,** 837–849.

24. Medkova, M. and Cho, W. (1999) Interplay of C1 and C2 domains of protein kinase C-alpha in its membrane binding and activation. *J. Biol. Chem.* **274,** 19,852–19,861.

25. Hurley, J. H., Newton, A. C., Parker, P. M., and Nishizuka, Y. (1997) Taxonomy and function of C1 protein kinase C homology domains. *Protein Sci.* **6,** 477–480.

26. Bolen, E. J. and Sando, J. J. (1992) Effects of phospholipid unsaturation on protein kinase C activation. *Biochemistry* **31,** 5945–5951.

27. Epand, R. M., Epand, R. F., Leon, B. T. C., Menger, F. M., and Kuo, J. F. (1991) Evidence for the regulation of the activity of protein kinase C through changes in membrane properties. *Biosci. Rep.* **11,** 59–64.

28. Senisterra, G. and Epand, R. M. (1993) Role of membrane defects in the regulation of the activity of protein kinase C. *Arch. Biochem. Biophys.* **300,** 378–383.

29. Slater, S. J., Kelly, M. B., Taddeo, F. J., Ho, C., Rubin, E., and Stubbs, C. D. (1994) The modulation of protein kinase C activity by membrane bilayer structure. *J. Biol. Chem.* **269,** 4866–4871.

30. Giorgione, J. R., Huang, Z., and Epand, R. M. (1998) Increased activation of protein kinase C with cubic phase lipid compared with liposomes. *Biochemistry* **37,** 2384–2392.

31. Das, S. and Rand, R. P. (1984) Modification by diacylglycerol of the structure and interaction of various phospholipid bilayer membranes. *Biochem. Biophys. Res. Commun.* **124,** 491–496.

32. Epand, R. M. (1985) Diacylglycerols, lysolecithin, or hydrocarbons markedly alter the bilayer to hexagonal phase transition temperature of phosphatidyletha-nolamines. *Biochemistry* **24,** 7029–7095.

33. Goldberg, E. M., Lester, D. S., Borchardt, D. B., and Zidovetzki, R. (1994) Effects of diacylglycerols and Ca^{2+} on structure of phosphatidylcholine/phosphatidylserine bilayers. *Biophys. J.* **66,** 382–393.

34. Dibble, A. R. G., Hinderliter, A. K., Shen, Y. A., Sando, J. J., and Biltonen, R. L. (1996) Lipid lateral heterogeneity in phosphatidylcholine/phosphatidylserine/ diacylglycerol vesicles and its influence on Protein Kinase C activation. *Biophys. J.* **71,** 1877–1890.

35. Shen, Y.-M. A., Chertihin, O. I., Biltonen, R. L., and Sando, J. J. (1999) Lipid-dependent activation of protein kinase C-α by normal alcohols. *J. Biol. Chem.* **274,** 34,036–34,044.

36. Mosior, M. and Epand, R. M. (1994) Characterization of the calcium-binding site that regulates association of protein kinase C with phospholipid bilayers. *J. Biol. Chem.* **269,** 13,798–13,805.

37. Boni, L. T. and Rando, R. R. (1985) The nature of protein kinase C activation by physically defined phospholipid vesicles and diacylglycerols. *J. Biol. Chem.* **260,** 10,819–10,825.

38. Hinderliter, A. K., Dibble, A. R. G., Biltonen, R. L., and Sando, J. J. (1997) Activation of protein kinase C by coexisting diacylglycerol-rich and diacylglycerol-poor lipid domains. *Biochemistry* **36**, 6141–6148.
39. Walker, J. M., Homan, E. C., and Sando, J. J. (1990) Differential activation of protein kinase C isozymes by short chain phosphatidylserines and phosphatidylcholines. *J. Biol. Chem.* **265**, 8016–8021.
40. Walker, J. M. and Sando, J. J. (1988) Activation of protein kinase C by short chain phosphatidylcholines. *J. Biol. Chem.* **263**, 4537–4540.
41. Hannun, Y. A., Loomis, C. R., and Bell, R. M. (1985) Activation of protein kinase C by Triton X-100 mixed micelles containing diacylglycerol and phosphatidylserine. *J. Biol. Chem.* **260**, 10,039–10,043.
42. Maurer, M. C., Sando, J. J., and Grisham, C. M. (1992) High-affinity Ca- and substrate-binding sites on protein kinase C α as determined by nuclear magnetic resonance spectroscopy. *Biochemistry* **31**, 7714–7721.
43. Vinton, B. B., Wertz, S. L., Steere, J., Grisham, C. M., Cafiso, D. S., and Sando, J. J. (1998) Influence of lipid on the structure and phosphorylation of Protein kinase C α substrate peptides. *Biochem. J.* **330**, 1433–1442
44. Bazzi, M. D. and Nelsestuen, G. L. (1987) Association of protein kinase C with phospholipid vesicles. *Biochemistry* **26**, 115–122.
45. Bazzi, M. D., Youakim, M. A., and Nelsestuen, G. L. (1992) Importance of phosphatidylethanolamine for association of protein kinase C and other cytosolic proteins with membranes. *Biochemistry* **31**, 1125–1134.
46. Sando, J. J. and Chertihin, O. I. (1996) Activation of protein kinase C by lysophosphatidic acid. *Biochem. J.* **317**, 583–588.
47. Cooney, R. V., Pung, A., Harwood, P. J., Boynton, A. L., Zhang, L.-X., Hossain, M. Z., et al. (1992) Inhibition of cellular transformation by triphenylmethane: a novel chemopreventive agent. *Carcinogenesis* **13**, 1107–1112.
48. Rao, Y.-P., Stravitz, R. T., Vlahcevic, Z. R., Gurley, E. C., Sando, J. J., and Hylemon, P. B. (1997) Activation of protein kinase C α and β by bile acids: correlation with bile acid structure and diacylglycerol formation. *J. Lipid Res.* **38**, 2446–2454.
49. Sando, J. J., Chertihin, O. I., Owens, J. M., and Kretsinger, R. H. (1998) Contributions to maxima in Protein Kinase C activation. *J. Biol. Chem.* **274**, 34,022–34,027.
50. Nishikawa, K., Toker, A., Johannes, F.-J., Songyang, Z., and Cantley, L. C. (1997) Determination of the specific substrate sequence motifs of protein kinase C isozymes. *J. Biol. Chem.* **272**, 952–960.
51. House, C. and Kemp, B. E. (1987) Protein kinase C contains a pseudosubstrate prototype in its regulatory domain. *Science* **238**, 1726–1728.
52. Mosior, M. and McLaughlin, S. (1992) Peptides that mimic the pseudosubstrate region of protein kinase C bind to acidic lipids in membranes. *Biophys. J.* **60**, 149–159.
53. Oancea, E. and Meyer, T. (1998) Protein kinase C as a molecular machine for decoding calcium and diacylglycerol signals. *Cell* **95**, 307–318.
54. Yang, L. and Glaser, M. (1995) Membrane domains containing phosphatidylserine and substrate can be important for the activation of protein kinase C. *Biochemistry* **34**, 1500–1506.

55. Bazzi, M. and Nelsestuen, G. L. (1991) Extensive segregation of acidic phospholipids in membranes induced by protein kinase C and related proteins. *Biochemistry* **30**, 7961–7969.
56. Bazzi. M. D. and Nelsestuen, G. L. (1987) Role of substrate in imparting calcium and phospholipid requirements to protein kinase C activation. *Biochemistry* **26**, 1974–1982.
57. Huang, S.-M., Leventhal, P. S., Wiepz, G. J., and Bertics, P. J. (1999) Calcium and phosphatidylserine stimulate the self-association of conventional protein kinase C isozymes. *Biochemistry* **38**, 12,020–12,027.
58. Hall, C., Monfried, C., Smith, C., Lim, H. H., Kozma, R., Ahmed, S., et al. (1990) Novel human brain cDNA encoding a 34 000 Mr protein, n-chimaerin, related to both the regulatory domain of protein kinase C and BCR, the product of breakpoint cluster region gene. *J. Mol. Biol.* **211**, 11–16.
59. Kazanietz, M. G. (2000) Eyes wide shut: protein kinase C isozymes are not the only receptors for the phorbol ester tumor promoters. *Mol. Carcinogenesis* **28**, 5–11.
60. Arnould, T., Sellin, L., Benzing, T., Tsiokas, L., Cohen, H. T., Kim, E., et al.(1999) Cellular activation triggered by the autosomal dominant polycystic kidney disease gene product PKD2. *Mol. Cell. Biol.* **19**, 3423–3434.
61. Suzuma, I., Suzuma, K., Ueki, K., Hata, Y., Feener, E.P., King, G. L., et al. (2002) Stretch-induced retinal vascualr endothelial growth factor expression is mediated by phosphatidylinositol 3-kinase and protein kinase C (PKC)-ζ but not by stretch-induced ERK1/2, Akt, Ras, or classical/novel PKC pathways. *J. Biol. Chem.* **277**, 1047–1057.
62. Lee, H. W., Smith, L., Pettit, G. R., and Smith, J. B. (1997) Bryostatin 1 and phorbol ester down-modulate protein kinase C-alpha and -epsilon via the ubiquitiin/ proteasome pathway in human fibroblasts. *Mol. Pharmacol.* **51**, 439–447, 1997
63. Kang, B. S., French, O. G., Sando, J. J., and Hahn, C. S. (2000) Activation-dependent degradation of protein kinase Cη. *Oncogene* **19**, 4263–4272.
64. Konishi, H., Yamaguchi, E., Taniguchi, H., Wamamoto, T., Matsuzaki, H., Ono, Y., Kikkawa, U., and Nishizuka, Y. (2000) Phosphorylation sites in protein kinase Cδ in H_2O_2-treated cells and its activation by tyrosine kinase in vitro. *Proc. Natl. Acad. Sci. USA* **98**, 6587–6592.
65. Parekh, D. B., Ziegler, W., and Parker, P. J. (2000) Multiple pathways control protein kinase C phosphorylation. *EMBO J.* **19**, 496–503.
66. Makowske, M. and Rosen, O. M. (1989) Complete activation of protein kinase C by an antipeptide antibody directed against the pseudosubstrate prototope. *J. Biol. Chem.* **264**, 16,155–16,159.
67. Leventhal, P. S. and Bertics, P. J. (1993) Activation of protein kinase C by selective binding of arginine-rich polypeptides. *J. Biol. Chem.* **268**, 13,906–13,913.
68. Mochly-Rosen, D. and Kauvar, L. M. (1998) Modulating protein kinase C signal transduction. *Adv. Pharmacol.* **44**, 91–145.
69. Jaken, S. and Parker, P. J. (2000) Protein kinase C binding partners. *BioEssays* **22**, 245–254.

6

Enzyme Assays for Protein Kinase C Activity

Julianne J. Sando and Jacquelyn K. Beals

1. Introduction

As protein kinase C (PKC) isozymes are implicated in increasing numbers of signal transduction pathways and disease processes, more investigators from different fields have the need to assay PKC activity. A simple in vitro assay suitable for measuring PKC activity in soluble cell extracts is provided here. Like other kinases, PKC catalyzes the transfer of the terminal phosphate from ATP to a protein or peptide substrate. Use of γ-^{32}P-labeled ATP allows this transfer to be monitored by scintillation counting or by autoradiography of the substrate following its removal from the excess ATP.

The PKC phosphorylation site in a substrate protein is a Ser or Thr residue surrounded by basic amino acids in a pattern that varies somewhat for different PKC isozymes. Nishikawa et al. (1) have characterized ideal linear recognition motifs for different isozymes. Recognition of some secondary structural motifs such as an α-helix also has been proposed (2). A great deal of overlap in substrate recognition exists, with several PKC isozymes being capable of phosphorylating the same substrate. Many PKC substrates also can be phosphorylated by kinases other than PKC. Lysine-rich histone can serve as an inexpensive in vitro substrate for many PKC isozymes (and other kinases). Myelin basic protein (MBP) is better for some calcium-independent PKCs.

PKC activity is distinguished from that of other kinases by its lipid dependence. Assays are conducted in the presence and in the absence of activating phospholipid vesicles or micelles, with the difference defined as PKC activity. Thus, the activation occurs on a two-dimensional membrane surface rather than in solution and this feature significantly complicates interpretation of PKC assays. All PKC isozymes (reviewed in refs. 3–6) are activated best by

From: *Methods in Molecular Biology, vol. 233: Protein Kinase C Protocols*
Edited by: A. C. Newton © Humana Press Inc., Totowa, NJ

phosphatidylserine (PS). Physical properties of the lipid, in addition to the head group structure, are critical to the activation (reviewed in **refs. *7–10***). The classic or cPKCs (α, βI, βII, and γ) are calcium-dependent because of their calcium-binding C2 domains (see the previous chapter for a cartoon illustrating structural differences in the isozymes). cPKCs and the calcium-independent novel or nPKCs (δ, ε, η, and θ) are further activated by diacylglycerol (DAG) or phorbol ester tumor promoters, which bind to one of two cysteine-rich C1 subdomains. Atypical or aPKCs (ζ, λ/ι, μ) are not activated by calcium or phorbol esters.

It is the molar ratio of lipid-soluble modulators like DAG with respect to the total lipid, rather than their concentration with respect to the aqueous solution, that is relevant to PKC activation. As the molar ratio of one lipid constituent increases, the molar ratio of some other constituent(s) necessarily decreases. Lipids can move in the plane of the membrane and segregate laterally into domains of different composition, size, and lifetime with varying abilities to activate PKC. Any substance that binds to or partitions preferentially into certain domains may alter the domain formation and thus PKC activity, even if it does not interact directly with PKC. PKC and many of its substrates are among the molecules that can bind to and alter domain formation. It is possible to have too many as well as too few activating domains for optimal PKC activation *(11)*. Thus, the lipid and calcium requirements for PKC activation are interdependent and vary with the type and concentration of the substrate and PKC isozyme as well. It is essential that in vitro PKC assays be optimized for the enzyme, lipid, and substrate conditions used. For a fuller discussion of these complexities in PKC assays, see Chapter 5 in this volume.

2. Materials

1. Phosphatidylserine, either mixed acyl chain PS, such as bovine brain PS (bbPS; from any supplier), or PS with defined acyl chains, such as dioleoyl-PS (DOPS) or 1-palmitoyl, 2-oleoyl-PS (from Avanti Polar-Lipids, Inc.). Store at $-20°$ or $-70°C$ in chloroform or chloroform–methanol under Ar or N_2 gas and protect from light.
2. DAG, such as 1,2-dioleoyl-sn-glycerol (DO; Avanti Polar-Lipids, Inc. or other supplier). Store at -20 or $-70°C$ in chloroform or chloroform–methanol and protect from light.
3. Argon or nitrogen gas for solvent evaporation.
4. An appropriate nonphosphate neutral range buffer. Tris-HCl (20 mM, pH 7.4) is least expensive. 3-N-(Morpholino)propanesulfonic acid (MOPS, 20 mM, pH 7.4) will remain at a more constant pH over varying temperatures (*see* **Note 1**).
5. CaCl$_2$, 75 mM in appropriate nonphosphate buffer at pH 7.4. Store at 4°C.
6. MgCl$_2$, 250 mM in appropriate nonphosphate buffer at pH 7.4. Store at 4°C.
7. ATP, 5 mM in water or appropriate nonphosphate buffer. Store at $-20°C$.

8. [γ-^{32}P]ATP, ~7000 Ci/mmol (ICN Biomedicals, Inc., Costa Mesa, CA or other supplier). Store in freezer (–20°C) with plexiglass shielding.
9. Lysine-rich Histone (Sigma HIIIS) 5 mg/mL or MBP or specific peptide substrate; store frozen at –20°C.
10. P-81 ion exchange paper (Whatman International, Maidstone, United Kingdom).
11. PKC source: cytosol or purified enzyme. [*See* **part II** of this book for PKC purification methods.]
12. 50 m*M* NaCl wash solution.
13. Shaking water bath and stir plate.
14. Beta scintillation counter, ideally one capable of Cerenkov counting, and Geiger counter for quantitating and monitoring ^{32}P.

3. Methods

The methods described below outline the following; 1) planning and set-up for the assay; 2) preparation of the lipids; 3) preparation of the aqueous reaction mixture; 4) preparation of the enzyme; and 5) initiation and termination of the reaction.

3.1. Planning and Set-Up

The experimental variable in a kinase assay may be one of the following: 1) the concentration of PKC; 2) the concentration or composition of lipids; 3) the concentration of additional cofactors such as calcium; 4) the incubation temperature; 5) the substrate; 6) the time course of the phosphorylation; or 7) the presence of PKC activators or inhibitors. Alternatively, identical conditions may be used to compare extracts from cells that had been treated differently. However, in the latter case, activity in the extract *does not* usually reflect activation of PKC in the cell (see the preceding introductory chapter for a discussion of PKC activation in a cell).

In all cases, it is essential to ensure that the assay is linear with time and enzyme concentration under the conditions used. It often is necessary to test several dilutions of an enzyme source to achieve conditions in which a linear relationship exists between enzyme concentration and activity. Decreasing the enzyme concentration or reaction time and/or increasing the substrate concentration may be necessary to establish linear assay conditions. Note also that *excess* activating lipid or substrate may decrease PKC activity. See Chapter 5 for a more complete discussion of these issues.

The planning involves deciding first what conditions will be examined and then how many replicates one wants with lipid and without lipid for each condition. For pilot experiments, we use a minimum of two plus-lipid tubes and one no-lipid tube per condition. In many experiments the no-lipid tubes do not vary among the conditions tested and can be averaged. Note that if lipid concentration is the variable, only one set of no-lipid tubes is required.

Table 1
Sample Protocol for Assay of PKC Activity
as a Function of PKC Concentration

Tube #	Condition	Ingredients (μL/tube)				
		PL[a]	PKC[b]	Buffer	Rxn Mix[c]	Total
1,2,3	0 PKC + PL	25	—	25	25	75
4,5	" – PL	—	—	50	25	75
6,7,8	1 × PKC + PL	25	5	20	25	75
9,10	" – PL	—	5	45	25	75
11,12,13	2 × PKC + PL	25	10	15	25	75
14,15	" – PL	—	10	40	25	75
16,17,18	3 × PKC + PL	25	15	10	25	75
19,20	" – PL	—	15	35	25	75
21,22,23	5 × PKC + PL	25	25	—	25	75
24,25	" – PL	—	25	25	25	75

[a]PL = Phospholipid vesicle preparation usually containing PS and DAG. If lipid composition is a variable, it will be necessary to prepare different lipid vesicle preparations for each condition.
[b]If PKC concentrations span a larger range, it will be necessary to make serial dilutions of the PKC stock rather than to pipet varying volumes directly into the assay as indicated here.
[c]Rxn Mix = aqueous reaction mixture containing ATP, γ-^{32}P-ATP, $MgCl_2$, protein, or peptide substrate, $CaCl_2$ (if required), and any other water-soluble ingredient common to all the tubes.

For publication quality experiments, we increase the number of replicates in addition to performing several independent replicate experiments. We find it convenient to carry out the assays in 75 μL of final volume with buffer diluting or replacing a variable constituent. **Table 1** shows a sample protocol in which enzyme concentration is the variable.

To set up for the experiment, the following should be performed.

1. Number enough small tubes for all conditions/replicates in the assay (*see* **Note 2**).
2. Number, in pencil, pieces of P81 paper ion exchange paper approx 3 cm square to correspond to each tube and arrange these for quick and orderly pick-up (*see* **Note 3**).
3. Set the temperature of the water bath to 30°C or other desired incubation temperature.
4. Prepare several liters of 50 m*M* NaCl wash solution, place approx 400 mL in a beaker with stirring, and shield this assembly for its use with ^{32}P.

Table 2 shows the final concentrations and volumes of appropriate stock solutions per assay tube for a standard assay. It is, of course, possible to vary

Table 2
Concentration of Each Ingredient in Standard PKC Assay

	Final concentration in assay	Stock concentration	Volume (µL) of stock/ 75 µL total assay volume	
Lipids				
bbPS	40 µg/mL	10 mg/mL in $CHCl_4$	0.3	These are dried and resuspended as
DO	1.6 µg/mL	1 mg/mL in $CHCl_4$	0.12	MLVs in 25 µL MOPS/tube
Aqueous Reaction Components				
$MgCl_2$	5 mM	250 mM in MOPS	1.5	
$CaCl_2$	300 µM	75 mM in MOPS	0.3	
ATP	40 µM	5 mM in MOPS	0.6	
Histone	0.2 mg/mL	5 mg/mL in MOPS	3.0	
MOPS	20 mM, pH 7.4	20 mM, pH 7.4	19.6	
[γ-^{32}P]ATP	1–4 Ci/mmol	1–5 mCi/30 µL	[1–5 µL per entire assay; volume/tube is negligible]	
Enzyme Source				
PKC	0.2–20 nM	0.6–60 nM in MOPS	25	

the stock concentrations and volumes to obtain the same final concentrations and it often is necessary to change the final concentrations of individual items to optimize the assay for specific enzyme, lipid, and substrate conditions. We find it convenient to group the lipids, the aqueous reaction components (Reaction Mixture) and the enzyme constituents separately. Their preparation will be discussed individually.

3.2. Preparation of the Lipid Vesicles

Lipid composition and concentration in the reaction mixture greatly affect enzyme activity. When the goal is to determine the maximal activity of a PKC preparation, it is essential to optimize the lipid conditions for the enzyme and substrate conditions used. If one simply wants to identify PKC-containing fractions from a purification column, an easier, cheaper lipid preparation with only bbPS and DAG may be sufficient. This simpler preparation will be described first.

3.2.1. Preparation of Multilamellar Vesicles Containing bbPS and DO

Table 2 provides an example of the volumes of bbPS and DO stocks needed per lipid-containing tube in a simple assay. The lipid vesicles for the entire experiment are prepared in a single tube and then apportioned into those individual assay tubes that require lipid. Thus, in the example shown in **Table 1**, there are 25 total tubes but only 15 that will contain lipid. It is necessary to prepare some extra volume of all stocks because it is not possible to remove quantitatively all of the liquid volume from a tube. Thus, in this sample experiment, we would prepare enough lipid vesicles for at least 18 tubes. At 0.3 µL of 10 mg/mL bbPS per tube, we will need 5.4 µL for 18 tubes. Similarly, at 0.12 µL of 1 mg/mL DO per tube, we will need 2.16 µL for 18 tubes. The lipid stocks are in chloroform (or chloroform:methanol) and must be pipetted into a glass tube. The chloroform will destroy PKC activity and therefore it must be removed completely by evaporation under an inert gas such as nitrogen or argon. The dried lipid mixture is then resuspended in the assay buffer (e.g., MOPS, pH 7.4) with vortexing to make multilamellar vesicles (MLVs) or with sonication to make small unilamellar vesicles (SUVs) (*see* **Note 4**). Other vesicle preparations can be made as described in the preceding introductory chapter. We find it convenient to resuspend the vesicles in a volume of 25 µL/assay tube; thus they will comprise one-third of the final volume (75 µL) per assay tube. The following summarizes the steps in this process:

1. Determine the number of lipid-containing tubes in the entire assay. Add 2–3 to this number and multiply the total by the volume per tube of bbPS and DO stocks (0.3 µL of 10 mg/mL bbPS/tube and 0.12 µL of 1 mg/mL DO/tube).
2. Pipet both lipids into a small glass tube (e.g., 13×75 mm) and evaporate the organic solvent under a stream of N_2 or Ar gas while rotating the tube so that the lipid evenly coats the bottom of the tube as the solvent evaporates. Leave the tube under the gas stream for another 5–15 min (e.g., while preparing the reaction mixture) to ensure that absolutely all of the solvent is removed.
3. Add assay buffer (e.g., MOPS, pH 7.4) to the dried lipid mixture in a volume of 25 µL × the number of assay tubes for which lipid is being prepared. Vortex this solution vigorously to prepare MLVs (*see* **Note 4**).
4. Pipet 25 µL of the lipid vesicle suspension into each plus-lipid tube in the assay.

3.2.2. Preparation of MLVs with Different Lipid Molar Ratios

Both the total lipid concentration and the molar ratio of particular lipids may be variables in PKC activity assays. **Table 3** shows a sample calculation for preparation of a lipid mixture containing 1 mM total lipid per tube with 25 mol% DOPS, 70 mol% dioleoylphosphatidylcholine (DOPC), and 5 mol% DO. The proportion of one component, such as the PS, can be varied at the expense of all of the others together or at the expense of a single component such as the PC. If, for example, one wants to keep the mol% of DO constant at 5%, PS will be the variable X mol%) and PC will be ($95–X$) mol%. The lipid mixtures are dried and resuspended as described above.

Table 3
Calculation for Preparation of 1 m*M* Lipid with Molar Ratio
DOPC:DOPS:DO (70:25:5)

	mmol/ 1000 mL	=	μmoles / 0.075 mL	×	mw (μg/μmol)	=	μg/tube	÷	[Stock] (in CHCl$_4$)	=	vol/tube (μL)
DOPC	0.7		0.0525		786.15		41.3		50 mg/mL		0.83
DOPS	0.25		0.1875		810.03		151.9		10 mg/mL		15.2
DO	0.05		0.00375		620.99		2.33		10 mg/mL		0.23
Total	1 mmol/L		0.075 μmol/75 μL								

3.3. Preparation of the Aqueous Reaction Mixture

We find it convenient to combine the aqueous ingredients common to all tubes in the assay into a single reaction mixture that can be apportioned into the individual tubes. This minimizes the pipetting and permits the use of smaller volumes of more concentrated stocks because the volumes to be pipetted are multiplied out over the full assay. Thus, more of the total assay volume is reserved for additional experimental variables or for components such as enzyme that may be available only at low concentration. Note that buffer volume compensates for variations in volume of other constituents. If calcium-independent PKCs are assayed, calcium is omitted and comparable amounts of MBP or appropriate peptide (30 μ*M*) may serve as a better substrate than histone. We find it convenient for many assays to allot 25 μL (or one-third of the final reaction volume) to the aqueous reaction mixture (*see* **Note 5**). Volumes per tube of specific stock solutions are listed in **Table 2**. As with the lipid preparations, enough reaction mixture is prepared for several extra tubes.

The specific activity of the [γ-^{32}P]ATP in the reaction mixture is set in each experiment by adding a small amount of very high specific activity [γ-^{32}P]ATP (~7000 Ci/mmol) to the reaction mixture. Because the specific activity of the stock [γ-^{32}P]ATP is so high, the volume added and its contribution to the final ATP concentration is negligible; thus, the ATP concentration is set by the concentration of unlabeled ATP in the tube (40 μ*M*). The final specific activity for each assay is determined directly by counting an aliquot of the reaction mixture and using the standard formula:

$$\text{cpm/efficiency of the scintillation counter for counting } ^{32}\text{P} = \text{dpm}$$
$$2.22 \times 10^6 \text{ dpm} = 1 \text{ μCi}$$

Because of the short half-life of ^{32}P (14.28 d), the specific activity of the stock [γ-^{32}P]ATP decreases significantly each day. Thus, as the ^{32}P-ATP ages, greater volumes must be added to the reaction mixture to achieve the same final specific activity in the assay. When the ^{32}P-ATP stock is new, addition of 1 μL to an aqueous reaction mixture adequate for a 40- to 80-tube assay may be more than enough. If small assays are performed when the ^{32}P-ATP is fresh, it may be convenient to dilute the stock a few-fold with water in order to pipet measurable amounts (*see* **Note 6**). We aim for a specific activity of 0.3–3.0 Ci/mmol in a typical assay or approx 1–10×10^6 cpm per assay tube (*see* **Note 7**). A sample of the aqueous reaction mixture (e.g., 2.5 μL) is counted to confirm that enough [γ-^{32}P]ATP has been added and to determine the achieved specific activity. With the established specific activity, the cpm transferred to substrate per unit enzyme can be converted back to pmol.

The efficiency of scintillation counters for detecting ^{32}P varies and can be determined by counting a commercial standard under the conditions used in the typical assay. With scintillation fluor, efficiencies approaching 100% may be achieved for ^{32}P. Cerenkov counting without fluor decreases efficiency by about half; but this trade in sensitivity is usually worth the savings in cost of fluor and its disposal.

A summary of the steps in preparation of a typical reaction mixture follows.

1. Combine the water-soluble ingredients (MgCl$_2$, ATP, substrate, CaCl$_2$ (for cPKCs), buffer) needed for the entire kinase assay into one disposable tube. Calculate the amounts by multiplying the volume/tube of appropriate stock concentrations of each component (**Table 2**) by the total number of tubes in the assay (plus a few extras) to obtain the total volume for each substance.
2. Spike the reaction mixture with [γ-^{32}P]ATP by adding 1–5 μL of the stock [γ-^{32}P]ATP for the entire assay (*see* **Note 8**). Because of the short half-life of ^{32}P, the volume added to the reaction mixture must increase during its several weeks of gradually decreasing activity. Allow the stock vial to thaw at room temperature for 5–10 min behind plexiglas shielding and similarly keep the spiked reaction mixture behind plexiglass shielding.
3. Count an aliquot (1–5 μL) of the reaction mixture. If necessary, add more [γ-^{32}P]ATP to the reaction mixture to achieve an adequate minimal specific activity (1–10×10^6 cpm per 25 μL reaction mixture per assay tube).

3.4. Preparation of Enzyme

3.4.1. Sources of PKC

The source of PKC may be anything from a crude cellular extract (*see* **Note 9**) to purified PKC isozymes. Purification of PKC is described in **part II** of this volume.

3.4.2. Lability of PKC

PKC is extremely labile. In crude preparations, it is susceptible to degradation by contaminating proteases. Some proteases can generate the catalytic fragment of PKC, known as protein kinase M (PKM). PKM is a cofactor-independent kinase so its activity will be evident in the no-lipid tubes in the assays along with the activity of any other lipid-independent kinases (*see* **Note 10**).

Highly purified preparations of PKC are typically of lower concentration and, thus, are more prone to significant decrease in concentration with time due to adherence to the surface of the tubes in which they are stored. PKC also undergoes self-aggregation with loss of activity, a problem that may be encountered during attempts to concentrate the enzyme.

Repeated freeze-thaw cycles damage PKC, as they do many enzymes, so it is best to freeze the enzyme in multiple small aliquots appropriate for individual experiments. Quick freezing in dry ice-acetone or liquid nitrogen followed by storage at −70°C is recommended. Addition of 30% glycerol to the preparations helps to stabilize the enzyme during freezing. PKC solutions can be thawed slowly on ice or the tubes can be immersed into room temperature water but they must be returned immediately to ice as thawing is completed. Proteolysis and degradation of PKC are much more rapid at higher temperatures. However, with careful handling, a good PKC preparation can be maintained for many months.

3.4.3. Concentration of PKC

It is essential for the investigator to establish that the PKC assay conditions provide linear increases in activity with increasing enzyme concentration. This can be performed with enzyme preparations of unknown concentration by including several dilutions of PKC as illustrated in the example assay (**Table 1**). Because the enzyme preparations may not be stable over long time periods, especially if they have been repeatedly frozen and thawed, it is appropriate to re-assay preparations that have been stored for some months.

Ideally, one will use known concentrations of PKC in the assay. Given the dramatic variations in activity that can be observed in the same enzyme preparation under different lipid conditions, we find it best to quantitate the enzyme by phorbol ester binding (e.g., **ref. *13***). Although the existence of a second, low affinity phorbol ester binding site on some isozymes has been reported *(12)*, only one high affinity site can be detected on cPKCs and nPKCs with ^{3}H-phorbol 12,13-dibutyrate *(14*; for review, *see* **ref. *6***). Thus, the number of ^{3}H-phorbol 12,13-dibutyrate binding sites reflects the number of PKC molecules. This assay, too, is lipid-dependent and must be optimized for specific lipid conditions. It is not useful for quantitation of aPKCs, which do not bind phorbol esters, and it is not suitable for use in crude preparations that contain non-PKC phorbol ester binding molecules such as N-chimaerin (reviewed in **ref. 6**). However, the phorbol ester binding assay is subject to many fewer complications than is the activity assay and we find it to be more accurate than many other methods for estimating PKC concentration (*see* **Note 11**). The use of phorbol esters for studying PKC function is discussed more fully in Chapter 12 and in **part IX** of this book.

If one has a PKC preparation of known concentration, it also is possible to estimate the concentration of unknown samples by quantitative Western blotting. In this procedure, serial dilutions of the PKC standard are compared with serial dilutions of the unknown preparation subjected to gel electrophoresis and Western blotting at the same time. A standard curve can be constructed from densitometry of bands in the known sample and used to estimate the concentration of the unknown. This method should provide an estimate within twofold of the true concentration and provides confirmation that the enzyme is intact, not necessarily that is it active either for phorbol ester binding or for kinase activity.

With the standard assay described here, we find that activity is linear somewhere in the range of 0.2–20 nM PKC. Precisely where linearity is achieved completely depends on the lipid and substrate conditions used in the assay. *See* Chapter 5 for a fuller discussion of these issues.

3.5. Initiation and Termination of the Reaction

Kinase assays are incubated for timed intervals that must be long enough to obtain sufficient ^{32}P incorporation into the substrate for adequate sensitivity but not so long that some component other than the enzyme becomes limiting. The assays can be initiated with either the substrate(s) [aqueous reaction mixture] or the enzyme. We find it easier to start the reactions with a constituent that does not vary over the assay. Thus, if different enzyme sources are being compared, it may be easier to pipet them leisurely into different assay tubes and to start the reactions with the common reaction mixture. If different reaction conditions are being compared, it may be easier to start the reactions with a common enzyme addition. Sometimes one may or may not want one particular component to pre-incubate with the lipid.

The reactions are initiated by pipetting the final component into the first assay tube, vortexing, placing the tube in the water bath, and immediately starting a stopwatch. At timed intervals (15–30 sec depending on the skill of the researcher), subsequent tubes are started.

After the desired incubation time (e.g., 4 min), an aliquot of the reaction mixture is removed from the first tube and blotted onto the correspondingly numbered square of P81 paper. The removed aliquot should be less than the total volume of the tube because it is not possible to remove accurately the total contents of the tube. Typically, we spot 60 µL of the 75 µL total in the tube. The paper is dropped immediately into the beaker of NaCl wash solution with stirring (*see* **Note 12**). The tube is placed behind plexiglass or back in the water bath to provide shielding from the ^{32}P until it is discarded as radioactive waste. Successive tubes are stopped at the appropriate timed intervals. If more tubes are assayed than can be started and stopped in the appropriate reaction time (e.g., 4 min), the assays will have to be run in sets so that they can be terminated at appropriate times. Alternatively, a second person may assist in the assay and begin stopping the reactions at the appropriate time while the first person continues to start new tubes. The papers are washed with at least three changes of the NaCl wash solution, dried, and counted in a beta scintillation counter.

Alternative methods for stopping the reaction include the following: 1) addition of 1 mL of 25% trichloroacetic acid to the reaction tube followed by separation of the precipitated proteins from ATP by filtration or 2) addition of sodium-dodecyl sulfate-containing sample buffer and boiling, followed by separation of the radiolabeled protein from free ATP by gel electrophoresis.

A summary of the initiation and termination of a typical P81 assay follows.

1. At timed intervals (10–30 sec depending on the skill of the researcher), pipet 25 µL of reaction mixture into assay tubes already containing lipid and enzyme. Vortex the mixture and place the tube in the water bath.
2. At the end of the incubation time (e.g., 4 min), begin stopping the reactions at timed intervals by spotting 60 µL from each tube onto the correspondingly numbered pieces of P81 paper and depositing these into the beaker of NaCl wash solution. Add two blank P81 squares to the wash to serve as controls for adequacy of the wash.
3. Change the NaCl wash solution at least three times at intervals of at least 20 min, pouring the waste into a liquid ^{32}P waste container (*see* **Note 13**).
4. Rinse the papers briefly with acetone to aid drying and allow them to dry thoroughly behind plexiglass shielding in a fume hood. The acetone may be reused in other assays if it is not too radioactive (*see* **Note 14**).
5. Count the papers in a beta scintillation counter. They may be counted dry (Cerenkov counting) to eliminate the cost of fluor. The vials, too, can be reused after discarding the papers in a dry ^{32}P waste container.
6. Subtract the counts on the blank P81 papers from the rest of the values. Average the plus-lipid and minus-lipid replicates for each condition and calculate the difference, which represents PKC activity.

4. Notes

1. HEPES, or PIPES buffers also are acceptable. Phosphate buffers cannot be used for kinase assays, which involve a phosphate transfer.

2 The tubes should be small (e.g., 12 × 75 mm) to facilitate delivery of µL volumes to the bottom of the tubes and their subsequent mixing and reaction. If calcium is a variable, it is necessary to use polypropylene tubes to avoid calcium contamination from the glass.

3. We find it convenient to order the P81 papers in the wells of a flat Styrofoam packaging rack for 50-mL conical test tubes.

4. MLVs are prepared by vigorous vortexing of the lipids in the hydrating buffer. Small unilamellar vesicles are prepared by sonicating. A probe sonicator may be more efficient but care must be taken to prevent the tip of the probe from touching the sides of the tube containing the lipids or from being exposed to air during the sonication. We find a minimum of 2 mL of lipid solution in a 12 × 75 mm glass tube is required to accommodate the probe tip. A bath sonicator may be more convenient for smaller volumes and multiple lipid samples. Times and conditions required for sonication may vary with the sonicator. On a Heat Systems Ultrasonics, Inc. W-375 sonicator, we use a pulsed sonication at 30% duty cycle with a power setting of 3 for 3 min. The power setting can be increased with a bath sonicator and the time usually must be increased (to about 10 min). After successful vortexing or sonicating, no lipid scum should be visible on the bottom of the tube and the solution, if not too dilute, will have a translucent appearance somewhat like scintillation fluor as opposed to water.

5. By planning for addition of a standard volume such as 25 µL, performance of the assay may be simplified by use of a repeat pipettor.

6. As the [γ-^{32}P]ATP ages beyond 3 wk (or in the event of a rare bad preparation from a supplier), some ^{32}P-labeled inorganic phosphate may contaminate the preparation. If calcium is included in the assay as for cPKCs, ^{32}P-labeled calcium phosphate may precipitate out of solution. This may be a source of considerable error if radioactive calcium phosphate precipitates on the P81 papers or other filters used to separate phosphorylated substrate from ATP. ^{32}P-labeled calcium phosphate precipitates are not such a big problem if phosphorylation reactions are monitored by autoradiography after gel electrophoresis because they are separated from the protein in the electrophoresis.

7. If PKC is serving as substrate and enzyme, that is, in an autophosphorylation reaction, it usually is necessary to increase the specific activity of the ^{32}P-ATP at least 10-fold to improve the sensitivity because the concentration of substrate, in this case, is so low.

8. Appropriate gloves, lab coats, and monitoring with a Geiger counter are used for all work with radioactivity and all items contaminated with ^{32}P are disposed of in appropriate dry or liquid radioactive waste containers stored behind plexiglass. Aerosol tips are ideal for pipetting radioactive materials. Because the micropipettors may become contaminated with ^{32}P during these assays, it is convenient to have a separate set of micropipettors reserved for ^{32}P use and stored behind plexiglass.

9. A crude 100,000g cytosolic extract in hypotonic buffer such as 20 m*M* Tris-HCl or MOPS with EDTA (2 m*M*)/EGTA (0.5 m*M*) and protease inhibitors, including leupeptin (20 µg/mL) usually has adequate amounts of the calcium-dependent

cPKCs from that cell and some calcium-independent nPKCs and aPKCs. The calcium chelators help to extract some calcium-dependent PKCs from membrane but some membrane-bound PKCs, especially some of the nPKCs like PKCη, are especially difficult to extract from membranes without detergents. Detergent (e.g., 1% Triton X-100) will release more enzyme from the membrane and also can help stabilize PKC preparations during freezing. However, as discussed in Chapter 5, detergents can both activate and inhibit PKC, depending on their concentration, and they have significant effects on the structure of lipid vesicles. For studies on PKC lipid requirements, it is better to avoid detergents.

10. The presence of PKM (and PKC) in the preparation can be assessed by Western blotting. Most commercially available PKC isozyme antibodies recognize a sequence in the carboxy-terminus of the enzyme and thus react with intact enzyme and the catalytic fragment.

11. The phorbol ester-binding assay is not based on the difference in activity with vs without PS-containing lipids, but on binding in the presence vs the absence of excess unlabeled phorbol ester. Thus, by including exogenous activating lipid in all tubes, cellular fractions of varying endogenous lipid can be compared. It should be noted, however, that demonstration of phorbol ester binding in a preparation does not indicate whether the catalytic portion of the enzyme is active or even present. The regulatory fragment of PKC binds phorbol esters with lipid dependence very similar to that of the whole enzyme.

12. A screen in the bottom of the beaker protects the P81 papers from damage by the stir-bar.

13. More washes may be necessary when the specific activity of the ATP is especially high. It is acceptable to leave the papers washing overnight. They should be processed within a day, however, because of the progressive decay of ^{32}P.

14. Drying may be speeded with the aid of a hair dryer. Place a screen over the beaker of papers to prevent them from blowing away and aim the dryer at them until they flutter freely.

References

1. Nishikawa, K., Toker, A., Johannes, F.-J., Songyang, Z., and Cantley, L. C. (1997) Determination of the specific substrate sequence motifs of protein kinase C isozymes. *J. Biol. Chem.* **272,** 952–960.
2. Vinton, B. B., Wertz, S. L., Steere, J., Grisham, C. M., Cafiso, D. S., and Sando, J. J. (1998) Influence of lipid on the structure and phosphorylation of Protein kinase C α substrate peptides. *Biochem. J.* **330,** 1433–1442
3. Stabel, S. and Parker, P. (1991) Protein kinase C. *Pharmacol. Ther.* **51,** 71–95.
4. Newton, A. C. (1995) Protein kinase C: structure, function and regulation. *J. Biol. Chem.* **270,** 28,495–28,498.
5. Mellor, H. and Parker, P. J. (1998) The extended protein kinase C superfamily. *Biochem. J.* **332,** 281–292.
6. Ron, D. and Kazanietz, M. G. (1999) New insights into the regulation of protein kinase C and novel phorbol ester receptors. *FASEB J.* **13,** 1658–1676.

7. Sando, J. J., Maurer, M. C., Bolen, E. J., and Grisham, C. M. (1992) Role of cofactors in protein kinase C activation. *Cell. Signal.* **4,** 595–609.
8. Zidovetski, R. and Lester, D. S. (1992) The mechanism of activation of protein kinase C: a biophysical perspective. *Biochem. Biophys. Acta* **1134,** 261–272.
9. Epand, R. M. (1994) In vitro assays of protein kinase C activity. *Anal. Biochem.* **218,** 241–247.
10. Stubbs, C. D. and Slater, S. J. (1996) The effects of non-lamellar forming lipids on membrane protein-lipid interactions. *Chem. Physics Lipids* **81,** 185–196.
11. Sando, J. J., Chertihin, O. I., Owens, J. M., and Kretsinger, R. H. (1998) Contributions to maxima in Protein Kinase C activation. *J. Biol. Chem.* **274,** 34022–34027.
12. Slater, S. J., Ho, C., Kelly, M B., Larkin, J. D., Taddeo, F. J., Yeager, M. D., et al. (1996) Protein kinase Cα contains two activator binding sites that bind phorbol esters and diacylglycerols with opposite affinities. *J. Biol. Chem.* **271,** 4627–4631.
13. Sando, J. J. and Young, M. C. (1983) Identification of a high affinity phorbol ester receptor in cytosol of EL4 thymoma cells: requirement for calcium, magnesium and phospholipids. *Proc. Natl. Acad. Sci. USA* **80,** 2642–2646.
14. Kikkawa, U., Takai, Y., Tanaka, Y., Miyake, R., and Nishizuka, Y. (1983) Protein kinase C as a possible receptor for tumor-promoting phorbol esters. *J. Biol. Chem.* **258,** 11,442–11,445.

7

Subcellular Translocation of Protein Kinase C

Julianne J. Sando, Jacquelyn K. Beals, and Isa M. Hussaini

1. Introduction

To implicate protein kinase C (PKC) or a specific PKC isozyme in the response of a cell to an agonist, one would like to demonstrate activation of PKC in the cell in response to the agonist and to show that specifically blocking this activation blocks the response. However, because of the lipid-dependence of PKC and the paucity of specific PKC activators, inhibitors, and substrates, it is very difficult to assay for the activation of PKC in a cell. In the early years of PKC investigation, it was observed that stimulation of cells with PKC-activating phorbol esters caused a loss of PKC activity from the cytosol *(1)* and an accompanying gain of activity in the membrane *(2)*. Since then, subcellular translocation of PKC has been observed in response to a variety of stimuli and assay of this translocation can serve as a surrogate for assessment of PKC activation in a cell. Changes in total cellular PKC can be estimated by assay of phosphatidylserine (PS)-dependent PKC activity or of phorbol ester binding capacity in subcellular fractions at varying times after a stimulus; however, some limitations of these assays must be recognized.

PKC in cytosolic fractions can be quantitated using the activity assay described Chapter 6. It is essential to assay several dilutions of each fraction to ensure that linear assay conditions are achieved uniformly. It is more difficult to determine activity of PKC in membrane-containing fractions because of the need to remove PKC from the lipid in order to assay activity with and without activating PS. Most PKCs can be removed from membrane with appropriate detergents (e.g., 1% Triton X-100); however, it then is necessary to dilute out the detergent or to remove it (e.g., with small ion exchange columns) so that it does not interfere with the assay. Still, the conditions for assaying the membrane

From: *Methods in Molecular Biology, vol. 233: Protein Kinase C Protocols*
Edited by: A. C. Newton © Humana Press Inc., Totowa, NJ

enzyme will not be equivalent to those for assaying the cytosolic enzyme (and the isozymes present in each fraction are likely to differ, too); therefore, comparison of absolute levels of PKC activity in cytosol and membrane usually cannot be made. Under some conditions it may be possible to show a reciprocal relationship between decrease in activity in one fraction and increase in activity in another, even though percentages may not add up to 100. Additional reasons that these percentages may not add up to 100 at each time point include possible degradation of some PKC isozymes following their activation, or degradation or poor recovery during preparation of subcellular fractions.

Assays for the specific binding of ^{3}H-labeled phorbol 12,13-dibutyrate (PDBu) to subcellular fractions have an advantage over kinase activity assays for PKC in that there is no need to extract membrane-associated PKCs from the lipid. The specificity of the assays depends not on the presence vs. the absence of lipid but on the presence vs. the absence of unlabeled PDBu competitor. Appropriate activating lipid vesicles can be added to all fractions and the assays can be performed in crude suspensions of particulate fractions. However, additional C1 domain-containing proteins, such as N-chimaerin, also bind to phorbol esters and will be included in assessment of PKC translocation via phorbol ester binding (reviewed in **refs. *3*** and ***4***), and regulatory domain fragments of PKC will not be distinguished from intact enzyme in this assay. In addition, PDB binding requires association/insertion of PKCs with appropriate activating lipid domains so this assay, too, will be dependent on the nature of the lipids present; these will be different in the membrane fractions of the cell. Methods for assessment of phorbol ester binding are described in a later section of this book.

Translocation assays also can be used to implicate specific isozymes in a response. These assays involve the following: 1) separating proteins from subcellular fractions by sodium dodecyl sulfate polyacrylamide gel electrophoresis (SDS-PAGE) followed by Western blotting for individual isozymes or 2) immunohistochemistry of intact cells. The Western blotting method has the advantage of allowing a better estimate of the concentration of the individual isozymes and confirmation that the isozymes are intact. Immunohistochemistry permits more rapid acquisition of early time points and more spatial resolution of the subcellular redistribution; however, detection of PKC fragments rather than intact enzyme may confuse interpretation of the results.

Critical to any of the methods for assessment of translocation is the time course over which the event is monitored. Phorbol esters cause a much more prolonged translocation of responsive PKCs than do agonists that generate diacylglycerols (DAGs) because most cells (macrophages are an exception) lack the esterases necessary for rapid degradation of phorbol esters. Translocation in response to physiological agonists typically is very transient. The slower generation of some DAGs from phosphatidylcholine (PC) hydrolysis versus phosphatidylinositol (PI) hydrolysis can affect the time course of membrane association; and the generation of calcium transients contributes to the association and longevity of classical PKC-membrane association *(5)*.

It should be noted that some isozymes, especially some of the calcium-independent isozymes like PKCη, tend to be found in membrane fractions under control conditions in many cells and may not undergo visible translocation in response to activation. Thus, the absence of translocation cannot be interpreted as a lack of activation.

2. Materials

1. Cell homogenization buffer: 20 mM N-(morpholino)propanesulfonic acid (MOPS), 2 mM ethylenediaminetetraacetic acid (EDTA); 0.5 mM ethylene glycol bis, 2,2'-tetraacetic acid (EGTA); 0.3 M sucrose; 20 μg/mL leupeptin; 0.5 mM phenylmethane sulfonyl fluoride (PMSF) (*see* **Note 1**).
2. Glass Dounce homogenizers or other suitable homogenizers appropriate for cell type.
3. Ultracentrifuge or microcentrifuge appropriate for desired cell fractionation.
4. 3X gel sample buffer: 1 mL of 0.5 M Tris-HCl, 3 mL of 10% (w/v) SDS; 3 mL glycerol; 1 mL β-mercaptoethanol; 0.5 mL of 0.1 % (w/v) bromophenol blue; store at room temperature.
5. SDS-PAGE Molecular Weight Standards: broad range or prestained broad range (Bio-Rad or other supplier); store at –20°C.
6. Gel electrophoresis apparatus and power supply.
7. 10% SDS (Bio-Rad).
8. 1.5 M Tris-HCl, pH 8.8, at room temperature; 0.5 M Tris-HCl, pH 6.8, at room temperature.
9. 30% Acrylamide/Bis solution (Bio-Rad).
10. 0.1% (w/v) ammonium persulfate (prepared fresh).
11. $N,N,N'N'$-Tetramethylethylenediamine (TEMED) (Sigma).
12. Gel running buffer (5X: Tris base (15 g), glycine (72 g), SDS (5 g) in 1 L of distilled water, pH 8; for use, dilute to 1X to yield 25 mM Tris, 192 mM glycine, 0.1% (v/v) SDS.
13. Transphor apparatus and power supply for Western blots.
14. Extra-thick filter paper (Bio-Rad).
15. Nitrocellulose membrane, 0.45 μm (Bio-Rad).
16. Transfer buffer (5X: Tris base (37.5 g), glycine (180 g) in 2 L of deionized water, pH 8.3; dilute to 1X then dilute 80:20 with 100% methanol to yield 25 mM Tris, 192 mM glycine, 20% (v/v) methanol for 1X.
17. Boiling waterbath with rack for Microfuge tubes.
18. PKC isozyme-specific antibodies (Santa Cruz, Transduction Laboratories, BioMol or other supplier) and antibodies to proteins unique to specific subcellular fractions of the cells under investigation.

19. Phosphate-buffered saline (PBS; 10X): $Na_2HPO_4H_2O$(45.8 g), NaH_2PO_4 (7.96 g), NaCl (350.72 g) in 4 L of deionized water; Dilute to 1X (9.7 mM phosphate, pH 7.4, 150 mM NaCl) for use.
20. PBS-Tween: 0.1% (v/v) Tween-20 in 1X PBS.
21. Nonfat dry milk (any brand).
22. Plastic bags for antibody incubation (e.g., Kapak heat-sealable pouches).
23. Heat sealer.
24. Enhanced Chemiluminescent kit (ECL or ECL+Plus [Amersham-Pharmacia]).
25. Film (e.g., Kodak X-Omat, $8 \times 10''$ or other size appropriate for gels).
26. Autoradiography film cassettes (Fisher or other supplier).
27. Developer and Fixer for film (usually in an automated film developer).

3. Methods

3.1. Cell Rupture and Subcellular Fractionation

3.1.1. Time Course of Agonist Stimulation

The time course for PKC translocation varies with the agonist, the isozyme, and the cell type. In EL4 mouse thymoma cells, total cytosolic PKC was lost within 20 min of phorbol ester treatment at 30° or within 5 min at 37°C (*1*). In human astrocytic lines U1241 and U252, we observe translocation of many, but not all, isozymes between 1 and 20 min after phorbol ester challenge (*unpublished obversations*).

3.1.2. Fractionation Method and Verification

Both the homogenization buffer and the cell rupture method must be appropriate for the cell type studied and the subcellular fractions desired. For separation of crude particulate and cytosolic fractions from many mammalian cells, we prepare 100,000g supernatant and pellet fractions after dounce homogenization in hypotonic buffer (*see* **Note 1**). For lymphocytes, buffer volumes that will yield at least 10^6 cells/mL usually will provide enough protein (in the 50–100 µL of cytosol loaded onto a gel) for detection of most isozymes. More concentrated preparations may be required for other cell types or for detection of less abundant isozymes. For direct comparison of cytosolic and membrane fractions, it is convenient, but not essential, to resuspend the 100,000g pellet fraction in a volume equal to that of the cytosol. The translocation assays should compare equal cell equivalents, not equal protein in the various fractions.

Preparation of specific organelle fractions (nuclei, mitochondria) varies with the cell type and may involve centrifugation through sucrose gradients. When movement of PKC to specific subcellular sites is investigated, it is essential to verify the purity of the fractions with appropriate assays for molecules unique to each fraction. Again, the cell rupture and fractionation procedures and the assays selected may be unique to a specific cell type (*see* **Note 2**). It may not be ideal to pursue translocation to specific organelles by subcellular fractionation because of the time required for generation of such fractions, with accompanying proteolysis, and the reversibility of the PKC translocation. Immunohistochemistry and more sophisticated microscopic methods described elsewhere in the volume better serve this purpose.

Once subcellular fractions have been prepared, appropriate aliquots should be boiled promptly in gel sample buffer to halt any proteolysis of the PKCs. A 3X sample buffer allows minimal dilution of the sample; a 1:1 (v:v) mixture of sample with 1X sample buffer is adequate. Samples then can be stored frozen until gels are run.

3.2. Gel Electrophoresis and Western Blotting for Isozymes

For all steps of gel electrophoresis, transfer of proteins to nitrocellulose, and Western blotting, it is essential to wear gloves to avoid marring the gels and blots with fingerprints.

3.2.1. Gel Electrophoresis

1. Prepare enough 10% running gels, each with 4% stacking gels and appropriately sized combs to make the wells, for all samples to be assayed (*see* **Note 3**). A recipe for preparation of 11.5- × 13.5-cm gels is provided in **Table 1**. Although blots can be stripped and reprobed with different antibodies, quality may be improved with a single probing of replicate gels for each isozyme (*see* **Note 4**). Fill the gel wells and the upper and lower buffer chambers with gel running buffer.
2. Load molecular weight standards and samples by underlaying appropriate volumes in the wells. For prestained molecular weight standards, 20 µL containing approx 40 µg of each standard usually is adequate (*see* **Note 5**). For PKC samples, we find that approx 10^6 cell equivalents per lane usually are adequate for detection of most endogenous isozymes in most cells. Fewer cell equivalents are needed for neurons, astroglial cells, and lymphocytes (e.g., **refs. 6** and **7**), which express high concentrations of many PKCs. For detection of PKCs in *Xenopus* oocytes, 1–10 cell equivalents usually are adequate (*8,9*).
3. Run the gels under conditions appropriate for their size. For 11.5 × 13.5 × 1.5-mm thick gels, we use 180 V for 3–4 h or 60 V overnight with stirring and cooling with circulating water (*see* **Note 6**). For mini or midi gels, follow the instructions of the manufacturer.

3.2.2. Transfer of Proteins to Nitrocellulose

1. Prepare transfer buffer by diluting the 5X stock to 1X with distilled water and adding methanol to a final concentration of 20% (v/v).
2. When electrophoresis is complete, remove the gels from the plates, cut off the stacking gel, and cut one corner of each gel for future orientation. Equilibrate

Table 1
Recipe for Preparation of Gels

Stock Solution	Volume of Stock for Different % Acrylamide Running Gels[a] (mL)				
	7%	8%	9%	10%	12%
Distilled water	15.1	14.1	13.1	12.1	10.1
1.5 M Tris, pH 8.8	7.5	7.5	7.5	7.5	7.5
10% SDS	0.3	0.3	0.3	0.3	0.3
30% Acrylamide-Bis	7.0	8.0	9.0	10.0	12.0
10% Ammonium persulfate	0.1	0.1	0.1	0.1	0.1
TEMED[b]	7.5 µL	7.5 µL	7.5 µL	7.5 µL	7.5 µL

	Volume of Stock for Different % Acrylamide Stacking Gels[a] (mL)	
	3%	4%
Distilled water	6.3	5.9
0.5 M Tris, pH 6.8	2.5	2.5
10% SDS	0.1	0.1
30% Acrylamide-Bis	1.0	1.33
10% Ammonium persulfate	0.1	0.1
TEMED[b]	5 µL	5 µL

[a]Volumes will provide enough solution for one gel of 13.5 cm × 11.5 cm × 1.5 mm.
[b]Add the reagents in order on ice and degas the mixture for 15 min before adding the TEMED. Note that TEMED volumes are in µL. Swirl gently after adding the TEMED and immediately pour the gel.

the gels, filter papers, fibrous sponges, and nitrocellulose sheets (sized to match the gel) in transfer buffer.

3. Assemble the gel and papers for transfer in the following order: fibrous sponge, extra thick filter paper, gel, nitrocellulose membrane, filter paper, fibrous sponge. Make sure there are no air bubbles between the gel and the nitrocellulose. Orient the gel toward the negative pole of the transfer tank (*see* **Note 7**). Transfer conditions may vary with the apparatus. We use 1.5–2 h at approx 0.21 amps at 4°C for big gels. Another protocol calls for transfer at 70 volts for at least 4 h at 4°C. Four liters of transfer buffer are required for big gels; half a liter is required for mini gels.

3.2.3. Western Blotting of the Transferred Proteins

Cut one corner of the nitrocellulose to match the cut gel and perform the following incubations, in separate dishes for each gel, on a rocker table at room temperature.

1. Wash the membranes in blocking solution (25 mL of PBS with 0.1% (v/v) Tween-20 and 5% (w/v) dry milk; *see* **Note 8**) for 1 h.

2. Rinse the membrane 3×5 min in PBS-Tween. If ECL will be used for detection, use 50-mL washes; if ECL+Plus will be used, use 200- to 300-mL washes.
3. Incubate the nitrocellulose with primary antibody in PBS-Tween containing 5% (w/v) fresh powdered milk for 1 h. Conducting the incubations in heat-sealable plastic bags can minimize the volume of solution (thus antibody) required. For 11.5×13.5 cm blots, 25 mL is suitable. The dilution of the primary antibody depends on the concentration of the antibody and how abundantly the isozyme is expressed in the cell system. We start with the dilution suggested by the supplier. For PKC polyclonal antibodies (e.g., from Santa Cruz), 50 µL of primary antibody is often adequate. For monoclonal antibodies, a higher concentration may be required (*see* **Note 9**). Eliminate air bubbles before sealing the bag.
4. Return the nitrocellulose to a dish and wash 3×5 min in PBS-Tween without milk.
5. Incubate the nitrocellulose with secondary antibody in PBS-Tween with 5% dry milk for 1–3 h at room temperature (in a fresh heat-sealable bag). Follow the manufacturer's recommendation for the antibody dilution. When using goat anti-rabbit secondary for detection of polyclonal anti-PKC antibodies prepared in rabbits, 5 µL in 25 mL often is adequate.
6. Return the nitrocellulose to a clean dish and wash it 2×5 min in PBS-Tween followed by a minimum of 2×5 min in PBS without Tween.

3.2.4. Detection of Reactive Proteins by Enhanced Chemiluminescence

Follow the procedure from the supplier of the ECL reagents. For the Amersham reagents, the steps are as follows.

1. Mix solutions A and B (or 1 and 2) just before use in appropriate ratios to provide enough mixture to cover the blot completely (e.g., for a 11.5- $\times$ 13.5-cm gel: 10 mL + 10 mL for ECL; 20 mL + 0.5 mL for ECL+Plus) and incubate the nitrocellulose for 1 min (ECL) or 5 min (ECL+Plus).
2. Drain off the solution and immediately place the damp nitrocellulose blot between two sheets of plastic taped in place in a film cassette. Alternatively, wrap the blot in plastic wrap and tape that into the film cassette to prevent movement. In either case, avoid air bubbles between the blot and the plastic.
3. In the dark, or with a safelight, place a sheet of film over the blot and allow it to expose for an appropriate time, typically varying from 30 sec to 30 min (*see* **Note 10**).
4. Develop the film. Expose more film for different times if needed to optimize the signal.

4. Notes

1. Different buffers can be used depending on the desired subcellular fractions and the particular cells. Detergent-containing buffers should be avoided at the initial cell rupture so that PKCs will not be extracted from membrane preparations and collected with the cytosolic enzyme. Hypotonic buffers may provide better

cell swelling and rupture but sucrose-containing isosmotic buffers may be more appropriate for maintenance of organelle integrity.

2. We found that dounce homogenization was not suitable for preparation of nuclei from lymphocytes because of adherence of plasma membrane fragments to the nuclei in these cells with relatively little cytosol *(10)*. Nitrogen cavitation proved to be a suitable cell rupture method to generate purer preparations of nuclei from those cells.

3. Mini and Midi-gels are convenient and can be prepared according to the manufacturers' instructions. However, the wells on these gels are small and may not accommodate enough sample from dilute PKC preparations for adequate detection.

4. To prevent misinterpretation due to spill-over of concentrated samples into adjacent lanes, do not over-fill the wells. An empty lane between unlike samples is often worth the "wasted" space.

5. Use of prestained standards enables immediate visualization of the standards on the nitrocellulose filter. With unstained standards, it is necessary to stain the blot, for example, with 0.2% Ponceau Red in 30% trichloroacetic acid, after completion of the ECL exposure. To confirm appropriate alignment of the standards (an issue if some bands have not entered the running gel, have been run off the running gel, or have degraded, generating more bands), we load a single known protein standard such as bovine serum albumin in an adjacent lane.

6. If the gel is run overnight, an additional hour at 180 V may be required to complete the run the next morning. Be sure that the water source for cooling is independent of plumbing cut-offs, or overheating may damage both the gel and the electrophoresis equipment.

7. To minimize bubbles between the membrane and the gel, and to stabilize the gel during its move to the transfer apparatus, slip the nitrocellulose membrane under the gel in the buffer, and then place the membrane plus gel on one filter paper and cover with the second filter paper.

8. If one intends to probe for phosphotyrosine, dry milk must be replaced with bovine serum albumin or gelatin because milk contains phosphotyrosine. It is convenient to make up 1–2 L of PBS-Tween, 1 L of PBS without Tween for final rinses, and multiple tubes containing 1.25 g powered milk for the entire blotting procedure.

9. Note that commercially available antibodies to mammalian PKC isozymes may not necessarily cross-react with similar isozymes in other species. For example, the Santa Cruz PKCα antibody does not recognize a *Xenopus* PKC. Check that the epitope to which the antibody reacts is present in the PKC isozyme under investigation. Some antibodies raised against native proteins will not recognize denatured proteins in a Western blot.

10. Usually several exposure times are necessary for optimal detection of the PKC bands. The use of intensifying screens can increase the signal but decrease the linearity of signal intensity; thus Western blotting with ECL detection is seldom quantitative. Semi-quantitative assessment of PKC can be obtained by running

serial dilutions of a standard PKC preparation of known concentration along with serial dilutions of the experimental samples. It is essential, in this case, to expose all the lanes to ECL reagents and to the film at the same time so that the intensities can be compared. Direct comparison of crude cytosol and membrane fractions to determine relative changes is possible when the samples are run and exposed together, provided the film has not been burned out as a result of overexposure in some lanes.

References

1. Kraft, A. S., Anderson, W. B., Cooper, H. L., and Sando, J. J. (1982) Decrease in cytosolic calcium/phospholipid-dependent protein kinase activity following phorbol ester treatment of EL4 thymoma cells. *J. Biol. Chem.* **257,** 13193–13196.
2. Kraft, A. S. and Anderson, W. B. (1983) Phorbol esters increase the amount of Ca^2+, phospholipid-dependent protein kinase associated with plasma membrane. *Nature* **301,** 621–623.
3. Ron, D. and Kazanietz, M. G. (1999) New insights into the regulation of protein kinase C and novel phorbol ester receptors. *FASEB J.* **13,** 1658–1676.
4. Kazanietz, M. G. (2000) Eyes wide shut: protein kinase C isozymes are not the only receptors for the phorbol ester tumor promoters. *Mol. Carcinog.* **28,** 5–11.
5. Oancea, E. and Meyer, T. (1998) Protein kinase C as a molecular machine for decoding calcium and diacylglycerol signals. *Cell* **95,** 307–318.
6. Hussaini, I. M., Karns, L. R., Vinton, G., Carpenter, J. E., Redpath, G. T., Sando, J. J., et al. (2000) Phorbol 12-myristate, 13-acetate induces protein kinase C η-specific proliferative response in astrocytic tumor cells. *J. Biol. Chem.* **275,** 22348–22354.
7. Resnick, M. S., Luo, X., Vinton, E. G., and Sando, J. J. (1997) Selective up-regulation of protein kinase C η in phorbol ester-sensitive and -resistant EL4 mouse thymoma cells. *Cancer Res.* **57,** 2209–2215.
8. Stith, B. J., Woronoff, K., Espinoza, R., and Smart, T. (1997) *sn*-1,2-diacylglycerol and choline increase after fertilization in Xenopus laevis. *Mol. Biol. Cell* **8,** 755–765.
9. Johnson, J. and Capco, D. G. (1997) Progesterone acts through protein kinase C to remodel the cytoplasm as the amphibian oocyte becomes the fertilization-competent egg. *Mech. Devel.* **67,** 215–226.
10. Jensen, D. E. and Sando, J. J. (1987) Absence of protein kinase C in nuclei of EL4 mouse thymoma cells. *Cancer Res.* **47,** 686–693.

IV

MEASURING THE INTERACTION OF PROTEIN KINASE C WITH MEMBRANES

8

Measuring the Interaction of Protein Kinase C with Membranes

An Introduction

Alexandra C. Newton

1. Translocation: The Hallmark for Protein Kinase C (PKC) Activation

The translocation of PKC to membranes has served as the hallmark for PKC activation since the historic discovery that phorbol esters cause PKC to redistribute from the cytosol to the membrane *(1,2)*. In the 20 years since this discovery, the literature has been flooded with reports on the intracellular redistribution of PKC after agonist or phorbol ester stimulation *(3)*. In recent years, this translocation has been beautifully captured in real time in live cells with the advent of fluorescent technologies *(4,5)*.

Extensive biophysical and biochemical studies have elucidated much of the molecular detail regarding the mechanism of translocation (reviewed in **ref. 6**). **Figure 1** shows a model based on these studies for the translocation of conventional PKC isozymes. In resting cells, PKC adopts a conformation in which the autoinhibitory pseudosubstrate (light grey rectangle; *see* **Fig. 1** in Chapter 1) is bound to the substrate-binding cavity; this inactive species bounces on and off the membrane by a diffusion-controlled mechanism. In the absence of ligands for its membrane-targeting modules, the C1 and C2 domains (*see* Chapter 1), the kinase is not retained at the membrane, and it diffuses back to the cytosol. (Note that scaffold or anchoring can retain PKC at specific intracellular locations, including the membrane, via protein:protein interactions; *see* Chapter 26.) The generation of signals that cause hydrolysis of phosphatidylinositol bis phosphate create two second messengers: inositol

From: *Methods in Molecular Biology, vol. 233: Protein Kinase C Protocols*
Edited by: A. C. Newton © Humana Press Inc., Totowa, NJ

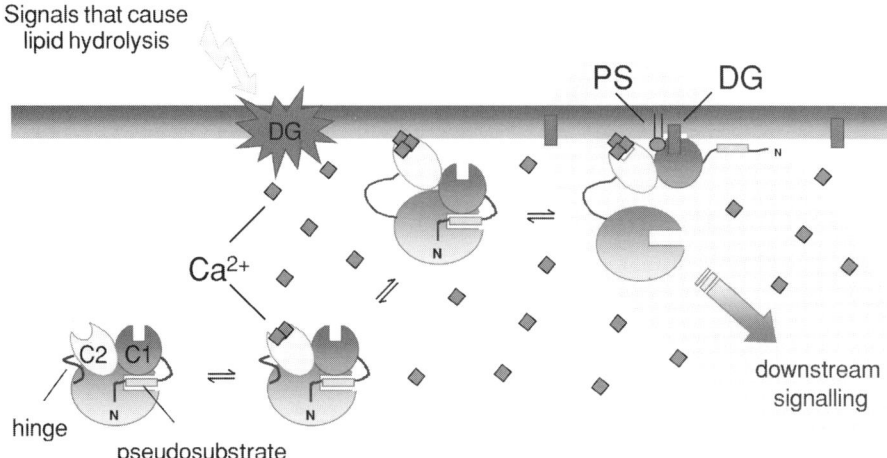

Fig. 1. Model describing the mechanism of translocation of conventional PKCs to membranes in response to signals that cause elevation of intracellular Ca^{2+} and diacylglycerol. In the absence of these stimuli, PKC localizes to the cytosol in a conformation in which the pseudosubstrate (light grey rectangle) blocks the substrate-binding cavity and a proteolytically labile hinge connecting the kinase domain (large circle with cleft) with the membrane targeting regulatory moieties is masked (species on far left). Elevation of intracellular Ca^{2+} (diamonds) recruits cytosolic PKC to the membrane by engaging the Ca^{2+}-bound C2 domain with anionic phospholipids, an event that unmasks the hinge but, because the interaction is of low affinity, does not result in release of the pseudosubstrate (middle). The membrane-bound enzyme then diffuses in the two-dimensional plane of the membrane searching for diacylglycerol; an encounter with this second messenger engages the C1 domain on membranes, with additional binding energy supplied by specific interaction of the C1 domain with phosphatidylserine (right). Engaging both domains on the membrane results in a high-affinity interaction that provides the energy to release the pseudosubstrate from the substrate-binding cavity, allowing phosphorylation and downstream signaling.

tris phosphate, which mobilizes intracellular Ca^{2+} (diamonds in **Fig. 1**), and diacylglycerol (dark grey rectangles in **Fig. 1**). Ca^{2+} binds the C2 domain of conventional PKCs (**Fig. 1**, second species from left), causing the affinity of this module for membranes to increase dramatically. Thus, on the next collision with the membrane, the kinase is retained at the membrane (**Fig. 1**, third species from left). Studies with the isolated C2 domain of a conventional PKC suggest that two Ca^{2+} ions bind cytosolic PKC and that a third Ca^{2+} ion bridges the C2 domain with anionic phospholipids in the membrane-bound species *(7)*. The enzyme then diffuses in the two-dimensional plane of the membrane until it encounters its membrane-localized ligand, diacylglycerol.

Binding of diacylglycerol to the C1 domain results in a high-affinity membrane interaction with a specific interaction between the diacylglycerol-bound C1 domain and phosphatidylserine, increasing the binding affinity of this module an additional 10-fold *(8)*. The engagement of the C1 and C2 domains on the membrane provides the energy to release the autoinhibitory pseudosubstrate sequence from the substrate-binding cavity, allowing substrate binding and phosphorylation, propagating downstream signaling.

Novel PKCs do not have the advantage of Ca^{2+}-triggered pretargeting to membranes because their C2 domains do not bind Ca^{2+}. Thus, the encounter of these isozymes with diacylglycerol is considerably less efficient because it occurs from the cytosol rather than the plane of the membrane. Kinetic measurements of green fluorescent protein-tagged PKCs suggest that the rate of translocation of novel PKCs is an order of magnitude slower than that of conventional PKCs in vivo *(9)*. Atypical PKCs have neither a C2 domain nor a ligand-binding C1 domain and thus respond to neither Ca^{2+} nor diacylglycerol.

Part IV describes the major techniques for studying the membrane interaction of PKC in vivo using fluorescence imaging (Chapter 9), and in vitro using model membranes for equilibrium (Chapter 10) and kinetic (Chapter 11) measurements. This part also describes the classic assay for measuring the binding of phorbol esters to PKC developed by Blumberg and co-workers (Chapter 12), whose synthesis of the relatively hydrophophilic phorbol dibutyrate was instrumental in studying PKC *(10)*.

References

1. Kraft, A. S., Anderson, W. B., Cooper, H. L., and Sando, J. J. (1982) Decrease in cytosolic calcium/phospholipid-dependent protein kinase activity following phorbol ester treatment of EL4 thymoma cells. *J. Biol. Chem.* **257,** 13,193–13,196.
2. Kraft, A. S. and Anderson, W. B. (1983) Phorbol esters increase the amount of Ca^{2+}, phospholipid-dependent protein kinase associated with plasma membrane. *Nature* **301,** 621–623.
3. Nishizuka, Y. (1995) Protein kinase C and lipid signaling for sustained cellular responses. *FASEB J.* **9,** 484–496.
4. Sakai, N., Sasaki, K., Ikegaki, N., Shirai, Y., Ono, Y, and Saito, N. (1997) Direct visualization of the translocation of the gamma-subspecies of protein kinase C in living cells using fusion proteins with green fluorescent protein. *J. Cell. Biol.* **139,** 1465–1476.
5. Oancea, E. and Meyer, T. (1998) Protein kinase C as a molecular machine for decoding calcium and diacylglycerol signals. *Cell* **95,** 307–318.
6. Newton, A. C. and Johnson, J. E. (1998) Protein kinase C: a paradigm for regulation of protein function by two membrane-targeting modules. *Biochim. Biophys. Acta* **1376,** 155–172.

7. Nalefski, E. A. and Newton A. C. (2001) Membrane binding kinetics of protein kinase C betaII mediated by the C2 domain. *Biochemistry* **40,** 13,216–13,229.

8. Johnson, J. E., Giorgione, J., and Newton, A. C. (2000) The C1 and C2 domains of protein kinase C are independent membrane targeting modules, with specificity for phosphatidylserine conferred by the C1 domain. *Biochemistry* **39,** 11,360–11,369.

9. Schaefer, M., Albrecht, N., Hofmann, T., Gudermann, T., and Schultz, G.. (2001) Diffusion-limited translocation mechanism of protein kinase C isotypes. *FASEB J.* **15,** 1634–1636.

10. Driedger, P. E. and Blumberg, P. M. (1980) Specific binding of phorbol ester tumor promoters. *Proc. Natl. Acad. Sci. USA* **77,** 567–571.

9

Fluorescence Imaging of Protein Kinase C Translocation in Living Cells

Naoaki Saito

1. Introduction

The existence of multiple isozymes of protein kinase C (PKC) raised a new question of whether each PKC isozyme has a specific function (*1–7*). The activation mechanism of each PKC isozyme is clearly different among the three subgroups, conventional PKC, novel PKC, and atypical PKC, but that of each isozyme in a subgroup has not been clarified (*1–8*). Many reports have demonstrated that a certain cellular response is regulated by a specific PKC isozyme, although the molecular mechanisms of PKC function in defined cellular responses has not been clarified.

In 1997, the visualization of PKC in living cells with confocal laser scanning fluorescent microscopy by using green fluorescent protein (GFP)-tagged PKCs was first described (*9*). The live imaging of PKC showed dynamic movement of PKCs within the cells and strongly suggested isozyme-specific and stimulus-specific targeting of PKC (*10–16*). It is also suggested that the multiple functions of PKC in signal transduction are regulated by targeting of PKC to specific intracellular compartments. This imaging technique has revealed the functional differences between PKC isozymes in a PKC subgroup and also cast doubt on the use of phorbol esters as a physiological PKC activator (*17,18*). Not only conventional biochemical or molecular biological analysis but also live imaging techniques are necessary to understand the physiological function of each PKC isozyme.

2. Materials

1. pEGFP-N vectors, pDsRed-N1 vector (Clontech).
2. cDNAs for PKC isozymes.

From: *Methods in Molecular Biology, vol. 233: Protein Kinase C Protocols*
Edited by: A. C. Newton © Humana Press Inc., Totowa, NJ

3. Oligonucleotide primers.
4. Restriction enzymes, *Taq* polymerase.
5. DNA sequencing equipment.
6. Cultured cell lines (COS-7, HeLa, CHO-K1 cells, etc.) or primary cultured cells.
7. Culture media (DMEM, F-12, etc.) as appropriate for each cell type.
8. Ringer's solution: 135 mM NaCl, 5.4 mM KCl, 1 mM MgCl$_2$, 1.8 mM CaCl$_2$, 5 mM HEPES, and 10 mM glucose, pH 7.3.
9. FuGENE™ 6 Transfection Regent (Roche, Molecular Biochemicals, IN), TransIT™-LT2 (Mirus Co., Madison, WI).
10. Glass-bottomed culture dishes (MatTek, Ashland, MA).
11. Fluorescent microscope.
12. Confocal laser scanning fluorescence microscope (LSM 510 Invert, CarlZeiss, Jena, Germany).
13. 12-*O*-Tetradecanoylphorbol 13-acetate (TPA; Sigma Chemical Co., St. Louis, MO).
14. Specific antibodies against each PKC isozyme (anti-α-, δ-, and ε-PKC monoclonal antibodies from Transduction Laboratories, Lexington, KY; anti-δ- and η-PKC polyclonal antibodies from Santa Cruz Biotechnology Inc., CA; anti-ζ-PKC polyclonal antibody from Upstate Biotechnology, Lake Placid, NY).
15. Anti-GFP polyclonal antibody (Molecular Probes, Leiden, Netherlands).

3. Methods
3.1. Construction of PKC-GFP Plasmids

1. cDNAs for GFP are now commercially available (Clontech, etc.; **Fig. 1**). The pEGFP-N (N1, N2, N3) vectors have a Cytomegalovirus promoter and a multiple cloning site upstream of the GFP gene, whereas the pEGFP-C (C1, C2, C3) vectors have a multiple cloning site downstream of the *GFP* gene. The protein can be expressed as a fusion protein having GFP at its C-terminus, or its N-terminus, when we use EGFP-N or pEGFP-C, respectively. GFP fusion to the C-terminus of PKC isozymes (use of pEGFP-N) is preferable, because the influence of GFP fusion on PKC activity and on PKC translocation was examined and revealed to be negligible *(9,12)*. Fusion of PKC to the N terminus of EGFP usually does not alter the fluorescence properties of native EGFP. GFP-fusion to the N-terminus of PKC, however, does not apparently inhibit its translocation. In some cases such as diacylglycerol kinase-gamma, N-terminal fusion is necessary, because GFP-fusion to the C-terminus of diacylglycerol kinase-gamma abolishes the kinase activity of this lipid kinase *(19)*. A cDNA fragment of each PKC isozyme with appropriate cloning sites at the 5′- and 3′-terminus is produced by a polymerase chain reaction with the cDNA of each PKC isozyme as a template by standard recombinant DNA methods.
2. Primers are designed to have appropriate cloning sites for subcloning into pEGFP-N vectors (*see* **Note 1**). To fuse GFP to the C-terminus of PKC, the stop codon of each PKC isozyme should be deleted and the target gene should be cloned into pEGFP-N1(N2, N3) in frame with the EGFP coding sequence, with

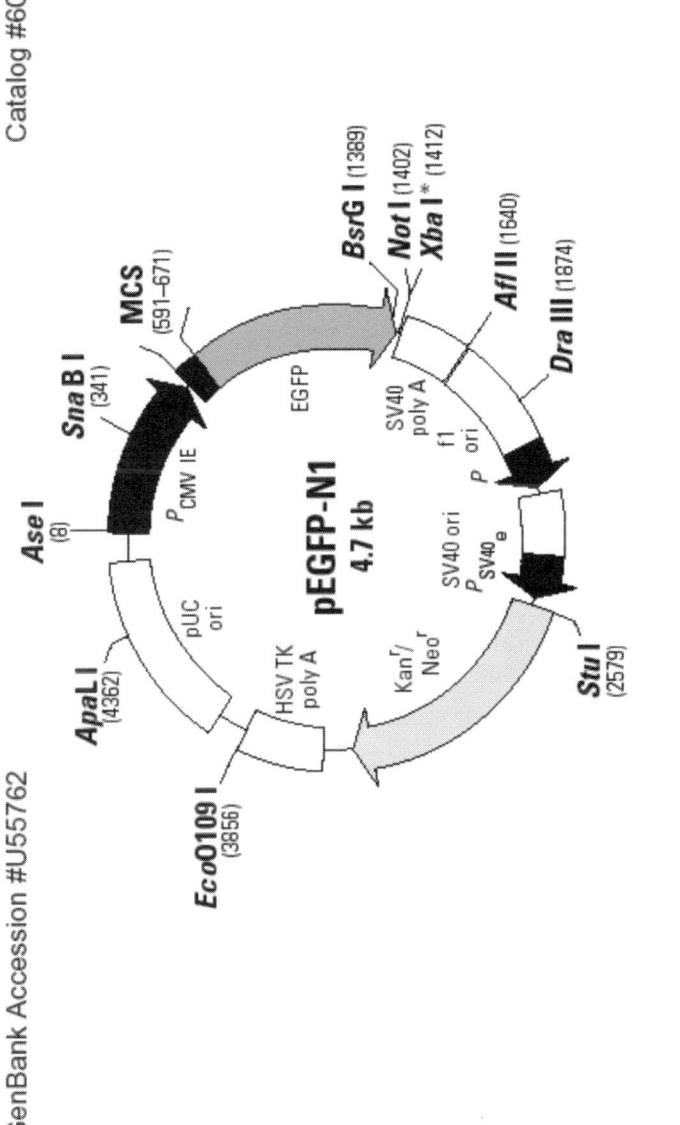

pEGFP-N1 Vector Information

GenBank Accession #U55762

PT3027-5

Catalog #6085-1

Fig. 1. Restriction map and multiple cloning site of pEGFP-N1. Cited from Clontech Catalog (#6085-1).

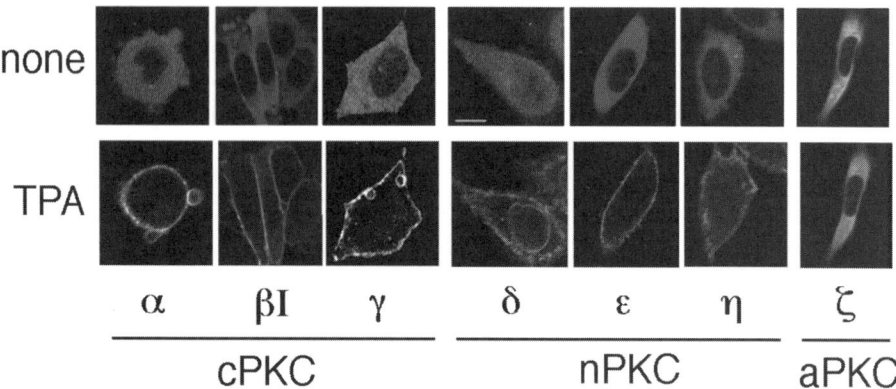

Fig. 2. Localization of 7 PKC isozymes expressed in CHO-K1 cells and its translocation by TPA treatment. Seven PKC isozymes (α, βI, γ, δ, ϵ, η, and ζ) were expressed in CHO-K1 cells after transfection by lipofection. Most isozymes were abundantly and homogeneously localized in cytoplasm and low in the nucleus. Only δPKC was abundant in the nucleus (upper row). TPA (1 μM) induced the irreversible translocation of all PKC isozymes except for ζPKC to the plasma membrane in 5–10 min (lower row).

no intervening in-frame stop codons. The sequences of the PCR product are checked by DNA sequencing.

3.2. Expression of PKC-GFP

1. Plasmids (approx 5.5 µg) encoding PKC-GFPs are transfected in 5×10^6 CHO-K1 cells by lipofection using TransIT-LT2 or FuGENE 6 Transfection Regent according to the manufacture's standard protocol.
2. The transfected cells are cultured at 37°C to obtain optimal GFP fluorescence then used for the observation of translocation. Transfection can be performed on the plastic culture dish or on glass-bottomed culture dish, but the fluorescence of GFP-tagged PKC is visible 8 h after the transfection and the fluorescence reaches maximum at 16–48 h. Fluorescence of cPKCs is usually bright and that of ϵPKC is slightly darker. In case of DsRed-tagged PKC, it takes longer (more than 24 h) to obtain the red fluorescence. The intracellular localization of PKC-GFP subtly differs between the isozymes and also differs in each cell line used for expression (*see* **Note 2**; **Fig. 2**). Longer culture (more than 72 h) sometimes induces changes in the shape of cells expressing some isozymes of PKC, such as ϵPKC *(20)*.

3.3. Characterization of GFP-Tagged PKC

The influence of GFP fusion on enzymological characteristics of each PKC isozyme should be examined.

1. Cells are transiently transfected by electroporation (*see* **Note 3**) with plasmids encoding GFP-tagged PKC then cultured at 37°C for 1 to 2 d.
2. GFP-tagged PKC are immunoprecipitated with anti-GFP antibody (Molecular Probes), and their kinase activities are assayed as described in other chapters. In brief, the immunoprecipitate is suspended in Dulbecco's phosphate-buffered saline [PBS(-)], and a part of the suspended pellet is used for the kinase assay. Kinase activity measurements of the immunoprecipitated PKC-GFP are based on the incorporation of ^{32}Pi into a fragment of myelin basic protein (Sigma) from [γ-^{32}P]ATP in the presence of 8 μg/mL phosphatidylserine (PS) and 0.8 μg/mL diolein. Basal activity is measured in the presence of 0.5 m*M* ethylenebis(oxyethylenenitrilo)tetraacetic acid instead of phosphatidylserine, diolein.
3. Immunoblot analysis should be performed to show the appropriate molecular size of the GFP-tagged PKC and the absence of degraded product (*see* **Note 4**). As we monitor the GFP fluorescence, it is important to know that all the fluorescence is derived from GFP tagged to PKC and not from GFP alone. For immunoblotting, the transfected cells are harvested with or without stimulation and homogenized by sonication with homogenate buffer (250 m*M* sucrose, 10 m*M* ethylenebis(oxyethylenenitrilo)tetraacetic acid, 2 m*M* ethylenediamine tetraacetic acid, 20 m*M* Tris-HCl, 200 μg/mL leupeptin, 1 m*M* phenylmethylsulfonyl fluoride, pH 7.4) with or without 1% Triton-X 100. After centrifugation at 19,000g for 15 min, the supernatant is collected and subjected to sodium dodecyl sulfate polyacrylamide gel electrophoresis and transferred to polyvinylidene difluoridefilter. The transferred protein is immunostained by anti-GFP antibody or anti PKC isozyme-specific antibody (**Fig. 3**).

3.4. Observation of PKC Translocation (Fig. 4)

1. Plate cells expressing PKC-GFPs onto glass-bottomed culture dishes (MatTek, Ashland, MA) at least 16 h before observation.
2. Replace the culture medium with 900 μL of Ringer's solution before the observation at 37°C (*see* **Note 5**).
3. Find appropriate cells expressing GFP under an inverted confocal laser scanning fluorescence microscope at 488-nm argon excitation laser with a 515-nm-long pass barrier filter (*see* **Note 6**).
4. Start scanning the image continuously, then add 100 μL of various stimulators at 10 times higher concentrations in Ringer's solution to obtain the appropriate final concentration (*see* **Note 7**). (Video of εPKC translocation by diacylglycerol in living cells can be viewed at http://www.biosig.kobe-u.ac.jp/biosignal/shinpro/image/video1.gif.)
5. After translocation of PKC, cells are fixed with a fixative containing 4% paraformaldehyde, 0.2% picric acid, and 0.01 *M* PBS (pH 7.4). The detailed localization of PKC can be examined by double staining with an organelle-specific marker. For example, Texas red-conjugated wheat germ agglutinin was used to monitor the Golgi network *(12,21)*. Fixed cells are treated with 0.3%

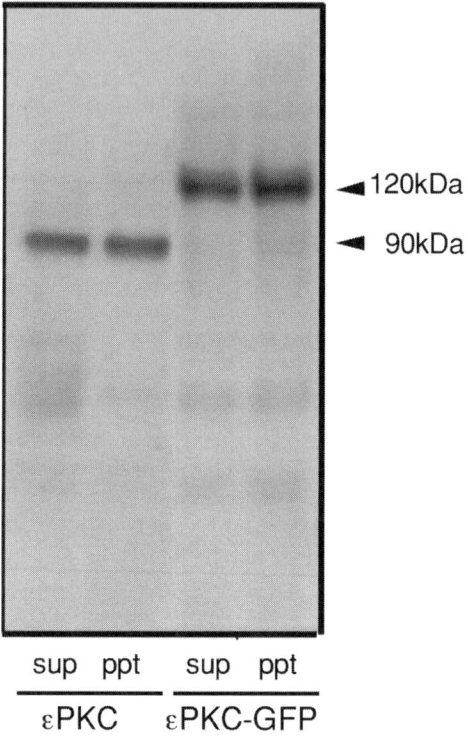

Fig. 3. Immunoblot of εPKC and εPKC-GFP.

Triton-X in PBS and 10% NGS for 1 h. Then, the cells are incubated with 0.5 μg/mL Texas red-conjugated wheat germ agglutinin (Molecular probes, Leiden, Netherlands) in PBS for 30 min. Finally, the fluorescence of Texas red and GFP are observed under a confocal laser scanning fluorescent microscope, the former at 543-nm argon excitation using a 590-nm-long pass barrier filter and the latter at 488-nm argon excitation using a 510- to 525-nm band pass barrier filter.

3.5. Simultaneous Observation of Two Different Fluorescent-Tagged PKC Isozymes

Various mutants of EGFP having different colors such as blue fluorescent protein, cyan fluorescent protein, and yellow fluorescent protein have been generated by the combination of point mutations. DsRed is a fluorescent protein isolated from an IndoPacific sea anemone-relative, *Discosoma sp*, (**Fig. 5** and **ref. 22**). Using these mutants, it is possible to monitor the spatiotemporal correlation of two different proteins. Among these fluorescent proteins, the combination of EGFP/DsRed is useful when we would like to observe the movement of proteins simultaneously. The fluorescence of EGFP can be seen using

ATP

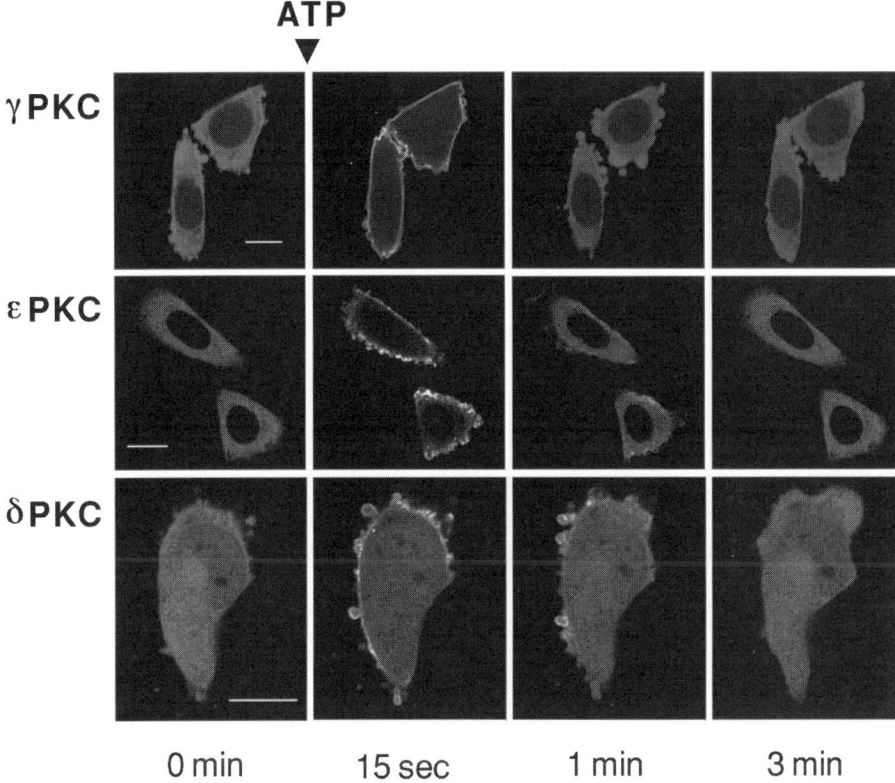

Fig. 4. Translocation of γPKC-, δPKC-, and εPKC-GFP by purinergic stimulation in CHO-K1 cells. All three PKC isozymes showed rapid and reversible translocation to the plasma membrane.

the usual filter-sets for FITC and that of DsRed is seen by Rhodamine filters under a fluorescent microscope. Although DsRed is a useful fluorescent protein, DsRed has major drawbacks, such as strong oligomerization and slow maturation *(23)*. Red fluorescence of DsRed can be seen more than 16 h after transfection, and sufficient fluorescence is obtained more than 36 h after transfection. Furthermore, oligomerization of DsRed sometimes inhibits the natural movement of DsRed-tagged PKC isozymes (*see* **Note 8**). Fusion of DsRed to nPKCs (δ and εPKC) and aPKC (ζPKC) significantly blocked their movement, whereas the movement of cPKCs (α, β, and γPKC) was not altered.

1. For the simultaneous observation, two cDNAs (DsRed-tagged and GFP-tagged proteins) were transfected as described above. When both red and green fluorescence are bright enough to be seen (usually 36 h after transfection), the movement of two proteins can be monitored.

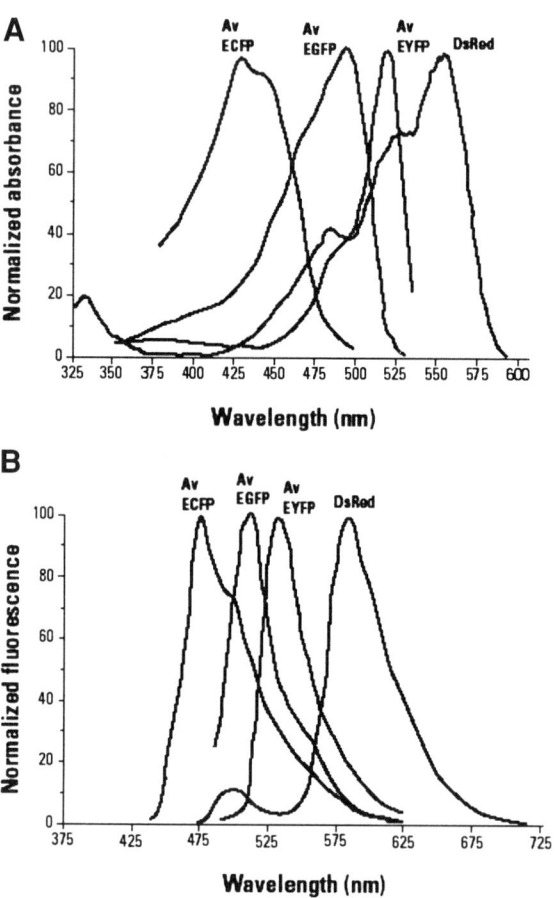

Fig. 5. DsRed, EGFP, ECFP, and EYFP have distinct absorption and emission spectra. (**A**) Absorbance. (**B**) Emission. Cited from Clontech Protocol (#PT 2040-1) *Living Colors Users Manual.*

2. DsRed is visualized using a 543-nm laser for excitation and a 560-nm-long pass filter for emission.

4. Notes

1. pEGFP and DsRed vectors from Clontech have kanamycin-resistant gene. Colony selection of recombinant plasmids should be performed by Kan. but not Amp.
2. Most PKC isozymes are localized throughout the cytoplasm homogeneously. However, GFP-tagged functional domains or deletion mutants of PKC are sometimes localized in specific intracellular compartment or aggregated. Kinase-

negative mutants of PKC usually show different localization from that of wild-type PKC. Most kinase-negative mutants of PKC are localized on the plasma membrane. This may be caused by the lack of maturation of PKC by autophosphorylation *(24)*.

3. Different transfection methods are selected depending on the purpose. For the imaging, lipofection is preferable because lipofection does not alter the shape of the cells. In contrast, electroporation sometimes changes the shape of the cells although its transfection-efficiency is better than that by lipofection. Electroporation is preferred for the biochemical assays to obtain larger amount of the recombinant protein.

4. Although it is known that cPKCs and nPKCs are degraded by phorbol ester treatment, GFP-tagged PKCs seem to be resistant to TPA-induced degradation. The reason is not clear.

5. Use a stage heater and lens heater to keep the cells at 37°C. Temperature is very important for live imaging of PKC translocation. At lower temperatures, translocation is usually undetectable or very slow *(9)*.

6. Appropriate cells are moderately bright cells that show homogenous GFP fluorescence in the cytoplasm. In very bright cells, PKC-GFP is sometimes aggregated and untranslocatable.

7. Because PKC is, as far as we know, translocated by diffusion, fast scanning at msec/image is not necessary. Scanning at an image/1–10 s seems reasonable.

8. Monomeric RFP (mRFP1) was produced by random or directed mutations (33 substitutions) in DsRed *(25)*. The maturation of mRFP1 is rapid and has minimal emission when excited at wavelengths of optimal for GFP. However, mRFP1 has several disadvantages and needs to be further improved. Currently, mRFP1 is not commercially available.

References

1. Nishizuka, Y. (1984) The role of protein kinase C in cell surface signal transduction and tumour promotion. *Nature* **308,** 693–698.
2. Nishizuka, Y. (1986) Studies and perspectives of protein kinase C. *Science* **233,** 305–312.
3. Nishizuka, Y. (1992) Intracellular signalling by hydrolysis of phospholipids and activation of protein kinase C. *Science* **258,** 607–614.
4. Ono, Y., Kikkawa, U., Ogita, K., Fujii, T., Kurokawa, T., Asaoka, Y., et al. (1987) Expression and properties of two types of protein kinase C determined by alternative splicing from a single gene. *Science* **236,** 1116–1120.
5. Ono, Y., Fujii, T., Ogia, K., Kikkawa, U., Igarashi, K., and Nishizuka, Y. (1988) The structure, expression, and properties of additional members of the protein kinase C family. *J. Biol. Chem.* **263,** 6927–6932.
6. Ono, Y., Fujii, T., Ogita, K., Kikkawa, U., Igarashi, K., and Nishizuka, Y. (1989) Protein kinase C ζ subspecies from rat brain: its structure, expression and properties. *Proc. Natl. Acad. Sci. USA* **86,** 3099–3103.

7. Osada, S., Mizuno, K., Saido, T. C., Akita, Y., Suzuki, K., Kuroki, T., et al. (1990) A phorbol ester receptor/protein kinase, nPKCη? a new member of the protein kinase C family predominantly expressed in lung and skin. *J. Biol. Chem.* **265,** 22,434–22,440.

8. Nishizuka, Y. (1995) Protein kinase C and lipid signaling for sustained cellular responses. *FASEB J.* **9,** 484–496.

9. Sakai, N., Sasaki, K., Ikegaki, N., Shirai, Y., and Saito, N. (1997) Direct visualization of translocation of γ-subspecies of protein kinase C in living cells using fusion proteins with green fluorescent protein. *J. Cell Biol.* **139,** 1465–1476.

10. Oancea, E., Teruel, M., Quest, A., and Meyer, T. (1998) Green fluorescent protein (GFP)-tagged cysteine-rich domains from protein kinase C as fluorescent indicators for diacylglycerol signaling in living cells. *J. Cell Biol.* **140,** 485–498.

11. Oancea, E. and Meyer, T. (1998) Protein kinase C as a molecular machine for decoding calcium and diacylglycerol signals. *Cell* **95,** 307–318.

12. Shirai, Y., Kashiwagi, K., Yagi, K., Sakai, N., and Saito, N. (1998) Distinct effects of fatty acids on translocation of γ- and ε-protein kinase C. *J. Cell. Biol.* **143,** 511–521.

13. Ohmori, S., Shirai, Y., Sakai, N., Fujii, M., Konishi, H., Kikkawa, U., et al. (1998) Three distinct mechanisms for translocation and activation of δ-subspecies of protein kinase C. *Mol. Cell. Biol.* **18,** 5263–5271.

14. Kajimoto, T., Ohmori, S., Shirai, Y., Sakai, N., and Saito, N. (2001) Subtype-specific translocation of the δ subtype of protein kinase C and its activation by tyrosine phosphorylation induced by ceramide in HeLa cells. *Mol. Cell. Biol.* **21,** 1769–1783

15. Almholt, K., Arkhammar, P. O. G., Thastrup, O., Tullin, S. (1999) Simultaneous visualization of the translocation of protein kinase Cα-green fluorescent protein hybrids and intracellular calcium concentrations. *Biochem. J.* **337,** 211–218

16. Feng, X., Zhang, J., Barak, L. S., Meyer, T., Caron, M. G., and Hannun, Y. A. (1998) Visualization of dynamic trafficking of a protein kinase C βII/green fluorescent protein conjugate reveals differences in G protein-coupled receptor activation and desensitization *J. Biol. Chem.* **273,** 10,755–10,762.

17. Wang, Q. J., Bhattacharyya, D., Garfield, S., Nacro, K., Marquez, V. E., and Blumberg, P. M. (1999) Differential localization of protein kinase Cδ by phorbol esters and related compounds using a fusion protein with green fluorescent protein. *J. Biol. Chem.* **274,** 37,233–37,239.

18. Wang, Q. J., Fang, T-W, Fenick, D., Garfield, S., Bienfait, B., Marquez, V. E., et al. (2000) The Lipophilicity of phorbol esters as a critical factor in determining the pattern of translocation of protein kinase Cδ fused to green fluorescent protein. *J. Biol. Chem.* **275,** 12,136–12,146.

19. Shirai, Y., Segawa, S., Kuriyama, M., Goto, K., Sakai, N., and Saito, N. (2000) Subtype-specific translocation of diacylglycerol kinase α and γ and its correlation with protein kinase C. *J. Biol. Chem.* **275,** 24,760–24,766.

20. Zeidman, R., Löfgren, B., Påhlman, S., and Larsson, C. (1999) PKCε, via its regulatory domain and independently of its catalytic domain, induces neurite-like processes in neuroblastoma cells. *J. Cell Biol.* **145,** 713–726.

21. Kashiwagi, K., Shirai, Y., Kuriyama, M., Sakai, N., and Saito, N. (2002) Importance of C1B domain for lipid messenger-induced targeting of PKC. *J. Biol. Chem.* **277,** 18,037–18,045.
22. Matz, M. V., Fradkov, A. F., Labas, Y. A., Savitsky, A. P., Zaraisky, A. G., Markelov, M. L., et al. (1999) Fluorescent proteins from nonbioluminescent Anthozoa species. *Nat. Biotechnol.* **17,** 969–973.
23. Baird, G. S., Zacharias, D. A., and Tsien, R. Y. (2000) Biochemistry, mutagenesis, and oligomerization of DsRed, a red fluorescent protein from coral. *Proc. Natl. Acad. Sci. USA* **97,** 11,984–11,989.
24. Newton, A. C. (1997) Regulation of protein kinase C. *Curr. Opin Cell Biol.* **9,** 161–167.
25. Campbell, R. E., Tour, O., Palmer, A. E., Steinbach, P. A., Baird, G. S., Zacharias, D. A., et al. (2002) A monomeric red fluorescent protein. *Proc. Natl. Acad. Sci. USA* **99,** 7877–7882.

10

Measuring the Binding of Protein Kinase C to Sucrose-Loaded Vesicles

Jennifer Giorgione and Alexandra C. Newton

1. Introduction

The binding of protein kinase C (PKC) to model membranes provides useful information about the mechanisms dictating the translocation of PKC to biological membranes. The sucrose-loaded vesicle (SLV) assay is useful for this purpose. It was originally designed to measure the binding of phospholipase C to vesicles *(1)* but later modified to be useful for the study of PKC *(2–4)*. Large unilamellar vesicles filled with sucrose are used as model membranes that can be readily pelleted by low-speed centrifugation (*see* **Subheading 3.1.**). PKC is incubated with the vesicles and other cofactors required for binding, and the mixture is centrifuged to pellet the vesicles along with any PKC bound to them. The supernatant containing the unbound PKC is separated from the vesicle pellet containing bound PKC (*see* **Subheadings 3.2.** and **3.3.**). The relative amounts of enzyme in the supernatant and pellet are determined by a kinase assay (*see* **Subheading 3.4.**). The percentage of PKC bound can be calculated from these values (*see* **Subheadings 3.5.** and **3.6.**). This assay is useful for determining various properties of PKC. For example, different compositions of lipids can be used to make the SLVs and the binding of PKC to each of them can be compared, to determine which lipids or combinations of lipids are preferred by the enzyme as well as optimal conditions for binding. Also, the binding of mutant PKCs can be compared to that of wild-type to determine the differences created by mutating specific residues.

From: *Methods in Molecular Biology, vol. 233: Protein Kinase C Protocols*
Edited by: A. C. Newton © Humana Press Inc., Totowa, NJ

2. Materials
2.1. SLVs

1. Lipids can be purchased from Avanti Polar Lipids.
2. 1-Palmitoyl, 2-oleoyl-phosphatidylserine (POPS) in chloroform (store at $-20°C$ under argon or nitrogen).
3. 1-Palmitoyl, 2-oleoyl-phosphatidylcholine (POPC) in chloroform (store at $-20°C$ under argon or nitrogen).
4. 2:1 (v:v) chloroform:methanol (store at room temperature in dark).
5. [^{3}H]1,2-dipalmitoylphosphatidylcholine (DPPC), 60 Ci/mmol, 1 µCi/µL (store at $-20°C$).
6. 1 mM phorbol myristate acetate (PMA) in dimethyl sulfoxide. Caution: PMA is a carcinogen (store at $-20°C$ in dark).
7. Sucrose buffer: 170 mM sucrose, 5 mM MgCl$_2$, 20 mM HEPES, pH 7.5. Store at 4°C.
8. KCl buffer: 100 mM KCl, 20 mM HEPES, 5 mM MgCl$_2$.
9. Microextruder from Avanti or Avestin.
10. Polycarbonate membranes (100-nM pore size).

2.2. PKC Dilution

1. Purified PKC, stored at $-80°C$.
2. 1 M dithiothreitol (DTT).
3. 1 M HEPES, pH 7.5.
4. 3 mg/mL bovine serum albumin (BSA).

2.3. Binding

1. Binding buffer: 0.36 mg/mL BSA, 120 mM KCl, 240 µM CaCl$_2$, 6 mM MgCl$_2$, 24 mM HEPES, pH 7.5, 1.2 mM DTT.
2. Buffer A: 0.3 mg/mL BSA, 100 mM KCl, 200 µM CaCl$_2$, 5 mM MgCl$_2$, 20 mM HEPES, 1 mM DTT.
3. Buffer B: 0.5 mg/mL BSA, 167 mM KCl, 333 µM Ca^{2+}, 33 mM HEPES, 8.3 mM MgCl$_2$, 1.7 mM DTT.

All three are stable for several weeks at 4°C.

2.4. Kinase Assay to Determine the Amount of PKC in Supernatant and Pellet

1. "GO": 25 mM MgCl$_2$, 20 mM HEPES, pH 7.5, 500 µM cold ATP, 100 µCi γ-^{32}P ATP, 1 mg/mL protamine sulfate.
2. "STOP": 0.1 M ATP, 0.1 M EDTA, pH 8.0.
3. Whatman P-81 ion exchange paper.

3. Methods

Methods will be described using specific volumes and concentrations for a standard assay, but these can be varied as needed (variations will be described in **Subheading 4.**). This section will describe how to make SLVs from pure lipid (**Subheading 3.1.**), how to dilute an aliquot of PKC for use in the binding assay (**Subheading 3.2.**), how to bind PKC to the vesicles and separate the bound from the unbound form (**Subheading 3.3.**), how to assay both the bound and unbound fractions to determine the amount of PKC in each of them (**Subheading 3.4.**), how to determine the percentage of SLVs that sediment (**Subheading 3.5.**), and finally how to calculate the fraction of PKC bound (**Subheading 3.6.**).

3.1. Making SLVs

This section will describe how to prepare the SLVs for use in **Subheading 3.3.**

1. Dilute POPS and POPC lipid solutions to 12.5 mM in 2:1 (v/v) chloroform:methanol. (If purchased in powder form, these can be weighed out and dissolved.)
2. Add 80 μL of 12.5 mM POPS and 120 μL of 12.5 mM POPC to a glass test tube.
3. Add 1 μL of [^{3}H]DPPC to test tube.
4. Vortex well.
5. Evaporate chloroform:methanol under a stream of nitrogen gas to leave a thin film of lipid around the bottom of the tube.
6. Dry under vacuum for 2 h.
7. Add 500 μL of sucrose buffer.
8. Vortex well.
9. Seal tube with Parafilm, but poke a hole in it.
10. Hold the tube with long tweezers, and place the portion of tube containing sucrose in liquid nitrogen, shaking gently. Tube will make a hissing noise, which indicates the solution inside has frozen.
11. Thaw under hot water.
12. Dry the outside of the tube completely and repeat the freeze-thaw cycle four times.
13. Extrude 300 μL of lipid using a microextruder (follow directions from manufacturer for assembly).
14. Use two 100-nm pore-size polycarbonate filters for extrusion.
15. Pass about 200 μL of sucrose buffer through extruder to wet the inside and ensure there are no leaks.
16. Discard, and then load vesicles.
17. Twenty full passes (back and forth) of the lipid through the filters should produce vesicles of uniform size. Remove vesicles from the opposite side to which they were added to ensure all have passed through the filter at least once.
18. Lipid solution becomes clearer after extrusion but remains slightly turbid.
19. Place 250 μL of extruded vesicles in a 1.5-mL ultracentrifuge tube.

20. Add 800 µL of KCl buffer.
21. Centrifuge at 100,000*g* for 30 min at 25°C (100°K; 50,000 rpm in a Beckman ultracentrifuge model TLA 120 rotor).
22. Remove 800 µL of KCl buffer.
23. Seal centrifuge tube with parafilm and vortex remaining lipid well (it should become slightly cloudier).
24. Add 5 µL of extruded and centrifuged lipid to 2 mL of scintillation fluid.
25. Add 5 µL of unextruded lipid to 2 mL of scintillation fluid.
26. Incubate both overnight since ^{3}H counts tend to increase slightly over the first few hours.
27. Determine counts per minute.
28. Use ratio of counts before and after extrusion to determine concentration of extruded vesicles.
29. Use KCl buffer to dilute extruded vesicles to 1 m*M*. The original vesicles were 5 m*M*, and are typically 3–4 m*M* after extrusion and sedimentation. There will be much more than needed for the one binding assay described below, but usually more than one assay is done at once.

3.2. PKC Dilution

This section will describe how to dilute the PKC for use in the binding assay. The amount of PKC used for each assay will vary slightly depending on the PKC preparation, the isoform used, and the specific activity of the enzyme. Typically, about 150–200 ng of PKC is used per assay tube. Because PKC obeys the law of mass action, the absolute amount of enzyme used will not change the fraction of it that binds.

Sample dilution:
  2 µL of purified PKC (about 150–200 ng).
  3 µL of 1 *M* DTT.
  3 µL of 3 mg/mL BSA.
  0.6 µL of 1 *M* HEPES.
  21.4 µL of H2O.

The total volume to be added to each tube is 30 µL (*see* **Subheading 3.3., step 6**).

3.3. Binding Assay

This section will describe the binding of PKC to the SLVs and separation of bound from unbound PKC; binding to isolated domains of PKC can also be measured (*see* **Note 1, Fig. 1**).

1. Add 4 µL of 1m*M* PMA to 200 µL of 1m*M* extruded vesicles. PMA should be added to the side of the Eppendorf tube, and vortexed vigorously for 30 s. If added directly to the vesicles, it may come out of solution.

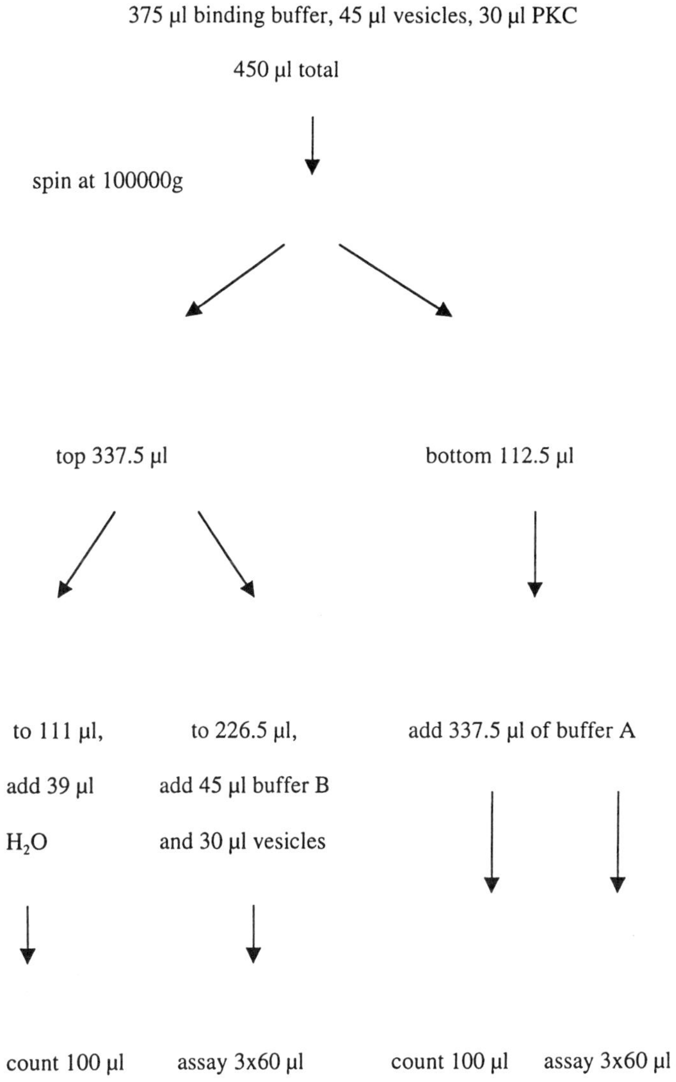

375 µl binding buffer, 45 µl vesicles, 30 µl PKC

450 µl total

spin at 100000g

top 337.5 µl   bottom 112.5 µl

to 111 µl,   to 226.5 µl,   add 337.5 µl of buffer A

add 39 µl   add 45 µl buffer B

H₂O   and 30 µl vesicles

count 100 µl   assay 3x60 µl   count 100 µl   assay 3x60 µl

Fig. 1. Schematic of binding assay.

2. Incubate the PMA and vesicles at room temperature for at least 30 min, vortexing occasionally. These vesicles contain 2 mol% PMA.
3. Add 375 µL of 1.2X binding buffer to each of two centrifuge tubes (*see* **Note 2**).
4. Add 45 µL of vesicles to one tube.
5. Add 45 µL of KCl buffer to the other tube. This tube is a control, and referred to as the "no lipid" tube. Always substitute KCl buffer in this tube whenever lipid is called for.

6. Add 30 µL of diluted PKC to each tube.
7. Mix well with a pipet.
8. Incubate for 10 min at room temperature.
9. Centrifuge at 100,000*g* for 30 min at 25°C.
10. Remove 337.5 µL of supernatant and store it in an Eppendorf tube. Label this the "kinase tube."
11. Add 337.5 µL of buffer A to pellet fraction, parafilm, and vortex well. This sample will be used in the kinase assay below, as well as for counting the amount of ^{3}H in the sample.
12. Remove 111 µL from the "kinase tube" and add it to 39 µL of H_2O. Save this for counting ^{3}H (do not use for kinase assay below).
13. Add 45 µL of buffer B and 30 µL of vesicles to remaining 226 µL of supernatant in the "kinase tube."
14. Substitute KCl buffer for vesicles for the "no vesicles" tube.
15. Mix all tubes well.
16. Add 60 µL from the "kinase tube" in **step 13** to each of three plastic culture tubes (to obtain results in triplicate).
17. Add 60 µL of pellet fractions from **step 11** to each of three plastic culture tubes (save the remainder of the fraction for counting ^{3}H).
18. Do the same for the "no lipid" controls.
19. Add a set of three "blanks" containing only 60 µL of buffer A and no PKC.

3.4. Kinase Assay

This section will describe the kinase assay, which is used to determine the amount of PKC in supernatant and pellet fractions. The activity obtained for each fraction is proportional to the amount of enzyme present (*see* **Note 3**).

This assay is timed, so prepare everything before starting (*see* **Note 4**).

1. Add 15 µL of "GO" solution to each tube.
2. Vortex quickly and incubate at 30°C for 6 min.
3. Add 25 µL of "STOP" solution to each tube, in the same order that the "GO" solution was added.
4. Mix briefly and place on ice.
5. Add 80 µL of each reaction to a numbered 2-cm^2 piece of P-81 ion exchange paper. The positively charged protamine sulfate will bind to the papers.
6. Let all papers dry.
7. Wash for 5 min in 200 mL of 0.4% phosphoric acid. This step washes away all of the ATP that is nonspecifically bound to the ion exchange papers, while leaving the protamine sulfate.
8. Remove papers then dispose of radioactive phosphoric acid appropriately.
9. Repeat **steps 7** and **8** three times.
10. Use 200 mL of ethanol for the fifth wash.
11. Remove papers and separate into scintillation vials.

12. Determine the counts per minute for each paper in a scintillation counter set to count ^{32}P (*see* **Note 5**).

3.5. Determination of Percentage of Vesicles Sedimented

The amount of ^{3}H in each supernatant and pellet fraction is counted, to determine the percentage of vesicles that have sedimented (*see* **Note 6**).

1. Add 100 μL of supernatant fractions (from **Subheading 3.3.**, **step 12**) to 2 mL of scintillation fluid.
2. Add 100 μL of pellet fractions (from **Subheading 3.3.**, **step 17**) to 2 mL of scintillation fluid.
3. Determine counts per minute in a scintillation counter set to count ^{3}H.

3.6. Calculations

This section will explain the calculations used to determine the fraction of PKC bound to the SLVs. From the amount of bound PKC, one can then calculate the apparent membrane affinity.

First, the percentage of vesicle associated activity (A_v) is determined using **Eq. 1**:

$$A_v = \frac{\beta A_b + (\beta - 1) A_t}{\alpha + \beta - 1} \tag{1}$$

where A_b and A_t are the measured activities of the bottom and top fractions, respectively. α is the fraction of sedimented vesicles and is calculated from the distribution of ^{3}H-labeled PC between the bottom and top fractions. β is fraction of kinase activity measured in the supernatant in the absence of lipid; within the limits of experimental error, it should be equal to the value expected for a nonsedimenting protein (i.e., 0.75 under the experimental conditions described; *see* **Note 7**).

Second, the fraction of bound PKC is calculated by taking the ratio of the vesicle-associated activity over the sum of the activities in the top and bottom fractions as in **Eq. 2**:

$$\% \text{ bound} = \frac{A_v}{A_b + A_t} \tag{2}$$

The apparent membrane affinity of the enzyme (apparent K_A) can now be calculated. It is defined as the ratio of membrane-bound to free enzyme divided by the total lipid concentration. Note that for accurate determination of K_A, it is important to measure binding under conditions where approx 50% of the PKC is bound.

4. Notes

1. A variation in the assay is to measure the membrane binding of either C1 or C2 domains of PKC rather than full length enzyme *(4)*. In this case, total protein is detected by SDS-PAGE analysis of bound vs free protein, followed by either silver staining or Western blot analysis. The amount of protein used in the binding step needs to be increased slightly so that it is enough to see on a gel but keeping the vesicle lipid in large excess of the protein. If using either a silver stain or a Western to determine the amount of PKC in the supernatant and pellet, simply separate the 337 µL of supernatant from the 112 µL of pellet, and then run one-third of each sample on a gel. There is no need to go through **steps 11–19** of **Subheading 3.3.** The gel replaces **Subheading 3.4.** (there is no need to add vesicles back to the supernatant fraction after separation because the amount of PKC is being determined by silver stain or a Western rather than by kinase activity). The fraction of PKC bound is the same regardless of the method used to determine it.

   The amount of PKC in the supernatant and pellet need to be determined by densitometry in the case of a silver stain, or by autoradiography in the case of a Western. Use the absolute amounts determined in place of counts per minute in **Subheading 3.6.** Instead of counting 100 µL of supernatant and pellet in **Subheading 3.5.**, count 33.7 µL of supernatant, and 11.2 µL of pellet fraction.

2. The volumes used for the binding steps above can be varied. The important part is to keep all the proportions the same. The rotor in the method above uses centrifuge tubes that hold volumes of up to about 1 mL. It is recommended that reaction volumes be at least 450 µL; smaller volumes result in small pellets that tend to float up when the supernatant fraction is pipetted off.

3. It is important that the kinase assay is linear over the entire 6-min reaction. This can be tested easily by setting up six reaction tubes and adding the "STOP" at different time intervals up to 6 min. The activity obtained should be linear over the entire 6 min. If not, the amount of PKC used needs to be reduced.

4. The assay described had only six tubes during the kinase portion. If there are many different conditions for an assay, the kinase portion of the procedure will get very cumbersome. It is easier to do the kinase **Subheading 3.4., steps 1–4**, in groups of 24 tubes, leaving the remaining tubes on ice.

5. Ideally, the sum of the activity in each supernatant and pellet fraction should be the same under each condition, since the same amount of PKC is added to each centrifuge tube. Realistically, this is not always the case, because there is some variation in the activation *seen* with different lipid compositions and amounts. The important part is that both the supernatant and pellet are treated similarly, in that they have the same amount and composition of vesicles added to them. Therefore, the ratio of the supernatant and pellet will be meaningful.

6. The percentage of vesicles in the pellet (**Subheading 3.6.**) should be at least 95% to avoid anomalies in the equation. If values are too low, it is possible that vesicles are being pipetted into the supernatant. Care should be taken not to pipet too close to the bottom of the tube, or to pipet the supernatant up too quickly.

Also, different phospholipids may sediment differently. Standard POPS/POPC vesicles should have about 96–98% in the pellet.

7. Different PKC preparations may have a different value for "no lipid," depending on whether there is some aggregation/denaturing affecting solubility. Ideally, this should be about 0.75 because this is the fraction of the total volume in the supernatant, but it is sometimes lower if there is some insoluble PKC in the preparation. Values for domains are sometimes lower.

Acknowledgment

This work was supported by NIH GM43154.

References

1. Rebecchi, M., Peterson, A., and McLaughlin, S. (1992) Phosphoinositide-specific phospholipase C-δ_1 binds with high affinity to phospholipid vesicles containing phosphatidylinositol 4,5-bisphosphate. *Biochemistry* **31,** 12,742–12,747.

2. Mosior, M. and Epand, R. M. (1993) Mechanism of activation of protein kinase C: roles of diolein and phosphatidylserine. *Biochemistry* **32,** 66–75.

3. Mosior, M. and Newton, A. C. (1995) Mechanism of interaction of protein kinase C with phorbol esters. Reversibility and nature of membrane association. *J. Biol. Chem.* **270,** 25,526–25,533.

4. Johnson, J. E., Giorgione, J., and Newton, A. C. (2000) The C1 and C2 domains of protein kinase C are independent membrane targeting modules, with specificity for phosphatidylserine conferred by the C1 domain. *Biochemistry* **39,** 11,360–11,369.

11

Use of Stopped-Flow Fluorescence Spectroscopy to Measure Rapid Membrane Binding by Protein Kinase C

Eric A. Nalefski and Alexandra C. Newton

1. Introduction

In investigating the binding of ligands to proteins, many studies have been concerned with answering the fundamental questions proposed by Scatchard *(1)*, that is, "How many" and "How tightly bound?" Although these questions are of utmost importance, the kineticist also wishes to know "How rapidly?" The development of rapid kinetic methodologies has provided many of the tools to answer the latter question *(2–4)*. Suitably designed kinetic experiments may, in many instances, be able to provide answers to the former questions as well. Moreover, kinetic studies may yield mechanistic insight that cannot be elucidated from other approaches.

The appearance and disappearance of transient second messengers during cell signal transduction places rigid constraints on the kinetic rate constants of signaling proteins that bind them. The time course of binding and release of these messengers must match their rise and fall to ensure that these signaling proteins serve as effective "on-off" switches. These constraints appear to operate on conventional isoforms of protein kinase C (PKC), a molecular sensor that translocates to cellular membranes and is activated upon binding two second messengers, Ca^{2+} and diacylglycerol (DG) *(5)*. Levels of both these signaling molecules transiently rise during cellular activation and then fall to a resting state *(6,7)*. Binding of Ca^{2+} to the PKC C2 domain triggers its engagement with anionic phospholipids *(8)*. Full activation of the enzyme, however, requires additional contact with the membrane through binding of its

From: *Methods in Molecular Biology, vol. 233: Protein Kinase C Protocols*
Edited by: A. C. Newton © Humana Press Inc., Totowa, NJ

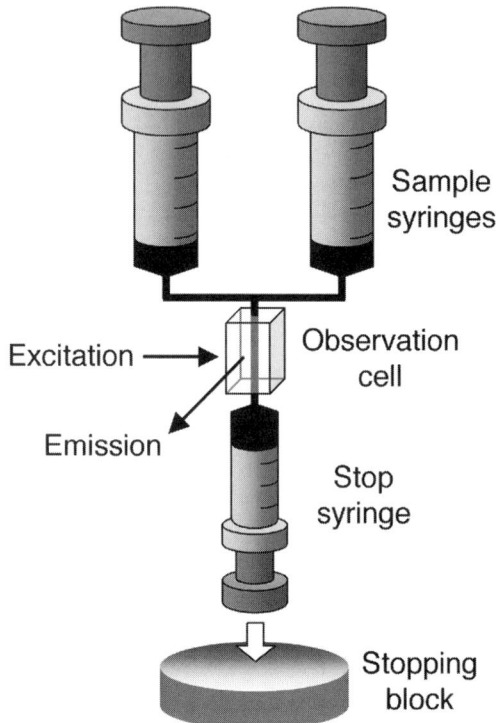

Fig. 1. Schematic of the stopped-flow apparatus and mechanism. Syringes containing two sample solutions are compressed, forcing the mixing of the reactants and their entry into an observation cell. The contents of the observation cell, consisting of the previous reaction solution, are forced out and pass into the stop syringe, whose plunger hits the stopping block, terminating the mixing process. The progress of the reaction is monitored in the observation cell using excitation light and emission measured at right angles to the excitation beam.

C1 domain(s) to membrane-sequestered DG and stereoselective recognition of phosphatidylserine *(5)*.

This chapter will provide strategies for determining the rate constants for binding of PKC and its isolated regulatory domains to synthetic phospholipid vesicles using rapid stopped-flow fluorescence spectroscopy. This technique measures the observed rate of binding of two reactants, in this case, protein and phospholipid vesicles of defined size and composition, under pseudo first-order conditions. These reactants are rapidly mixed together by the action of a stop syringe (**Fig. 1**). The reaction is monitored by recording fluorescence changes in the reactants or products upon formation of products. By fitting the spectroscopic data to simple exponential equations, quantitative information about the extent of binding, from the fluorescence amplitudes, and the observed rate of the approach to binding equilibrium (k_{obs}) is obtained. The dependence of k_{obs} on reactant concentrations for a simple one-step binding mechanism allows determination of kinetic constants k_{on} and k_{off}, from which the apparent dissociation constant (K_d) can be calculated ($K_d = k_{off}/k_{on}$). The dependence of the observed fluorescence change amplitude on reactant concentration provides an independent measure of the K_d value.

2. Materials

1. Several milliliters of purified protein in the micromolar range are required for the assay. Methods for purification of full-length PKC have been described *(9)* as have methods for production of recombinant PKC regulatory C2 and C1 domains as glutathione S-transferase fusion proteins expressed in *Escherichia coli (8)*. Alternatively, His$_6$-fusion proteins for purification over nickel resins can be used *(10)*.

2. Stocks of the phospholipids phosphatidylserine, phosphatidylcholine (PC), and DG are diluted in chloroform and stored at –20°C. Stock concentrations of commercial phospholipids should be determined as described below. Stocks of phorbol esters are diluted in dimethylsulfoxide and stored at –20°C. For the fluorescence resonance energy transfer (FRET) application *(11)*, fluorescent or nonfluorescent acceptor phospholipids are required. A suitable fluorescence acceptor of tryptophans in proteins that bind phospholipid vesicles is *N*-(5-dimethylaminonapthalene-1-sulfonyl)-1,2-dihexadecanoyl-*sn*-glycero-3-phosphoethanolamine (dansyl PE or dPE) *(12)*. Fluorescent phospholipid stocks are stored in the dark to avoid photobleaching. A method for preparing phospholipid vesicles of defined size by extrusion is described below. A visible light absorbance spectrophotometer is required for phospholipid quantification.

3. A stopped-flow fluorescence spectrophotometer capable of UV excitation and collecting and storing emission data is required. Hardware capable of reliably mixing together relatively small volumes (<100 µL) will provide efficient use of samples. Because fluorescence is highly dependent on temperature, thermostatted chambers surrounding sample syringes and the observation cell are required to maintain constant temperature. Depending on the instrument, filters (cut-off or band pass) may be required to collect fluorescence emission at the desired wavelengths.

4. Computer software capable of rapid least-squares analysis of data using exponential equations is required *(3)*. If possible, data can be analyzed directly by the software supplied with the stopped-flow apparatus. Alternatively, data may be processed using the method of global analysis *(13)*.

5. Buffered solutions for the experiment must be individually optimized for the protein and should be free of detergents and fluorescent contaminants. Suitable buffers include HEPES, PIPES, MOPS, but not Tris. Phosphate buffers must not be used because they interfere with phospholipid quantification and are

incompatible with assays involving the addition of Ca^{2+}. Assay buffer consisting of 150 mM NaCl, 20 mM HEPES (pH 7.4), and 5 mM dithiothreitol has been successfully used with full-length PKC βII and its isolated C2 domain *(14)*. Either ethylenediaminetetraacetic acid or Ca^{2+} may be included, depending on whether a Ca^{2+}-triggered binding event is being measured.

3. Methods
3.1. Protein Stock Preparation

Protein should be stored in buffers that stabilize protein activity. Protein stocks may be quantified using molar extinction coefficient and absorbance spectroscopy *(15)* or bichinoninic acid using bovine serum albumin as a standard *(16)*. Proteins must be able to survive the conditions of the assay for extended periods of time, for example, up to an hour at the temperature chosen. Immediately before use, proteins are rapidly thawed from frozen stocks maintained at –80°C, diluted into assay buffer, and kept on ice until analysis.

3.2. Phospholipid Vesicle Stock Preparation

Vesicles are prepared, as described in **Subheading 3.2.1.**, from mixtures of 20–40% PS, with the remaining phospholipid consisting of PC. Such vesicles have been used successfully with full-length PKC βII and its isolated C2 domain *(14)*. For the FRET application, the anionic phospholipid dPE may be incorporated at 2–5%, and the PS composition is adjusted to maintain the desired percentage of total anionic phospholipid. The concentration of vesicles is determined by a modified phosphate assay *(17)*, as described in **Subheading 3.2.2.**

3.2.1. Generation of Phospholipid Vesicles

1. Phospholipid stocks in chloroform are dispensed into glass test tubes, mixed, and dried down under a gentle stream of nitrogen gas into a thin transparent film.
2. Assay buffer is added to reconstitute phospholipids at approx 1–10 mM phospholipid.
3. The suspension is agitated vigorously by vortexing until a cloudy solution is formed and the film is completely removed from the sides of the tube.
4. The solution is transferred to a plastic vessel and subjected to at least three freeze-thaw cycles at –80°C and room temperature, respectively. Reconstituted phospholipids may be stored at –80°C until extrusion.
5. A suspension of 100-nm diameter vesicles is formed by extruding the thawed solution through double-stacked 100-nm pore size polycarbonate filters according to manufacturer's instructions.

3.2.2. Phospholipid Vesicle Quantification

1. A series of the unknown vesicles, ranging from 1 to 10 μL, and phosphate standards, ranging from 1 to 10 nmol (1 to 10 μL of 10 mM phosphate) are dispensed into disposable borosilicate glass test tubes along with a blank tube lacking phosphate.
2. 100 μL of 5 M sulfuric acid is added, and the mixture is heated to 150°C for at least 3 h.
3. 30 μL of 30% hydrogen peroxide (stored at 4°C) is added, and the mixture is heated to 150°C for an additional 90 min.
4. 920 μL of 0.22% ammonium molybdate and 50 μL of 10% ascorbic acid are added, and the mixture solution is heated to 100°C for 15 min.
5. Solutions are diluted fourfold into water, and absorbance is recorded at 830 nm.
6. After subtracting the blank value, measured from the reaction tube lacking phosphate, absorbance is plotted against the mass of the phosphate standard and the volume of the unknown. Slopes are determined using linear regression of data points in the linear range of the assay. The concentration of phosphate in the vesicle stock (in nmol per μL) is calculated by dividing its slope (in absorbance units per μL unknown) by the standard slope (in absorbance units per nmol phosphate). Phospholipid concentration is taken as the calculated phosphate concentration, from which vesicle concentration is calculated, assuming 90,000 phospholipids per vesicle of 100 nm diameter (*18*).

3.3. Sample Analysis

3.3.1. Sample Dilutions

A series of 2X vesicle solutions, 1–2 mL each, and a single protein solution, equal in volume to the total volume of vesicle solutions, are prepared in assay buffer immediately prior to analysis. The concentration of protein, generally 0.1–1 μM, required to provide an adequate fluorescence signal must be determined empirically. Vesicles are diluted over a wide range (e.g., 0.2–2.4 nM) gravimetrically. To assure pseudo first-order conditions of binding sites with respect to protein, the total concentration of target phospholipids must be in vast excess (at least 10-fold) of protein. Vesicles containing fluorescent phospholipids should be kept in the dark to avoid photobleaching.

3.3.2. Fluorometer Preparation

While the apparatus is allowed to equilibrate, stopped-flow plumbing lines and sample syringes are washed out extensively to remove any residual impurities from previous use. A good regimen would include successive washes with 10% SDS, 1 N NaOH, water, 1 N HCl, and then copious amounts of water. Finally, the plumbing lines are flushed extensively with assay buffer.

3.3.3. Instrument Settings

Because tryptophan emission varies considerably between proteins (λ_{max} = 310–350 nm), intrinsic tryptophan fluorescence may be monitored using 280 nm excitation and collecting emission with a 325 nm high-pass filter. For the FRET application, tryptophan is excited using 280 nm light and dPE emission is recorded using a 475 nm high-pass filter.

Although other instrument settings must be optimized carefully for each fluorescence application and instrument, certain guidelines may be followed; these are detailed in **ref. 3**. For instance, voltage and gain settings must be optimized for each application to achieve suitable signal to noise ratios. Excitation and emission slit widths should be adjusted, for example, to ≤5 nm, to allow sufficient light detection without photobleaching of sensitive materials. For short time collections (<1 s), pressure should be maintained on the stop syringe during data collection, if possible, to avoid artifacts caused by abrupt pressure changes. To conserve sample materials, the stopped-syringe settings should be adjusted to minimize the mixing volume without compromising the ability of the mechanism to flush out the contents of the previous reaction from the observation cell.

3.3.4. Data Collection

Protein solution is applied to one sample syringe, and the series of vesicle dilutions are applied, beginning with the lowest vesicle concentration, to the other. This approach obviates washing out the lines after application of a new vesicle stock. Samples are allowed to equilibrate to the thermostatted temperature for 5 min before collecting data. Individual test traces must be collected after application of a new vesicle stock to establish when the plumbing lines are completely cleared of previous reaction mixtures. For each time course, 1000–2000 data points are collected. Five such time courses are collected and averaged together, and then three sets of these averaged experiments are collected. If possible, each trace should be viewed prior to averaging so that aberrant traces, if any, can be omitted. The time base required for each experiment must be determined iteratively: a test run is analyzed to estimate the rate constant for the reaction, and then the final experiment is carried out at a collection time equal to five times the half-life to observe the reaction proceed to 97% completion (*see* **Note 1**).

3.4. Data Analysis

The goal of the kinetic approach presented here is to formulate the simplest mechanism that describes the steps in a binding reaction and to determine the observed rate constants for these steps. Analysis of the time courses described presently is simplified by conducting the experiments under pseudo first-order conditions, which reduces the observable steps to one or more exponential processes. Hence, the primary data are first fitted with simple exponential equations to determine the number of slowest observable steps in the binding process. This analysis provides the observed pseudo first-order rate constants and amplitudes for each reactant concentration. The dependence of these rate constants on reactant concentrations is then determined to ascertain the forward and reverse rate constants, from which the dissociation constant can be calculated. The dependence of the observed amplitudes on reactant concentrations is determined to independently evaluate the dissociation constant.

The following will describe kinetic experimentation in which intrinsic tryptophan emission is recorded and the time-dependent changes in fluorescence intensity are analyzed to extract apparent rate and equilibrium constants. Nevertheless, this analysis procedure may be applied to primary FRET data as well *(14)*.

3.4.1. Fitting Primary Fluorescence Data

Only data collected beyond the experimental dead time, which must be determined empirically (*see* **Note 2**), are to be included in the data analysis. Using the method of least-squares analysis *(19)*, the fluorescence intensities are fitted with exponential equations of the general form *(3)*:

$$F(t) = F_0 + \sum_{i=1}^{n} A_{obs(i)} e^{-k_{obs(i)}t} \tag{1}$$

where $F(t)$ represents the observed fluorescence at time t, F_0 is a fluorescence offset representing the final fluorescence, $A_{obs(i)}$ represents the amplitude and $k_{obs(i)}$ is the observed rate constant for the ith of n phases. Time courses are first fitted with a monoexponential equation to test whether the binding process may be approximated by a simple bimolecular interaction between the reactants (*see* **Note 3**).

The fitted value of F_0 may be subtracted from $F(t)$ to display each time course as an "offset fluorescence" in arbitrary units (U), which runs from a value of 0 to A_{obs} (**Fig. 2A**). This is particularly useful because the parameter that best reflects fractional binding saturation at equilibrium is the magnitude of A_{obs}, not the final observed fluorescence, which may vary between experiments due to instrument drift. Alternatively, the progression to binding equilibrium in different binding reactions may be represented by dividing the offset fluorescence by A_{obs} to obtain time courses that proceed from 0 to 1.

Two straightforward tests for the adequacy of data fitting involve simple inspection of the fitted parameters and analysis of residual plots. The fitted

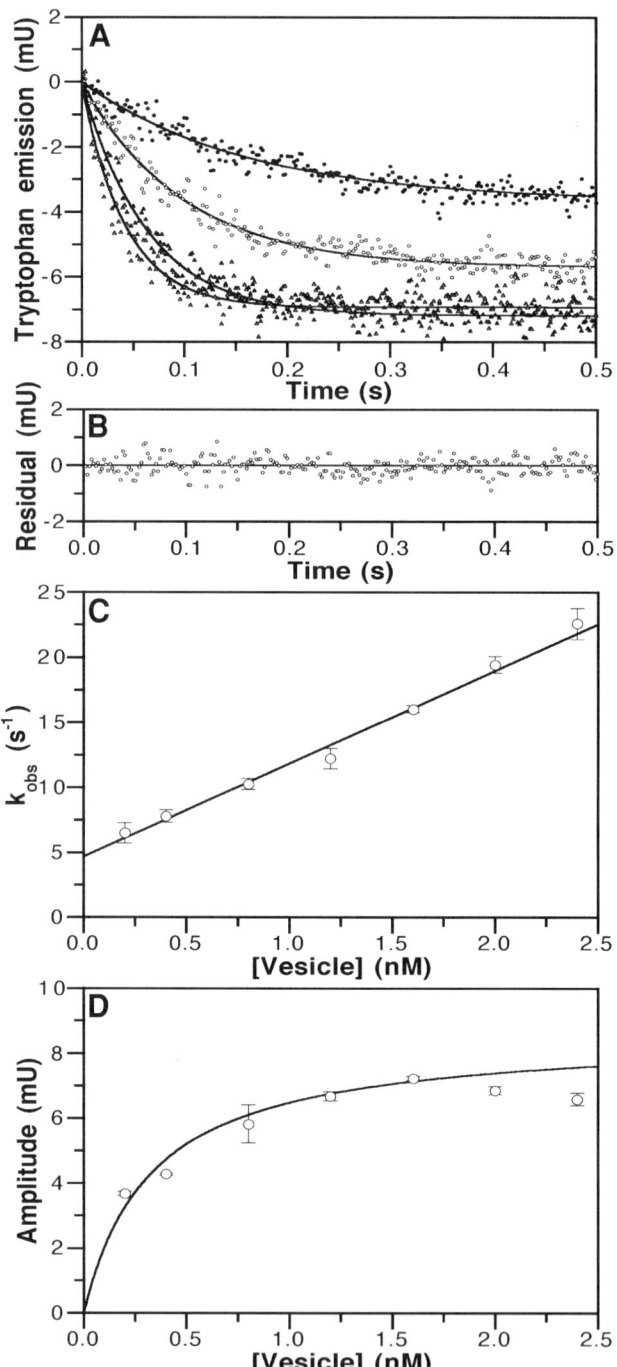

parameters must be reasonable (e.g., give rate constants that are consistent with the time base of data collection) and have small errors associated with them. For each vesicle concentration, residuals are calculated as the difference between $F(t)$ and $F_{calc(t)}$, the theoretical value based on the fitted parameters (*see* **Fig. 2B**). Residuals that are small and randomly distributed about zero are indicative of adequate fitting of the data.

The data are then fitted with biexponential equations to determine whether significant improvement in the fitting results (*see* **Note 4**). Caution is advised when judging the significance of the improvement in curve fitting by the addition of a second exponential term, since the inclusion of additional parameters almost always improves the performance of a given equation. Guidelines for ascertaining the "goodness of fit" are described in detail elsewhere (*2,20,21*).

3.4.2. Determining Apparent Forward and Reverse Rate Constants

For simple monophasic time courses representing the reversible binding of proteins to vesicles carried out under pseudo first-order conditions, the observed rate constants are plotted as a function of vesicle concentration (**Fig. 2C**). The data are fitted with a linear equation represented by **Eq. 2**:

$$k_{obs} = k_{on}[v] + k_{off} \tag{2}$$

Here, k_{on} (the slope) represents an apparent second-order association rate constant (in units of $M^{-1} s^{-1}$), and k_{off} (the y-intercept) represents an apparent first-order dissociation rate constant (in units of s^{-1}). Because three determina-

---

Fig. 2. (*see facing page*) Kinetics of PKC βII C1B domain binding to anionic phospholipid vesicles containing DG. Purified PKC βII C1B domain variant Y123W (0.5 μ*M*) was rapidly mixed with an equal volume of solutions containing increasing concentration of DG-PS-PC (5:35:60) vesicles. Tryptophan was excited with 280 nm excitation light and emission was recorded with a 325 nm high-pass filter. (A) Representative traces showing time course of tryptophan emission decreases for C1B-Y123W domain mixed with 0.2, 0.8, 1.6, and 2.4 n*M* (final) vesicles (top to bottom). Solid lines show monoexponential fit with **Eq. 1**, which yielded k_{obs} and amplitude values. (B) Representative residual plot from monoexponential fit of time course taken from data employing 0.8 n*M* vesicles shown in A. (C) Dependence of k_{obs} on vesicle concentration. Bars indicate standard deviation of three determinations from a single experiment. Weighted least-squares analysis using **Eq. 2** revealed a slope value for k_{on} equal to 0.71 (±0.03) × 10^{10} $M^{-1} s^{-1}$ and a y-intercept value for k_{off} equal to 4.7 (±0.4) s^{-1}. From these values, a K_d^{calc} equal to 0.66 (±0.06) n*M* can be estimated. (D) Dependence of amplitude on vesicle concentration. Bars indicate standard deviation of three determinations from a single experiment. Weighted least-squares analysis using Equation 3 yielded a K_d^{obs} equal to 0.33 (±0.02) n*M*.

tions are performed for each vesicle concentration, the data should be weighted by the uncertainties in k_{obs} values *(19)*. The ratio of the fitted parameters, k_{off} to k_{on}, provides the calculated apparent vesicle dissociation constant (K_d^{calc}, in units of M). For the more complex case of biphasic time courses generated by protein binding to vesicles, estimating the rate constants is more involved (*see* **Note 5**).

3.4.3. Determining Apparent Dissociation Constants

The observed amplitudes (A_{obs}) are plotted as a function of vesicle concentration and fitted with a hyperbolic equation represented by **Eq. 3**:

$$A_{obs} = A_{max}\left(\frac{[v]}{[v] + K_d^{obs}}\right) \tag{3}$$

Here, A_{max} represents the calculated maximal amplitude and K_d^{obs} represents the observed vesicle dissociation constant (*see* **Fig. 2D**). In cases where vesicle concentrations may be tested both well below and above the K_d value, K_d^{obs} should agree reasonably well with the K_d^{calc} value based on the measured kinetic constants.

4. Notes

1. Because the half-life ($t_{1/2}$) is equal to $\ln 2/k$, the amplitude at $t = 5t_{1/2}$ equals $A_{obs}e^{(-5\ln 2)}$, which in turn equals $0.03A_{obs}$.
2. A general method for determination of the mixing time of a stopped-flow apparatus is based on the reduction of 2,6-dichlorophenolindophenol (DCIP) by L-ascorbic acid (AA) *(22)*, which can be followed spectroscopically. This procedure involves the determination of the dependence of the observed amplitude on the observed chemical reaction rate constant. A stock solution of 500 μM DCIP is prepared by dissolving 7.25 mg of DCIP into 5 mL of isopropanol containing 0.58 g NaCl, and the volume is raised to 50 mL with water. A 200 mM solution of AA is prepared by dissolving 0.88 g AA into 25 mL of an acid-salt solution comprised of 0.02 N HCl and 0.2 M NaCl. A series of AA dilutions, ranging from 2 to 100 mM, are then prepared in 3 mL of the acid-salt solution. Beginning with the lowest concentration, the AA solutions are rapidly mixed with the DCIP solution in the stopped-flow apparatus in absorbance mode, recording absorbance at 524 nm. The absorbance is plotted against time, and the average value observed within the first few milliseconds of each time course is noted. Because the absorbance of reduced DCIP is very small, this value represents ΔA_{obs}, the change in absorbance observable after mixing. Then, with the first several milliseconds masked, the data are fitted with monoexponential equations to obtain k_{obs}. Because the reaction is carried out under pseudo first-order conditions of AA with respect to DCIP, a plot of k_{obs} against AA concentration should yield a straight line. Because deviation from linearity is expected at high

AA concentrations, where reactions are completed within the mixing time, only those rate constants in the linear range are considered further. The $\ln(\Delta A_{obs})$ is plotted as a function of k_{obs}, and the absolute vale of the resulting slope is taken as the dead time. The inverse natural log of the y-intercept represents the theoretic amplitude of the reaction, that is, the absorbance change that results from complete reduction of DCIP by AA under these conditions.

3. The simple bimolecular interaction may be represented by **Scheme I**:

$$P + V \underset{k_{off}}{\overset{k_{on}}{\rightleftarrows}} P{\bullet}V$$

Scheme I

where P represents protein, V represents the phospholipid vesicle, and P•V represents the protein–vesicle complex.

4. Significantly improved fitting with biexponential equations suggests that the binding process involves two phases, which may be represented either by **Scheme II**:

$$P + V \underset{k_{-1}}{\overset{k_1}{\rightleftarrows}} P{\bullet}V \underset{k_{-2}}{\overset{k_2}{\rightleftarrows}} [P{\bullet}V]^*$$

Scheme II

or by **Scheme III**:

$$P + V \underset{k'_{-1}}{\overset{k'_1}{\rightleftarrows}} P^*{+}V \underset{k'_{-2}}{\overset{k'_2}{\rightleftarrows}} P^*{\bullet}L$$

Scheme III

Scheme II illustrates a mechanism whereby the bimolecular interaction between protein and vesicle is then followed by a unimolecular isomerization of the P•V complex. In contrast, **Scheme III** shows how unimolecular isomerization of the protein (or the vesicle) may precede its binding to vesicles. Because either may take place, in principle, neither mechanism may be excluded solely on the basis of analysis of apparent rate constants *(3)*.

5. For **Scheme II**, based on equations derived for general two-step mechanisms *(3)*, the following approximations for the two phases are shown in **Eqs. 4** and **5**:

$$k_{obs(1)} \simeq k_1[v] + k_{-1} + k_2 + k_{-2} \tag{4}$$

$$k_{obs(2)} \simeq \left(\frac{k_1[v](k_2 + k_{-2}) + k_{-1}k_{-2}}{k_1[v] + k_{-1} + k_2 + k_{-2}} \right) \tag{5}$$

The $k_{obs(1)}$ values for the fast phase are plotted as a function of vesicle concentration and fitted with a linear equation represented by **Eq. 6**:

$$k_{obs(1)} = k_1 [v] + C \tag{6}$$

In this case, k_1 (the slope) represents the apparent second-order rate constant for a bimolecular step. C (the y-intercept) represents the sum $k_{-1} + k_2 + k_{-2}$, where k_{-1} equals the apparent dissociation rate constant for the bimolecular step and $k_2 + k_{-2}$ is the sum of the forward and reverse rate constants for a first-order transition step. For the slow phase, $k_{obs(2)}$ is plotted as a function of vesicle concentration and fitted with a hyperbolic equation represented by **Eq. 7**:

$$k_{obs(2)} = k_{max(2)} \left(\frac{[v]}{[v] + C} \right) \tag{7}$$

where C is a constant and $k_{max(2)}$ represents the calculated asymptote, which is equal to the sum $k_2 + k_{-2}$. Hence, the difference in the y-intercept **(Eq. 4)** and the asymptote **(Eq. 5)** yields an estimate for k_{-1}. Estimates for the remaining rate constants k_2 and k_{-2} can in certain circumstances be determined by application of a trapping experiment *(14,23)*. Different approaches have been used to determine these constants in other binding systems *(24,25)*.

For the alternative **Scheme III**, the two phases are approximated by rearrangement of **Eqs. 4** and **5** *(3)*:

$$k'_{obs(1)} \simeq k'_1 + k'_{-1} + k'_2 [v] + k'_{-2} \tag{8}$$

$$k'_{obs(2)} \simeq \frac{k'_1 (k'_2 [v] + k'_{-2}) + k'_{-1} k'_{-2}}{k'_1 + k'_{-1} + k'_2 [v] + k'_{-2}} \tag{9}$$

Here, $k'_{obs(1)}$ values are plotted as a function of vesicle concentration and fitted with a linear equation given by **Eq. 10**:

$$k'_{obs(1)} = k'_2 [v] + C' \tag{10}$$

where the slope represents the apparent second-order rate constant k'_2 and the y-intercept C' represents the sum $k'_1 + k'_{-1} + k'_{-2}$. For the slow phase, $k'_{obs(2)}$ is plotted as a function of vesicle concentration and fitted with a hyperbolic equation represented by **Eq. 11**:

$$k'_{obs(2)} = k'_{max(2)} \left(\frac{[v]}{[v] + C'} \right) \tag{11}$$

where C' is a constant and $k'_{max(2)}$ represents the calculated asymptote, which is equal to k'_1. As before, estimates for the remaining constants k'_{-2} and k'_{-1} can be elucidated in certain circumstances from a trapping experiment *(14)*.

Acknowledgment

This work was supported by NIH GM43154.

References

1. Scatchard, G. (1949) The attractions of proteins for small molecules and ions. *Ann. N. Y. Acad. Sci.* **51,** 660–672.
2. Johnson, K. A. (1986) Rapid kinetic analysis of mechanochemical adenosinetriphosphatases. *Methods Enzymol.* **134,** 677–705.
3. Johnson, K. A. (1992) Transient-state kinetic analysis of enzyme reaction pathways in *The Enzymes*, 3rd ed, Vol. 20 (Sigman, D. S., ed.), Harcourt Brace Jovanovich, San Diego, pp. 1–60.
4. Gutfreund, H. (1999) Rapid-flow techniques and their contributions to enzymology. *Trends Biochem. Sci.* **24,** 457–460.
5. Newton, A. C. and Johnson, J. E. (1998) Protein kinase C: a paradigm for regulation of protein function by two membrane-targeting modules. *Biochim. Biophys. Acta* **1376,** 155–172.
6. Berridge, M. J. (1990) Calcium oscillations. *J. Biol. Chem.* **265,** 9583–9586.
7. Werner, M. H., Bielawska, A. E., and Hannun, Y. A. (1992) Multiphasic generation of diacylglycerol in thrombin-activated human platelets. *Biochem. J.* **282,** 815–820.
8. Johnson, J. E., Giorgione, J., and Newton, A. C. (2000) The C1 and C2 domains of protein kinase C are independent membrane targeting modules, with specificity for phosphatidylserine conferred by the C1 domain. *Biochemistry* **39,** 11,360–11,369.
9. Orr, J. W. and Newton, A. C. (1992) Interaction of protein kinase C with phosphatidylserine. 1. Cooperativity in lipid binding. *Biochemistry* **31(19),** 4661–4667.
10. Ananthanarayanan, B., Das, S., Rhee, S. G., Murray, D., and Cho, W. (2002) Membrane targeting of C2 domains of phospholipase C-δ isoforms. *J. Biol. Chem.* **277,** 3568–3575.
11. Nalefski, E. A. and Falke, J. J. (2002) Use of fluorescence resonance energy transfer to monitor Ca^{2+}-triggered membrane docking of C2 domains. *Methods Mol. Biol.* **172,** 295–303.
12. Bazzi, M. D. and Nelsestuen, G. L. (1987) Association of protein kinase C with phospholipid vesicles. *Biochemistry* **26,** 115–122.
13. Beechem, J. M. (1992) Global analysis of biochemical and biophysical data. *Methods Enzymol.* **210,** 37–53.
14. Nalefski, E. A. and Newton, A. C. (2001) Membrane binding kinetics of protein kinase C βII mediated by the C2 domain. *Biochemistry* **40,** 13,216–13,229.
15. Gill, S. C. and von Hippel, P. H. (1989) Calculation of protein extinction coefficients from amino acid sequence data. *Anal. Biochem.* **182(2),** 319–326.

16. Smith, P. K., Krohn, R. I., Hermanson, G. T, and Klenk, D. C. (1985) Measurement of protein using bicinchoninic acid. *Anal. Biochem.* **150,** 76–85.
17. Bartlett, G. R. (1958) Phosphorus assay in column chromatography. *J. Biol. Chem.* **234,** 466–468.
18. Arbuzova, A., Wang, J., Murray, D., Jacob, J., Cafiso, D. S., and McLaughlin, S. (1997). Kinetics of interaction of the myristoylated alanine-rich C kinase substrate, membranes, and calmodulin. *J. Biol. Chem.* **272,** 27,167–27,177.
19. Taylor, J. R. (1997) *An Introduction to Error Analysis*, 2nd ed, University Science Books, Sausalito, CA.
20. Mannervik, B. (1982). Regression analysis, experimental error, and statistical criteria in the design and analysis of experiments for discrimination between rival kinetic models. *Methods Enzymol.* **87,** 370–390.
21. Straume, M. and Johnson, M. L. (1992) Analysis of residuals: criteria for determining goodness-of-fit. *Methods Enzymol.* **210,** 87–105.
22. Tonomura, B., Nakatani, H., Ohnishi, M., Yamaguchi-Ito, J., and Hiromi, K. (1978) Test reactions for a stopped-flow apparatus. Reduction of 2,6-dichlorophenol-indophenol and potassium ferricyanide by L-ascorbic acid. *Anal. Biochem.* **84,** 370–383.
23. Torok, K. and Trentham, D. R. (1994) Mechanism of 2-chloro-(ε-amino-Lys75)-[6-[4-(N,N- diethylamino)phenyl]-1,3,5-triazin-4-yl]calmodulin interactions with smooth muscle myosin light chain kinase and derived peptides. *Biochemistry* **33,** 12,807–12,820.
24. Zhao, Z., Rothery, R. A., and Weiner, J. H. (1999) Stopped-flow studies of the binding of 2-n-heptyl-4-hydroxyquinoline-N-oxide to fumarate reductase of Escherichia coli. *Eur. J. Biochem.* **260,** 50–56.
25. Lacourciere, K. A., Stivers, J. T., and Marino, J. P. (2000) Mechanism of neomycin and Rev peptide binding to the Rev responsive element of HIV-1 as determined by fluorescence and NMR spectroscopy. *Biochemistry* **39,** 5630–5641.

12

[³H]Phorbol 12,13-Dibutyrate Binding Assay for Protein Kinase C and Related Proteins

Nancy E. Lewin and Peter M. Blumberg

1. Introduction

In vivo, *sn*-1,2-diacylglycerol (DAG) is a lipophilic second messenger, generated as one arm of the phosphoinositide turnover pathway as well as indirectly though phospholipase D activity, that plays a central role in cellular signaling. DAG binds with high affinity to the C1 domains currently found in six receptor families: the protein kinase C (PKC), protein kinase D (PKD), chimaerin, munc-13, RasGRP, and DAG kinase families *(1)*. The phorbol esters, potent tumor promoters, function as ultrapotent analogs of DAG for binding to PKC and related receptors. The structure of the bimolecular complex between phorbol ester and the C1b domain of PKC δ has been solved by X-ray crystallography *(2)*, and our understanding has been extended to DAG, to other ligands and to homologous C1 domains by computer modeling *(3)*. A remaining, significant limitation in our understanding, however, is that high affinity binding requires the presence of phospholipid, and the details of the role played by the phospholipid in the ternary complex of lipid–ligand–C1 domain remain unresolved.

Around the time that PKC was first identified, our laboratory developed a specific binding assay for phorbol ester receptors *(4)*. The similarities in the characteristics of these receptors and of PKC contributed to the subsequent demonstrations that the phorbol esters stimulated PKC enzymatic activity *(5)*, bound to the enzyme *(6)*, and were competitively inhibited in their binding activity by DAG *(7)*. The binding assay has proven to be of great value in analysis of ligand–C1 domain interactions. The details of this assay, as refined over the years, are described below.

From: *Methods in Molecular Biology, vol. 233: Protein Kinase C Protocols*
Edited by: A. C. Newton © Humana Press Inc., Totowa, NJ

Phorbol 12-myristate 13-acetate is the usual phorbol ester used in biological studies *(8)*. However, its lipophilicity renders it highly problematic for binding studies. A critical aspect in the development of our binding assay for PKC therefore was the use of [20-^{3}H]phorbol 12,13-dibutyrate. This derivative, now commercially available, represents the phorbol derivative with the optimal ratio between binding affinity and nonspecific partitioning into the membrane *(4)*. Although PDBu will bind to PKC in the absence of phospholipid, the binding affinity is greatly enhanced in its presence *(9)*. We therefore routinely analyze the binding in the presence of a saturating amount of phosphatidylserine (*see* **Note 1**). For assays measuring competition of PDBu binding by other ligands, it is not essential that these other ligands be soluble in water *(10)*. We describe variants of the assay in which the competing ligands are added to the assay either in aqueous solution or else are incorporated directly into the lipid phase. In addition, we describe precautions that we routinely take to prevent losses of our ligands, which typically are amphipathic, to surfaces and the air–water interface. Finally, we describe aspects of data analysis that contribute to the robustness of the assay. Perhaps an unusual feature of our assay from the modern perspective is that we use centrifugation rather than filtration to separate bound and free ligand. Although less suitable for mass screening, for "sticky" ligands such as the phorbol esters this approach contributes to low backgrounds and low variation.

2. Materials

2.1. Reagents

1. [^{3}H]phorbol 12,13-dibutyrate, Perkin-Elmer Life Sciences/New England Nuclear (Boston, MA), cat. no. NET 692, 13–22 Ci/mmol.
2. Phosphatidylserine, Avanti Polar Lipids (Alabaster, AL), cat. no. 840032, concentration 10 mg/mL in CHCl$_3$.
3. Calcium chloride: suggested stock concentration 10 mM in deionized water.
4. Ethylenebis(oxyethylenenitrilo)tetraacetic acid (EGTA): suggested stock concentration 0.5 M in deionized water neutralized with NaOH.
5. Phorbol 12,13-dibutyrate, LC Laboratories (Woburn, MA), cat. no. P-4833, suggested stock concentration 40 mM in dimethyl sulfoxide (DMSO).
6. Immune gamma globulins (IGG), bovine from Cohn fraction II, III, Sigma-Aldrich (St. Louis, MO), cat. no. G-5009, suggested stock concentration 5 mg/mL in 50 mM Tris-HCl, pH 7.4.
7. Tris-HCl: 1 M, pH 7.4.
8. Polyethylene glycol 6000: suggested stock concentration 35.0% wt/vol in 50 mM Tris-HCl, pH 7.4.
9. Protein kinase C isoforms, Panvera Corp. (Madison, WI), cat. no. (for 20 µg): PKC α, P2227; PKC β 1, P2281; PKC β II, P2251; PKC γ, P2233; PKC δ, P2287; PKC ε, P2282; PKC ζ, P2268.

2.2. Equipment

1. Microcentrifuge racks, Research Products International Corp. (Mt. Prospect, IL), cat. no. H18905. A two-piece rack, lower tray holds crushed ice, top tray holds tubes.
2. Gilson Microman positive-displacement pipettors, Rainin (Emeryville, CA), cat. no. M-10, M-25, M-50, M-100, M-250, or M-1000. Recommended for working with volatile solutions.
3. Eppendorf Repeater Plus pipettor, VWR Scientific Products (So. Plainfield, NJ), cat. no 21516-002.
4. Gilson Pipetman, Rainin (Emeryville, CA), or equivalent micropipettors, range 2–1000 µL.
5. Sonicator with microtip, Misonix Incorporated (Farmingdale, NY), cat. no. XL2020 with microtip 420.
6. Microcentrifuge, refrigerated, VWR International (Bridgeport, NJ), Beckman 367570 Allegra 21 refrigerated centrifuge with Beckman 363000 rotor assembly, bowl H6002. We recommend using a horizontal drum rotor, which spins 60 microfuge tubes and gives a flat pellet.
7. Scintillation counter, beta counter, Perkin-Elmer Life Sciences (Boston, MA).

3. Methods

The precipitation binding assay currently has three versions. The first version, for Scatchard analysis, determines the K_d (dissociation constant) of the receptor protein for [³H]PDBu. The second version is for determining the K_i of a competing ligand that is hydrophilic and can be added in DMSO to the aqueous reaction mixture. The third version is for determining the K_i of a competing ligand that is hydrophobic and is dissolved in chloroform and added to the assay after its incorporation into the lipid phase. The same basic assay protocol is used for all three versions, but the versions differ in how the various assay components (mixes) are assembled and added to the assay tubes.

The basic assay protocol is detailed in **Subheading 3.1.** The subsequent three sections detail the preparation of the individual assay components (mixes) for the versions of the assay.

3.1. Basic Assay Protocol

The assay is performed in 1.5-mL Microfuge tubes (Sarstedt with screw caps). Each assay condition, whether the concentration of [³H]PDBu for Scatchard analysis or the concentration of competing ligand for determining the K_i, is performed in triplicate with a fourth tube to establish the nonspecific binding under that condition. The assay mixture is based on a volume of 0.25 mL per assay tube. A typical experiment will have eight different concentrations of ligand (including the control) for a total of 32 microfuge tubes in the assay. The arrangement of these tubes in the microfuge rack is important

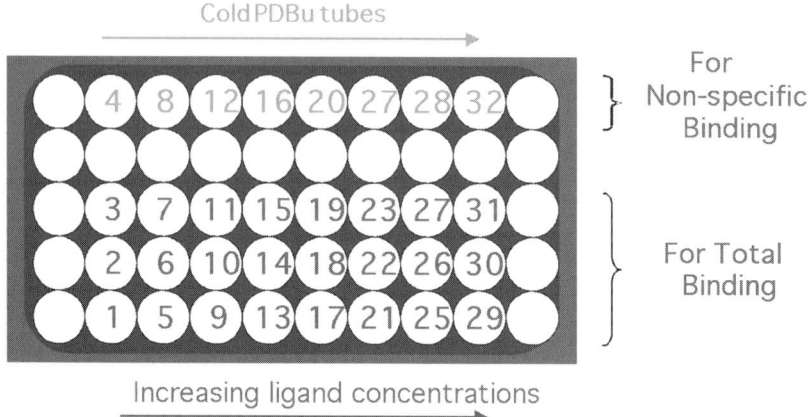

Fig. 1. Placement of tubes and direction of additions to assay tubes. Solutions are added to the control tubes first, proceeding from left to right, then to the tubes for the nonspecific binding.

for guiding the addition of the assay components. We use a 50-hole microfuge rack, 5 rows by 10 columns, with a lower tray in which ice can be placed (**Fig. 1**).

Experience suggests leaving the first and last columns of holes in the rack empty. For an assay with eight ligand concentrations, the tubes are placed in the second to ninth columns of holes in order of increasing concentration. The triplicates for each ligand concentration are placed in rows 1–3 with the fourth tube for determining the nonspecific binding in row 5. Label the tubes 1–32 as indicated. The microfuge tubes are located in the rack so that additions will be made from left to right, adding first to the triplicate rows (24 tubes) and finally to the last row (tubes 4, 8, 12, etc.). The specified microfuge tube placement and the method of addition are designed to prevent confusion or possible cross contamination of reaction components.

1. With the microfuge tubes labeled and placed on ice, the individual assay components are assembled and added to the microfuge tubes. The details for making the assay solutions and their order of addition will be given in **Subheadings 3.2.** and **3.3.** After the assay solutions are prepared, they are added to the tubes using a combitip repipettor. The last addition to the tubes is the receptor protein, which begins the reaction.
2. After all the assay solutions are pipetted into the tubes, each tube is vortexed and placed back on ice.
3. The top of the microfuge rack containing the assay tubes is then lifted off the ice-filled rack base and transferred to a water bath, where it is placed on another

microfuge rack base, which is half-filled with water. Both the bath and the rack base are at 37°C. (A covered water bath is preferred because it keeps the temperature constant.) When the top rack is placed on the microfuge rack base, the height of the water should come up to the bottom of the rack holes but not so high that the microfuge tubes float or that water can get into the tubes. The tubes are incubated for 5–60 min (5 min is the usual time; *see* **Note 2**).

4. After incubation, the tubes are removed from the water bath and placed on ice for 10 min.

5. After 10 min, 200 µL of refrigerator-chilled 35.0% wt/vol polyethylene glycol (PEG) is added to each tube. The function of the PEG is to precipitate both the PKC and the carrier protein. The PEG addition is made with a combitip barrel, which does not have a pipet tip placed on its end. The PEG is viscous and a tip will block the viscous flow and make the addition less accurate. After the PEG is added to all the tubes on ice, the tubes are vortexed. (The PEG sinks to the bottom of the tube and forms two layers, careful vortexing will produce one cloudy layer.) Failure to vortex adequately will lead to variable results and is a common mistake for inexperienced workers. The tubes are now capped with the screw caps and further incubated on ice for 10–15 min to allow complete precipitation of the protein. Under usual assay conditions, the release of [³H]PDBu from PKC is very slow at 4°C. If binding is being measured to a variant C1 domain for which the binding affinity is weak, with a corresponding rapid off-rate of the [³H]PDBu, then the addition of PEG will change the volume at which final equilibrium occurs and will thereby decrease the concentration of [³H]PDBu and of the competing ligands. These changes need to be considered when calculating the experiment and will be discussed in **Subheading 3.4.**

6. After precipitation of the PKC and carrier protein, the tubes are spun in the cold for 15 min at 12,400g to pellet the precipitate. For this we use Beckman 12 microfuges that are kept in a cold room. The rotor holds six microfuge racks of 10 tubes each, and the tubes spin horizontally so that flat pellets form at the bottom of the tubes. Because the Beckman 12 microfuges are no longer manufactured, we list in the equipment section a newer model centrifuge, which should be equivalent.

7. After the tubes are centrifuged, they are placed on ice, and racks of scintillation vials are prepared for counting. For an assay containing 32 tubes we need 32 scintillation vials for measurement of the radioactivity in the supernatants and 32 scintillation vials for measurement of the radioactivity in the pellets. (Minivials reduce the amount of scintillation fluid that is needed; we use Wheaton or Kimble 8 mL low-background glass vials).

8. After setting up the scintillation vial racks for the supernatants and pellets, the caps are removed from the microfuge tubes and 100 µL of each supernatant is transferred to the corresponding scintillation vial for a total of 32 vials. The supernatant is viscous and a pipet tip of adequate bore size (Rainin RT-20) and a suitable pipettor (200 µL Gilson Pipetman) is needed for consistent pipetting. Because of the viscosity of the supernatant, any problems in the quality of the

seal between pipet tip and pipettor shaft or between the pipettor piston and the sealing ring will be reflected in inaccurate transfer of the supernatant and thus in variation in the measured supernatant values. We routinely use a new tip for every tube, but at a minimum one tip for each ligand concentration (four tubes) should be used. The remaining supernatant is now carefully aspirated from the centrifuge tubes, leaving just the meniscus of liquid above the pellet. To aspirate we use a 7-inch Pasteur pipet attached to a vacuum flask.

9. The pellet is then dried and the bottom of the microfuge tube containing the pellet is cut off. We have two methods for drying the pellet. The first is to wind a Kimwipe to a point and use it to adsorb the remaining supernatant from the pellet in the tube. The pellet is then cut off and placed in the scintillation vial. The second method is quicker but less tidy. After aspiration of the supernatant, the pellet is cut off and the pellet is turned over on an absorbent paper towel to remove excess supernatant. The pellet is then placed in the scintillation tube. Either method works well. To cut the tubes, we use a single edge razor blade mounted in a holder; the razor blade holders are sold at local stores for scraping wallpaper.

10. After the supernatants and pellets have been added to the scintillation vials, 3–4 mL of Cytoscint ES are added to each vial. The vials are then capped and vortexed. The scintillation vials are put in the counter with the supernatants (which equilibrate the fastest) counted first and the pellets counted after them. We find that 8 h is necessary for complete equilibration of the pellets, so we program the counter appropriately so that 8 h elapse before the pellets begin to count.

The details for making the solutions for each of the versions for this protocol will now be discussed.

3.2. Solutions for Scatchard Version of Binding Assay

The Scatchard assay *(11)* is used to determine the affinity (K_d) of the receptor for the radioactive ligand [^{3}H]PDBu. The K_d value for [^{3}H]PDBu binding to its receptor is necessary for computing the K_i for competing ligands. For the Scatchard assay, the following solutions are combined to make the assay mix:

Solution 1: Serial dilutions of the [^{3}H]PDBu ligand for each set of four assay tubes; volume, 50 µL per tube.

Solution 2: A solution containing all the other necessary ingredients for binding; volume, 100 µL per tube.

Solution 3: Nonradioactive (cold) PDBu for determining the nonspecific binding, added to the fourth tube of each increasing [3H]PDBu concentration, and a control solution containing the components minus the nonradioactive PDBu for addition to the triplicate specific binding tubes; volume, 50 µL per tube.

Solution 4: Solution containing the receptor protein; volume, 50 µL per tube.

The assay is designed with a final volume for each tube of 0.25 mL. All the four solutions include 1–5 mg/mL IGG dissolved in 50 mM Tris-HCl, pH 7.4 (IGG-Tris). The additions are made in amounts of 50 μL or 100 μL using the 2.5 mL combitip repipettor.

3.2.1. Solution 1

The highest concentration of [³H]PDBu in the Scatchard assay is in the range of 80 times the anticipated K_d of the receptor, using approx 0.2 pmol receptor (typically 10–30 ng/tube receptor protein) in the assay. All the receptors, PKC, RasGRP, Chimaerin, PKD, and munc-13 have high affinity for PDBu (0.2–1 nM), so the use of [³H]PDBu of the specific activity commercially available gives a convenient level of dpms in the assay. Starting with the highest concentration of [³H]PDBu, seven 1:2 dilutions are made. For PKC α with an anticipated K_d of 0.2–0.3 nM, the range of [³H]PDBu concentrations is thus: 16 nM, 8 nM, 4 nM, 2 nM, 1 nM, 0.5 nM, 0.25 nM, and 0.125 nM. If this range of concentrations proves not to be appropriate, it can be adjusted as needed.

The serial dilutions are made in eight small glass tubes (12-× 75-mm Kimble). [³H]PDBu tends to stick to plastic over a period of time even in the presence of protein, so glass is preferred for diluting. Calculate the amount of [³H]PDBu needed for the assay as follows. A volume of 50μL [³H]PDBu solution is added to each of four assay tubes for each concentration of [³H]PDBu. The required volume of solution is thus 200 μL; an amount of 250 μL allows extra volume. The dilutions are made in 1 mg/mL IGG-Tris. The highest [³H]PDBu concentration (16 nM, final concentration in the assay tube) needs double the final volume (500 μL) for the following dilutions: 16 nM [³H]PDBu stock = 16 nM × 0.5 mL (volume) × (250 μL, final assay vol./50 μL, addition vol.) × (0.17) μL/pmol (determined from specifications sheet of [³H]PDBu, see calculation in **Subheading 3.4.**) = 6.8 μL [³H]PDBu added to glass tube. 493.2 μL (500 – 6.8 μL) of IGG-Tris is added to bring volume to 500 μL.

Serial 1:2 dilutions are made to provide the stock concentrations of [³H]PDBu ranging from 8 nM down to 0.125 nM [³H]PDBu. 250 μL of 1 mg/mL IGG-Tris is placed in each glass tube (2–8), then 250 μL of the 16 nM [³H]PDBu is added to tube 2 and vortexed; 250 μL of tube 2 is added to tube 3 and vortexed, and so on, until all dilutions have been made. The dilutions are then added to the assay tubes, 50 μL/tube, using a micropipettor. Add the lowest concentration (0.125 nM [³H]PDBu) to tubes 1–4, 0.25 nM [³H]PDBu to tubes 5–8, and so on until 16 nM [³H]PDBu is added to tubes 28–32.

3.2.2. Solution 2

A solution is made that contains all the necessary ingredients for binding other than the receptor and the [³H]PDBu. The solution contains the following:

calcium (for classic PKCs) or EGTA (for novel PKCs or other receptors lacking a calcium interacting domain), phosphatidylserine, and IGG-Tris. To allow extra volume, enough solution for 40 assay tubes is prepared.

The solution composition is calculated for addition of 100 μL to each assay tube. The final assay volume is 250 μL. The needed volume is 4 mL (100 μL × 40 tubes). The final assay volume after addition of the solutions to the assay tubes is 10 mL (250 μL × 40 tubes) total volume.

The components of solution 2 are calculated as follows: Ca^{2+} = 0.1 mM final conc. × 10 mL final vol./10 mM $CaCl_2$ in deionized water = 100 μL.

For receptor protein not having a C2 domain, use 1 mM EGTA instead of the 0.1 mM calcium. 1 mM final conc. × 10 mL final volume/0.5 M EGTA stock solution in deionized water = 20 μL).

The phosphatidylserine (PS) is dissolved in chloroform at 10 mg/mL. It is dried down to remove the chloroform and sonicated in a 1.2 mL volume of 50 mM Tris-HCl. One milliliter is added to the mix for solution 2. A glass scintillation vial (8 mL) is recommended for the drying and sonication of the PS.

PS = 0.1 mg/mL final conc. × 10 mL final vol. × (1.2 mL/1.0 mL)/10 mg/mL stock sol. = 120 μL (use positive displacement pipettor). 120 μL of PS is dried down under nitrogen. Then, 1.2 mL 50 mM Tris-HCl is added and the PS is sonicated with a microtip at 5–7% power for a total of 15 s, performed in three bursts of 5 s each. If all the PS is not dispersed it can be sonicated longer but attention should be paid that the sample does not overheat.

IGG-Tris = All four assay solutions contain IGG-Tris, so it is only necessary to calculate the amount needed in the volume for this solution: 1 mg/mL IGG-Tris × 4 mL mix vol./5 mg/mL stock conc. = 0.8 mL.

1 M Tris-HCl = The components of solution 2 are diluted to a final volume of 4 mL in 50 mM Tris-HCl. The amount of 1 M Tris-HCl necessary to add to deionized water for the final 50 mM Tris-HCl concentration is determined by subtracting from the final 4 mL volume of solution 2 those components, which were already in buffered solutions:

(4 mL vol. – 1 mL PS sol. – 0.8 mL IGG-Tris) = 2.2 mL unbuffered volume.

2.2 mL × 0.05 M Tris-HCl/1 M Tris-HCl = 110 μL 1 M Tris-HCl.

Calculate the amount of deionized water necessary to bring up the total volume to 4 mL by subtracting all the components of solution 2 from the total volume of solution 2.

4 mL – (Ca^{2+} or EGTA) vol. – (PS vol.) – (IGG-Tris vol.) – (1 M Tris-HCl vol.) = for the version of solution 2 containing calcium (4 mL – 0.1 mL – 1 mL – 0.8 mL – 0.11 mL) = 1.99 mL deionized water.

100 μL of solution 2 is now added to all the tubes.

3.2.3. Solution 3 (Same for All Assay Versions)

Cold PDBu is added to every fourth assay tube to determine the nonspecific binding for each [³H]PDBu concentration. The PDBu is added in 5 mg/mL IGG-Tris at a concentration greater than 10,000 times the receptor K_d. The cold PDBu effectively dilutes the specific activity of the [³H]PDBu so that negligible radioactivity is bound to receptor binding sites and any measured dpms are nonspecifically bound to the pellet. A routine assay concentration of 30 μM PDBu is used, with a stock made up in DMSO at a concentration of 40 mM. A solution of 5 mg/mL IGG-Tris + DMSO (mock) is made up to add to the triplicate tubes, so the volumes in all the tubes are kept equivalent.

> PDBu sol: For eight tubes, 50 μL per tube = 0.4 mL needed. Prepare 0.6 mL (0.6 mL) 5 mg/mL IGG-Tris × 30 μM PDBu × (250 μL final vol./50 μL addition vol.)/40 mM PDBu stock in DMSO = 2.25 μL PDBu
> mock sol: For 24 tubes, 50 μL per tube = 1.2 mL. Prepare 1.4 mL.
> Add the same amount of DMSO per unit volume as was added to the PDBu solution.
> 1.4 mL 5mg/mL IGG-Tris × (2.25 μL/0.6 mL) = 5.25 μL DMSO
> Add 50 μL of the mock solution to the triplicate assay tubes. Then add 50 μL PDBu solution to the fourth set of tubes (last row of tubes).

3.2.4. Solution 4 (Same for All Assay Versions)

The receptor protein is diluted in 1 mg/mL IGG-Tris in a volume of 50 μL per tube. A protein solution for 35 tubes, or 1.75 mL, provides adequate volume for the 32-tube assay. Thirty nanograms per tube protein × 35 tubes = 1.05 μg/ (1 mg/mL or your stock conc.) = 1.05 μL added to 1.75 mL 1 mg/mL IGG-Tris.

A 50-μL aliquot of receptor is now added to the assay tubes. It is important to add the receptor to the triplicate tubes first and then to the tubes containing cold PDBu. Vortex all the tubes and continue the Scatchard assay by referring back to the general assay protocol (*see* **Subheading 3.1.**) for the remaining steps of the assay.

3.3. K_i *Version of Binding Assay*

Once the K_d of the receptor protein has been determined, the K_i of a competing ligand can be determined using the same basic assay. A [³H]PDBu concentration of 2.5–5 times the determined K_d for the receptor protein is recommended for the assay. The competing ligand is added in logarithmically increasing amounts to the assay. No competing ligand is added to the first set of assay tubes to determine the control or maximal amount of [³H]PDBu bound.

The competing ligand is added to the assay in one of two ways, depending on the solubility characteristics of the compound. If the compound is soluble in DMSO it can be diluted and added directly in 1 mg/mL IGG-Tris. If the compound is lipophilic and soluble in chloroform it is added to the PS, also dissolved in chloroform; the solvent is evaporated under nitrogen; and the mixture of compound and PS is sonicated in the presence of 50 mM Tris-HCl.

3.3.1. K_i Version of Assay with Soluble Competing Ligand

The assay is similar to the Scatchard assay. The assay is designed with a final volume for each tube of 0.25 mL. All the components are added in 1-5 mg/mL IGG-Tris . The assay solutions are as follows.

> Solution 1: Increasing ligand (compound) concentrations, volume, 50 μL per tube.
> Solution 2: Solution containing assay binding components, volume, 100 μL per tube.
> Solution 3: PDBu or mock solution, volume, 50 μL per tube.
> Solution 4: Receptor protein solution, volume, 50 μL per tube.

3.3.1.1. SOLUTION 1

The basic assay is designed for 32 tubes, consisting of a control and seven concentrations of ligand. The compound/ligand is dissolved in ethanol, methanol, or DMSO at a concentration 10–50 mM. The solvent of choice is DMSO. The final concentration of solvent in the assay is kept below 0.2%. A final volume of 0.2% allows a maximum volume of 0.5 μL solvent per assay tube, so concentrated stock solutions (10–50 mM) are needed to be able to reach high μM concentrations in the assay. Occasionally, when very high μM concentrations are needed, this low percentage of solvent is unattainable, but the maximum addition should not exceed 0.5%, and, if such concentrations of solvent are used, the lack of effect of the solvent should be verified. If the affinity of the compound is unknown, a smaller assay of 10-fold dilutions spanning the nM to μM range can be performed to determine the range of activity. For example, concentrations could be: 0 nM, 10 nM, 100 nM, 1000 nM, and 10,000 nM. If there are several compounds to test, screening the compounds at 1000 nM or (100 nM and 10,000 nM) will indicate whether the compound is active in the nM or μM range.

Several steps are needed to prepare the dilutions of the compound. First, the compound is dissolved (in DMSO or other solvent) to make a stock solution of the appropriate concentration (10–50 mM). Fresh dilutions in DMSO or other solvent are made from the stock solution for each experiment, so small dilution volumes are used each time to conserve compound. These dilutions are then further diluted into 1 mg/mL IGG-Tris for addition to the reaction tubes. To test a compound dissolved in DMSO for the 10 nM–10 μM range, the compound is diluted in DMSO down to 10 μM from a concentration of the stock solution of 10–50 mM.

e.g., = [20 mM (dil 1:20, 4 μL into 76 μL) – 1000 μM (dil 1:10, 4 μL into 36 μL) – 100 μM (dil 1:10, 4 μL into 36 μL) – 10 μM].

Microfuge tubes work well for this purpose. Keep them on ice if ethanol or methanol is the solvent. The 4 μL volumes of compound are measured with a 1–10 μL pipettor (Rainin, Pipetman) if DMSO is the solvent or with a 10 μL positive displacement pipettor (Gilson, Microman) if ethanol/methanol is the solvent.

Set up 5 glass 12- × 75-mm tubes (Kimble) to make the actual dilutions that are added to the assay tubes. Each compound concentration needs a volume of (50 μL × four assay tubes) or 0.2 mL; 0.6–0.8 mL is typically chosen to give additional volume and improved accuracy in pipetting the compound.

Tube 1 (0 nM): control, add 3 μL solvent to 0.6 mL 1 mg/mL IGG-Tris.
Tube 2 (10 nM): (10 nM final) × 0.6 mL IGG-Tris × (250 μL, assay vol./50 μL addition vol.)/10 μM = 3 μL of 10 μM dilution.
Continue is same manner until tube 5.
Tube 5 (10,000 nM): (10 μM final) × 0.8 mL IGG-Tris × (250 μL, assay vol./50 μL addition vol.)/20 mM = 2 μL.

These dilutions are now added to the assay tubes in a 50 μL volume: tube 1, to assay tubes 1–4; tube 2, to assay tubes 5–8; continue in same manner until tube 5 is added to assay tubes 17–20.

After the active range of the compound has been determined using 10-fold dilutions, spanning 3 log units, an assay is set up with one concentration between each of the 10-fold dilutions (e.g., 3 nM between 1 nM and 10 nM) to better define the binding curve (the Log [3] = 0.477 and is thus well spaced between 1 and 10 on a Log scale). Likewise, the selection of the appropriate range of concentrations is important so that the concentrations give inhibition values 5–95%. An inhibition curve with seven points (100% point = the control value) should ideally have three points above the ID_{50} and three points below. If the compound is found to inhibit in the range of 1–1000 nM, an eight-point assay might be designed to have the following concentrations: 0 nM, 1 nM, 3 nM, 10 nM, 30 nM, 100 nM, 300 nM, 1000 nM, which are optimally spaced for a compound with an ID_{50} of 30 nM.

3.3.1.2. SOLUTION 2

Solution 2 contains all the necessary ingredients for binding other than PKC. The solution is similar to the corresponding Scatchard solution 2 except that it also contains the [^{3}H]PDBu. The solution contains the following: calcium (for classic PKCs) or EGTA (for novel PKCs or other receptors lacking a calcium interacting domain), [^{3}H]PDBu, phosphatidylserine, and IGG-Tris.

To allow extra volume, enough solution for a 40-tube assay is calculated. The solution volume is calculated for 100 µL per assay tube, and a final volume per assay tube of 250 µL. The needed volume of solution 2 = (100 µL × 40 tubes) = 4 mL. The final volume (after addition of all solutions to assay tubes) is (250 µL × 40 tubes =) 10 mL total volume.

The components of solution 2 are calculated as follows:

Ca^{2+} = 0.1 mM final conc. × 10 mL final vol./10 mM CaCl$_2$ in deionized water = 100 µL. For receptor proteins not having a calcium interacting domain, substitute 1 mM EGTA for the 0.1 mM calcium. (1 mM final conc. × 10 mL final volume/ 0.5 M stock sol. in deionized water = 20 µL).

[^{3}H]PDBu = The amount of radioactive ligand in the assay should be approx 2.5–5 times the K_d of the protein receptor. For example, for PKCα, which has a K_d of 0.3 nM, a final free concentration of 0.75 nM [^{3}H]PDBu is needed in the assay. To achieve this concentration in the assay, a value of 1–2 nM is used in the calculations for the solution.

e.g., 2 nM [^{3}H]PDBu × 10 mL (final volume) × (0.17) µL/pmol or your stock concentration (*see* **Subheading 3.4.** on calculations) = 3.4 µL

PS is in chloroform at 10 mg/mL and needs to be dried down and sonicated in a 1.2 mL volume of 50 mM Tris-HCl; 1.0 mL is added to the mix. A scintillation vial is recommended for use in the drying and sonication of the PS.

PS = 0.1 mg/mL final conc. × 10 mL final vol. × (1.2 mL/1.0 mL)/10 mg/mL stock sol. = 120 µL (use a positive displacement pipettor). PS (120 µL) is dried down under nitrogen, then 1.2 mL 50 mM Tris-HCl is added and the PS is sonicated with a microtip at 5–7% power for a total of 15 s, performed in three bursts of 5 s each. If all the PS is not dissolved, it can be sonicated longer but attention should be paid that the sample does not overheat.

IGG-Tris = All the four solutions contain IGG-Tris, so it is only necessary to calculate the amount needed in the volume of solution 2: 1 mg/mL IGG-Tris × 4 mL solution vol./5 mg/mL stock conc. = 0.8 mL.

1 M Tris-HCl = The components of solution 2 are diluted to a final volume of 4 mL in 50 mM Tris-HCl, pH 7.4. The amount of 1 M Tris-HCl necessary to add to deionized water for the final 50 mM Tris-HCl concentration is determined by

subtracting from the final 4 mL solution volume those components, which were added in buffered solutions:

(4 mL vol. − 1 mL PS sol. − 0.8 mL IGG-Tris) = 2.2 mL unbuffered volume.

2.2 mL × 0.05 M Tris-HCl/1 M Tris-HCl = 110 μL 1 M Tris-HCl.

Deionized water = The solution is brought up to the final volume in deionized water, by subtracting from the 4 mL total volume of solution 2 all the added components.

4 mL −(Ca^{2+} or EGTA vol.) − (PS vol.) − (IGG-Tris vol.) − (1 M Tris-HCl vol) = for the solution containing calcium (4 mL − 0.1 mL − 1 mL − 0.8 mL − 0.11 mL) = 1.99 mL deionized water.

100 μL of solution 2 is now added to all the tubes.

3.3.1.3. Solution 3 (Same as for Scatchard; *see* Subheading 3.2. for Details)

A solution containing 5 mg/mL IGG-Tris + 30 μM nonradioactive PDBu is added to every fourth assay tube to determine the nonspecific binding of [³H]PDBu in the assay. A solution of 5 mg/mL IGG-Tris + DMSO (mock) is made up to add to the triplicate tubes, so the volumes in all the tubes are kept equivalent. Add 50 μL of the mock solution to the triplicate assay tubes. Then, add 50 μL PDBu solution to the fourth set of tubes (last row of tubes).

3.3.1.4. Solution 4 (Same as for Scatchard; *see* Subheading 3.2. for Details)

The receptor protein is suspended in 1 mg/mL IGG-Tris in a volume of 50 μL per tube. Fifty microliters of this receptor solution (in IGG-Tris) is now added to the assay tubes. It is important to add the protein to the triplicate tubes first, and then to the tubes containing cold PDBu. Vortex all the tubes and continue by referring back to the assay protocol for the next steps of the protocol.

3.3.2. K$_i$ Version of Assay with Lipophilic Competing Ligand

For compounds soluble in chloroform, the assay is modified to include addition of the lipophilic ligand in the lipid, PS. The assay is designed with a final volume for each tube of 0.25 mL. All the components except the compound/PS dilutions are added in 1–5 mg/mL IGG-Tris. The assay components are as follows.

Solution 1:   Increasing compound concentrations in PS, volume 100 μL per tube.

Solution 2:   Solution containing assay binding components other than PS, volume 50 μL per tube.

Solution 3:   PDBu and mock solution, volume 50 μL per tube.

Solution 4:   Receptor protein solution, volume 50 μL per tube.

3.3.2.1. SOLUTION 1

The basic assay is designed for 32 tubes, with control and seven concentrations of ligand. The addition of the ligand to the assay varies from the direct addition of a soluble ligand in that the lipophilic ligand is dissolved in chloroform, added to the PS in chloroform (to ensure complete mixing), the chloroform is then dried down, and mixed lipid dispersions in buffer are prepared by sonication. The first four tubes are control tubes, with PS alone, the rest of the tubes contain seven increasing concentrations of ligand with the PS. The compound is dissolved in chloroform at a concentration of 10–20 mM. Because the ligand/PS mixture is dried down and sonicated there is no maximum concentration limit imposed by the amount of solvent in the assay. As a practical matter, however, we would expect the binding behavior of [^{3}H]PDBu to be distorted when the proportion of ligand in the PS becomes an appreciable fraction of the total. Under these conditions, inhibition of [^{3}H]PDBu binding may result from interference with the ability of the PS to support [^{3}H]PDBu binding rather than from direct competition. The highest concentration of ligand that is recommended is a 1:1 ratio of ligand to PS but preferably it should be 0.1:1 or less. The final concentration of the PS is 0.1 mg/mL, thus the highest recommended concentration of compound is 0.1 mg/mL or approx 300 µM. If the active range of the compound is unknown, a small initial experiment using the concentrations of 10 nM, 100 nM, 1000 nM, and 10,000 nM can be performed to determine the appropriate concentration range for the assays. If there are several compounds to be assayed, the compounds can be screened at 1000 nM (or 100 nM and 10,000 nM) to determine whether they are active in the nM or µM range.

Several steps are needed to prepare the dilutions of the compound. First, the compound is dissolved in chloroform to make a stock solution of the appropriate concentration (10–50 mM). Fresh dilutions in chloroform are made from the stock solution for each experiment, so small dilution volumes are used each time to conserve compound. The 1.5 mL (Sarstedt) microfuge tubes placed on ice work well for these dilutions. The dilutions are made using a positive displacement pipettor. Because of its volatility and density, chloroform cannot be pipetted accurately unless a positive displacement pipet is used. Chloroform is added to the microfuge tubes and serial dilutions are made until the lowest concentration needed is obtained.

e.g., = 10 mM stock (dil 1:10, 4 µL into 36 µL) – 1000 µM (dil 1:10, 4 µL into 36 µL) – 100 µM to 1.0 µM lowest concentration.

The next step for preparing the dilutions is to add the compound to aliquots of the lipid. Because the compound and the lipid are both in chloroform, glass works best for this step. An 8-mL glass scintillation vial works well, especially during the sonication step because the thicker vial does not easily shatter. For the smaller range-finding assay, five scintillation vials are placed in a scintillation rack. First the amount of PS that is needed will be calculated and added to the vials. Then the compound will be added to the same vials. The final volume of compound/PS to be added to each assay tube is 100 µL. Each dilution is added to four assay tubes, giving a needed volume of 0.4 mL. An amount of 0.6 mL is chosen to provide an adequate volume for sonication. The amount of PS, 0.1 mg/mL final concentration, is calculated as follows.

PS = 0.6 mL (volume) × 0.1 mg/mL (final PS concentration) × (250 µL, final assay volume/100µL, volume added)/10 mg/mL (PS stock concentration) = 15 µL PS.

With a positive displacement pipettor, add 15 µL 10 mg/mL PS to each of the five scintillation vials. The first vial will be the control, so no ligand is added to it. The next four tubes receive increasing concentrations of compound, with the amount calculated in a similar manner.

Compound = 10 nM, 100 nM, 1000 nM, 10,000 nM.

10 nM = 0.6 mL (volume of dilution) × 0.01µM (final concentration) × (250 µL, final assay volume/100µL, volume added)/1.0 µM compound diluted in chloroform) = 15 µL

100 nM = 0.6 mL × 0.1µM × 250µL/100µL/10 µM compound = 15 µL

1000 nM = 0.6 mL × 1.0µM × 250µL/100 µL/100 µM compound = 15 µL.

10,000 nM = 0.6mL × 10µM × 250µL/100µL/1000 µM compound = 15 µL.

Fifteen microliters of the appropriate chloroform dilution of compound is added to vials 2–5 containing the PS. The compound/PS mixtures are then dried under nitrogen to remove the chloroform. Chloroform denatures the protein if added directly to the assay, so it is essential that the chloroform is completely dried down. Add 0.6 mL 50 mM Tris-HCl to each of the vials. Each vial is now sonicated at 5–7% power using a microtip for 15 s using three bursts of 5 s. Sonicate in order of increasing compound concentration.

The dilutions are now ready for addition to the assay tubes. The PS alone is added to tubes 1–4 as the control, the 10 nM compound concentration to tubes 5–8, 100 nM to tubes 9–12, and so on. After the active range of the compound has been determined (about three log values) using 10-fold dilutions of compound, an assay is set up with one concentration between each of the 10-fold dilutions (e.g., 3 nM between 1 and 10 nM) to better define the binding curve (the Log [3] = 0.477 and is thus well spaced between 1 and 10 on a Log scale). Likewise, the selection of the appropriate range of concentrations is important so that the concentrations give inhibition values between 5–95%. An inhibition curve with seven points (100% point = the control value) should ideally have three points above the ID_{50} and three points below. If the compound is found to inhibit in the range of 1–1000 nM, an eight-point assay might be designed to have the following concentrations: (0 nM, 1 nM, 3 nM, 10 nM, 30 nM, 100 nM, 300 nM, 1000 nM) with 30 nM equaling the ID_{50}.

3.3.2.2. SOLUTION 2

A solution is made that contains all the necessary components for binding other than PS, with the volume based on addition of 50 µL per assay tube. The solution contains the following: calcium (for classic PKCs) or EGTA (for novel PKCs or other receptors lacking a calcium interacting domain), [³H]PDBu, and IGG-Tris. To allow extra volume, enough solution for a 40-tube assay is calculated. The needed solution volume is calculated for 50 µL per assay tube and a final volume per assay tube of 250 µL. The solution volume = (50 µL × 40 tubes = 2 mL). The final volume (after addition of all solutions to assay tubes) is (250 µL × 40 tubes = 10 mL total volume). The components of solution 2 are calculated as follows.

Ca^{2+} = 0.1 mM final conc. × 10 mL final vol./10 mM $CaCl_2$ in deionized water = 100 µL.

For receptor proteins not having a calcium-binding domain, include 1 mM EGTA in the assay instead of the 0.1 mM calcium.

1 mM (final concentration EGTA) × 10 mL (final volume)/0.5 M (EGTA stock in deionized water) = 20 µL.

[³H]PDBu = The amount of radioactive ligand in the assay should be approx 2.5–5 times the K_d of the protein receptor. For example, for PKCα, which has a K_d of 0.3 nM, a final free concentration of 0.75 nM [³H]PDBu is needed in the assay. To achieve this concentration in the assay, a value of 1–2 nM is used in the calculations for the solution.

e.g., 2 nM [³H]PDBu × 10 mL (final volume) × 0.17 µL/pmol or your stock concentration (*see* **Subheading 3.4.** on calculations) = 3.4 µL

IGG-Tris = In the previous assays all the assay components were added in IGG-Tris, but in this version the compound/PS is added in 50 mM Tris-HCl, so it is necessary to compensated for the lack of IGG in the compound/PS mix by adding extra IGG to solution 2. Calculate the amount of IGG needed in solution 2 as follows:

The assay volume including the compound/PS portion is (100 µL compound/PS + 50 µL solution 2) = 150 µL per assay tube. Thus, for 40 tubes, a volume of (40 tubes × 150 µL/tube) = 6 mL is used to calculated the amount of 1 mg/mL IGG-Tris needed in solution 2.

1 mg/mL IGG-Tris × 6 mL (needed volume)/5 mg/mL stock conc. = 1.2 mL

1 M Tris-HCl = The components of solution 2 are diluted to a final volume of 2 mL in 50 mM Tris-HCl. The amount of 1 M Tris-HCl necessary to add to

deionized water for the final 50 mM Tris-HCl concentration is determined by subtracting from the final 2 mL solution volume those components that were added in buffered solutions.

(2 mL vol. – 1.2 mL IGG-Tris) = 0.8 mL unbuffered volume.

0.8 mL × 0.05 M Tris-HCl/1 M Tris-HCl = 40 µL 1 M Tris-HCl.

Deionized water = The mix is brought up to volume in deionized water.
2 mL – (Ca²⁺ or EGTA) vol. – (IGG-Tris vol.) –(1 M Tris-HCl vol) =
for the solution 2 containing calcium (2 mL – 0.1 mL – 1.2 mL – 0.04 mL) = 0.66 mL deionized water.
50 µL of this solution is now added to all the assay tubes.

3.3.2.3. SOLUTION 3 (SAME AS FOR SCATCHARD; *SEE* SUBHEADING **3.2.** FOR DETAILS)

A solution containing 5 mg/mL IGG-Tris + 30 µM nonradioactive PDBu is added to every fourth assay tube to determine the nonspecific binding of [³H]PDBu in the assay. A solution of 5 mg/mL IGG-Tris + DMSO (mock) is made up to add to the triplicate tubes, so the volumes in all the tubes are kept equivalent. Add 50 µL of the mock solution to the triplicate assay tubes. Then add 50 µL PDBu solution to the fourth set of tubes (last row of tubes).

3.3.2.4. SOLUTION 4 (SAME AS FOR SCATCHARD; *SEE* SUBHEADING **3.2.** FOR DETAILS)

The receptor protein is suspended in 1 mg/mL IGG-Tris in a volume of 50 µL per tube. A 50-µL aliquot of this protein is now added to the assay tubes. It is important to add the protein to the triplicate tubes first, and then to the tubes containing PDBu. Vortex all the tubes and continue by referring back to the assay protocol section for the next steps in the protocol.

3.4. Evaluation of the Data

After the experiment has been counted in the scintillation counter, there is a set of 32 values for dpms representing the amounts of free [³H]PDBu in the assay tubes and a set of 32 values for dpms representing the total amounts of [³H]PDBu bound in the pellet fractions. Every fourth value represents the nonspecific binding in the assay tube, which contains a vast excess of nonradioactive PDBu. The dpms representing the free ligand for this fourth tube should be higher than the triplicate values for the other three tubes because the nonradioactive PDBu will have displaced the specifically bound [³H]PDBu from the PKC, causing the [³H]PDBu to shift to the supernatant. In the competition versions of the assay, as the amount of competing ligand increases the free dpms in the triplicates will also increase as less [³H]PDBu is bound to the PKC in the pellet because of competition. The pellet values should have triplicates which are very close to each other in dpms, whereas the fourth tube should have a much lower value, perhaps 5% of the other three. The number of dpms in the pellet is related to the amount of PKC in the assay. The suggested amount of PKC should give about 4000–8000 dpm in the control triplicates. If the pellet values are much higher or much lower the amount of PKC added in subsequent assays should be adjusted. If there are 8000 dpm for the triplicate values of the control, the nonspecific value for bound ligand will be around 500 dpm. If the triplicate pellet counts are similar to the nonspecific pellet counts in the control tubes, then the initial interpretation is that there is no specific binding, most likely because of a problem with the receptor. Initially, it is valuable to label your dpms so that you get a visual overview of the results. Label each set of four tubes for the concentration of ligand used and mark every fourth value as a reminder that this value is for the calculation of the nonspecific binding.

One consideration in the calculations of the assay is the type of protein that was used in the assay. The full-length PKCs bind [^{3}H]PDBu sufficiently tightly so that the complex of protein and [^{3}H]PDBu does not re-equilibrate after the PEG is added to the chilled assay solution. This is not true, for example, if the isolated C1b domain or its mutants are used. In this case, the calculations for the concentrations of both [^{3}H]PDBu and added ligand will need to be based on an assay volume corresponding to the volume after addition of PEG.

3.4.1. Calculation of the Amount of [^{3}H]PDBu to Be Added to the Assay

For calculating the amount of [^{3}H]PDBu to add to the assay (μL/pmol), you need the following information from the specifications sheet:

From the sheet, obtain the Ci/mmol. For example: 17.0 Ci/mmol
From the packaging section, obtain the concentration. For example: 0.1 mCi/mL.
   17 Ci/mmol = 17 μCi/nmol × 10^{-3} nmol/pmol = 0.017 μCi/pmol.
   0.1 mCi/mL = 0.1 μCi/μL.
   0.017 μCi/pmol/0.1 μCi/μL = 0.17 μL/pmol.

3.4.2. Calculation of the Data

3.4.2.1. VALUES NEEDED FOR CALCULATIONS

For calculating the pmol/mg bound in your assay, the following information is needed.

1. The final volume of the assay, 250 μL or (for proteins that re-equilibrate) 450 μL.
2. The volume ratio, 4.5; the amount of supernatant in the assay tube divided by the amount of supernatant counted.

450 µL supernatant (250 µL assay volume + 200 µL PEG)/
100 µL counted volume = 4.5

3. Receptor protein concentration. The receptor protein added to the assay is usually 10–30 ng/tube or 0.01–0.03 µg/tube and should represent about 0.2 pmol binding activity.

   Divide by the assay volume (0.25 mL or 0.45 mL) to get µg/mL protein.

If the protein concentration is unknown, the final calculations can be made using µL of protein added to the assay for pmol bound/µL protein stock solution.

3.4.2.2. CALCULATING NONSPECIFIC BINDING

The first calculation of the results is to determine the nonspecific binding in the assay. The typical approach in binding assays is to determine specific binding by subtracting the binding in the presence of the excess nonradioactive ligand from the total binding in the absence of the nonradioactive ligand. Unfortunately, this typical approach is imprecise and, although it is satisfactory under circumstances of low nonspecific binding and low ratios of receptor to radioactive ligand, becomes unsatisfactory if those conditions do not prevail. To calculate nonspecific binding more accurately, we recognize that in our system the level of nonspecific binding is proportional to the concentration of free [³H]PDBu. We determine a proportionality constant (partition coefficient) between the concentration of the free [³H]PDBu and the measured nonspecific binding. We then use the actual, measured free concentration of [³H]PDBu in each experimental tube times this proportionality constant to calculate the predicted nonspecific binding at that concentration of [³H]PDBu. This approach corrects for the change in free [³H]PDBu concentration as nonradioactive PDBu displaces the [³H]PDBu from the receptor.

The partition coefficient (proportionality constant) is calculated as follows. For the 32-tube assay there are eight supernatants for the nonspecific values and eight pellets (every fourth value). To demonstrate the calculations, the dpms for a control point of an experiment (triplicate control values + one nonspecific control value in bold) are presented as an example.

Control Supernatant, dpms	Control Pellet, dpms
3420	7514
3301	7510
3466	7511
4815	**495**

To determine the partition coefficient, first calculate total free [³H]PDBu in the nonspecific tube.

e.g., 4815 (supernatant dpms) × 4.5 (volume ratio) = 21,667.5 dpms

From the pellet value for this assay tube for nonspecific binding, calculate the percentage of free [³H]PDBu bound nonspecifically.

e.g., 495 (pellet dpms)/21,667.5 total supernatant dpms = 0.02285 partition coefficient. 2.3% of the dpms are nonspecifically bound to the protein.

An eight-point assay will have eight values for the partition coefficient. The eight partition coefficient values in a competition experiment to determine a K_i should be very close and the average of all eight points will give the best estimate of the partition coefficient to be used to calculated the nonspecific binding in the assay. Occasionally, the partition coefficients in an assay will increase with increasing ligand concentration as a result of large amounts of ligand dissolving in the PS, thereby changing its ability to nonspecifically bind [³H]PDBu. In that case, use the individual partition coefficients to calculate nonspecific binding at each concentration of compound. Likewise, in the Scatchard protocol, the assay tubes in which the concentrations of [³H]PDBu are very low may yield partition coefficients which are greater than for those points that have higher amounts of [³H]PDBu. In this case the individual partition coefficients determined for each concentration of [³H]PDBu should be used to calculated the nonspecific binding.

3.4.3. Specific Binding Calculations

The calculations for specific binding are performed for the triplicate values. The control values given previously will be used to calculate total dpms in the supernatant and the dpms specifically bound in the pellet. Other calculations, such as total dpms, percent specific bound, and percent bound of total, help troubleshoot the assay. For example,

1 Set	2 Supernatant	3 Total sup	4 Pellet	5 Nonspecific	6 Specific pellet	7 Total dpms	8 % Spec bd	9 % Bound of total
1	**3420**	15390	**7514**	315.7	7162.3	22904	95.3	31.3
1	**3301**	14854.5	**7510**	339.4	7170.6	22364.5	95.5	32.1
1	**3466**	15597	**7511**	356.4	7154.6	23,108	95.3	31.0

1. The number of sets in the usual assay is eight. The first set is the control.
2. The supernatant values (bold) are taken directly from the list of dpms generated by the scintillation counter.

3. The total supernatant dpms are calculated by multiplying the measured supernatant dpms by the volume ratio.

   e.g., 3420 dpms × 4.5 (vol. ratio) = 15390 total supernatant dpms.

4. The pellet values (bold) are taken directly from the list of dpms generated by the scintillation counter.

5. The nonspecific dpms are determined by multiplying the total supernatant dpms (3) by the partition coefficient determined for each set or by the average partition coefficient calculated from all eight sets.

   e.g., 15390 (total dpms) × 0.02285 (calculated partition coefficient) = 351.7 dpms.

6. The specific pellet dpms are calculated by subtracting the nonspecific dpms from the pellet dpms.

   e.g., 7514 (dpms) – 351.7 (nonspecific dpms) = 7162.3 dpms.

The next calculations are not necessary for determining the pmol/mg bound but are helpful to verify that the assay conditions were appropriate.

7. Total dpms are calculated by adding the total supernatant dpms to the total pellet (bold) dpms.

   e.g., 15390 dpms + 7514 dpms = 22904 dpms.

The total dpms should be equal in all the sets of data in the competition (K_i) assay.

In the Scatchard assay, the values within a set of triplicates should be similar and should increase by approximate factors of 2. If the total dpms are not the same, the reason should be determined. A bad pipettor can give varying supernatant dpms, or dpms can be lost by sticking to dilution tubes or assay tubes if they sit for long periods of time.

8. Percent of specific binding is calculated by dividing the specific pellet dpms by the total pellet dpms.

   e.g., 7162.3/7514 = 95.3%

The specific binding should be high (>90%). Obviously, as the percent of specific binding decreases, the values become increasingly sensitive to any uncertainties in the determination of nonspecific binding. If the percent specific binding is low, the usual correction is to increase the amount of receptor in the assay. We normally consider values with less than 30% specific binding unreliable.

9. The percent bound of total is determined by dividing the specific pellet dpms by the total dpms in the assay.

   e.g., = 7162.3 (specific pellet dpms)/22904 (total dpms) = 31.3%

It is important for several reasons that sufficient free ligand is maintained. First, there may be radioactive impurities in the radioactive ligand. Because the impurities will remain in the supernatant as the actual ligand binds to the receptor, the removal of much of the free ligand from the supernatant will reciprocally increase the proportion of radioactivity in the supernatant, which is an impurity and will thereby magnify the inaccuracy in the calculation of free ligand. In Scatchard analysis, this problem will be reflected in the points at the upper left end of the curve (near the y-intercept) falling below the straight line, because the ratio of bound/free ligand is reduced by an overestimate of the free ligand. Second, in competition experiments the shape of the inhibition curve will be distorted because the free concentration of [^{3}H]PDBu will vary dramatically over the competition curve as the [^{3}H]PDBu is displaced from the receptor by the competitor. This increasing concentration of [^{3}H]PDBu will compete more effectively with the competitor. Third, in saturation binding experiments the amount of [^{3}H]PDBu present will obviously limit the amount that can bind to the receptor, giving the appearance of an artifactually low B_{max}. We typically aim for 30% or less of total [^{3}H]PDBu to be specifically bound. We normally consider it unacceptable for more than 65% of total [^{3}H]PDBu to be bound. The remedies for an excessive amount of total [^{3}H]PDBu being bound are either to decrease the amount of receptor in the assay or to increase the amount of [^{3}H]PDBu. Because either remedy decreases the ratio of specific binding to total binding, conditions may need to be optimized to appropriately balance these conflicting problems in the case of assays at the limits of what is feasible.

3.4.4. Calculation of pmol/mg

The values of free and specifically bound ligand can now be converted from dpms to molar units. The number of dpms in a pmol of the hot ligand is calculated from the specific activity of the [^{3}H]PDBu obtained from the specifications sheet. If the [^{3}H]PDBu specific activity is 17 Ci/mmol.

e.g., 17 Ci/mmol $\times$ (2.22×10^{12}) dpms/Ci $\times 10^{-9}$ mmol/pmol = 37,740 dpms/pmol.

1. Calculate the nM of free supernatant as follows:
   Total dpms supernatant/0.25 mL or 0.45 mL (assay volume) = dpms/mL/dpms/pmol = pmol/mL = nM
   e.g., 15390 dpms/0.25 mL = 61,560 dpms/mL/37740 dpms/pmol = 1.63 nM.
2. Calculate the bound pmol/mg.
   dpms specific pellet/0.25 mL or (0.45 mL) assay volume = dpms/mL/dpms/pmol = pmol/mL = nM.
   e.g., 7162.3 dpms/0.25 mL = 28,649.2 dpms/mL/
   37740 dpms/pmol = 0.759 nM(pmol/mL).
   The amount of the protein in the assay is 0.03 µg/tube/0.25 mL/tube or (0.45 mL) =
   e.g., = 0.25 mL/tube final assay volume = 0.120 µg/mL.
   0.759 pmol/mL/0.12 µg/mL = 6.325 pmol/µg $\times 10^3$ µg/mg = 6325 pmol/mg.

From the calculated individual bound values determine the mean ± SEM value for each triplicate set. The SEM = SD divided by the square root of the number of replicates $((3)^{1/2} = 1.73)$. Calculating the pmol/mg of the given triplicate values you get: 6325, 6333, and 6317 pmol/mg.

The mean ± SEM: mean = 6325 pmol/mg
SD = 8.0 pmol/mg
SEM = 8.0/1.73 = 4.62 pmol/mg

A spreadsheet program, such as Excel or Quattro Pro, is sufficient for performing the above calculations.

3.4.5. Scatchard Analysis of Calculated Data

A Scatchard plot is a traditional way to display binding of a ligand to a receptor *(11)*. The amount of ligand bound is plotted on the x-axis; the ratio of bound/free ligand is plotted on the y-axis. In the absence of cooperativity, the points fall on a straight line, for which the x-intercept indicates the level of binding at saturation and the negative reciprocal of the slope indicates the dissociation constant. A line is readily fitted to the points using linear regression.

Despite its simplicity, the Scatchard plot has several flaws. The line is dominated by the points at the two extremes, near the x- and y-axes, although these points in fact are least well determined by the data. The points near the y-axis represent those values for which free and bound ligand are lowest; the points near the x-axis represent the points for which the nonspecific binding is greatest relative to total binding. In addition, statistically the analysis is complicated because the x- and y-axes are not independent. An error in measurement of specifically bound ligand is reflected both in the value of bound on the x-axis and in the value of bound/free on the y-axis.

Because of these defects in the Scatchard analysis, it is preferable to fit the data to the standard binding equation.

$$\text{Bound} = (B_{\max} \times \text{L}) / (K_{\text{d}} + \text{L})$$

where $B_{\max}$ represent binding at saturation, L represents free ligand, and K_{d} represents the dissociation constant. This can be performed either using the spreadsheet program or using a plotting program such as Origin. We then use the parameters derived from this analysis for graphical display of the data and for comparison of the fit between the points and the predicted curve.

Inappropriate assay conditions can lead to distortions in the Scatchard plot. If the percentage of specific to total binding becomes too small, then errors in the calculation of nonspecific binding become significant and the points at the highest ligand concentrations become unreliable. The line may appear to curve back to lower bound values as the curve approaches the x-axis (if the correction for nonspecific binding is excessive) or the curve may flatten and appear to parallel the x-axis (if the correction for nonspecific binding is insufficient). In either case, the remedy is to increase the amount of receptor or to reduce the maximum amount of ligand used. The opposite problem is that the proportion of total ligand that is bound is too large. In this case, as already discussed, the concentration of the radioactive impurities in the apparent free ligand will lead to an overestimate of the concentration of free ligand. The ratio of bound/free ligand will therefore appear artifactually low and the points will fall below the line near the y-intercept. The solution is to reduce the amount of receptor in the assay. Through attention to the appropriate conditions for the assay, it should be possible to obtain a straight line of six to eight points with an $r < 0.95$.

3.4.6. Determination of K_i from Calculated Data for Competition Experiments

The calculations of the data from an eight-point competition assay yield a control value for specific binding plus seven values for specific binding at the increasing concentrations of the competing ligand. The binding is expressed in units of pmol/mg.

1. To plot the data, the pmol/mg values are converted to percent of control for each concentration of compound by dividing the pmol/mg bound for each concentration of compound by the pmol/mg bound for the control and multiplying by 100. The SEM values are likewise normalized by dividing by the value of the control in pmol/mg and multiplying by 100. The inhibitor concentrations are plotted on the x-axis and the corresponding percent control values are plotted on the y-axis. The inhibitor concentrations:

    e.g., 10 nM, 30 nM, 100 nM, 300 nM, 1000 nM, 3000 nM, 10,000 nM

    are plotted on the x-axis using a Log scale. It is important to remember that, with a re-equilibrating receptor, the concentrations of the compound decrease because of the volume change from 0.25 mL to 0.45 mL. The inhibitor concentrations for an assay in which re-equilibration occurs are as follows.

    e.g. 10 n$M \times$ (250μL/450μL) = 5.56 nM, 30 nM =16.67 nM, 100 nM = 55.6 nM etc.

    The percent of control ± SEM for each ligand concentration is plotted on the y-axis on a linear scale from 0–100%.
2. The points are fitted to the equation.

$$y = 100 \times (1 + L/K_d)/(1 + I/K_i + L/K_d)$$

where y represents binding as percent of control, L represents the measured concentration of free [³H]PDBu for that point, K_d represents the measured dissociation constant of [³H]PDBu for the receptor, I represents the concentration of inhibitory compound, and K_i represents the dissociation constant of the inhibitor. All the values have either been measured in the assay (y, L) or previously (K_d), except for K_i, which is the value that is being determined. It is easy to solve the equation for K_i using the numerical analysis tools in either the Excel or Quattro Pro spreadsheets. The approach is to find the value of K_i which minimizes the sum of the squares of the differences between the experimental values and those predicted from the binding equation.

To plot the inhibition curve, it is useful to determine the predicted ID50 (50% inhibitory dose).

$$ID_{50} = K_i (1 + L / K_d)$$

where L represents the concentration of [³H]PDBu at the concentration of I which gives 50% inhibition. An approximate, predicted curve for the inhibition of binding with the parameters determined above is then given by the following equation:

$$y = 1 - (I / (ID_{50} + I))$$

We routinely compare the data points with the predicted inhibition curve based on the determined parameters. Almost always, a good fit is obtained unless there is a high amount of receptor present, which, as discussed previously, will cause L to vary markedly as a function of I. A second possible cause for deviation is aberrant behavior of the competing ligand, either reflected in problems in dilution or else in perturbation of the lipid environment for the receptor.

If there is any reason to suppose that the binding should reflect cooperative binding kinetics, then the Hill equation *(11)* can be used for the fitting in place of the noncooperative model described above.

3.5. Troubleshooting the Assay

As stated earlier, these assays are very reproducible and values of specific binding are attainable with an SEM of approx 2.0%. This section discusses problems that may arise, how to identify the problems, and how to correct them.

1. The assay depends on accurate pipetting. One of the reasons Rainin pipet tips are chosen for the assay is that there are lines on the tips for the 50 μL and 100 μL volumes so that the amount pipetted can be visually checked for accuracy. The assay relies heavily on the use of the combitip repipettor. The major problems encountered with this pipettor are using a pipettor that is defective, pipetting bubbles, and using the last dispelled amount. The combitip repipettor is easily checked for accuracy. Set the repipettor on a setting of five and draw the fluid lever up to maximum fill. The delivery lever should click exactly nine times and stop. If there is room for a partial click after nine clicks the repipettor is out of

adjustment and will deliver variable amounts. Get rid of any bubbles that collect in the tip before pipetting and fill the combitip barrel slowly to avoid bubbles. We have seen data where the amount of bound was low in every 24th tube. We traced the problem to totally emptying the combitip barrel rather than refilling before the last position was reached.

2. Another potential problem for inexperienced personnel is cross-contamination of tubes with nonradioactive PDBu or with competing ligands. It is very important to use fresh pipet tips when going into stock bottles. Likewise, we have seen data where the first pellet of every set of four was much lower than the other two replicates. We found that this occurred when the final addition of protein was made to all four tubes in the series rather than to all the triplicate tubes followed by the tubes with the nonradioactive PDBu. A small amount of PDBu was being carried over from the tube containing nonradioactive PDBu to the next tube, which was then contaminated by this competing ligand.

3. Inaccurate pipetting while making dilutions can also cause problems. If the competition curve appears as a series of steps with every second point deviating from the curve, then the problem is with the dilutions. If the problem occurs with a volatile solvent, then the remedy is to carefully keep the solvent dilutions on ice and pipet these dilutions immediately. Avoid pipetting very small volumes. We have routinely used 2–4 μL amounts for diluting with good results.

4. It is important that all the tubes be kept cold and adequately vortexed before incubation. Make sure when incubating the assay tubes that the bottoms of the tubes are evenly covered with water. We have often seen lower dpms in tubes that have been against a wall in the tray and not adequately covered by the water. In the same manner it is important that the assay tubes are adequately chilled on ice before the protein is precipitated with PEG. The on and off rates of the ligands are rapid at 37°C. At 4°C, the equilibration is very slow, and the samples will not re-equilibrate with the usual receptor after the samples are diluted with PEG. Make sure that the assay tubes are adequately vortexed after addition of the PEG. Because of the viscosity of the PEG and its density, the mixing is problematic. Inadequate mixing can be recognized because agitation will result in schlieren lines being visible in the solution. If the PEG is not well mixed, the precipitation will be incomplete, which will be reflected in variable triplicates.

5. The carrier protein should not be neglected. IGG is added to the assay to stabilize the receptor protein and to give a pellet that is large enough to be worked with. We have tried several proteins, including bovine serum albumin, and none of them precipitate as well as IGG. Furthermore, in the absence of carrier protein the [^{3}H]PDBu will be variably lost at the air–water interface and on surfaces.

6. When the tubes are spun, make sure that they are spun in the cold. Our microfuges in the cold room do warm up toward the end of the spin cycle, but, to avoid re-equilibration, it is important that the tubes do not warm up while the initial pelleting is occurring. Make sure the pellets are firmly precipitated. A soft pellet will give variable results. A 15-min spin at 12,400g is adequate to firmly precipitate the pellet.

7. Once the protein is precipitated, the manipulations of the supernatant must be done carefully so as not to disturb the pellet. Carefully remove the supernatant aliquot and for extra caution keep the tubes on ice. In the same manner care must be taken not to disturb the pellet while aspirating the supernatant and drying the pellet. The tube should be cut as close to the pellet as possible. A single-edge razor blade works the best. (Animal nail) tube cutters will leave a lot of tube above the pellet. The extra length of tube wall will markedly retard the rate at which the [³H]PDBu is extracted from the pellet into the scintillation fluid. We have determined that complete extraction can require more than 24 h if the tube is not appropriately cut. Make sure to vortex the supernatants and the pellets. The vortexing helps the pellets equilibrate. The first thing to check after the samples are counted in the scintillation counter is the agreement among the triplicate values. If the pellets are not well equilibrated there will be sporadic, low values among the triplicates for the pellets. This will tend to be more noticeable among the first sets of pellets counted. Confirmation that this is the explanation comes from recounting the samples. If the problem is equilibration of the pellets, then the triplicates will agree after the recount.

8. If there is variability among the pellets which is not due to lack of equilibration before counting, then the next diagnostic step is to examine the corresponding value for the supernatant for that tube. If the precipitation of the receptor was inadequate, then a low pellet value will be reflected in a correspondingly elevated value of the supernatant. If the supernatant is also low, then the probable explanation is a problem during pipetting of the initial mix with the [³H]PDBu. If the supernatant is neither elevated nor low, then the problem may be partial loss of the pellet either during aspiration or drying. If the triplicates for the supernatants show variability but the triplicates for the pellets do not, then the problem is the pipetting of the viscous, 100 μL aliquot of supernatant for measurement.

4. Notes

1. Because the phospholipid contributes to the binding affinity of the complex, both the amount and composition of the phospholipid will influence the measured binding affinity of PDBu. Moreover, different ligands/receptors will show different dependence on the phospholipid, influencing structure activity relations *(12)*.

2. The length of time for incubation depends on the time needed for equilibration and on stability of the receptor. In our experience, equilibration can be problematic for sufficiently lipophilic ligands if added to the aqueous phase of the assay *(10,13)*. Stability may be a problem for isolated C1 domains, depending on their structure and on whether they have been mutated *(14)*. In case of uncertainty, it is important to verify at the beginning of a series of experiments that the choice of incubation time and temperature is appropriate. For example, we routinely incubate the C1b domain of PKCδ for 10–30 min at 18°C to preserve adequate stability and time for equilibration.

References

1. Kazanietz, M. G. (2002) Novel "nonkinase" phorbol ester receptors: the C1 domain connection. *Mol. Pharmacol.* **61,** 759–767.

2. Zhang, G., Kazanietz, M .G., Blumberg, P. M., and Hurley, J. H. (1995) Crystal structure of the cys2 activator-binding domain of protein kinase C delta in complex with phorbol ester. *Cell* **81,** 917–924.

3. Pak, Y., Enyedy, I., Varady, J., Kung, J. W., Kang, J. H., Lewin, N. E., et al. (2001) Structural basis of binding of high affinity ligands to protein kinase C: prediction of the binding modes through a new molecular dynamics method and evaluation by site-directed mutagenesis. *J. Med. Chem.* **44,** 1690–1701.

4. Driedger, P. E. and Blumberg, P. M. (1980) Specific binding of phorbol ester tumor promoters. *Proc. Natl. Acad. Sci. USA* **77,** 567–571.

5. Castagna, M., Takai, Y., Kaibuchi, K., Sano, K., Kikkawa, U., and Nishizuka, Y. (1982) Direct activation of calcium-activated, phospholipid-dependent protein kinase by tumor-promoting phorbol esters. *J. Biol. Chem.* **257,** 7847–7851.

6. Niedel, J. E., Kuhn, L. J., and Vandenbark, G. R. (1983) Phorbol diester receptor copurifies with protein kinase C. *Proc. Natl. Acad. Sci. USA* **80,** 36–40.

7. Sharkey, N. A., Leach, K. L., and Blumberg, P. M. (1984) Competitive inhibition by diacylglycerol of specific phorbol ester binding. *Proc. Natl. Acad. Sci. USA* **81,** 607–610.

8. Hecker, E. (1968) Cocarcinogenic principles from the seed oil of Croton tiglium and from other Euphorbiaceae. *Cancer Res.* **28,** 2338–2349.

9. Kazanietz, M. G., Barchi, J. J., Jr., Omichinski, J. G., and Blumberg, P. M. (1995) Low affinity binding of phorbol esters to protein kinase C and its recombinant cysteine-rich region in the absence of phospholipids. *J. Biol. Chem.* **270,** 14,679–14,684.

10. Sharkey, N. A. and Blumberg, P. M. (1985) Highly lipophilic phorbol esters as inhibitors of specific [^{3}H]phorbol 12,13-dibutyrate binding. *Cancer Res.* **45,** 19–24.

11. Matthews, J. C. (1993) *Fundamentals of Receptor, Enzyme, and Transport Kinetics.* CRC Press, Inc., Boca Raton, FL.

12. Lorenzo, P. S., Beheshti, M., Pettit, G. R., Stone, J. C., and Blumberg, P. M. (2000) The guanine nucleotide exchange factor RasGRP is a high-affinity target for diacylglycerol and phorbol esters. *Mol. Pharmacol.* **57,** 840–846.

13. Wang, Q. J., Fang, T. W., Fenick, D., Garfield, S., Bienfait, B., Marquez, V. E., et al. (2000) The lipophilicity of phorbol esters as a critical factor in determining the pattern of translocation of protein kinase Cα fused to green fluorescent protein. *J. Biol. Chem.* **275,** 12,136–12,146.

14. Shindo, M., Irie, K., Nakahara, A., Ohigashi, H., Konishi, H., kkawa, U., et al. (2001) Toward the identification of selective modulators of protein kinase C (PKC) isozymes: establishment of a binding assay for PKC isozymes using synthetic C1 peptide receptors and identification of the critical residues involved in the phorbol ester binding. *Bioorg. Med. Chem.* **9,** 2073–2081.

V

MEASURING THE PHOSPHORYLATION OF PROTEIN KINASE C

13

Protein Kinase C Phosphorylation

An Introduction

Peter J. Parker

The constitutive, lipid-dependent protein kinase activity of purified protein kinase C (PKC) delayed the realization that phosphorylation of the PKCs themselves played a fundamental role in their catalytic activities. With the clarity of hindsight, it is evident that the proteins originally purified from tissue sources and indeed those recombinant proteins derived from baculovirus infection of insect cells were already highly phosphorylated *(1)*. The requirement for modification was thus not obvious.

Some of the first studies to provide insights into this problem came with the identification in tissue culture of a particulate, fast-migrating, apparently dephosphorylated form of PKCα, that was inactive *(2)*. More directly, it was also shown that purified PKCα was inactivated by protein phosphatase treatment *(3)*. Subsequently, there has been a substantial effort in mapping the phosphorylation sites and determining their effects on activity and other properties of these kinases. This has been pursued by combinations of mutagenesis, phospho-specific antisera, and mass spectrometry. These studies have shown that the serine/threonine phosphorylation sites that confer optimum catalytic activity on PKCs are broadly conserved in members of the protein kinase A, protein kinase G, protein kinase C(AGC) kinase subfamily of protein kinases (*see* http://pkr.sdsc.edu for detailed protein kinase superfamily alignments and also domain structures, vide infra). As indicated in the alignment (**Fig. 1**), one phosphorylation site is in the activation loop, with two at the C-termini of the PKCs. For aPKCs and also PRKs/PKNs, the most C-terminal site is replaced by an acidic amino acid albeit within the same hydrophobic FXXFXF/Y context.

From: *Methods in Molecular Biology, vol. 233: Protein Kinase C Protocols*
Edited by: A. C. Newton © Humana Press Inc., Totowa, NJ

Activation Loops

```
PKCα        ...DFGMCKEHMMDGVTTRTFCGTPDYIAPEII...
PKCβ1       ...DFGMCKENIWDGVTTKTFCGTPDYIAPEII...
PKCβ2       ...DFGMCKENIWDGVTTKTFCGTPDYIAPEII...
PKCγ        ...DFGMCKENVFPGSTTRTFCGTPDYIAPEII...
PKCδ        ...DFGMCKENIFGENRASTFCGTPDYIAPEIL...
PKCθ        ...DFGMCKENMLGDAKTNTFCGTPDYIAPEIL...
PKCε        ...DFGMCKEGIMNGVTTTTFCGTPDYIAPEIL...
PKCη        ...DFGMCKEGICNGVTTATFCGTPDYIAPEIL...
PKCζ        ...DYGMCKEGLGPGDTTSTFCGTPNYIAPEIL...
PKCι        ...DYGMCKEGLRPGDTTSTFCGTPNYIAPEIL...
PRK1        ...DFGLCKEGMGYGDRTSTFCGTPEFLAPEVL...
PRK2        ...DFGLCKEGMGYGDRTSTFCGTPEFLAPEVL...
PKNβ           DFGLCKEGIGFGDRTSTFCGTPEFLAPEVL...
               *  *  ***        *******    ***
```

```
PKCα        ...FTRGQPVLTPPDQL-VIANIDQSDFEGFSYVNPQFVHPILQSAV
PKCβ1       ...FTRHPPVLTPPDQE-VIRNIDQSEFEGFSFVNSEFLKPEVKS
PKCβ2       ...FTRQPVELTPTDKL-FIMNLDQNEFAGFSYTNPEYVINV
PKCγ        ...FTRAAPALTPPDRL-VLASIDQAEFQGFTYVNPDFVHPDARSPISPTPVPVM
PKCδ        ...FLNEKPQLSFSDKN-LIDSMDQTAFKGFSFVNPKYEQFLE
PKCθ        ...FLNEKPRLSFADRA-LINSMDQNMFRNFSFMNPGMERLIS
PKC         ...FTREEPILTLVDEA-IIKQINQEEFKGFSYFGEDLMP
PKCη        ...FIKEEPVLTPIDEG-HLPMINQDEFRNFSYVSPELQL
PKCζ        ...FTSEPVQLTPDDED-VIKRIDQSQFEGFEYTNPLLLSAEESV
PKCι        ...FTNEPVQLTPDDDD-IVRKIDQSEFEGFEYINPLLMSAEECV
PRK1        ...FTGEAPTLSPPRDARPLTAAEQAAFLDFDFVAGGC
PRK2        ...FTSEAPILTPPREPRILSEEEQEMFRDFDYIADWC
PKNβ        ...FTGLPPALTPPAPHSLLTARQQAAFRDFDFVSERFLEP...
               *              *   *   *
```

C-termini

Fig. 1. Alignment of PKC superfamily conserved catalytic domain phosphorylation sites. The phosphorylation sites are as indicated by the outlined single letter code amino acids. The substitution of acidic residues in the most C-terminal site is indicated for the atypical PKCs and for the related PRK/PKN kinases.

The predicted structures of cPKC catalytic domains *(4,5)* are consistent with the notion that these domains of PKCs can exist in open and closed conformations, the latter of which are stabilized by phosphorylation and probably reflects the catalytically active conformer. It is possible that the more open conformation is required for substrate binding and product release; this would be consistent with the prediction of buried nucleotide. Irrespective of catalytic mechanism, it is clear that these phosphorylations and the effects they have on conformation and activity have profound implications for the functions of PKCs and by inference with respect to the diverse regulatory inputs that control steady state phosphorylation.

The only well-established upstream kinase involved in the control of PKC is the PDK-1 *(6,7)*. PDK-1 comprises a kinase domain and a phosphatidylinositol 3,4,5 P_3-binding pleckstrin homology domain *(8,9)*. It serves a general role in the phosphorylation of multiple AGC family protein kinases in their catalytic loops *(10,11)*.

In addition to these conserved sites of phosphorylation, further serine/threonine and tyrosine phosphorylation sites have been identified in different PKCs. These reside in both regulatory and catalytic domains. The properties conferred by these phosphorylations are varied and include: lipid independence *12,13)*, increased/decreased specific activity or substrate selectivity *(14,15)*, and modified susceptibility to degradation *(16)*. Certain of these phosphorylations appear to be autophosphorylation events, whereas others, for example, tyrosine phosphorylations, are clearly not.

A consequence of the multisite phosphorylation of PKC is that our ability to predict the actions of pathways and PKC functionality will not be guided by monitoring lipid binding (the process of translocation) or any one dominant phosphorylation event but requires a more comprehensive view. The means to achieve this through the use of site-specific antisera is the subject matter of this chapter.

References

1. Keranen, L. M., Dutil, E. M., and Newton, A. C. (1995) Protein kinase C is regulated in vivo by three functionally distinct phosphorylations. *Curr. Biol.* **5,** 1394–1403.
2. Borner, C., Filipuzzi, I., Wartmann, M., Eppenberger, U., and Fabbro, D. (1989) Biosynthesis and posttranslational modifications of protein kinase C. *J. Biol. Chem.* **264,** 13,902–13,909.
3. Pears, C., Stabel, S., Cazaubon, S., and Parker, P. J. (1992) Studies on the phosphorylation of protein kinase C-alpha. *Biochem. J.* **283,** 515–518.
4. Orr, J. W. and Newton, A. C. (1994) Requirement for negative charge on "activation loop" of protein kinase C. *J. Biol. Chem.* **269,** 27,715–27,718.

5. Srinivasan, N., Bax, B., Blundell, T. L., and Parker, P. J. (1996) Structural aspects of the functional modules in human protein kinase-C alpha deduced from comparative analyses. *Proteins* **26,** 217–235.
6. Le Good, J. A., Ziegler, W. H., Parekh, D. B., Alessi, D. R., Cohen, P., and Parker, P. J. (1998) Protein kinase C isotypes controlled by phosphoinositide 3-kinase through the protein kinase PDK1. *Science* **281,** 2042–2045.
7. Dutil, E. M., Toker, A., and Newton, A. C. (1998) Regulation of conventional protein kinase C isozymes by phosphoinositide-dependent kinase 1 (PDK-1). *Curr. Biol.* **8,** 1366–75.
8. Alessi, D. R., Deak, M., Casamayor, A., Caudwell, F. B., Morrice, N., Norman, D. G., et al. (1997) 3-Phosphoinositide-dependent protein kinase-1 (PDK1): structural and functional homology with the Drosophila DSTPK61 kinase. *Curr. Biol.* **7,** 776–789.
9. Stephens, L., Anderson, K., Stokoe, D., Erdjument-Bromage, H., Painter, G. F., Holmes, A. B., et al. (1998) Protein kinase B kinases that mediate phosphatidylinositol 3,4,5-trisphosphate-dependent activation of protein kinase B. *Science* **279,** 710–714.
10. Williams, M. R., Arthur, J. S., Balendran, A., van der Kaay, J., Poli, V., Cohen, P., et al. (2000) The role of 3-phosphoinositide-dependent protein kinase 1 in activating AGC kinases defined in embryonic stem cells. *Curr. Biol.* **10,** 439–448.
11. Vanhaesebroeck, B. and Alessi, D. R. (2000) The PI3K-PDK1 connection: more than just a road to PKB. *Biochem. J.* **346,** 561–576.
12. Konishi, H., Tanaka, M., Takemura, Y., Matsuzaki, H., Ono, Y., Kikkawa, U., et al. (1997) Activation of protein kinase C by tyrosine phosphorylation in response to H2O2. *Proc. Natl. Acad. Sci. USA* **94,** 11,233–11,237.
13. Konishi, H., Yamauchi, E., Taniguchi, H., Yamamoto, T., Matsuzaki, H., Takemura, Y., et al. (2001) Phosphorylation sites of protein kinase C delta in H2O2-treated cells and its activation by tyrosine kinase in vitro. *Proc. Natl. Acad. Sci. USA* **98,** 6587–6592.
14. Gschwendt, M., Kielbassa, K., Kittstein, W., and Marks, F. (1994) Tyrosine phosphorylation and stimulation of protein kinase C delta from porcine spleen by src in vitro. Dependence on the activated state of protein kinase C delta. *FEBS Lett.* **347,** 85–89.
15. Haleem-Smith, H., Chang, E. Y., Szallasi, Z., Blumberg, P. M., and Rivera, J. (1995) Tyrosine phosphorylation of protein kinase C-delta in response to the activation of the high-affinity receptor for immunoglobulin E modifies its substrate recognition. *Proc. Natl. Acad. Sci. USA* **92,** 9112–9116.
16. Gatti, A. and Robinson, P. J. (1996) Unique phosphorylation of protein kinase C-alpha in PC12 cells induces resistance to translocation and down-regulation. *J. Biol. Chem.* **271,** 31,718–31,722.

14

Pulse-Chase Analysis of Protein Kinase C

Mikiko Takahashi and Yoshitaka Ono

1. Introduction

Pulse-chase experiments allow study of the fate of proteins after synthesis, such as processing, intracellular transport, secretion, degradation, and physical chemical properties of proteins. Pulse-chase protocols have been used to analyze the phosphorylation of endogenous and transiently expressed protein kinase C (PKC) family proteins (*1–5*). To radiolabel newly synthesized proteins, the cells are incubated with radiolabeled amino acid for short periods (pulse labeling). Then the pulse is followed by a chase in which cells are further incubated with the excess amount of unlabeled counterpart of the precursor used for labeling. The radiolabeled protein of interest is analyzed by electrophoresis after isolation from other cellular proteins by immunoprecipitation.

Labeling of cellular proteins is metabolically achieved by placing cells in a medium containing all components necessary for growth of cells in culture, except for one (or two) amino acid(s) that are substituted by its radiolabeled form. The radiolabeled amino acids are transported across the plasma membrane and incorporated into newly synthesized proteins. The common means of labeling amino acids is substituting radioisotopes, such as ^{35}S, ^{3}H, or ^{14}C in place of their nonradioactive counterparts. Methionine and cysteine are conveniently labeled with ^{35}S and most often used in pulse-chase experiments (*6*). Because the average abundance of methionine and cysteine in proteins is relatively low, it is important to check the contents of methionine and cysteine in the protein of interest (PKC family proteins contain sufficient amount of methionine/cysteine). For labeling proteins that contain little methionine and cysteine, other radiolabeled amino acids (e.g., leucine) should be used.

From: *Methods in Molecular Biology, vol. 233: Protein Kinase C Protocols*
Edited by: A. C. Newton © Humana Press Inc., Totowa, NJ

Immunoprecipitation protocols consist of several steps: solubilization of the antigens, formation of antigen-antibody complex, and adsorption of the complex to a matrix, such as Protein A- and Protein G-Sepharose beads, to allow separation and washing of the complex by low-speed centrifugation (*see* **Note 1**). Extraction with nondenaturing detergents such as Triton X-100 and Nonidet P-40, or in the absence of detergent allows immunoprecipitation with antibodies to epitopes that are exposed on native proteins. If the antigen is in the particulate fraction insoluble by nondenaturing detergents (e.g., cytoskeletal structures, chromatin, membrane "rafts"), extraction under denaturing conditions is applied. Further, immunoprecipitated antigens can be dissociated from the antibodies and reimmunoprecipitated by a protocol of sequential immunoprecipitation to identify components of multisubunit complexes or to study protein–protein interactions.

Posttranslational modifications, such as phosphorylation and glycosylation, of newly synthesized proteins often cause a mobility shift of the proteins in sodium dodecyl sulfate polyacrylamide gel electrophoresis (SDS-PAGE). These modifications can be confirmed by treatment of the proteins in the immunoprecipitates with specific enzymes (phosphatases, endoglycosidases), which results in shift of the bands to the original mobility.

2. Materials
2.1. Pulse-Chase Labeling

The following list is for the experiments with adherent cells maintained in Dulbecco's modified essential medium (DME) supplemented with 10% fetal bovine serum (FBS). The medium and other supplements may differ according to the cells used.

1. Dialyzed FBS (dFBS): FBS is dialyzed against saline or phosphate-buffered saline (PBS) overnight at 4°C to remove amino acids, and then sterilized by filtration.
2. Methionine/cysteine-free DME (available from Invitrogen, Sigma, etc.).
3. [^{35}S]-labeled hydrolysate (*see* **Note 2**): EasyTagEXPRESS PROTEIN LABELING MIX, [^{35}S] (Perkin-Elmer Life Sciences), which is hydrolysate of *E. coli* grown in the presence of carrier-free $^{35}SO_4$ and contains approx 73% methionine and 22% cysteine. Specific activity: 43.5 TBq/mmol. Similar products are available from other manufacturers such as Amersham Bioscience. Handle the vial in a fume hood as described in **Subheading 2.3**. This product may be stored in a refrigerator during frequent use (the rate of decomposition is approx 1% per week at 4°C). If used infrequently, freezing is recommended to minimize the risk of microbial contamination. The half-life of [^{35}S] is 87.4 d.
4. Pulse-labeling medium: methionine/cysteine-free DME supplemented with 5–10% dFBS prewarmed at 37°C.

5. Labeling cocktail: [^{35}S]-labeled hydrolysate (7.4 MBq/mL) in pulse-labeling medium prepared just before use by adding [^{35}S]-labeled hydrolysate to pulse-labeling medium prewarmed at 37°C.
6. Chase medium: DME supplemented with 10% FBS and excess amount of cold methionine (90 μg/mL) and cysteine•2 HCl (187.8 μg/mL) prewarmed at 37°C.
7. Vacuum aspirator with trap for liquid radioactive waste.
8. Disposable cell scraper.

2.2. Immunoprecipitation

1. Basal buffer: 50 mM Tris-HCl at pH 7.5, 150 mM NaCl, 0.5 mM ethylene-diaminetetraacetic acid , 0.5 mM ethylenebis(oxyethylenenitrilo)tetraacetic acid, 1 mM dithiothreitol (DTT), 1.5 mM MgCl$_2$, protease inhibitors (1 mM phenylmethylsulfonyl fluoride, 20 μg/mL aprotinin, 10 μg/mL leupeptin), and phosphatase inhibitors (10 mM NaF, 1 mM Na$_3$VO$_4$). Inhibitors are added just before use. Keep on ice.
2. Lysis buffer 1: 1% Triton X-100 in basal buffer.
3. Lysis buffer 2: 1% SDS in basal buffer.
4. Wash buffer: 0.1% SDS, 0.1% NaDOC, and 1% Triton X-100 in basal buffer.
5. Dissociation buffer: 1% SDS and 1 mM DTT in basal buffer.
6. Dilution buffer: 1% Triton X-100 and 1 mM iodoacetamide in basal buffer.
7. Protein A-Sepharose or Protein G-Sepharose (Amersham Bioscience).
8. Specific and control antibodies: monoclonal antibody is the first choice for immunoprecipitation because the level of nonspecific binding is usually low compared with that of polyclonal antibodies. Affinity purification is recommended in case polyclonal antibodies are employed. Control antibody should be in the same biochemical form (*see* **Note 3**).
9. Rotator.

2.3. Precautions for Handling ^{35}S-Labeled Compounds

Solutions containing ^{35}S-labeled compounds have been found to release volatile radioactive substances. In addition to the usual safety practices for handling radioactive materials, some extra precautions should be taken when using ^{35}S-labeled amino acids.

1. Fume hood equipped with an activated charcoal filter: vials containing ^{35}S-labeled compounds should always be handled in this type of hood. Handling includes thawing the solution, opening the vial, and adding the radiolabeled amino acid to the labeling medium.
2. A needle attached to a syringe packed with activated charcoal: before opening, vials should be vented with this needle.
3. A tray containing a layer of activated charcoal: place this tray in the CO$_2$ incubator to reduce the amount of ^{35}S-labeled compounds released to the air during cell labeling.

3. Methods

3.1. Pulse-Chase Labeling of Overexpressed Protein

The following procedure is for pulse-chase labeling of PKC overexpressed in adherent cells. Labeling of endogenous PKC may need higher concentrations of radiolabeled amino acids and scale up of the culture.

1. Grow adherent cells in a 6 cm-diameter dish, and transfect the cells with expression plasmid for PKC.
2. Incubate for 2 d.
3. Wash the cells three times with PBS prewarmed at 37°C.
4. Add 3 mL of pulse-labeling medium prewarmed at 37°C and incubate for 1.5 h to deplete intracellular pools of methionine and cysteine.
5. Remove the medium by using aspirator, add 1 mL of labeling cocktail (containing [^{35}S]methionine/cysteine) prewarmed at 37°C and incubate for 5 min (pulse labeling).
6. Quickly remove the cocktail, and wash the cells twice with chase medium prewarmed at 37°C.
7. Add the same medium and incubate at 37°C for various time periods (chase).
8. Quickly remove the medium and wash the cells twice with ice-cold PBS.
9. Lyse the cells by direct addition of lysis buffer (1 mL) containing appropriate detergent to the cell layer, and scrape the cells off the dish with a scraper and transfer them into a 1.5-mL tube. The lysate is ready for immunoprecipitation.
10. If labeled cells are not to be used immediately, the cells can be stored before lysis: after washing the cells with PBS, add 1 mL of ice-cold PBS and scrape the cells off the dish with a scraper and transfer them into a 1.5-mL tube, centrifuge 3 min at 500g at 4°C, and discard supernatant. The cell pellet can be stored frozen at −80°C for several days.

3.2. Immunoprecipitation

The methods described below outline the following: (1) immunoprecipitation of labeled proteins extracted by nondenaturing lysis buffer containing Triton X-100; (2) immunoprecipitation of labeled proteins solubilized from the remaining pellet by denaturing lysis buffer containing SDS; (3) sequential immunoprecipitation protocol of detergent-soluble proteins to detect protein–protein interaction; and (4) protocol for phosphatase treatment of the immunoprecipitate.

3.2.1. Immunoprecipitation of Detergent-Soluble Proteins

This protocol is for the cells labeled in a 6-cm dish (*see* **Subheading 3.1., step 9**).

1. Lyse the cells with 1 mL of lysis buffer 1 (containing 1% Triton X-100) and transfer the lysate to a 1.5-mL tube.

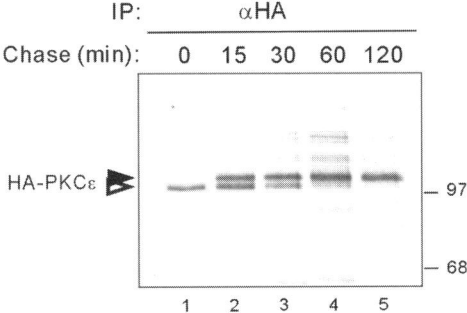

Fig. 1. Pulse-chase labeling of PKCε expressed in COS7 cells. COS7 cells expressing HA-tagged PKCε were pulse labeled with [^{35}S]methionine/cysteine for 5 min and chased for the indicated time periods. Detergent-soluble cell extracts were immunoprecipitated with anti-HA antibody, and then resolved on SDS-PAGE followed by fluorography. Black and white arrowheads indicate the positions of slow and fast migrating species of PKCε, respectively. [Reproduced, with permission, from Takahashi et al. *(4)*]

2. Disrupt the cells and chromosomal DNA by passing through a 26-gauge needle 15–20 times.
3. Centrifuge the lysate at 17,800*g* in a microcentrifuge for 5 min at 4°C.
4. Transfer the supernatant to a fresh 1.5-mL tube as detergent-soluble extract. Optionally, preclear procedure is employed (*see* **Note 4**). Pellet is discarded or processed for further lysis described in **Subheading 3.2.2., step 1**.
5. Mix the extract with 1 μg of specific antibody or negative control antibody and incubate at 4°C for 2 h (*see* **Note 5**). Amount of antibody differs case by case depending on the quality of specific antibody.
6. Add 10 μL of Protein A-Sepharose bead slurry (50% v/v) and incubate at 4°C for a further 30 min by using a rotator.
7. Centrifuge at 4000*g* for 1 min (or 17,800*g* for 2 sec) in a microcentrifuge at 4°C to sediment the beads, and then discard the supernatant.
8. Add 1 mL of lysis buffer 1 and resuspend the beads. To minimize nonspecific binding, wash buffer (containing 0.1% SDS, 0.1% NaDOC and 1% Triton X-100) can be used in this step.
9. Repeat **steps 7** and **8** twice, and then **step 7** once.
10. Spin down the beads again to remove residual supernatant.
11. Resusupend the beads with SDS-PAGE sample buffer and boil for 3 min.
12. Separate the sample on SDS-PAGE and process the gel for fluorography by using Amplify (Amersham Bioscience) according to manufacturer's instruction (*see* **Fig. 1**).

3.2.2. Immunoprecipitation of Detergent-Insoluble Proteins

This protocol describes immunoprecipitation of (nondenaturing) detergent-insoluble proteins in the pellet obtained by **step 4** of **Subheading 3.2.1.**

1. Add 100 µL of lysis buffer 2 (containing 1% SDS) to the pellet and resuspend it by pipetting.
2. Add 900 µL of lysis buffer 1 (containing 1% Triton X-100, *see* **Note 6**), and disrupt the pellet by passing through a 26-gauge needle 15–20 times.
3. Centrifuge the lysate at 17,800g in a microcentrifuge for 5 min at 4°C.
4. Transfer the supernatant to a fresh 1.5-mL tube as detergent-insoluble extract.
5. Follow the same procedure as described in **Subheading 3.2.1., steps 5–12**. For the washing step, wash buffer (containing 0.1% SDS, 0.1% NaDOC and 1% Triton X-100) should be used.

3.2.3. Sequential Immunoprecipitation

This protocol is used to detect a labeled protein (e.g., PKC) associated with another protein (binding protein): immunoprecipitation is performed with the first antibody (to the binding protein), and then the immunoprecipitate is dissociated by boiling in the presence of SDS and DTT. The protein of interest (PKC) is then reimmunoprecipitated with the second antibody. Cells labeled in a 10-cm dish with higher concentrations of radiolabeled amino acids (14.8 MBq/mL or more) may be needed because of the lower yield of the labeled protein.

1. Prepare 1 mL of detergent-soluble extract from the cells labeled in a 10-cm dish according to the procedure described in **Subheading 3.2.1., steps 1–4**.
2. Mix the detergent soluble extract with the first antibody (to the binding protein) or negative control antibody and incubate at 4°C for 2 h.
3. Add 20 µL of Protein A-Sepharose bead slurry (50% v/v) and incubate at 4°C for 30 min by using rotator.
4. Spin down the beads and then wash them with lysis buffer 1 as described in **Subheading 3.2.1., steps 8–10**.
5. Resuspend the beads with 100 µL of dissociation buffer (containing 1% SDS and 1 mM DTT) and boil the suspension for 5 min to dissociate the immunoprecipitate.
6. Spin down the beads, and then transfer the supernatant to a fresh 1.5-mL tube.
7. Add 900 µL of dilution buffer (containing 1% Triton X-100 and 1 mM iodoacetamide, *see* **Notes 6** and **7**).
8. Add the second antibody (to PKC) or negative control antibody and perform immunoprecipitation as described in **Subheading 3.2.1., steps 5–12**.

3.2.4. Alkaline Phosphatase Treatment of the Immunoprecipitates

This protocol is to examine the phosphorylation status of labeled protein in the immunoprecipitate prepared by procedures described in **Subheadings 3.2.1. to 3**.

1. After washing the beads carrying immunoprecipitate with appropriate wash buffer, resuspend the beads with 1 mL of phosphatase reaction buffer (50 mM Tris-HCl at pH 8.5 containing 1 mM MgCl$_2$).

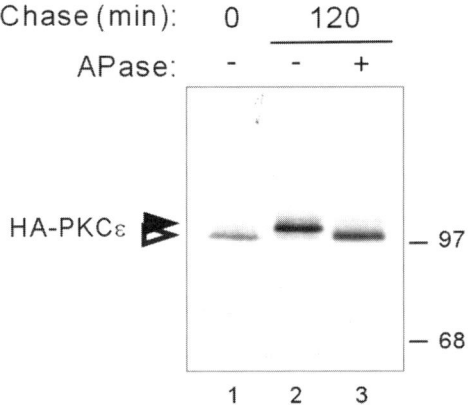

Fig. 2. Alkaline phosphatase treatment of the pulse-chase labeled PKCε. COS7 cells expressing HA-tagged PKCε were pulse-labeled with [^{35}S]methionine/cysteine for 5 min and then chased for 0 or 120 min. Detergent-soluble cell extracts were immunoprecipitated with anti-HA antibody, and then the immunoprecipitates were treated with or without alkaline phosphatase (APase) as indicated. This result indicates that newly synthesized PKCε was phosphorylated at 120-min chase. Black and white arrowheads indicate the positions of slow and fast migrating species of PKCε, respectively. [Reproduced, with permission, from Takahashi et al. (*4*)]

2. Spin down the beads, and discard the supernatant. This step removes detergent and other materials that may interfere phosphatase reaction.
3. Resuspend the beads with 30 µL of the same buffer containing 30 U of alkaline phosphatase.
4. Incubate at 37°C for 1 h by using rotator.
5. Spin down the beads and discard the supernatant.
6. Wash the beads and process the sample for SDS-PAGE and fluorography as described in **Subheading 3.2.1., steps 8–12**. (*see* **Fig. 2**).

4. Notes

1. An alternative protocol can also be used: immobilize the antibody to Protein A-Sepharose beads first, and then the immobilized antibody is incubated with solubilized antigen. This protocol is preferable when impure antibody source is employed such as ascites and antiserum to exclude the proteins other than antibody.
2. Purified [^{35}S]-methionine can be used for the labeling substituting [^{35}S]-protein hydrolysate. The merit of this form is less production of volatile decomposition products than the hydrolysate.
3. It is critical to include appropriate nonspecific controls in immunoprecipitation. For instance, parallel experiment should be used using control antibody in the same biochemical form as the experimental antibody (e.g., serum, ascites,

affinity-purified immunoglobulin) and belonging to the same species and immunoglobulin subclass.

4. To preclear the lysate, the lysate is mixed with 10 μL of Protein A-Sepharose bead slurry (50% v/v) with or without 1 μg of control antibody, and incubate at 4°C for 30–60 min by using rotator. After microcentrifuge at 4000g for 1 min, the supernatant is processed for immunoprecipitation. This step is to remove proteins that bind to Protein A-Sepharose beads or to antibody other than antigen recognition site.

5. If affinity of the antibody is relatively low, longer incubation may increase the efficiency of immunoprecipitation. However, there is an increased risk of protein degradation, formation of aggregates, and appearance of nonspecific bands especially when using polyclonal antibody.

6. Dilution of the buffer containing 1% SDS with excess amount of Triton X-100 sequesters SDS into Triton X-100 micelles, which allow binding of antibody to antigen.

7. DTT in the dissociation buffer is neutralized with excess amount of iodoacetamide.

References

1. Borner, C., Filipuzzi, I., Wartmann, M., Eppenberger, U., and Fabbro, D. (1989) Biosynthesis and posttranslational modifications of protein kinase C in human breast cancer cells. *J. Biol. Chem.* **264,** 13902–13909.

2. Zhang, J., Wang, L., Schwartz, J., Bond, R. W., and Bishop, W. R. (1994) Phosphorylation of Thr642 is an early event in the processing of newly synthesized protein kinase C beta 1 and is essential for its activation. *J. Biol. Chem.* **269,** 19578–19584.

3. Bornancin, F. and Parker, P. J. (1997) Phosphorylation of protein kinase C-alpha on serine 657 controls the accumulation of active enzyme and contributes to its phosphatase-resistant state. *J. Biol. Chem.* **272,** 3544–3549.

4. Takahashi, M., Mukai, H., Oishi, K., Isagawa, T., and Ono, Y. (2000) Association of immature hypophosphorylated protein kinase c epsilon with an anchoring protein CG-NAP. *J. Biol. Chem.* **275,** 34592–34596.

5. Sonnenburg, E. D., Gao, T., and Newton, A. C. (2001) The phosphoinositide-dependent kinase, PDK-1, phosphorylates conventional protein kinase C isozymes by a mechanism that is independent of phosphoinositide 3-kinase. *J. Biol. Chem.* **276,** 45289–45297.

6. Meisenhelder, J. and Hunter, T. (1988) Radioactive protein-labelling techniques. *Nature* **335,** 120.

15

PDK-1 and Protein Kinase C Phosphorylation

Alex Toker

1. Introduction

The discovery of the phosphoinositide-dependent kinase-1 (PDK-1) as the upstream kinase for protein kinase C (PKC) represented an important step in the understanding of the regulation of this crucial lipid-signaling enzyme. Three laboratories simultaneously described PDK-1 as the activation loop upstream kinase for conventional (PKCα and PKCβII; **ref. _1_**), novel (PKCδ and PKCε; **ref. _2_**), and atypical (PKCζ; **refs. _2_** and _3_) isozymes. It is now well established that PDK-1 is the upstream kinase for all PKC family members, and numerous studies have addressed the detailed biochemical mechanisms by which PDK-1 phosphorylates PKC thereby regulating its function in cells (for reviews, _see_ **refs. _4–6_**). PDK-1 phosphorylates PKCs at a critical Thr residue in the so-called activation loop sequence of the highly conserved catalytic kinase domain (_see_ Chapter 13), and this event is required for PKC to gain catalytic competency. Phosphorylation of the activation loop Thr correctly aligns residues within the active site and this permits transfer of the gamma phosphate of ATP to an exogenous substrate. This phosphorylation triggers two phosphorylations at the carboxyl-terminus required to stabilize the catalytically competent species of PKC. Therefore, phosphorylation is a rate-limiting step in the regulation of PKC and precedes other regulatory events, including binding of lipid activators _(7)_.

PDK-1 was originally discovered as the upstream kinase of Akt/PKB, also an AGC kinase _(8)_. In the case of Akt/PKB, phosphorylation of Akt/PKB by PDK-1 occurs in a phosphatidylinositol (PI) 3-K-dependent manner, and is dependent on the binding of the PI 3-K lipids PtdIns-3,4-P_2 or PtdIns-3,4,5-P_3 to the Akt/PKB and PDK-1 pleckstrin homology (PH) domains _(9)_. Phosphorylation of Akt/PKB at its activation loop Thr308 residue leads to

From: _Methods in Molecular Biology, vol. 233: Protein Kinase C Protocols_
Edited by: A. C. Newton © Humana Press Inc., Totowa, NJ

a 20-fold increase in the specific activity of Akt/PKB toward exogenous substrates, although full catalytic competency is achieved by an addition phosphorylation that occurs at the carboxyl-terminal site, Ser473 *(10)*. Therefore, phosphorylation of Akt/PKB at Thr308 represents an ON/OFF switch in the regulation of the kinase, and for this reason monitoring this event is often used as a marker for activation of Akt/PKB, and indeed PI 3-K. Although PDK-1 also phosphorylates PKCs at their activation loop Thr residues, the precise mechanism is somewhat different from that of Akt/PKB, depending on the PKC family member. Numerous studies have examined these differences, and there is an emerging consensus that only in the case of atypical PKCs (PKCζ, PKCι/λ) does phosphorylation also serve as a direct ON/OFF switch. For example, stimulation of cells with mitogens results in increased activation and phosphorylation of PKCζ at the activation loop Thr410 in a PI 3-K-dependent manner *(2,3)*. Thus, as with Akt/PKB, monitoring phosphorylation of Thr410 (hPKCζ) or Thr403 (hPKCι/λ) has often been used as a marker for activation of these atypical PKCs.

In the case of novel PKCs, phosphorylation by PDK-1 also results in a modest activation of the kinase, and again there are reports that this occurs in a PI 3-K-dependent manner. For example, stimulation of cells with either serum or PDGF results in phosphorylation of hPKCε at Thr566 and of hPKCδ at Thr505, and this is concomitant with their activation *(11–13)*. It is, however, extremely important to note that in the case of novel PKCs, the final step in the activation of the kinase is binding of lipid activators (phosphatidylserine and diacylglycerol), and this is required to release the pseudosubstrate from the active site, leading to full enzymatic activity. Thus, although phosphorylation of novel PKCs at their activation loops has been used as a marker for their activation, it is important to note that this is not the final activating step. In summary, there are two equally important inputs in the regulation of novel PKCs, the PI 3-K → PDK-1 signal leading to activation loop phosphorylation, and the PLCγ-1 → DG signal required for membrane translocation and ultimately, activation.

Finally, the regulation of conventional PKCs by PDK-1 is also different from that of novel and atypical family members, by two criteria; first, the phosphorylation of PKCα or PKCβII by PDK-1 in cells does not require a PI 3-K input *(14)*, and this is consistent with the fact that PDK-1 is not entirely dependent on PtdIns-3,4-P_2 or PtdIns-3,4,5-P_3 for its intrinsic protein kinase activity *(15)*, rather these lipids primarily regulate its membrane localization *(16)*. Second, phosphorylation of conventional PKCs at their activation loops is constitutive, such that these sites are near stoichiometrically phosphorylated even in serum-starved cells, and there is no appreciable increase in phosphorylation upon mitogenic stimulation of cells *(1,7)*. This is a critical issue because the phosphorylation of conventional PKCs cannot be used as a marker for second messenger-triggered activation. In fact, once phosphorylated at the carboxyl-terminal sites, in vitro studies reveal that the phosphorylation state of the activation loop does not result in a significant change in kinase activity *(1)*. Rather, phosphorylation at the activation loop of newly synthesized PKC is a trigger for subsequent autophosphorylations at the carboxyl terminus *(17)*. Consequently, the only mechanism that is responsible for quantitatively increasing the intrinsic kinase activity of conventional PKCs is diacylglycerol binding, which leads to pseudosubstrate release.

Phosphorylation of PKC at the activation loop is often used as a marker for activation, at least in the case of novel and atypical isozymes. Historically, phosphorylation of PKC was monitored by total phosphate incorporation using cells labeled with inorganic phosphate to label the intracellular ATP pool. However, this did not discriminate between the three conserved phosphorylation sites in the PKC kinase domain, and a tedious analysis using two-dimensional phosphopeptide mapping was required. Phosphorylation of PKC could also be monitored by observing the classical SDS gel mobility shifts that are readily observed upon phosphorylation of many proteins, including PKC, or by pulse-chase analysis (*see* Chapter 14). Although this served as a good guide for PKC phosphorylation because phosphorylation of each site was responsible for a discrete shift, this was not readily evident with all isozymes. With the advent of phospho-specific antibodies against the PKC activation loop, these technical challenges have been overcome. This chapter describes the methodology used to detect phosphorylation of PKC at the activation loop using phosphospecific antibodies and thus monitor the PDK-1-dependent step. In the case of novel and atypical PKCs, this can be used as a reasonable marker for activation upon cell stimulation. Procedures for detecting phosphorylation of both endogenous, as well as transfected, over-expressed PKCs are described. In addition, a procedure for detecting the phosphorylation of purified recombinant PKCs by PDK-1 in vitro is also described, again using phosphospecific antibodies. Finally, a recently developed procedure for detecting PDK-1 activity using an immune-complex kinase assay and synthetic peptide substrate is described.

2. Materials
2.1. Measuring Phosphorylation in Cells
2.1.1. Maintenance of Cells

1. HEK293, COS-7 or NIH 3T3 cells (American Type Culture Collection, Manassas, VA; http://www.atcc.org).
2. Fetal Bovine Serum (FBS; Irvine Scientific, Santa Ana, CA; http://www.irvinesci.com).

3. High-glucose Dulbecco's modified Eagle medium (DMEM; Invitrogen Life Technologies, Carlsbad, CA; http://www.lifetech.com).
4. Trypsin/ethylenediaminetetraacetic acid (EDTA) solution (Invitrogen Life Technologies).
5. Antibiotic/antimycotic (100× Invitrogen Life Technologies).
6. Sterile phosphate-buffered saline (PBS).
7. 37°C humidified CO_2 incubator.
8. Tissue culture dishes (100 mm) or 175-cm^2 flasks (BD Biosciences, Bedford, MA; http://www.bdbiosciences.com).
9. Sterile Pasteur pipets.
10. Sterile disposable 2-, 5-, 10-, and 50-mL pipets.

2.1.2. Cell Transfection

1. Tissue culture dishes, 60 mm or 100 mm.
2. Sterile, 5-mL polystyrene tubes (BD Biosciences).
3. Transfection reagent (LipofectAMINE, Invitrogen Life Technologies; Superfect, QIAGEN, Valencia, CA; http://www.qiagen.com).
4. Transfection medium (Optimem, Invitrogen Life Technologies).
5. Purified PKC cDNA in mammalian expression vector.

2.1.3. Cell Stimulation and Lysis

1. Growth factors (platelet-derived growth factor BB [PDGF-BB], insulin-like growth factor-1 [IGF-1], epidermal growth factor [EGF], FBS, Invitrogen Life Technologies).
2. Ice-cold PBS.
3. Lysis buffer A: 20 mM Tris-HCl, pH 7.5, 10% glycerol, 1% NP-40, 10 mM EDTA, 150 mM NaCl, 20 mM NaF, 5 mM sodium pyrophosphate, 1 mM sodium vanadate.
4. Protease inhibitor cocktail (Sigma-Aldrich, St. Louis, MO; http://www.sigmaaldrich.com).
5. Rubber policeman/cell scraper.

2.1.4. Immunoprecipitation

1. Protein Sepharose A/G beads (Santa Cruz Biotechnology, Santa Cruz, CA; http://www.scbt.com).
2. PKC isoform-specific antibodies (*see* **Table 1**).
3. Epitope tag antibodies (*see* **Table 2**).
4. Rocking platform.
5. Vacuum aspirator.
6. Wash buffer B: 1× PBS + 1% NP-40.
7. Wash buffer C: 10 mM Tris-HCl, pH 7.5, 0.5 M LiCl.
8. Wash buffer D: 10 mM Tris-HCl, pH 7.5, 100 mM NaCl, 1 mM EDTA.
9. Laemmli SDS sample buffer.

Table 1
PKC Isoform-Specific Antibodies

PKC Isoform	Antibody	Company	Species	Specificity	Epitope
PKCα	α-PKC(MC5)	Santa Cruz	Mouse	PKCα PKCβ PKCγ	Hinge
	α-PKCα (C-20)	Santa Cruz	Rabbit	PKCα (PKCβ/γ)	CT
	α-PKCα (M4)	Upstate	Mouse	PKCα	N.D.
	α-PKCα (clone 3)	BD TL	Mouse	PKCα(PKCβ)	Hinge
PKCβI	α-PKCβI (C-16)	Santa Cruz	Rabbit	PKCβI	CT
	α-PKCβI (E-3)	Santa Cruz	Mouse	PKCβI	CT
	α-PKCβ (clone 36)	BD TL	Mouse	PKCβI and PKCβI	C1 + C2
PKCβII	α-PKCβII (C-18)	Santa Cruz	Rabbit	PKCβII	CT
	α-PKCβ (clone 36)	BD Tl	Mouse	PKCβI and PKCβI	C1 + C2
PKCγ	α-PKCγ (C-19)	Santa Cruz	Rabbit	PKCγ	CT
	α-PKCγ (clone 20)	BD TL	Mouse	PKCγ	CT
PKCδ	α-PKCδ (C17)	Santa Cruz	Rabbit	PKCδ	CT
	α-PKCδ (G-9)	Santa Cruz	Mouse	PKCδ	CT
	α-PKCδ	Upstate	Rabbit	PKCδ	CT
	α-PKCδ (clone 14)	BD TL	Mouse	PKCδ	C1a + C1b
PKCε	α-PKCε (C-15)	Santa Cruz	Rabbit	PKCε	CT
	α-PKCε (E-5)	Santa Cruz	Mouse	PKCε	CT
	α-PKCε	Upstate	Rabbit	PKCε	CT
	α-PKCε (clone 21)	BD TL	Mouse	PKCε	NT
PKCη	α-PKCη (C-15)	Santa Cruz	Rabbit	PKCη	CT
	α-PKCη (clone 31)	BD TL	Mouse	PKCη	Hinge
PKCθ	α-PKCθ (C-19)	Santa Cruz	Rabbit	PKCθ	CT
	α-PKCθ (E-7)	Santa Cruz	Mouse	PKCθ	CT
	α-PKCθ (clone 27)	BD TL	Mouse	PKCθ	C1a + C1b
PKCζ	α-PKCζ (C-20)	Santa Cruz	Rabbit	PKCζ, PKCι/λ	CT
	α-PKCζ (N-17)	Santa Cruz	Goat	PKCζ	NT
	α-PKCζ	Upstate	Rabbit	PKCζ	CT
PKCι/λ	α-PKCι(N-20)	Santa Cruz	Rabbit	PKC/ι/λ	NT
	αPKCλ (N-17)	Santa Cruz	Rabbit	PKC/ι/λ	NT
	αPKCι (clone 23)	BD TL	Mouse	PKCι/λ, PKCζ	CT
	αPKCλ (clone 41)	BD TL	Mouse	PKCι/λ, PKCζ	Catalytic

This table should only be used as a guide when selecting PKC isoform-specific antibodies. Numerous other companies also market PKC isoform-specific antibodies. The manufacturer's instructions should be followed when evaluating the ability of each antibody to immunoprecipitate, Western immunoblotting or immunohistochemistry. The specificity of each antibody listed is according to the manufacturer's specifications, and the PKC isoforms in parentheses in the Specificity columns refer to weak cross-reactivity of the antibody against those PKCs. CT, epitope from the carboxyl terminus, NT, epitope from the amino terminus, N.D., not determined.

Table 2
Epitope Tag Antibodies

Epitope	Antibody	Company	Species
HA	α-HA (12CA5)	Roche	Mouse
	α-HA (3F10)	Roche	Mouse
	α-HA (12CA5)		
Myc	α-Myc (9E10)	Upstate	Mouse
	α-Myc	Invitrogen LT	Mouse
	α-Myc (9E10)	Roche	Mouse
FLAG	α-FLAG (M2)	Sigma-Aldrich	Mouse

The anti-HA, anti-Myc, and anti-FLAG antibodies are also available from several other companies. The anti-HA and anti-Myc antibodies are derived from the mouse hybridoma 12CA5 and 9E10, respectively, and the hybridomaa are available from ATCC. The hybridoma can be cultured and the supernatant can be collected for direct use in immunoblotting. Alternatively, we routinely purify the supernatant using HiTRAP protein G columns (Amersham-Pharmacia Biotech). The resulting antibody is used for immunoprecipitation, immunoblotting, and immunohistochemistry. The anti-Xpress antibody based on the epitope DLYDDDK (Invitrogen Life Technologies) is similar to the FLAG epitope with only two amino acid substitutions between the two epitopes. This epitope, and the corresponding antibody (α-Xpress) can be used in place of the anti-FLAG antibody with similar results.

2.1.5. Western Immunoblotting

1. SDS polyacrylamide gel electrophoresis apparatus (Bio-Rad, Hercules, CA; http://www.biorad.com. Amersham-Pharmacia Biosciences, Piscataway, NJ; http://www.electrophoresis.apbiotech.com).
2. Nitrocellulose (Bio-Rad) or Immobilon (Millipore Corporation, Bedford, MA; http://www.millipore.com).
3. Tris-buffered saline (TBS); 10 mM Tris-HCl pH 8.0, 150 mM NaCl.
4. Ponceau red stain (Sigma-Aldrich).
5. Tween-20 (Sigma-Aldrich).
6. Bovine serum albumin (BSA; Sigma-Aldrich).
7. PKC phospho-specific antibodies (*see* Chapter 19).
8. Secondary horseradish peroxidase-coupled antibodies (anti-mouse, anti-rabbit, or anti-goat, Chemicon International, Temecula, CA; http://www.chemicon.com).
9. Chemiluminescence reagents: Western Lightning PLUS (Perkin-Elmer Life Sciences, Boston, MA, http://lifesciences.perkinelmer.com) or Supersignal West PICO (Pierce Biotechnology Inc., Rockford, IL; http://www.piercenet.com).
10. Autoradiography film.

2.2. Measuring Phosphorylation in Vitro

1. Purified, recombinant PKC (*see* Chapter 3).
2. Sf9 cells (ATCC).

3. 500-mL spinner flask (Bellco Glass, Vineland, NJ; http://www.bellcoglass.com).
4. Baculovirus directing expression of His_6-PDK-1.
5. Ni^{2+}-NTA (nitriloacetate) column, 5 mL (Qiagen).
6. Mono Q column, 1 mL (Amersham-Pharmacia Biosciences).
7. Equilibration buffer E: 20 mM Tris-HCl, pH 8.0, 500 mM KCl, 5 mM 2-mercapto-ethanol, 10% glycerol.
8. Wash buffer F: 20 mM Tris-HCl, pH 8.0, 500 mM KCl, 5 mM 2-mercaptoethanol, 10% glycerol, 10 mM imidazole.
9. Elution buffer G: 20 mM Tris-HCl, pH 8.0, 500 mM KCl, 5 mM 2-mercapto-ethanol, 10% glycerol, 100 mM imidazole.
10. Mono Q column buffer H: 20 mM Tris-HCl, pH 8.0, 5 mM 2-mercaptoethanol.
11. Phosphatidylserine, diacylglycerol (Avanti Polar Lipids, Alabaster, AL; http://www.avantilipids.com).
12. L-α-Dipalmitoyl PtdIns-3,4,5-P_3 and L-α-dipalmitoyl PtdIns-3,4-P_2 (Calbiochem, La Jolla, CA; http://www.calbiochem.com).

2.3. Measuring PDK-1 Activity In Vitro

1. Anti-PDK-1 antibody (Upstate Biotechnology, Waltham, MA; http://www.upstatebiotech.com. Biosource International, Camarillo, CA; http://www.biosource.com).
2. cDNA for PDK-1 in a mammalian expression vector (e.g., pcDNA3).
3. Insulin (Sigma Aldrich).
4. 30% H_2O_2 solution (Pierce Chemical Company).
5. Sodium orthovanadate (Sigma-Aldrich).
6. Catalase (8 U/mg) (Sigma-Aldrich).
7. PDK-1 synthetic peptide substrate.
8. p81 phosphocellulose paper discs (Whatman, Clifton, NJ; http://www.whatman.com).
9. [γ-^{32}P]ATP, 3000 Ci/mmol (Perkin-Elmer Life Sciences).
10. 1% phosphoric acid solution.
11. 100% ethanol.
12. Liquiscint scintillation solution (National Diagnostics, Atlanta, GA; http://www.nationaldiagnostics.com).

3. Methods
3.1. Cell Culture

The procedures described below have been developed for detecting PKC phosphorylation in mouse fibroblasts (NIH 3T3), human embryonic kidney cells (HEK293), or African green monkey kidney cells (COS-7, SV40 transformed), and indeed numerous studies addressing the regulation of PKC by phosphorylation have used these cells. However, the methodologies described can also be applied to numerous other cell lines, with minor modifications to the growth conditions and media requirements depending on the cell type. Cells are grown in a humidified CO_2 (5%) incubator at 37°C in high-glucose DMEM supplemented with 10% FBS. We routinely also supplement with an antibiotic/antimycotic. Standard sterile tissue culture techniques should be used at all times. Cells are typically seeded in 100-mm dishes; alternatively, 175-cm_2 flasks can also be used to achieve higher cell densities. Cells are serially passaged prior to confluence, typically at 50–70% confluence (*see* **Note 1**). To determine stimulation-dependent phosphorylation of PKC, cells should be serum-starved before stimulation with growth factors. The conditions for serum-starvation vary greatly depending on the cell type, but as a general guide a minimum of 10–14 h starvation in DMEM in the absence of FBS should be used. 1% BSA can be used as a supplement, and some researchers also use 0.1% FBS. We have found that in the case of NIH 3T3 *(3)*, HEK293 *(3,13)*, or COS-7 cells *(1)*, starvation for 14 h in DMEM alone is ideal, although this can be increased to 24 h (*see* **Note 2**).

3.2. Transient Transfection of Cells with PKC cDNA

Although the phosphorylation of PKC can be monitored using endogenous proteins, it is often useful and advantageous to monitor the phosphorylation of heterologous PKC, which can be achieved by transient transfection. There are a number of reasons for this: the phosphorylation of PKC is more readily detected when present is larger quantities compared to the endogenous; the PKC cDNA can be tagged with a variety of epitopes and highly specific antiodies are available to such tags (this can circumvent the low specificity or affinity of some PKC isoform-specific antibodies); a given cell type may not express the particular PKC being studied; most importantly, when evaluating the role of a signaling pathway or molecule(s) on PKC phosphorylation, cotransfection allows the monitoring of only those cells which have been transfected.

Transient transfection with PKC cDNA has been widely used, and numerous mammalian expression vectors are available. The most commonly used vectors are those that use the cytomegalovirus promoter to achieve high overexpression in transient transfections. Examples of these are the pcDNA3 series (Invitrogen Life Technologies). Expression vectors for PKCα *(18)*, PKCβI *(19)* and PKCβII *(20)*, PKCγ *(21)*, PKCδ *(22)*, PKCε *(13)*, PKCη *(23)*, PKCθ *(24)*, PKCζ *(3)*, PKCι *(25)*, and PKCλ *(26)* have been widely published and their construction will not be further described here. We recommend the cloning of PKC cDNA in a CMV promoter-based vector, and the in-frame addition of an epitope tag to the amino or carboxyl-terminus of the PKC open reading frame (*see* **Note 3**). We routinely use the following epitope tags:

HA: MYPYDVPDYA
        5' ATG TAC CCT TAC GAC GTC CCG GAC TAT GCT 3'

Myc: MEQKLISEEDL
5′ ATG GAG CAG AAG CTG ATC AGC GAG GAG GAC CTG 3′
FLAG: MDYKDDDDK
5′ ATG GAC TAT AAG GAC GAT GAT GAC AAA 3′

The epitope tags are added to the PKC cDNA by PCR using oligonucleotide primers under standard PCR reaction conditions. The PKC cDNA vectors are then purified using conventional plasmid isolation techniques from bacterial cultures. The procedure below describes a transient transfection protocol for either NIH 3T3, HEK 293, or COS-7 cells using the LipofectAMINE procedure, adapted from the manufacturer's protocol (*see* **Note 4**). We also routinely use Superfect with similar results. There are numerous other transfection reagents and procedures, in each case the transfection efficiency of a given PKC cDNA should be determined empirically for each procedure and cell line.

1. Seed cells into 60-mm dishes at a density of 7×10^5 in DMEM + 10% FBS. Allow cells to reach a density of 40–50% confluence.
2. Wash cells twice with Optimem media. Leave in 2 mL of Optimem media.
3. In one sterile, 5-mL polystyrene tube, add 400 μL Optimem media and 6 μL Lipofectamine reagent. Mix by gently tapping the tube.
4. In a separate tube, add 400 μL Optimem media and 3–5 μg of PKC plasmid DNA. Mix by gently tapping the tube.
5. Add Optimem/Lipofectamine mixture to tube containing Optimem/DNA mixture dropwise. Mix by tapping gently. Do not vortex. Incubate for 20 min at room temperature.
6. Add 1.2 mL of Optimem to tube containing Lipofectamine/DNA complexes and mix by pipetting up and down gently.
7. Aspirate media from each dish and gently overlay 2 mL of transfection mixture to each dish.
8. Return cells to incubator and allow transfection to proceed for 5–6 h.
9. Aspirate transfection mixture from cells, and add 4 mL of complete media (DMEM + 10% FBS + antibiotic/antimycotic).
10. Cells are typically harvested 24–36 h after transfection, although the optimal time for maximal protein expression has to be determined for each DNA and cell line.
11. If cells are to be stimulated with agonists, they should be serum-starved in 4 mL of DMEM without serum, as described in **Subheading 3.1.**

3.3. Cell Stimulation, Lysis, and Immunoprecipitation

To determine the stimulation-dependent phosphorylation of PKCs, cells should first be serum-starved as described above. The procedure below describes the specific determination of the phosphorylation of either endogenous or heterologous transfected PKCs using a number of growth factors.

The procedure should be modified to take into account distinct experimental conditions depending on the cell line or agonist(s).

1. Add agonist to the dishes containing the 4 mL of DMEM; we routinely use FBS (10%) PDGF-BB (50 ng/mL), IGF-1 (100 ng/mL), or EGF (30 ng/mL). It is recommended that both dose-dependence and time-dependence be determined for each agonist and cell line.
2. At the appropriate time point, aspirate media, wash cells twice with ice-cold PBS, maintaining dishes on ice.
3. Add 1 mL of lysis buffer A to which fresh protease inhibitor cocktail has been added. This can also be replaced with leupeptin (1 μg/mL), pepstatin A (1 μg/mL), and phenylmethylsulfonyl fluoride (1 mM).
4. Scrape cells into a 1-mL centrifuge tube and vortex vigorously.
5. Centrifuge lysate for 15 min at 4°C at 16,000g.
6. Remove supernatant into a new centrifuge tube, discard pellet (*see* **Note 5**).
7. Take 10% of the lysate and immediately add SDS sample buffer, heat to 100°C for 10 min., and store at –20°C.
8. Add PKC isoform-specific or epitope tag antibody to lysate (*see* **Table 1** and **Table 2**). Typically, a concentration of 1–5 μg/mL for each antibody is sufficient, according to the manufacturer's suggestions.
9. Add 40 μg/mL of a 50% slurry of protein Sepharose A/G beads.
10. Incubate immunoprecipitates for 3 h with rocking at 4°C.
11. Pellet beads by centrifugation at 10,000g for 2 min at 4°C.
12. Aspirate supernatant, and wash twice with buffer B, twice with buffer C, and twice with buffer D. Centrifuge at 10,000g for 2 min at 4° at each step.
13. Add SDS sample buffer to the beads and heat to 100°C for 5 min.
14. Store at –20°C.

3.4. Western Immunoblotting with Phospho-Specific Antibodies

The immunoprecipitates recovered from **steps 1–14** in **Subheading 3.3.** are used to detect phosphorylation of the corresponding PKC with the appropriate phospho-specific antibodies. These antibodies are commercially available, or can also be obtained from other researchers. In many cases, the phospho-specific antibodies are not qualitatively specific for any given PKC isoform and will cross-react with other PKCs. The recommendations of the manufacturer should be followed. This is only an issue when monitoring PKC phosphorylation on total cell lysates. In immunoprecipitates, issues of cross-reactivity with PKC phospho-specific antibodies are not a concern. *See* Chapter 19 for more information on the selection and use of PKC phosphospecific antibodies. *See* **Fig. 1** for an example of PDK-1-mediated phosphorylation of PKCζ at Thr410 under transient transfection conditions. Generally, we recommend that both immunoprecipitates as well as total cell lysates be used by Western immunoblotting to monitor phosphorylation of PKC, as follows:

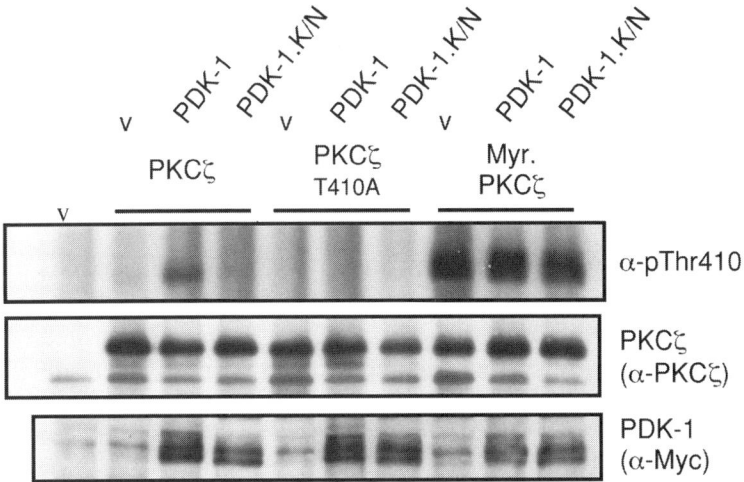

Fig. 1. PDK-1 phosphorylates PKCζ at Thr410 in transiently transfected cells. HEK293 cells were transiently co-transfected with FLAG-PKCζ, FLAG-PKCζ.T410A, or Myr.PKCζ.FLAG and either vector alone (v), Myc.PDK-1, or Myc.PDK-1.K110N, as indicated. Cell lysates were immunoprecipitated with FLAG antibody and immunoblotted with a phospho-specific antibody against pThr410. Levels of PKCζ and PDK-1 expression were detected by immunoblotting with anti-PKCζ or anti-Myc. Adapted from **ref. 3**.

1. Using standard sodium dodecyl sulfate polyacrylamide gel electrophoresis (SDS-PAGE) apparatus and reagents, prepare a 7.5% SDS polyacrylamide gel.
2. Centrifuge beads boiled in SDS sample buffer at 16,000g for 15 min at room temperature, and carefully load supernatant into stacking gel avoiding carry over of beads. Also load 10% of the total cell lysate recovered in **Subheading 3.3., step 7**. In the first lane, load appropriate prestained molecular mass standard proteins.
3. Separate proteins using constant current.
4. Transfer proteins to either nitrocellulose or PVDF membrane (Immobilon) in a semi-dry or wet-transfer apparatus.
5. Rinse membrane with TBS for 5 min.
6. Stain with Ponceau Red stain to reveal proteins (optional).
7. Rinse membrane for 30 min with TBS with rocking.
8. Block membrane with 3% BSA in TBS with rocking (*see* **Note 6**).
9. Rinse in TBST (TBS + 0.2% Tween-20) for 5 min.
10. Incubate with primary antibody at appropriate dilution (typically 1:1000) in TBST + 1% BSA ON at 4°C with rocking (*see* Chapter 19 and **Note 7**).
11. Rinse three times 15 min each with TBST.
12. Incubate with secondary antibody at appropriate dilution (typically 1:10,000) in TBST + 1% BSA for 1 h at room temperature, with rocking.
13. Rinse three times 15 min each with TBST.

14. Rinse twice in TBS for 5 min.
15. Expose to chemiluminescence reagents as recommended by the manufacturer, typically 1–5 min.
16. Expose to autoradiography film.

3.5. Measuring Phosphorylation In Vitro

3.5.1. Recombinant PKC Expression and Purification

For in vitro assays, recombinant insect-cell expressed and catalytically active PKCs are used. The expression and purification of PKCs using the baculovirus system is described in Chapter 3. Recombinant, purified PKCs are also commercially available and can be obtained from various companies, for example, PanVera Corporation (Madison, WI; http://www.panvera.com).

3.5.2. Recombinant PDK-1 Expression and Purification

Although PDK-1 can be heterologously expressed in active form in bacteria *(3,15)*, we recommend expression and purification of PDK-1 from baculovirus-infected Sf9 insect cells. The PDK-1 obtained by this technique is at least 20-fold more active than that which can be obtained from bacteria *(15)* and also does not suffer from the proteolytic degradation, which occurs in bacterially produced PDK-1. We recommend the use of the Bac-to-Bac baculovirus expression system (Invitrogen Life Technologies) to generate recombinant His_6-PDK-1, which can be purified using Nickel-NTA affinity purification *(13)*, essentially the same as described for PKC in Chapter 3.

1. The human PDK-1 cDNA (Genbank accession number 015530) is subcloned into one of the many baculovirus transfer vectors, with the initiator Met in-frame to the His_6 or GST in the vector. This is performed using standard polymerase chain reaction conditions. The following procedure outlines the use of the pFastBac HT (Bac-to-Bac) series of baculovirus transfer vectors (HTa, HTb, HTc) that permit subcloning of a given cDNA in one of three open reading frames. The resulting vector is used for the generation of recombinant baculovirus directing the expression of His_6-PDK-1, as described in Chapter 3 and according to the manufacturer's protocol.
2. Grow 500-mL Sf9 cells to a density of 1.5×10^6/mL in a spinner flask, with cell viability of >95%.
3. Infect cells with 50-mL high-titer (10 p.f.u./cell) His_6-PDK-1 baculovirus.
4. Allow to infect for 3 d.
5. Recover cells from spinner flask, centrifuge at 1500g for 5 min at 4°C.
6. Discard supernatant, cell pellet can be stored at –70°C until needed.
7. Lyse cells in 20 mL of lysis buffer A (important: omit EDTA) with protease inhibitor cocktail. Incubate at 4°C for 30 min with rocking.
8. Centrifuge lysate at 16,000g for 30 min at 4°C.

9. Discard pellet, apply supernatant to a 5-mL Nickel-NTA column equilibrated with buffer E.
10. Wash column with five column volumes of wash buffer F.
11. Elute His_6-PDK-1 with elution buffer G, collecting 1-mL fractions and monitoring absorbance at A_{280}.
12. Analyze fractions by SDS-PAGE and Coomassie staining to detect fractions containing His_6-PDK-1 (~70 kDa).
13. At this point, the resulting His_6-PDK-1 is approx 90% pure. We recommend a further column purification step using anion exchange chromatography to further purify the fusion protein.
14. Pool the fractions containing the His_6-PDK-1 protein and dilute fivefold into buffer H.
15. Apply to a MonoQ column equilibrated with buffer H.
16. Elute proteins with a 40-mL linear gradient of NaCl (0–0.5 M) in buffer H, at a flow rate of 1 mL/min, using Fast Protein Liquid Chromatography (FPLC). Monitor absorbance at A_{280} and collect 1-mL fractions.
17. Analyze resulting fractions by SDS-PAGE and pool fractions containing His_6-PDK-1 (*see* **Note 8**).
18. Store protein in 50% glycerol at –20°C in small aliquots. The protein is stable for up to 1 yr under these conditions. Avoid repeated freezing or thawing.

3.5.3. Preparation of Lipid Activators

To monitor phosphorylation of PKCs by PDK-1 in vitro, it is recommended that assays be carried both in the presence and absence of lipid activators to evaluate the contribution of the open conformation of the kinase induced by lipid binding *(27)*. Phosphatidylserine and diacylglycerol vesicles or mixed micelles should be prepared for this purpose, as described in Chapter 6. It may also be necessary to evaluate the contribution of the PI 3-K phosphoinositides PtdIns-3,4-P_2 and PtdIns-3,4,5-P_3 in the phosphorylation of PKCs under these in vitro conditions *(2,3)*. If this is the case, mixed vesicles/micelles should be prepared by inclusion of 5–10 mol% of synthetic L-α-DiC$_{16}$-PtdIns-3,4-P2 or L-α-DiC$_{16}$-PtdIns-3,4,5-P_3 before sonication (mixed vesicles) or mixed micelle preparation. For an example of PDK-1-mediated phosphorylation of PKCζ, *see* **Fig. 2**. Other phosphoinositides, such as PtdIns-3-P, PtdIns-4-P, PtdIns-4,5-P_2, and PtdIns-3,5-P_2 may also be included in the analysis to evaluate the selectivity of the phosphoinositides.

3.5.4. Kinase Assay

1. In a 1.5-mL microcentrifuge tube, combine the following assay components: 0.2 µg of the purified PKC, 0.2 µg of purified PDK-1, mixed lipid vesicles/micelles (PtdSer 140 µM, DAG 4 µM), in a total volume of 50 µL in kinase buffer comprising 20 mM Tris-HCl, pH 7.5, 25 mM MgCl$_2$, 80 µM cold ATP.

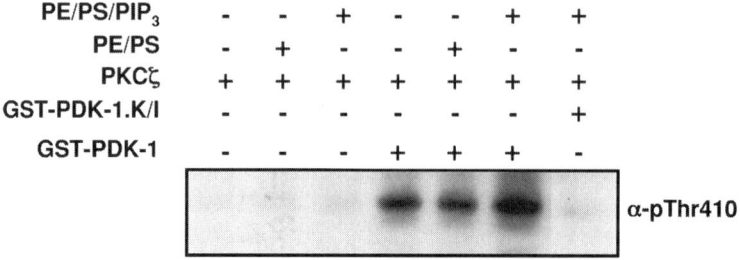

Fig. 2. PDK-1 phosphorylates PKCζ at Thr410 in vitro. Purified PKCζ was incubated in an in vitro kinase assay with recombinant PDK-1 and kinase inactive PDK-1 (PDK-1.K/I) (0.2 µg each protein) in the presence or absence of phosphatidylethanolamine/ phosphatidylserine (PE/PS, 100 µM) or PE/PS + PtdIns-3,4,5-P₃ (10 µM) vesicles. The phosphorylated PKCζ was resolved by SDS-PAGE, transferred to nitrocellulose, and immunoblotted with a phospho-PKCζ antibody against pThr410 (α-pThr410). Adapted from **ref. 3**.

2. Allow reaction to proceed for 20 min at 25°C.
3. Terminate reaction by addition of SDS sample buffer and heating to 100°C for 5 min.
4. Resolve proteins on a 7.5% SDS-PAGE, transfer to nitrocellulose, and immunoblot with the appropriate PKC phosphospecific antibodies as outlined in **Subheading 3.4.**

3.6. PDK-1 Assay

PDK-1 is often referred to as a constitutively active enzyme, such that increases in the intrinsic protein kinase activity of PDK-1 are not readily observed upon stimulation of cells with growth factors or hormones *(5,8)*. This is unlike its substrates, such as Akt/PKB or S6K1, whose activities are increased as a consequence of PDK-1-mediated phosphorylation and that can be measured using immune-complex kinase assays *(10)*. However, there are reports that in fact PDK-1 activity can be increased, albeit modestly, upon stimulation of certain cells with agonists such as insulin, or pervanadate *(28)*. This section describes an immune-complex assay for the analysis of changes in PDK-1 activity following stimulation of cells. This assay makes use of a specific synthetic peptide substrate based on the phosphorylation motif of Akt/PKB at Thr308. The peptide sequence we have found to work best is: Ser-Lys-Gln-Ala-Arg-Ala-Asn-Ser-Phe-Val-Gly-Thr-Ala-Gln-Tyr-Val-Ser-Arg-Ag-Lys-Arg. Similar assays using bacterially expressed glutathione S transform fusion proteins of Akt/PKB as PDK-1 substrates have also been described *(8)*.

1. HEK293, NIH 3T3, or COS-7 cells are grown and maintained as described in **Subheading 3.1.** Serum-starve cells for 12–24 h in DMEM without FBS.
2. Stimulate cells with appropriate agonists; as a control, stimulate cells with insulin (100 nM) or pervanadate (75 µM). To prepare pervanadate solution, mix 1 mL of a 20 mM sodium orthovanadate stock solution with 330 µL of 30% H_2O_2 and incubate for 10 min at room temperature. This yields a 15 mM solution of pervanadate and H_2O_2. Inactivate residual H_2O_2 by incubating with 10 µL catalase (8 U/mg) for 15 min. Fresh preparations should be made for each experiment.
3. Lyse cells in lysis buffer A and prepare lysates as described in **Subheading 3.3.**
4. Add anti-PDK-1 antibody at 3 µg/mL and 40 µL of a 50% slurry of protein A/G Sepharose beads.
5. Incubate immunoprecipitates and wash beads as described in **Subheading 3.3.**
6. Discard supernatant from the last wash, and to the beads add 25 µL of kinase assay buffer containing 50 mM Tris-HCl, pH 7.5, 20 mM $MgCl_2$, 10 mM dithiothreitol, 30 µM PDK-1 substrate peptide, 10 mM cold ATP + 10 µCi [γ-^{32}] ATP per reaction.
7. Incubate for 30 min at 37°C.
8. Pellet beads by centrifugation.
9. Recover 25 µL of the supernatant and spot onto p81 phosphocellulose discs.
10. Wash discs four times with 200 mL of a 1% phosphoric acid solution (*see* **Note 9**).
11. Wash discs once in 100% ethanol and allow to dry.
12. Add each disc to a scintillation vial containing 5 mL of scintillation fluid, and count on a scintillation counter.

4. Notes

1. The primary reason to maintain cells at low confluence is to avoid increases in the basal PI 3-K activity, which is often seen in cell lines that are serially passaged over time, and in particular when cells are allowed to reach 100% confluency. We have noticed that in HEK293 as well as NIH 3T3 cells this high basal PI 3-K activity translates into increased basal PKC phosphorylation even when cells are serum starved *(3)*. Thus, we recommend that cells are not allowed to grow in culture beyond 80% confluency and that a new aliquot of cells is thawed into a fresh culture after 15–20 serial passages. For the same reason, we recommend the use of early passage NIH 3T3 fibroblasts (<150 passages), which can be obtained from ATCC.
2. Many transformed cells lines display elevated PI 3-K levels even under serum-starved conditions. Therefore, extensive serum starvation in the absence of serum may be necessary for many transformed cell lines that display elevated basal PI 3-K activity even under these conditions. We have noted that with early passage NIH 3T3 cells, 14-h starvation with DMEM in the absence of serum is sufficient to eliminate basal PI 3-K activity *(3)*, and that this basal activity increases with

serial passaging of cells over time. In the case of HEK 293 or COS-7, we have not been able to completely or quantitatively eliminate PI 3-K activity even with extensive serum starvation (>24 h). One way to reduce this basal activity is to pretreat cells with PI 3-K inhibitors such as wortmannin (100 n*M* final concentration) or LY294002 (50 µ*M* final concentration) for 10 min before stimulation, as reported *(14)*. These issues are particularly important when evaluating the phosphorylation of novel and atypical PKCs, which is mediated by increased PI 3-K and PDK-1 activity.

3. We have not noticed any adverse affect in the activity or regulation of PKCε or PKCζ when an epitope tag is added to either the amino terminus or carboxyl terminus. Numerous studies have made use of epitope-tagged PKCs with the tag fused in frame to the initiator Met residue, and again this has not been reported to have a detrimental effect on the activity or regulation of the PKC in question.

4. The procedure described is for transfection into 60-mm dishes, which in our hands results in easily detectable PKC expression by western immunoblotting. This procedure can be scaled up or down by altering the amounts of Optimem, DNA, and Lipofectamine, accordingly. Note that the critical variables in this, and most other transfection techniques, is the concentration of DNA and amount of transfection reagents used, and this should be determined for each cell line and DNA under investigation. Additional detailed information, including troubleshooting, can be found in the manufacturer's protocols. We have found that both HEK293 cells, NIH 3T3 fibroblasts, as well as COS-7 cells are easily and reproducibly transfected with PKC mammalian expression vectors using the Lipofectamine protocol, with transfection efficiencies approaching 40–50% depending on the cell type.

5. It is important to note that the pellet that is obtained from this procedure contains not only unbroken nuclei but also the cytoskeletal fraction. There are numerous reports that PKCs are localized to the nucleus (reviewed in **ref. 29**), and equally importantly, translocation of PKC to the detergent-insoluble cytoskeletal fraction has been described *(30)*. It is therefore advisable to further analyze the NP-40 pellet further if one wishes to evaluate the status of PKC phosphorylation in the nuclear and cytoskeletal fraction.

6. Although lower backgrounds are often obtained on autoradiographs of immunoblots when using non-fat milk (5% in TBS) as the blocking agent, this is known to contain active phosphatases which can effectively strip the phosphate groups off many proteins. Although this is particularly troublesome when evaluating tyrosine phosphorylation, we favor the use of BSA as the blocking agent to circumvent this issue, particularly when immunoblotting with phospho-specific antibodies. Also, if one wishes to strip immunoblots for reprobing, we recommend that the first incubation be carried out with the phosphospecific antibody, followed by stripping and reprobing with PKC isoform-specific antibodies because the commonly used blot stripping procedures also tend to completely strip phosphate groups from proteins.

7. We find that an overnight incubation with the primary antibody provide better signals and lower backgrounds, provided the incubation is performed at 4°C. This is particularly true for phosphospecific antibodies. However, if time is an issue, comparable results can be achieved by carrying out the primary antibody incubation at room temperature for 2 h, with rocking.

8. It is advisable that the His_6 moiety be removed from the resulting His_6-PDK-1 fusion protein, although we have not found any detrimental effect of the His_6 tag on PDK-1 activity in vitro. The pFastBac HT series of vectors are designed with an rTev protease cleavage site between the His_6 sequence and the first codon of the cDNA of interest. The rTev protease is highly specific and can be obtained from various commercial sources (e.g., Invitrogen Life technologies). The manufacturer's directions should be followed for the use of this protease. Other baculovirus transfer vectors contain cleavage sites for other proteases, such as thrombin.

9. The appropriate institutional radiation safety procedures should be followed when performing the PDK-1 kinase assay. This is particularly important when handling the 200 ml wash solution containing the unbound radiolabeled ATP. The first wash will contain more than 95% of the radioactive ATP, and we routinely adsorb this into adsorbent material such as kitty litter. The subsequent washes contain low levels of radioactivity and can be disposed of down the sink, according to institutional guidelines and regulations.

References

1. Dutil, E. M., Toker, A., and Newton, A. C. (1998) Regulation of conventional protein kinase C isozymes by phosphoinositide-dependent kinase 1 (PDK-1). *Curr. Biol.* **8,** 1366–1375.
2. Le Good, J. A., Ziegler, W. H., Parekh, D. B., Alessi, D. R., Cohen, P., and Parker, P. J. (1998) Protein kinase C isotypes controlled by phosphoinositide 3-kinase through the protein kinase PDK1. *Science* **281,** 2042–2045.
3. Chou, M. M., Hou, W., Johnson, J., Graham, L. K., Lee, M. H., Chen, C. S., et al. (1998) Regulation of protein kinase C zeta by PI 3-kinase and PDK-1. *Curr. Biol.* **8,** 1069–1077.
4. Toker, A. and Newton, A. C. (2000) Cellular signaling: pivoting around PDK-1. *Cell* **103(2),** 185–188.
5. Storz, P. and Toker, A. (2002) 3′-phosphoinositide-dependent kinase-1 (PDK-1) in PI 3-kinase signaling. *Front. Biosci.* **7,** d886–d902.
6. Parekh, D. B., Ziegler, W., and Parker, P. J. (2000) Multiple pathways control protein kinase C phosphorylation. *EMBO J.* **19(4),** 496–503.
7. Keranen, L. M., Dutil, E. M., and Newton, A. C. (1995) Protein kinase C is regulated in vivo by three functionally distinct phosphorylations. *Curr. Biol.* **5(12),** 1394–1403.
8. Alessi, D. R., James, S. R., Downes, C. P., Holmes, A. B., Gaffney, P. R., Reese, C. B., et al. (1997) Characterization of a 3-phosphoinositide-dependent protein

kinase which phosphorylates and activates protein kinase Balpha. *Curr. Biol.* **7,** 261–269.

9. Stokoe, D., Stephens, L. R., Copeland, T., Gaffney, P. R., Reese, C. B., Painter, G. F., et al. (1997) Dual role of phosphatidylinositol-3,4,5-trisphosphate in the activation of protein kinase B. *Science* **277,** 567–570.

10. Alessi, D. R., Andjelkovic, M., Caudwell, B., Cron, P., Morrice, N., Cohen, P., et al. (1996) Mechanism of activation of protein kinase B by insulin and IGF-1. *EMBO J.* **15,** 6541–6551.

11. Moriya, S., Kazlauskas, A., Akimoto, K., Hirai, S., Mizuno, K., Takenawa, T., et al. (1996) Platelet-derived growth factor activates protein kinase C epsilon through redundant and independent signaling pathways involving phospholipase C gamma or phosphatidylinositol 3-kinase. *Proc. Natl. Acad. Sci. USA* **93(1),** 151–155.

12. Parekh, D., Ziegler, W., Yonezawa, K., Hara, K.. and Parker, P. J. (1999) Mammalian TOR controls one of two kinase pathways acting upon nPKCdelta and nPKCepsilon. *J. Biol. Chem.* **274,** 34,758–34,764.

13. Cenni, V., Doppler, H., Sonnenburg, E. D., Maraldi, N., Newton, A. C., and Toker, A. (2002) Regulation of novel protein kinase C epsilon by phosphorylation. *Biochem. J.* **363,** 537–545.

14. Sonnenburg, E. D., Gao, T., and Newton, A. C. (2001) The phosphoinositide-dependent kinase, PDK-1, phosphorylates conventional protein kinase C isozymes by a mechanism that is independent of phosphoinositide 3-kinase. *J. Biol. Chem.* **276,** 45,289–45,297.

15. Alessi, D. R., Deak, M., Casamayor, A., Caudwell, F. B., Morrice, N., Norman, D. G., et al. (1997). 3-phosphoinositide-dependent protein kinase-1 (PDK1): structural and functional homology with the Drosophila DSTPK61 kinase. *Curr. Biol.* **7,** 776–789.

16. Anderson, K. E., Coadwell, J., Stephens, L. R., and Hawkins, P. T. (1998) Translocation of PDK-1 to the plasma membrane is important in allowing PDK-1 to activate protein kinase B. *Curr. Biol.* **8,** 684–691.

17. Behn-Krappa, A. and Newton, A. C. (1999) The hydrophobic phosphorylation motif of conventional protein kinase C is regulated by autophosphorylation. *Curr. Biol.* **9,** 728–737.

18. Ohno, S., Konno, Y., Akita, Y., Yano, A., and Suzuki, K. (1990) A point mutation at the putative ATP-binding site of protein kinase C alpha abolishes the kinase activity and renders it down-regulation-insensitive. A molecular link between autophosphorylation and down- regulation. *J. Biol. Chem.* **265,** 6296–6300.

19. Zhang, J., Wang, L., Petrin, J., Bishop, W. R., and Bond, R. W. (1993) Characterization of site-specific mutants altered at protein kinase C beta 1 isozyme autophosphorylation sites. *Proc. Natl. Acad. Sci. USA* **90,** 6130–6134.

20. Feng, X. and Hannun, Y. A. (1998) An essential role for autophosphorylation in the dissociation of activated protein kinase C from the plasma membrane. *J. Biol. Chem.* **273,** 26,870–26,874.

21. Hata, A., Akita, Y., Suzuki, K., and Ohno, S. (1993) Functional divergence of protein kinase C (PKC) family members. PKC gamma differs from PKC alpha

and -beta II and nPKC epsilon in its competence to mediate-12-O-tetradecanoyl phorbol 13-acetate (TPA)- responsive transcriptional activation through a TPA-response element. *J. Biol. Chem.* **268,** 9122–9129.

22. Li, W., Yu, J. C., Shin, D. Y., and Pierce, J. H. (1995) Characterization of a protein kinase C-delta (PKC-delta) ATP binding mutant. An inactive enzyme that competitively inhibits wild type PKC- delta enzymatic activity. *J. Biol. Chem.* **270,** 8311–8318.

23. Ueda, E., Ohno, S., Kuroki, T., Livneh, E., Yamada, K., Yamanishi, K., et al. (1996) The eta isoform of protein kinase C mediates transcriptional activation of the human transglutaminase 1 gene. *J. Biol. Chem.* **271,** 9790–9794.

24. Chang, J. D., Xu, Y., Raychowdhury, M. K., and Ware, J. A. (1993) Molecular cloning and expression of a cDNA encoding a novel isoenzyme of protein kinase C (nPKC). A new member of the nPKC family expressed in skeletal muscle, megakaryoblastic cells, and platelets. *J. Biol. Chem.* **268,** 14,208–14,214.

25. Murray, N. R. and Fields, A. P. (1997) Atypical protein kinase C iota protects human leukemia cells against drug-induced apoptosis. *J. Biol. Chem.* **272,** 27,521–27,524.

26. Kotani, K., Ogawa, W., Matsumoto, M., Kitamura, T., Sakaue, H., Hino, Y., et al. (1998) Requirement of atypical protein kinase clambda for insulin stimulation of glucose uptake but not for Akt activation in 3T3-L1 adipocytes. *Mol. Cell. Biol.* **18,** 6971–6982.

27. Newton, A. C. (1997) Regulation of protein kinase C. *Curr. Opin. Cell. Biol.* **9(2),** 161–167.

28. Park, J., Hill, M. M., Hess, D., Brazil, D. P., Hofsteenge, J., and Hemmings, B. A. (2001) Identification of tyrosine phosphorylation sites on 3-phosphoinositide-dependent protein kinase-1 and their role in regulating kinase activity. *J. Biol. Chem.* **276(40),** 37,459–37,471.

29. Toker, A. (1998) Signaling through protein kinase C. *Front Biosci* **3,** D1134–D1147.

30. Edwards, A. S., Faux, M. C., Scott, J. D., and Newton, A. C. (1999) Carboxyl-terminal phosphorylation regulates the function and subcellular localization of protein kinase C betaII. *J. Biol. Chem.* **274,** 6461–6468.

16

In Vitro Autophosphorylation of Protein Kinase C Isozymes

Antonio M. Pepio and Wayne S. Sossin

1. Introduction

Protein kinase C (PKC) autophosphorylates at multiple sites. Two of these sites that are important for PKC activity are usually quantitatively phosphorylated in vivo *(1–8)*. Therefore, these sites are poorly detected in vitro autophosphorylation studies unless phosphatases are used first to remove endogenous phosphorylation. PKCs also autophosphorylate at other sites that are not quantitatively phosphorylated in cells and are not required for PKC activity *(7–14)*. These sites may nevertheless be important for modulating the activity of PKC *(10,12,13)*. These sites can be isoform specific and may also be important in determining differences in isoform activation.

Our work has concentrated on the role of autophosphorylation in the nervous system using the invertebrate model system of *Aplysia*. This is an excellent system for studying the role of PKCs in neuronal plasticity *(15,16)*. One simplification in this system is that there are only two phorbol ester-activated PKCs, the Ca^{2+}-activated PKC APL I and Ca^{2+}-independent PKC APL II *(17,18)*. We have focused on the role of two autophosphorylation sites. First, we studied a site in the carboxy-terminal that is well conserved in many PKCs and may play a role in the removal of PKC from membranes *(12)*. We have also discovered a site that is specific for the C2 domain of the Ca^{2+}-independent PKC and that plays a role in the lipid binding of the C2 domain *(13)*. The specific examples we show are for these isoforms. The techniques we use should be readily applicable to the study of PKC in any system.

Investigation of in vitro phosphorylation of PKC usually follows the following route. First, conditions for the in vitro kinase reaction are optimized

From: *Methods in Molecular Biology, vol. 233: Protein Kinase C Protocols*
Edited by: A. C. Newton © Humana Press Inc., Totowa, NJ

to increase molar incorporation of phosphate into PKC. Next, the autophosphorylated kinases are analyzed using one-dimensional or two-dimensional chromatography. These methods are useful for defining the region of PKC phosphorylated and the number of sites phosphorylated. Although they do not by themselves define the site of phosphorylation, they can be used to generate hypotheses concerning the sites and are very useful for confirming a putative site of phosphorylation. Once an autophosphorylation site has been discovered, it is still a challenge to determine its role. In this chapter, we describe the general techniques used to determine sites of in vitro phosphorylation and to investigate the function of these sites.

2. Materials
2.1. In Vitro Kinase Reaction

1. Purified PKC (*see* Chapter 3).
2. Phosphatidylserine (PS) (Avanti Polar lipids).
3. Phosphatidylcholine (PC) (Avanti Polar lipids).
4. Phorbol 12-myristate 13-acetate (TPA).
5. Amber glass vials with phenolic closure (12 × 35 m).
6. Nitrogen tank.
7. Bath-sonicator.
8. Dimethylsulfoxide.
9. ATP (1 mM, pH 7.0).
10. γ^{32}P-ATP.
11. Plexiglass shield.
12. Screw-top microcentrifuge tubes.

2.2. Phosphopeptide Mapping

1. Hunter thin-layer peptide mapping electrophoresis system.
2. Thin-layer cellulose plates (EM systems).
3. Tosyl Phenylalanyl Chlorometlylketone (TPCK)-treated trypsin (Worthington Biochemicals).
4. Polyvinylpolypyrrolidone 360 (PVP-360).
5. Nitrocellulose paper.
6. 5× sample buffer: 2% sodium dodecyl sulfate (SDS), 10% glycerol, 100 mM dithiothreitol, 0.06 M Tris-HCl (pH 6.8), 0.001% bromophenol blue.
7. Whatmann blotting paper.
8. 2D markers (DNP-lysine; Xylene cyanol).
9. Phospho-chromatography buffer: 37.5% 1-butanol, 25% pyridine, and 7.5% acetic acid.
10. Thin-layer chromatography (TLC) tank.
11. High-sensitivity autoradiography film (i.e., Kodak Biomax MS).

3. Methods

3.1. The In Vitro Autophosphorylation Reaction

Studies of in vitro phosphorylation begin by examining, in a purified preparation of PKC, how activation of PKC leads to autophosphorylation. Below we cover the standard protocols used in our lab for this procedure.

3.1.1. Preparation of Lipids

We have found that mixtures of 90% PS and 10% PC lead to more stable and more reproducible results than 100% PS vesicles *(19)*. To make lipids, first combine chloroform stocks of PS and PC to give a 9:1 molar ratio of PS. This stock solution should be aliquoted into small amounts (25 µL of a 10 mg/mL stock) and stored in brown glass vials. Seal the vials under nitrogen and store at –70°C. On the day of the experiment, dry down the material under nitrogen. Resuspend the dried lipid film in 500 µL of water. Vortex vigorously for 30 s and then incubate in the bath sonicator for 30 s; repeat this process three times. This should result in a cloudy suspension of lipid vesicles. Be sure to vortex immediately before use. In order to vary the amount of lipid in a systematic way, mixed micelles should be used *(20)*. Care should be taken with TPA because it is a tumor promoter. Store as aliquots in dimethylsulfoxide at a concentration (2-10 mM) greater than 1000× the final concentration used.

3.1.2. The In Vitro Kinase Reaction

Purified Ca^{2+}-independent PKC (10–100 pmol) (*see* **Note 1**) is incubated with vesicles of PS/PC (50 µg/mL), TPA (50 nM), and ATP (10 µM ATP, 1–5 µCi, γ^{32}P-ATP) at 37°C (*see* **Note 2**). Because of the radioactivity involved, this reaction must be shielded (plexiglas shield). We also use screw-top microcentrifuge tubes because they do not release radioactivity; this is much preferred to wrapping tubes. Usually we incubate for 30 min, but an appropriate first experiment is a time course to determine when the reaction begins to saturate. An important consideration is the concentration of total ATP. Usually, we find that 1–5 µCi γ^{32}P-ATP and 10 µM cold ATP is optimal. Using only hot ATP reduces ATP concentration well below the K_{max} of PKCs and leads to less total incorporation. Increasing the total ATP concentration to about 10 µM leads to a good compromise between specific activity of ATP and K_{max} considerations. The stock solution of cold ATP (1 mM) must be around neutral pH because ATP can acidify the reaction if not buffered. Also it is advisable to measure the counts per minute (cpm) of the hot ATP in a scintillation counter to facilitate later determination of moles of ATP incorporated in the in vitro phosphorylation reaction (*see* **Subheading 3.1.4.**).

The concentration of PS is also important and is not necessarily the same as for substrate reactions *(21)*. For PKC Apl I, a concentration of PS higher than normally used (e.g., 50 µg/mL as opposed to 10 µg/mL) is required for good autophosphorylation *(12)*. The concentration of PKC used may be important if there is an issue between *cis*-autophosphorylation (one molecule of PKC phosphorylating itself) vs *trans*-autophosphorylation (one molecule of PKC phosphorylating a different molecule of PKC). For *cis*-autophosphorylation, the amount of incorporation should increase linearly with the amount of kinase. In contrast, *trans*-autophosphorylation should increase with the square of the concentration of kinase since it is a bimolecular reaction. Thus, higher concentrations of PKC will be required for *trans*-phosphorylations (*see* **Note 3**).

3.1.3. Trans-*Phosphorylation of PKC Fusion Proteins*

Even sites that are normally *cis*-autophosphorylated may be phosphorylated in *trans* in an in vitro reaction *(13)*, although this is not the case for all PKC autophosphorylations *(22)*. In this case, a fragment of PKC or a PKC fusion protein is used as a substrate. The reaction is similar to that above, but usually less PKC (2–5 pmol) and a higher concentration of the PKC fusion protein substrate (100–500 nmol) is used. One can assure oneself that the site phosphorylated in the fusion protein (in *trans*) is the same as in the autophosphorylation using two-dimensional phosphopeptide mapping (*see* **Subheading 3.2.3.**) *(13)*.

3.1.4. *Measuring Autophosphorylation*

The kinase reaction can be stopped by the addition of 5× sample buffer to a final concentration of 1×. The reactions are then run on SDS-polyacrylamide gel electrophoresis (PAGE) gels. Because all the radioactive ATP is being run on the gel, the gel also needs to be run behind a plexiglas shield. It is important to run a percentage gel such that the dye (bromophenol blue in sample buffer) does not have to run out the bottom. ATP will run with the dye and clean-up is much easier if the ATP remains in the gel as opposed to running into the buffer. After running the gel, cut off the region near the dye and dispose in the radioactive waste. Usually, this region of the gel, the tubes, and the pipet tips used to add the ATP and to load the gel are the only radioactive waste generated.

We find that instead of drying the gel and exposing it to film, it is usually easier to transfer the proteins to nitrocellulose by electrophoresis (Western blot) and expose the nitrocellulose to film. Usually, once the dye has been cut off, the remaining radioactivity that enters the transfer buffer is of low enough specific activity that it can be disposed down a sink (check with the radioactive standards in your institution). After exposure of the nitrocellulose to film (*see*

Note 4), development of the film allows visualization of the incorporation of ATP into the kinase.

It is often useful to know the molar percentage of incorporation in the kinase reaction. This is important for determining the feasibility of identifying the phosphorylated residues by direct means (*see* **Subheading 3.3.2.**). Assuming that you have purified PKC it is simple to approximate the moles of PKC in the preparation, although this is probably an over estimate as it assumes that all the PKC protein is active. To determine the moles of ATP incorporated into PKC, one needs to separate the incorporated ATP from unincorporated. Probably the easiest method is to simply cut out the band from the nitrocellulose and count it in a scintillation counter. The moles of phosphate incorporated are the cpm in the band divided by the total cpm added to the reaction (*see* **Subheading 3.1.1.**) and multiplied by the moles of ATP in the reaction. However, this cannot be used to reliably determine the number of sites because the in vitro kinase reaction rarely goes to completion. Although incorporation of more than one mole of phosphate/mole of kinase is proof of multiple sites, even with multiple sites of phosphorylation we rarely *see* incorporation of greater than 0.2 mol of phosphate per mole of PKC.

3.2. Phosphopeptide Mapping

An important step in studying autophosphorylation of PKC is to determine which residues are the sites of autophosphorylation. Phosphopeptide mapping is a first step in this process. Even in the absence of definitive identification of one site, phosphopeptide mapping allows one to determine how many sites are being used, and how different sites may be regulated. Although sites with high molar incorporation can be detected immediately by mass spectrometry (*see* **Subheading 3.3.2.**), this is difficult to do directly with sites of low incorporation.

3.2.1. One-Dimensional Mapping

In one-dimensional mapping, the autophosphorylated PKC is subjected to partial proteolysis before separation by SDS-PAGE. One-dimensional mapping is useful to determine whether sites are in the catalytic or regulatory domain. If multiple antibodies are available to different regions of the PKC, they can restrict the location of the site to smaller domains (i.e., C2 domains, *see* **Fig. 1**). Purified PKCs are autophosphorylated as described above. To prevent confusion from phosphorylation of the protease, the reaction should be quenched either with ethylenediamine tetraacetic acid or an inhibitory peptide. The proteolysis reaction contains 10 pmol of phosphorylated PKC. Conditions for partial hydrolysis of proteases will need to be determined for each preparation. For our studies we use 2 µg/mL of TPCK-treated trypsin

Autorad Immunoblot

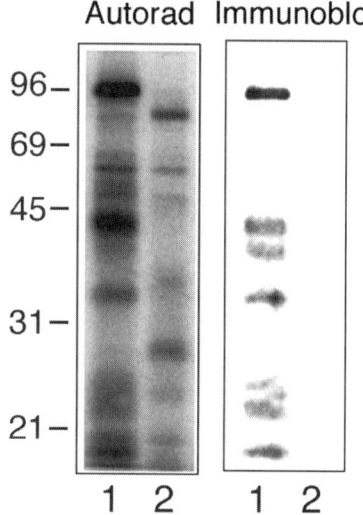

Fig. 1. One-dimensional map of autophosphorylation. In this case, an invertebate
Ca^{2+}-independent PKC (PKC Apl II) was autophosphorylated and subjected to partial
tryptic hydrolysis as described *(13)*. Because the goal of this experiment was to obtain
evidence for phosphorylation in the C2 domain, the kinase without the C2 domain (PKC
Apl IIΔC2) was treated in a similar manner. The digests were then run on a 9% SDS-gel,
transferred to nitrocellulose, and exposed for autoradiography (Autorad). PKC Apl II
(1), PKC Apl IIΔC2 (2). The nitrocellulose was then immunoblotted with an antibody
whose epitope was in the C2 domain (Immunoblat). Note that many of the major bands
seen in the autoradiogram in match with the bands detected by the C2 domain antibody
and are absent in the digest of the kinase that lacks the C2 domain.

for 5 min at 30°C, but initially a range either of trypsin concentration or
of time should be attempted to obtain a good partial digest. Reactions are
then quenched by addition of 20 μL of sample buffer and boiled. Proteins of
interest are then separated by SDS-PAGE and electrophoretically transferred to
nitrocellulose. Membranes containing proteins are subjected to autoradiography
to visualize phosphate incorporation and probed with the anti-PKC antibody
for identification of protein fragments containing a particular domain. Because
only fragments containing the epitope for the antibody will react, this can
identify phosphorylations within a domain (i.e., C2 domain; **Fig. 1**).

3.2.2. Two-Dimensional Phosphopeptide Mapping

There are a number of comprehensive reviews of two-dimensional phosphopeptide mapping *(23,24)*. The following is a detailed description of the
protocol used in our lab. Autophosphorylation reactions are quenched with

5× sample buffer. These reactions are then separated by SDS-PAGE. After electrophoresis, the proteins are transferred to nitrocellulose membranes again by electrophoresis. After autoradiography to identify the phosphorylated band (*see* **Note 5**), the radioactive band is cut out with a clear sterile scalpel. Use forceps to handle the membrane fragments. Place each radioactive membrane fragment in a screw-top microcentrifuge tube. Cerenkov count the tubes (no scintillation fluid) to get an approximate level of the counts incorporated into the PKC. Membrane fragments containing protein bands are next preblocked in 0.5% PVP-360 in 100 mM acetic acid for 30–60 min at 37°C in the same tube. Wash the fragments five times in deionized water. The membrane fragments are then sliced into very small strips (approx 2 by 5 mm) using forceps and dissecting scissors. Return these strips to a screw-top tube for digestion overnight at 37°C with 10 µg of TPCK-treated trypsin in 10 mM Tris-HCl (pH 7.5; *see* **Note 5**). To ensure complete digestion, an additional 10 µg of trypsin can be added for 3 h on the following day. Remove the supernatant containing the cleaved peptides and place it in a new clean screw-top microcentrifuge tube and then lyophilize to completion in a Speedvac. Next, it is important to oxidize the peptides. Because cysteines and methionines will oxidize to some extent during the procedure, it is preferable to oxidize all of them to retain consistency between different preparations. Oxidation is accomplished by resuspending the peptides in 50 µL of performic acid and letting them sit on ice for 1 h. Do not to let the oxidation reaction proceed for more than 1 h. The performic acid is made by mixing 900 µL of concentrated formic acid with 100 µL of hydrogen peroxide (33%) and leaving on ice for 1 h. Stop the oxidation by adding 1 mL of deionized H$_2$O and vortex. Immediately snap freeze on dry ice and ethanol to stabilize the residues. Again lyophilize to completion. Resuspend the peptides in 1mL of deionized H$_2$O and again lyophilize for the third time to completion. Peptides are then resuspended in pH 1.9 buffer (0.6 M formic acid and 1.4 M acetic acid; the same buffer used for the first dimension of 2-D analysis) *(23)*. Resuspend each radiolabeled protein tube in 10 µL (*see* **Note 7**). Centrifuge the resuspended sample briefly to sediment large particles. Prepare the TLC plates by marking with a stencil for the location of sample origin *(23)*. Apply only the top 7 µL of the centrifuged sample to the origin mark one drop at a time and dry each drop under a stream of nitrogen. Be sure to keep the total diameter of the spotted sample as small as possible to improve localization of radiolabeled peptides. In the marker origin apply a mixture of DNP-lysine (yellow) and xylene cyanol (blue) *(23)* to control for the separation you achieve in the electrophoretic dimension. Run the plate in the first dimension for 30 min at 1500 volts (*see* **Note 8**). The time and voltage will have to be optimized for separation of different peptides. When the procedure is complete, remove the plate from the apparatus and carefully allow it to dry in the fume hood until acetic acid can no longer be detected by smell (usually about 2 h). The plate is then ready for the second chromatography dimension. A TLC tank should be lined with Whatmann paper on the back and sides and filled with 100 mL of phosphochromatography buffer *(23)* and placed in the fume hood. This amount of buffer in a standard TLC tank remains just below the placement of the sample and the marker spot. Allow the strips of Whatmann paper to wet by capillary action and seal the tank airtight with silicone vacuum grease. Allow the tank to stand and equilibrate at least overnight and do not move the tank once equilibrated. Re-spot the TLC plate with a new marker dot of the same dye composition at the same horizontal height as the origin of the sample. Allow the marker spot to dry thoroughly. Place the TLC plate sample and marker spot down into the TLC tank such that the plate sits at a 75- to 80-degree angle with the surface of the counter. Close the tank and seal again with vacuum grease. Allow chromatography to proceed for 8 h undisturbed in the tank. When 8 h have passed, gently remove the plate from the tank and allow to dry overnight in the TLC hood. Once the plate is dry place dots of radiolabeled ink to align the plate with the film after autoradiography. Next wrap the completed TLC plate in saran wrap and place it in an autoradiography cassette with intensifying screens and BioMax MS film. It is important to use the most sensitive film available to decrease the required time of exposure. Wrap the cassette in foil to prevent stray light entering and place the cassette at $-70°C$ for 1–4 wk for optimum exposure depending on the counts loaded onto the plate. An example of a successful two-dimensional blot is seen in **Fig. 2**.

3.2.3. Analysis of Two-Dimensional Maps

There are a number of caveats to remember in analyzing these maps. First, not all spots will separate cleanly and one spot can represent multiple peptides. Second, it is difficult to rule out partial cleavage, such that two spots may result from differential cleavage products containing the same phosphorylated residue. Third, there may be multiple phosphorylation sites on the same peptide. Usually, these spots will run close to the cathode because of their negative charge. The reproducibility of these maps is a major challenge. If one wants to compare maps on two different plates (i.e., after altering a condition in the in vitro reaction or after mutating a putative site of phosphorylation), it is best to run them right after each other and develop them in the same tank. Smudging, smearing of spots, or failure to leave the origin are usually the result of high salt in the sample, not being careful to remove nonsoluble components, or loading too much sample.

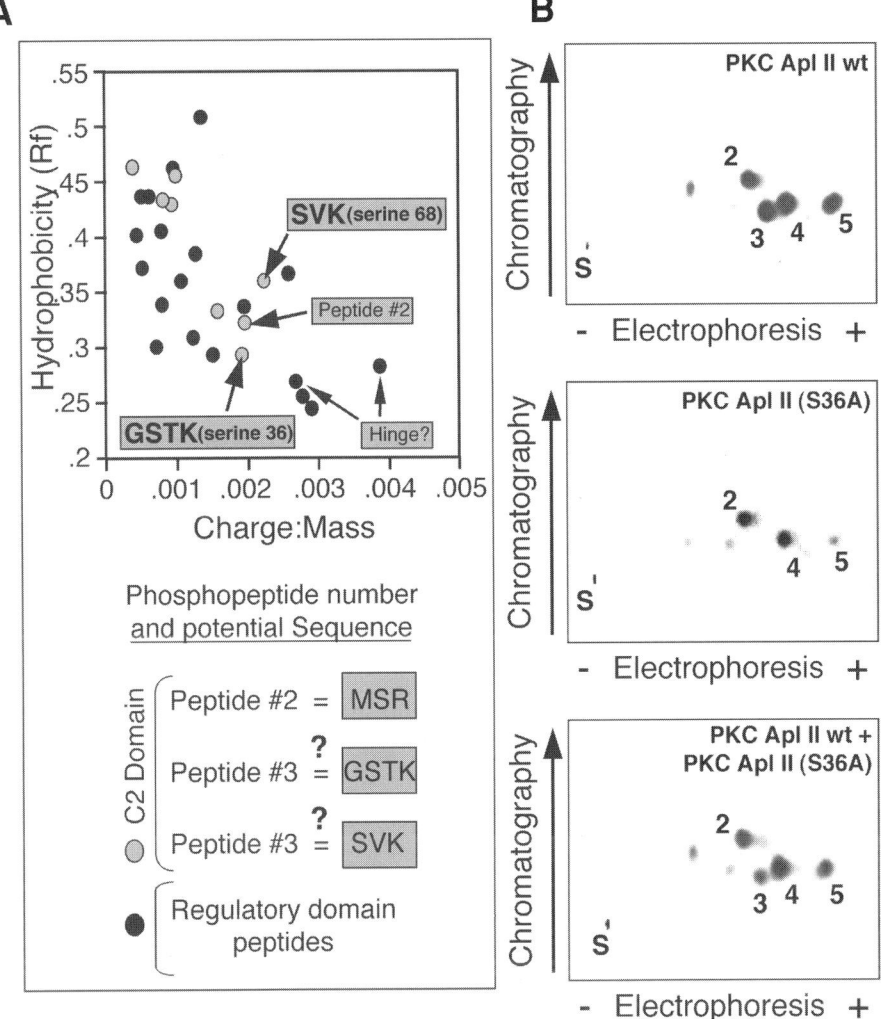

Fig. 2. Theoretical phosphopeptide maps can help define a phosphorylation site. (A) The regulatory region of PKC Apl II was divided into constituent tryptic peptides and their mass/charge ratio calculated based on singular phosphorylations. The relative hydrophobicity of each peptide was then calculated in the phospho-chromatography buffer. The resulting values for regulatory domain peptides (black circles) and C2 domain specific peptides (grey circles) were plotted against each other to generate a phosphopeptide "map" where the location of each peptide is relative to each other. Spot 2 had already been defined as MSR and helped root the map. The hypothetical map showed that the most likely site for spot 3 was serine 36, whereas serine 68 was less likely. Indeed in (B) one can see that the conversion of serine 36 to alanine specifically led to the loss of spot 3, whereas conversion of serine 68 did not (data not shown). + and – refer to the polarity of the primary electrophoresis dimension in buffer pH 1.9; vertical arrows indicate direction of liquid chromatography; S signifies origin of sample application.

3.3. Identification of Phosphopeptide

Phosphopeptide analysis allows one to initially analyze autophosphorylations and is useful in measuring changes in levels of autophosphorylation, but clearly does not replace identification of the phosphorylation site.

3.3.1. Hypothetical Phosphopeptide Maps

A hypothetical phosphopeptide map can help in limiting candidate regions for the phosphorylated amino acid. Divide the domain of interest into constituent tryptic peptides based on primary sequence analysis. Their mass/charge ratio is then calculated based on singular phosphorylations *(23)*. The relative hydrophobicity of each peptide is then calculated in the phospho-chromatography buffer *(23)*. The resulting values are plotted against each other to generate a phosphopeptide "map" where the location of each peptide is relative to the others (**Fig. 2**) *(13)*. The map enables one to limit the putative peptides to a small number that may then be tested by mutagenesis.

3.3.2. Direct Identification of the Peptide

Several analytical methods are available for the identification of phosphorylated peptides within proteins. Mass spectrometry is becoming common as a method for identifying these sites *(25–29)*. The peptide may be directly isolated from the two-dimensional phosphopeptide map although the yield from this is low *(30)*. Alternatively, the complete tryptic digest can be first fractionated by high-performance liquid chromatography, radioactive peptides eluted and then used for Edman degradation or mass spectrometry *(9,30)*. Tandem mass spectrometry is an additional step that not only gives the mass of a peptide but then by further analysis of the peptide can help identify phosphorylated residues *(26)*. One caveat of all these techniques is the low amount of phosphorylated peptide, compared to nonphosphorylated peptides. Thus, even after two-dimensional analysis or high-performance liquid chromatography, the peptide is usually not pure, or even a major constituent of the sample for mass spectrometry.

3.3.3. Testing the Identified Site by Mutation

The usual next step in determining phosphopeptide identity is to test a mutation of the kinase where the putative site identified above has been converted to an alanine. If there are multiple phosphorylation sites mutation of any one site is unlikely to be detected as a loss of incorporation. Depending on the reproducibility of the in vitro kinase reaction, it may be hard to see less than a 50% reduction in incorporation. The loss of a spot on the two-dimensional gel is good evidence (but not conclusive) that the site has been determined. The major caveat of this test is that the serine to alanine change could alter the conformation of the kinase and thus prevent autophosphorylation without actually being the site of autophosphorylation. An additional step of proof is the construction of a phosphopeptide antibody to the site (*see* Chapter 19). The combination of these two tests is conclusive.

3.4. Determining Role of Autophosphorylation

Autophosphorylations of PKC that are not quantitatively phosphorylated in cells do not appear to be critical for activity. Possible roles for these phosphorylations are modulation of kinase activity, modulation of lipid binding, and the modulation of protein–protein interactions (*10,12,13*).

3.4.1. Does Phosphorylation Affect Activity in Vitro?

A major problem in determining the effect of autophosphorylation is that because the molar incorporation is usually low, the overall level of kinase activity or binding will not be modified after the in vitro kinase reaction. Indeed, one needs assays in which one can directly measure the phosphorylated kinase. This can be performed if phosphorylation affects binding by comparing the binding of the phosphorylated protein to the total protein. Mutating the phosphorylated amino acid to glutamic acid or aspartatic acid to mimic phosphorylation can serve to make a kinase that acts as if it is totally phosphorylated. However, often these mutations do not mimic phosphorylation because there are important structural differences between the phosphorylated serine/threonine and either aspartic acid or glutamic acid.

3.4.2 Does Phosphorylation Affect Activity in Vivo?

Measuring changes in the in vivo activities of PKC will probably be the most convincing proof of the importance of an autophosphorylation site. Even transient sites of autophosphorylation may be important in rates of translocation that can now be measured by green fluorescent protein translocation experiments (*31–34*) using PKCs with mutations at the phosphorylation site. Phosphopeptide antibodies and immunoblotting can determine differences in the basal or activated location of the kinase and how this localization is altered by phosphorylation (*12,35*).

4. Notes

1. Some Ca^{2+}-activated or Ca^{2+}-dependent PKCs also required calcium (usually 1–300 μM). However, we have found that calcium actually inhibits activity from Ca^{2+}-independent PKCs at that concentration (*35*).
2. Vertebrate PKCs require 37°C. Many of our experiments are performed with invertebrate PKCs and they work better at room temperature.
3. The phosphorylation reaction does not occur in an aqueous media but rather on the PS vesicles, so the effective concentration of PKC on the vesicles is actually

much higher than what would be expected for soluble PKC. A better test to determine *cis*- vs *trans*-autophosphorylation is to determine whether kinase-dead mutants of PKC (e.g., catalytic lysine converted to arginine) can be phosphorylated. A positive result is definitive proof that *trans*-phosphorylation can occur; however, because of misfolding of kinase-dead PKCs, negative results should be treated with caution.

4. It is important to be able to align the radioactive band on the nitrocellulose with the autoradiogram without damaging the nitrocellulose in the process. Seal the nitrocellulose membrane in a plastic bag and affix the bag to a piece of Whatmann filter paper with magic tape such that there are no wrinkles in the bag or membrane. In order to align the film after audioradiography, label the Whatmann paper with three distinguishable radioactive ink dots such that the exposure time of the ink dots is comparable to signal from the protein of interest. The membrane is then placed in an autoradiography cassette with enhancing screens with BioMax MS film and stored for an appropriate period of time based on counts (1000 cpm - 1 d).

5. After exposing the film, affix the film to the Whatmann paper/membrane preparation with magic tape to make a "sandwich" such that the ink spots are aligned properly and trace the shape of the protein bands onto the back of the Whatmann paper. Place the bagged membrane and Whatmann paper assembly Whatmann side down on a glass plate and place the plate on top a light box. Use a clean sterile scalpel to excise the bands by cutting along the traced line visible through the bagged membrane

6. The TPCK-trypsin stock should be stored at –20°C in 0.1 mM HCL at 1 mg/mL. We use 10 mM Tris-HCl rather than 100 mM Tris-HCl because the 10-fold decrease in salt reduces streaking in the primary dimension. It is important to keep the total volume of trypsin added to a minimum in order to increase the efficiency of counts recovered. Typically, we add just enough trypsin solution to cover the top of the strips (approx 200 µL). One can also use chymotrypsin instead of trypsin. This will give a different digest pattern, and depending on the sequence of the PKC may be more instructive.

7. This volume will vary with the total number of counts in the tube. Usually we attempt to load 100–250 counts onto the plate so we would try to put the loadable 100–250 counts in 7–10 µL. For example suppose you have 1000 cpm in a given sample and want to load 1/10 of it on the TLC plate (100 cpm). You would then resuspend the total peptide count (1000 cpm) in 100 µL or 100 cpm/10 µL. You would then spot 7–10 µL of the total.

8. The set up of the plate on the Hunter apparatus is described in the manual that comes with the apparatus. However, we have found that the following detailed protocol is useful as most users have little experience with this machine. Allow both the marker and sample to dry. Prepare the Blotter and wicks from Whatmann paper. The blotter should be the thickness of two pieces of Whatmann paper taped together with 1- to 1.5-cm holes cut for the sample and marker to show through. The wicks for the Hunter-3000 apparatus should be 20 × 28 cm and

folded in half on the 28-cm length. Place approx 750 mL of pH 1.9 buffer in each electrophoresis tank of a Hunter apparatus. Place 40–50 mL of the 1.9 buffer in a clean polycarbonate tray. This buffer will be used to wet the Blotter prior to dampening the TLC plate for electrophoresis. Thoroughly wet the Blotter in pH 1.9 buffer and then gently lay the blotter over the TLC plate and allow the plate to wet by capillary action. If any area of the plate is not moist use gentle pressure with your fingers to transfer buffer to the surface of the plate. It should be noted that even pressure should be applied to the area around the holes that encircle the sample and marker application spots. This serves to concentrate the samples in a tight spot prior to electrophoresis and will maintain tight spots of radioactivity for visualization. You are now ready to start the first dimension electrophoresis. Quickly but gently place the moistened TLC plate in the Hunter apparatus with the sample origin on the left near the anode. Samples will travel toward the cathode during electrophoresis. Fold the wicks down from the buffer chamber onto each of the left and right edges of the plate. Then cover the plate gently with a piece of PVC plastic sheeting slightly larger than the TLC plate itself. Prepare the TLC sandwich by placing the Teflon cover sheet over the PVC plastic sheet and then place the black rubber pad over the Teflon sheet. Close the apparatus and secure with the two metal locking pins. Attach the inflow air tube to an air jet in the lab and adjust the pressure to 10 lb/in^2 and verify that the air bag has inflated to put pressure on the TLC plate. Next attach the wires for the anode and cathode and close the security cover of the Hunter TLC apparatus. Attach the cooling system hose to a cold water laboratory tap and turn on gentle pressure such that the system is cooled but the pressure is not so great that the hose is ejected from the tap. The final step for the first dimension is to connect the wires for the anode and cathode to a suitable power supply that will monitor voltage and preferably has a timer.

References

1. Bornancin, F. and Parker, P. J. (1996) Phosphorylation of threonine 638 critically controls the dephosphorylation and inactivation of protein kinase c-alpha. *Curr. Biol.* **6,** 1114–1123.
2. Bornancin, F. and Parker, P. J. (1997) Phosphorylation of protein kinase C-alpha on serine 657 controls the accumulation of active enzyme and contributes to its phosphatase-resistant state. *J. Biol. Chem.* **272,** 3544–3549.
3. Behn-Krappa, A. and Newton, A. C. (1999) The hydrophobic phosphorylation motif of conventional protein kinase C is regulated by autophosphorylation. *Curr. Biol.* **9,** 728–737.
4. Dutil, E. M., Keranen, L. M., DePaoli, R. A., and Newton, A. C. (1994) In vivo regulation of protein kinase C by trans-phosphorylation followed by autophosphorylation. *J. Biol. Chem.* **269,** 29,359–29,362.
5. Edwards, A. S. and Newton, A. C. (1997) Phosphorylation at conserved carboxyl-terminal hydrophobic motif regulates the catalytic and regulatory domains of protein kinase C. *J. Biol. Chem.* **272,** 18,382–18,390.

6. Edwards, A. S., Faux, M. C., Scott, J. D., and Newton, A. C. (1999) Carboxyl-terminal phosphorylation regulates the function and subcellular localization of protein kinase C beta II. *J. Biol. Chem.* **274,** 6461–6468.
7. Keranen, L. M., Dutil, E. M., and Newton, A. C. (1995). Protein kinase C is regulated in vivo by three functionally distinct phosphorylations. *Curr. Biol.* **5,** 1394–1403.
8. Tsutakawa, S. E., Medzihradszky, K. F., Flint, A. J., Burlingame, A. L., and Koshland, D. E., Jr. (1995) Determination of in vivo phosphorylation sites in protein kinase C. *J. Biol. Chem.* **270,** 26,807–26,812.
9. Flint, A. J., Paladini, R. D., and Koshland, D. E., Jr. (1990) Autophosphorylation of protein kinase C at three separated regions of its primary sequence. *Science* **249,** 408–411.
10. Orr, J. W., Keranen, L. M., and Newton, A. C. (1992) Reversible exposure of the pseudosubstrate domain of protein kinase C by phosphatidylserine and diacylglycerol. *J. Biol. Chem.* **267,** 15,263–15,266.
11. Ng, T., Squire, A., Hansra, G., Bornancin, F., Prevostel, C., Hanby, A., et al. (1999) Imaging protein kinase Calpha activation in cells. *Science* **283,** 2085–2089.
12. Nakhost, A., Dyer, J. R., Pepio, A. M., Fan, X., and Sossin, W. S. (1999) Protein kinase C phosphorylated at a conserved threonine is retained in the cytoplasm. *J. Biol. Chem.* **274,** 28,944–28,949.
13. Pepio, A. M. and Sossin, W. S. (2001) Membrane translocation of nPKCs is regulated by phosphorylation of the C2 domain. *J. Biol. Chem.* **276,** 3846–3855.
14. Sweatt, J. D., Atkins, C. M., Johnson, J., English, J. D., Roberson, E. D., Chen, S. J., et al. (1998) Protected-site phosphorylation of protein kinase C in hippocampal long-term potentiation. *J. Neurochem.* **71,** 1075–1085.
15. Byrne, J. H. and Kandel, E. R. (1996) Presynaptic facilitation revisited: state and time dependence. *J. Neurosci.* **16,** 425–435.
16. Byrne, J. H., Zwartjes, R., Homayouni, R., Critz, S. D., and Eskin, A. (1993) Roles of second messenger pathways in neuronal plasticity and in learning and memory. Insights gained from Aplysia. *Adv. Second Messenger Phosphoprotein Res.* **27,** 47–108.
17. Kruger, K. E., Sossin, W. S., Sacktor, T. C., Bergold, P. J., Beushausen, S., and Schwartz, J. H. (1991) Cloning and characterization of Ca(2+)-dependent and Ca(2+)-independent PKCs expressed in Aplysia sensory cells. *J. Neurosci.* **11,** 2303–2313.
18. Sossin, W. S., Diaz, A. R., and Schwartz, J. H. (1993) Characterization of two isoforms of protein kinase C in the nervous system of Aplysia californica. *J. Biol. Chem.* **268,** 5763–5768.
19. Sossin, W. S. and Schwartz, J. H. (1992) Selective activation of Ca(2+)-activated PKCs in Aplysia neurons by 5-HT. *J. Neurosci.* **12,** 1160–1168.
20. Hannun, Y. A., Loomis, C. R., and Bell, R. M. (1985) Activation of protein kinase C by Triton X-100 mixed micelles containing diacylglycerol and phosphatidylserine. *J. Biol. Chem.* **260,** 10,039–10,043.

21. Newton, A. C. and Koshland, D. E., Jr. (1990) Phosphatidylserine affects specificity of protein kinase C substrate phosphorylation and autophosphorylation. *Biochemistry* **29,** 6656–6661.
22. Newton, A. C. and Koshland, D. E., Jr. (1987) Protein kinase C autophosphorylates by an intrapeptide reaction. *J. Biol. Chem.* **262,** 10,185–10,188.
23. Boyle, W. J., van, der, Geer, P., and Hunter, T. (1991) Phosphopeptide mapping and phosphoamino acid analysis by two-dimensional separation on thin-layer cellulose plates. *Methods Enzymol.* **201,** 110–149.
24. Nagahara, H., Latek, R. R., Ezhevsky, S. A., and Dowdy, S. F. (1999) 2-D phosphopeptide mapping. *Methods Mol. Biol.* **112,** 271–279.
25. Resing, K. A. and Ahn, N. G. (1997) Protein phosphorylation analysis by electrospray ionization-mass spectrometry. *Methods Enzymol.* **283,** 29–44.
26. Zhou, H., Watts, J. D., and Aebersold, R. (2001) A systematic approach to the analysis of protein phosphorylation. *Nat. Biotechnol.* **19,** 375–378.
27. Shi, S. D., Hemling, M. E., Carr, S. A., Horn, D. M., Lindh, I., & McLafferty, F. W. (2001) Phosphopeptide/phosphoprotein mapping by electron capture dissociation mass spectrometry. *Anal. Chem.* **73,** 19–22.
28. Annan, R. S. and Carr, S. A. (1996) Phosphopeptide analysis by matrix-assisted laser desorption time-of-flight mass spectrometry. *Anal. Chem.* **68,** 3413–3421.
29. Carr, S. A., Huddleston, M. J., & Annan, R. S. (1996) Selective detection and sequencing of phosphopeptides at the femtomole level by mass spectrometry. *Anal. Biochem.* **239,** 180–192.
30. Quadroni, M. and James, P. (2000) Phosphopeptide analysis. *Exs* **88,** 199–213.
31. Feng, X. and Hannun, Y. A. (1998) An essential role for autophosphorylation in the dissociation of activated protein kinase C from the plasma membrane. *J. Biol. Chem.* **273,** 26,870–26,874.
32. Feng, X., Zhang, J., Barak, L. S., Meyer, T., Caron, M. G., & Hannun, Y. A. (1998) Visualization of dynamic trafficking of a protein kinase C betaII/green fluorescent protein conjugate reveals differences in G protein-coupled receptor activation and desensitization. *J. Biol. Chem.* **273,** 10,755–10,762.
33. Feng, X., Becker, K. P., Stribling, S. D., Peters, K. G., & Hannun, Y. A. (2000). Regulation of receptor-mediated protein kinase C membrane trafficking by autophosphorylation. *J. Biol. Chem.* **275,** 17,024–17,034.
34. Oancea, E. and Meyer, T. (1998) Protein kinase C as a molecular machine for decoding calcium and diacylglycerol signals. *Cell* **95,** 307–318.
35. Pepio, A. M., Fan, X., and Sossin, W. S. (1998) The role of C2 domains in Ca2+-activated and Ca2+-independent protein kinase Cs in Aplysia *J. Biol. Chem.* **273,** 19,040–19,048.

17

Tyrosine Phosphorylation of Protein Kinase C

Toshiyoshi Yamamoto, Emiko Yamauchi, Hisaaki Taniguchi, Hidenori Matsuzaki, and Ushio Kikkawa

1. Introduction

Protein kinase C (PKC) comprises a family of 10 isoforms divided into conventional, novel, and atypical groups according to the structural differences in their regulatory domains. The PKC isoforms are phosphorylated in the activation loop at a threonine residue conserved among the family. In addition, conventional and novel isoforms are phosphorylated at the two serine and/or threonine motif sites in the carboxyl-terminal end region, whereas atypical isoforms have a single threonine phosphorylation site in the carboxyl-terminal sequence *(1,2)*. Phosphorylation on these threonine and serine residues is important to render the enzyme catalytically active. Recently, most of PKC isoforms have been revealed to be phosphorylated on tyrosine in vivo *(3–9)*. Especially, the δ isoform of the novel group is phosphorylated on different tyrosine residues presumably by distinct stimuli, and the covalent modification reaction regulates its catalytic activity *(10–14)*. The tyrosine phosphorylation sites are also identified for the θ isoform of the novel group *(15)* and the λ isoform of the atypical group *(16)*. In contrast to serine and threonine phosphorylation, most of these tyrosine residues thus far detected are not conserved among the PKC family, suggesting that the tyrosine phosphorylation regulates the enzyme by a mechanism specific to each isoform rather than a common way for the whole family. Therefore, it is important to determine the tyrosine phosphorylation sites of each isoform induced by different stimuli to elucidate the control mechanism of the PKC family. For such a purpose, the phosphopeptide map method has been used to search the candidates of the phosphorylation sites. The mutated molecules are then expressed in cells to

From: *Methods in Molecular Biology, vol. 233: Protein Kinase C Protocols*
Edited by: A. C. Newton © Humana Press Inc., Totowa, NJ

confirm the phosphorylated residues. It is, however, difficult to identify the phosphorylation sites, if the molecule is phosphorylated on multiple residues on tyrosine as well as on threonine and serine residues. This chapter describes the identification of the phosphorylation sites of PKCδ isoform by mass spectrometric analysis recovered from the cells stimulated with hydrogen peroxide as an example of the advanced method for the determination of the phosphorylation sites *(14)*.

2. Materials

1. COS-7 cells cultured in Dulbecco's modified Eagle's medium supplemented with 10% fetal calf serum at 37°C in a 5% CO_2 incubator.
2. pECE expression vector of FLAG-epitope tagged PKCδ.
3. Electroporation apparatus and 0.4-cm cuvettes (Gene Pulsar, Bio-Rad; Richmond, CA).
4. Hydrogen peroxide.
5. Extraction buffer: 20 m*M* Tris-HCl, pH 7.5 containing 1 m*M* ethylenediaminetetraacetic acid (EDTA),
   1 m*M* ethylenebis(oxyethylenenitrilo)tetraacetic acid, 1 m*M* dithiothreitol, 150 m*M* NaCl, 10 m*M* NaF, 1 m*M* Na_3VO_4, 20 m*M* β-glycerophosphate, and 50 μg/mL phenylmethylsulfonyl fluoride.
6. Anti-FLAG antibody M2 affinity gel (Sigma; St. Louis, MO).
7. Elution buffer: 100 m*M* glycine-HCl, pH 3.0.
8. Lysyl endoprotease (Wako Pure Chemical Industries, Osaka).
9. Digestion buffer: 100 m*M* Tris-HCl, pH 9.0 containing 2 *M* urea.
10. Liquid chromatography/mass spectrometry instruments: the electrospray triple quadruple mass spectrometer (API-III, PE Sciex; Foster City, CA) equipped with a nanospray ion source (Protana; Odense, Demmark); a capillary high-performance liquid chromatography column (Pepmap C18, 75 μm × 150 mm, LC Packings, Amsterdam, The Netherlands) connected to a fused silica nanospray capillary (New Objective, Woburn, MA).
11. Poros R3 column (Applied Biosystems, Foster City, CA) self-packed in the pulled glass capillaries.
12. Tandem mass spectrometry instruments: the hybrid-type mass spectrometer equipped with a nanospray ion source (Micromass Q-TOF; Manchester, UK).
13. Phosphopeptides and control peptides.
14. Sf-9 cells cultured in Grace's insect medium containing yeastolate, lactoalbumin, and glutamine (GIBCO BRL; Carlsbad CA), supplemented with 10% fetal calf serum at 27°C.
15. The baculovirus vector for the expression of a fusion protein of PKCδ having glutathione S-transferase (GST) on its amino-terminal end.
16. Lysis buffer: 20 m*M* Tris-HCl, pH 7.5 containing 1 m*M* EDTA, 1 m*M* ethylenebis(oxyethylenenitrilo)tetraacetic acid, 10 m*M* 2-mercaptoethanol,

1% Triton X-100, 150 mM NaCl, 10 mM NaF, 1 mM Na$_3$VO$_4$, and 50 µg/mL phenylmethylsulfonyl fluoride.

17. Washing buffer: 20 mM Tris-HCl, pH 7.5 containing 150 mM NaCl and 1% Triton X-100.
18. Glutathione-Sepharose (Pharmacia; Piscataway, NJ).
19. Anti-FLAG antibody (Sigma).
20. Protein A-Sepharose (Pharmacia).
21. Sodium dodecyl sulfate-polyacrylamide gel electrophoresis (SDS-PAGE) and electroblot apparatuses.
22. Immobilon P membrane (Millipore; Bedford, MA).
23. Anti-PKCδ antibody (Transduction Laboratories, Lexington, KY).
24. Alkaline phosphatase-conjugated anti-rabbit or anti-mouse antibody (Promega; Madison, WI).
25. Color reaction reagents.

3. Methods

The methods described below outline (1) the recovery of tyrosine-phosphorylated PKCδ from hydrogen peroxide-treated COS-7 cells; (2) the determination of the phosphorylation sites by mass spectrometric analysis; (3) the generation of phosphorylation site-specific antibodies; and (4) immunoblot analysis of tyrosine-phosphorylated PKCδ.

3.1. Recovery of Tyrosine-Phosphorylated PKCδ

1. COS-7 cells grown in Dulbecco's modified Eagle's medium supplemented with 10% fetal calf serum were recovered, washed with phosphate-buffered saline, and suspended in phosphate-buffered saline containing 5 mM MgCl$_2$.
2. The cells (10^7 cells/cuvette) were transfected with the expression vector of FLAG-epitope tagged PKCδ (10 µg) by using Gene Pulsar (220 V, 975 µF), cultured for 48 h, and treated with 5 mM hydrogen peroxide for 10 min.
3. The following purification procedures were performed at 0–4°C. The cells were homogenized by sonication in the extraction buffer. After centrifugation for 10 min at 18,000g, the supernatant was subjected to FLAG M2 affinity column chromatography and FLAG-PKCδ was eluted by glycine-HCl buffer. The eluate is neutralized with 1 M Tris-HCl, pH 8.0, and NaF, Na$_3$VO$_4$, and β-glycerophosphate were added to the solution at the final concentrations in the extraction buffer. Approximately 1 µg of FLAG-PKCδ was purified from each 10-cm culture plate, and the sample was frozen at –80°C until use.

3.2. Determination of Phosphorylation Sites by Mass Spectrometry

1. Phosphorylation sites of PKCδ were analyzed by mass spectrometry in two steps. First, the immunoisolated protein was digested with lysyl endoprotease and the

resulting peptide mixture was analyzed on a triple-quadruple mass spectrometer operated in precursor scan mode. In this operation mode, only phosphopeptides producing a fragment ion specific for the phosphoryl group were detected and mass analyzed.

2. Second, the same peptide mixture was analyzed in a Q-TOF type mass spectrometer operated in product scan mode. In this mode, the phosphopeptides detected in the previous analysis were chosen as precursor and the fragment ions produced by the collision-induced dissociation were analyzed to determine the phosphorylation sites in the peptides.

3.2.1. Detection of Phosphopeptides in the Precursor Scan Mode

1. FLAG-PKCδ (6 μg) was digested by lysyl endoprotease (1 μg) in 100 mM Tris-HCl, pH 9.0 containing 2 M Urea at 37°C for 16 h, and 0.1% (final concentration) formic acid was added to stop the reaction.
2. Liquid chromatography/mass spectrometry experiments were performed on an API-III electrospray triple quadruple mass spectrometer equipped with a nanoelectrospray ion source *(17)*.
3. The peptide mixture was separated on a capillary high-performance liquid chromatography column by a linear gradient of H$_2$O-acetonitrile in the presence of 0.1% formic acid at a flow rate of 0.2 μL/min, and the column eluate was directly introduced into a fused silica nanospray capillary.
4. Phosphopeptides were selectively detected by the precursor-ion scanning for the phosphopeptide-specific fragment ion (PO$_3^-$, m/z 79) in the negative ion mode (**Fig. 1A,B**).

3.2.2. Determination of Phosphorylation Sites

1. The phosphopeptides identified were subjected to tandem mass spectrometric analysis to obtain sequence data. Protein digests were desalted on 1 μL of Poros R3 column self-packed in the pulled glass capillaries *(18)* and eluted directly into a nanoelectrospray needles *(19)*.
2. Tandem mass spectrometry experiments were performed in a Q-TOF-type mass spectrometer equipped with a nanoelectrospray ion source. Each phosphopeptide was selected as precursor and subjected to the collision-induced dissociation, in which sequence-specific fragment ions were produced. By comparing with the known peptide sequence, the phosphorylation sites in the peptide were unambiguously determined (**Fig. 1C**).

3.3. Generation of Phosphorylation Site-Specific Antibodies

Phosphorylation site-specific antisera were raised against the following phosphopeptides: CPETVGIpYQGFEK (amino acids 305-316), CIPDNNGT-pYGKIWEG (amino acids 325-338), and CGTPDpYIAPEILQG (amino acids 508-520), where pY indicates phosphotyrosine. These peptides have an additional cysteine residue at the amino-terminal end of the amino acid sequence of rat PKCδ. The phosphopeptides were coupled to keyhole limpet

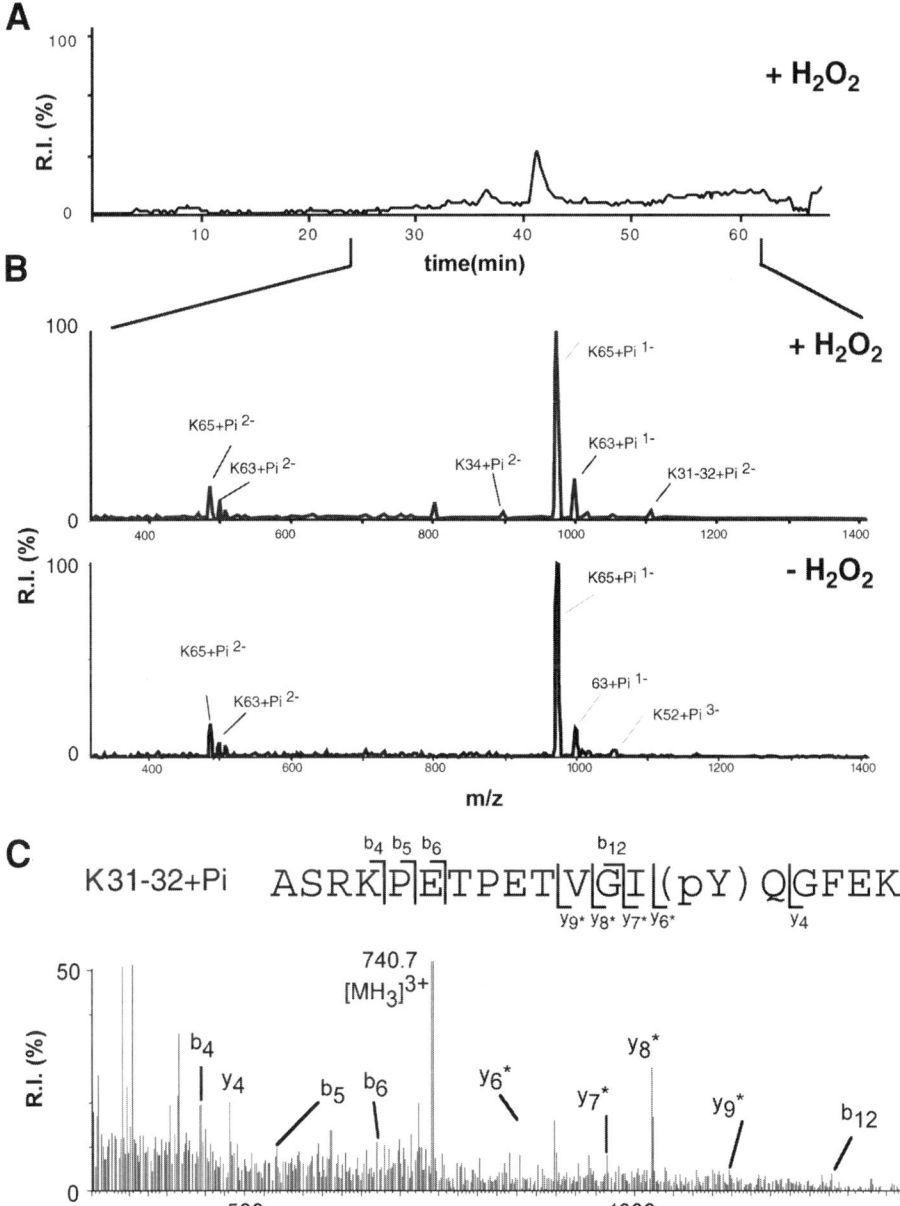

Fig. 1. Mass spectrometric analysis of phosphorylation sites. (A) Total ion chromatogram of liquid chromatography/mass spectrometry analysis operated on precursor-ion scan mode for PO_3^- (m/z –79). Data obtained with PKCδ digests prepared from cells treated with hydrogen peroxide were shown. (B) Precursor ion scan spectra of PKCδ isolated from control (upper) and treated (lower) cells. All scans containing significant peaks in the liquid chromatography/mass spectrometry experiment were combined. (C) Product ion spectrum of the phosphotyrosine containing peptide K31-32 (Arg298-Lys316) (m/z 740, 3+). Singly charged y6 to y9 ions carrying the phosphoryl group (marked with an asterisk) and y4 ion allow identification of Tyr311 as the phosphorylation site.

hemocyanin and used to immunize rabbits, and the phosphorylation site-specific antibodies were purified from antisera by successive column chromatographies using affinity resins coupled with each phosphopeptide and non-phosphopeptide essentially as described *(20,21)*. The resulting antibodies were designated as anti-pY311, anti-pY332, and anti-pY512, respectively. Where indicated, the antibodies were adsorbed with each phosphopeptide at 1 mg/mL at 4°C for 1 h.

3.4. Immunoblot Analysis of Tyrosine-Phosphorylated PKCδ

The tyrosine phosphorylation sites of PKCδ were confirmed by immunoblot analysis by using the phosphorylation site-specific antibodies described above. Tyrosine phosphorylation of the enzyme was first analyzed for GST-fusion protein of PKCδ expressed in Sf-9 cells, because a much amount of tyrosine-phosphorylated protein was obtained from insect cells treated with hydrogen peroxide. Tyrosine phosphorylation of FLAG-PKCδ recovered from COS-7 cells was then explored.

3.4.1. Immunoblot Analysis of GST-PKCδ from Sf-9 Cells

1. The baculovirus for the expression of a fusion protein of PKCδ having GST on its amino-terminal end was constructed by using pAcGHLT transfer vector (Pharmingen) and by the recombination into *Autographa californica* nuclear polyhedrosis virus using Bacvector 2000 transfection kit (Novagen) as described and was purified and amplified according to the manufacturer's protocol.
2. Sf-9 cells were infected with the recombinant virus and cultured for 48 h.
3. After the treatment with hydrogen peroxide, the GST-fusion protein was purified by affinity chromatography on glutathione-Sepharose.
4. The samples were subjected to SDS-PAGE and the proteins were transferred onto an Immobilon P membrane.
5. Immunoblot analysis was carried out using with anti-PKCδ and phosphorylation site-specific antibodies as the primary antibodies and alkaline phosphatase-conjugated anti-rabbit or anti-mouse antibodies as the secondary antibodies (**Fig. 2**).
6. GST-PKCδ was recognized by these phosphorylation site-specific antibodies and the immunoreaction was diminished by the adsorption of the antibodies by each phosphopeptide.

3.4.2. Immunoblot Analysis of FLAG-PKCδ from COS-7 Cells

FLAG-PKCδ was immunoprecipitated from the lysate of COS-7 cells. The following procedures were carried out at 0–4°C.

1. The cells were washed with phosphate-buffered saline and lysed in the lysis buffer.

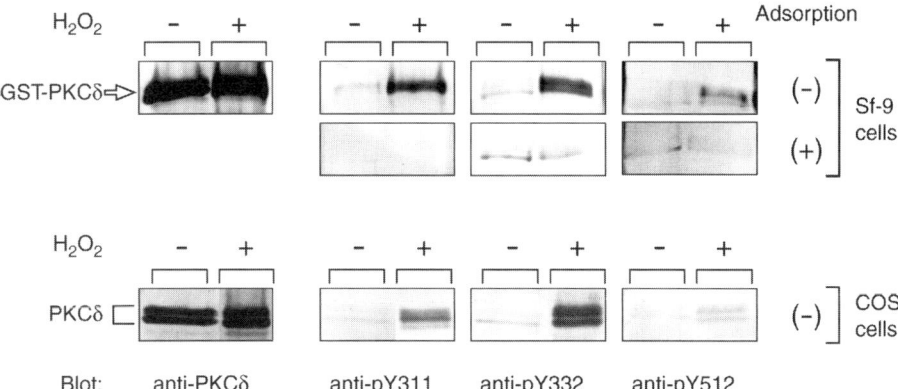

Fig. 2. Immunoblot analysis of PKCδ by the phosphorylation site-specific antibodies. Sf-9 and COS-7 cells expressing GST-PKCδ and FLAG-PKCδ, respectively, were cultured in the medium containing 0.1% bovine serum albumin instead of fetal calf serum for 18 h and then treated with hydrogen peroxide. The enzymes isolated were subjected to immunoblot analysis using each primary antibody. Where indicated, the antiphosphopeptide antibodies adsorbed with each phosphopeptide were used.

2. The lysate was centrifuged, and the supernatant was incubated for 1 h with an anti-FLAG monoclonal antibody.
3. Then, protein A-Sepharose was added to the mixture and incubated for 30 min.
4. The immunoprecipitates were collected by centrifugation and were washed four times each time with the washing buffer.
5. The samples were subjected to SDS-PAGE and verified by immunoblot analysis as described in **Subheading 3.4.1. (Fig. 2)**.
6. Tyrosine phosphorylation of the three residues was confirmed by immunoblot analysis with the anti-phosphotyrosine antibody.

4. Notes

1. Two tyrosine residues of Tyr311 and Tyr332 in K31-32 and K34, respectively, were identified as the phosphorylation sites by mass spectrometric analysis. These results are consistent with those of phosphopeptide map study. Namely, a fragment with the approximate molecular mass of 45 kDa had been detected by the antibody against phosphotyrosine as well as the antibody recognizing the carboxyl-terminal end of PKCδ after partial tryptic digestion of the enzyme recovered from the hydrogen peroxide-treated cells *(9)*. The 45-kDa fragment is large enough to include Tyr311 and Tyr332 because the calculated molecular mass of the peptide Arg298-Glu673 is 43.6 kDa, which ranges from K31-32 (Arg298-Lys316) to the carboxyl-terminal end of the enzyme.

2. Another tyrosine-phosphorylated fragment with the approximate molecular mass of 25 kDa had been found by phosphopeptide map study *(9)*, which should be phosphorylated on tyrosine other than Tyr311 and Tyr332 within the carboxyl-terminal end region. Among the phosphopeptides of K51, K63, and K65 detected by mass spectrometric analysis, Tyr512 in K51 is only one tyrosine residue. The calculated molecular mass of the peptide, which covers from K51 (Glu495-Lys522) to the carboxyl-terminal end of the enzyme, is 21.0 kDa. Therefore, Tyr512 is a candidate for the phosphorylation site.

3. The phosphorylation site-specific antisera were raised against synthetic phospho-peptides, including each tyrosine phosphorylation site such as Tyr311, Tyr332, and Tyr512. Hydrogen peroxide treatment induced tyrosine phosphorylation of PKCδ in Sf-9 cells, and thus the phosphorylation site-specific antibodies were first calibrated by using the enzyme recovered from the insect cells. It is useful to employ the sample prepared from Sf-9 cells especially for Tyr512, a trace phosphorylation site in mammalian cells.

4. Presently, it is ideal to determine the tyrosine phosphorylation sites directly by mass spectrometric analysis and to confirm them by immunoblot analysis using the phosphorylation site-specific antibodies. It is helpful to examine the mutated molecule replacing the phosphorylation sites *(14)*, although all of the multiple phosphorylation sites need to be determined before the construction of the mutant molecule.

Acknowledgments

This study was supported in part by research grants from the Scientific Research Funds of the Ministry of Education, Science, Sports and Culture of Japan and the Special Coordination Funds for Promoting Science and Technology of STA of Japan.

References

1. Parekh, D. B., Ziegler, W., and Parker, P. J. (2000) Multiple pathways control protein kinase C phosphorylation. *EMBO J.* **19,** 496–503.
2. Newton, A. C. (2001) Protein kinase C: structural and spatial regulation by phosphorylation, cofactors, and macromolecular interactions. *Chem. Rev.* **101,** 2353–2364.
3. Denning, M. F., Dlugosz, A. A., Howett, M. K., and Yuspa, S. H. (1993) Expression of an oncogenic rasHa gene in murine keratinocytes induces tyrosine phosphorylation and reduced activity of protein kinase C δ. *J. Biol. Chem.* **268,** 26079–26081.
4. Li, W., Mischak, H., Yu, J.-C., Wang, L.-M., Mushinski, J. F., Heidaran, M. A., et al. (1994) Tyrosine phosphorylation of protein kinase C-δ in response to its activation. *J. Biol. Chem.* **269,** 2349–2352.
5. Liu, F., and Roth, R. A. (1994) Insulin-stimulated tyrosine phosphorylation of protein kinase C α: evidence for direct interaction of the insulin receptor and protein kinase C in cells. *Biochem. Biophys. Res. Commun.* **200,** 1570–1577.

6. Gschwendt, M., Kielbassa, K., Kittstein, W., and Marks, F. (1994) Tyrosine phosphorylation and stimulation of protein kinase C δ from porcine spleen by src in vitro. Dependence on the activated state of protein kinase C δ. *FEBS Lett.* **347,** 85–89.

7. Haleen-Smith, H., Chang, E.-Y., Szallasi, Z., Blumberg, P. M., and Rivera, J. (1995) Tyrosine phosphorylation of protein kinase C-δ in response to the activation of the high-affinity receptor for immunoglobulin E modifies its substrate recognition. *Proc. Natl. Acad. Sci. USA* **92,** 9112–9116.

8. Soltoff, S. P., and Toker, A. (1995) Carbachol, substance P, and phorbol ester promote the tyrosine phosphorylation of protein kinase C δ in salivary gland epithelial cells. *J. Biol. Chem.* **270,** 13490–13495.

9. Konishi, H., Tanaka, M., Takemura, Y., Matsuzaki, H., Ono, Y., Kikkawa, U., et al. (1997) Activation of protein kinase C by tyrosine phosphorylation in response to H2O2. *Proc. Natl. Acad. Sci. USA* **94,** 11233–11237.

10. Szallasi, Z., Denning, M. F., Chang, E.-Y., Rivera, J., Yuspa, S. H., Lehel, C., et al. (1995) Development of a rapid approach to identification of tyrosine phosphorylation sites: application to PKCδ phosphorylated upon activation of the high affinity receptor for IgE in rat basophilic leukemia cells. *Biochem. Biophys. Res. Commun.* **214,** 888–894.

11. Li, W., Chen, X.-H., Kelley, C. A., Alimandi, M., Zhang, J., Chen, Q., et al. (1996) Identification of tyrosine 187 as a protein kinase C-δ phosphorylation site. *J. Biol. Chem.* **271,** 26404–26409.

12. Brodie, C., Bogi, K., Acs, P., Lorenzo, P. S., Baskin, L., and Blumberg, P. M. (1998) Protein kinase C δ (PKCδ) inhibits the expression of glutamine synthetase in glial cells via the PKCδ regulatory domain and its tyrosine phosphorylation. *J. Biol. Chem.* **273,** 30713–30718.

13. Blake, R. A., Garcia-Paramio, P., Parker, P. J., and Courtneidge, S. A. (1999) Src promotes PKCδ degradation. *Cell Growth Differ.* **10,** 231–241.

14. Konishi, H., Yamauchi, E., Taniguchi, H., Yamamoto, T., Matsuzaki, H., Takemura, Y., et al. (2001) Phosphorylation sites of protein kinase C δ in H_2O_2-treated cells and its activation by tyrosine kinase in vitro. *Proc. Natl. Acad. Sci. USA* **98,** 6587–6592.

15. Liu, Y., Witte, S., Liu, Y. C., Doyle, M., Elly, C., and Altman, A. (2000) Regulation of protein kinase Cθ function during T cell activation by Lck-mediated tyrosine phosphorylation. *J. Biol. Chem.* **275,** 3603–3609.

16. Wooten, M. W., Vandenplas, M. L., Seibenhener, M. L., Geetha, T., and Diaz-Meco, M. T. (2001) Nerve growth factor stimulates multisite tyrosine phosphorylation and activation of the atypical protein kinase C's via a src kinase pathway. *Mol. Cell. Biol.* **21,** 8414–8427.

17. Taniguchi, H., Manenti, S., Suzuki, M., and Titani, K. (1994) Myristoylated alanine-rich C kinase substrate (MARCKS), a major protein kinase C substrate, is an in vivo substrate of proline-directed protein kinase(s). A mass spectroscopic analysis of the post-translational modifications. *J. Biol. Chem.* **269,** 18299–18302.

18. Neubauer, G. and Mann, M. (1999) Mapping of phosphorylation sites of gel-isolated proteins by nanoelectrospray tandem mass spectrometry. *Anal. Chem.* **71,** 235–242.

19. Wilm, M. and Mann, M. (1996) Analytical properties of the nanoelectrospray ion source. *Anal. Chem.* **68,** 1–8.

20. Nishizawa, K., Yano, T., Shibata, M., Ando, S., Saga, S., Takahashi, T., et al. (1991) Specific localization of phosphointermediate filament protein in the constricted area of dividing cells. *J. Biol. Chem.* **266,** 3074–3079.

21. Czernik, A. J., Girault, J. A., Nairn, A. C., Chen, J., Snyder, G., Kebabian, J., et al. (1991) Production of phosphorylation state-specific antibodies. *Methods Enzymol.* **201,** 264–283.

18

Methods to Study Dephosphorylation of Protein Kinase C In Vivo

Karen England and Martin Rumsby

1. Introduction

The phosphorylation of protein kinase C (PKC) in vivo has been well documented and is known to be important in the maturation of the enzyme to a fully functional form localized correctly within the cell *(1–7)*. The subsequent dephosphorylation of PKC, however, has not been well studied, especially in vivo. PKC is phosphorylated, typically on three sites—the activation loop site, the turn motif, and the hydrophobic region *(1–7)*. Each of these phosphorylations has different effects on the properties of the enzyme as defined by in vitro studies. Phosphorylation of the activation loop site seems to be vital for enzyme activity *(1–3,5,8)* whereas phosphorylation of the other sites is not essential for catalytic activity but, instead, seem to affect the stability of the enzyme *(1–3,9–13)*. Because of the distinct effects of phosphorylation of PKC at different sites, it is important to look at the dephosphorylation of each site individually when studying dephosphorylation of PKC in vivo.

Our studies concentrate on the dephosphorylation of PKCε at site Ser729, the hydrophobic site (*see* Chapter 13), upon passage of fibroblast cells *(14,15)*. This chapter will detail methods involved in this study, including identification of site-specific dephosphorylation of PKC, measurement of its activity, the study of localization by cell fractionation, and Western blotting. Finally, the use of chemical inhibitors to identify pathways important in the dephosphorylation of PKCε on cell passage is described.

From: *Methods in Molecular Biology, vol. 233: Protein Kinase C Protocols*
Edited by: A. C. Newton © Humana Press Inc., Totowa, NJ

2. Materials

2.1. Equipment

2.1.1. Sample Preparation and Western Blotting

1. Sodium dodecyl sulfate-polyacrylamide gel electrophoresis (SDS-PAGE) and Western blotting equipment; we used the Bio-Rad mini-protean II system.
2. Probe sonicator for shearing DNA; if one is not available, DNA in total cell lysates can be sheared by passing through a 26-gauge needle.
3. Benchtop ultracentrifuge for cell fractionation.
4. Freeze dryer for cell fractionation.
5. Rocking table.
6. Heating block.
7. Rotating stirrer for immunoprecipitation.

2.1.2. Matrix-Assisted Laser Desorption/Ionization Time-of-Flight (MALDI-TOF) Mass Spectrometry

1. Speed vac.
2. MALDI-TOF mass spectrometer. (We used a Voyager DeStr from Applied Biosystems and the associated Data Explorer software.)
3. Internet access for database searching. From the ExPasy site, we used the Peptide mass and PeptIdent software (http://www.expasy.ch). We also used the protein prospector MS FIT software (http://www.prospector.ucsf.edu).

2.1.3. (^{35}S)-Methionine (Met) Labeling and PKC Activity Assay

1. Gel dryer.
2. Scintillation counter.

2.2. Reagents

2.2.1. Western Blotting and Cell Fractionation and Immunoprecipitation

1. Lysis buffer: 50 mM Tris-HCl, 0.5 mM ethylenediaminetetraacetic acid, 2 mM ethylenebis(oxyethylenenitrilo)tetraacetic acid (EGTA), pH 7.5, with 0.5% NP40, protease inhibitors: 10 µg/mL aprotinin, 10 µg/mL leupeptin, 2 mM 4-(2-Aminoethyl)benzenesulfonyl-fluoride-HCL (AEBSF), phosphatase inhibitors; 50 mM sodium fluoride, 5 mM sodium pyrophosphate, 10 µM sodium ortho-vanadate. This lysis buffer can be made and stored in the fridge and the inhibitors added fresh before use. Phosphatase inhibitors can be made up, aliquoted and frozen. Alternatively commercial phosphatase inhibitor mixes are available.
2. Homogenization buffer: 20 mM Tris-HCl, 5 mM EGTA, pH 7.5. Add protease and phosphatase inhibitors as before.
3. Laemmli protein loading buffer; 3.8 mL of distilled water, 1.0 mL of 0.5 M Tris-HCl, pH 6.8, 0.8, mL of glycerol, 1.6 mL of 10% (w/v) SDS, 0.4 mL of 2-β mercaptoethanol, 0.4 mL of 0.05% (w/v) Bromophenol blue, total volume 8 mL.

4. SDS-PAGE running buffer: 10 g of SDS, 58.2 g of Tris-HCl, and 148.6 g of glycine in 10 liters.
5. Western Blot Transfer buffer: 58.2 g of Tris-HCl and 29.3 g of glycine in 10 liters.
6. Ponceau S Stain solution (Sigma).
7. 0.5% Tween 20 in Tris-HCl buffered saline (TBST; 24.2 g of Tris-HCl, 80 g of NaCl in 10 liters).
8. Antibodies (*see* **Note 1**). The polyclonal PKCε antibody used for Western blotting and immunoprecipitations was generated to the C-terminal peptide sequence by Professor N Groome (Oxford Brookes University, UK). A commercially available polyclonal antibody to PKCε (Sigma) was also used for these purposes. The anti-phospho-Ser729-PKCε was from Upstate Biotechnology (Lake Placid, NY). The anti-phospho-Thr566-PKCε was a kind gift from Davey Parekh and Peter Parker, ICRF (London, UK). Antiphosphothreonine and antiphosphoserine antibodies were from Zymed (San Francisco, CA). The antiphosphotyrosine antibody was from Sigma. Peroxidase- and FITC-conjugated secondary antibodies were from Sigma.
9. Protein A Sepharose (PAS; Sigma product code P3391).

2.2.2. In Vitro Dephosphorylation

1. Alkaline phosphatase buffer: 50 mM Tris-HCl, pH 9.0, 1 mM MgCl$_2$, 20 mg/mL leupeptin, 2 µg/mL aprotinin.
2. Lambda phosphatase: 50 mM Tris-HCl, pH 7.8, 5 mM dithiothreitol, 2 mM MnCl$_2$, 100 µg/mL bovine serum albumin (BSA).

2.2.3. MALDI-TOF m/s

1. All chemicals must be of a high (mass spectrometry/HPLC) grade here.
2. Destain solution: 50% acetonitrile in 25 mM ammonium bicarbonate.
3. Trypsin solution: 0.1 µg/µL trypsin, sequencing grade, in 25 mM ammonium bicarbonate, 5 mM CaCl.
4. Peptide elution solution: 50% acetonitrile: 5% trifluoroacetic acid.
5. Matrix: 10 mg/mL α cyano-4-hydroxy-cinnamic acid in 50% acetonitrile, 5% trifluoroacetic acid.

2.2.4. [^{35}S]-Met Labeling of Cells

1. Methionine-free Dulbecco's modified eagle's medium (DMEM; Gibco-BRL).
2. [^{35}S]-Met/Cysteine (Cys; Promix, Amersham Radiochemicals, UK).
3. 25/20/65(v/v/v) isopropanol/acetic acid/H$_2$0 for 1 h.
4. Amplify (Amersham, UK).

2.2.5. PKC Activity Assay

1. PKC activity assay buffer: 20 mM Tris-HCl, pH 7.5, 25 mM MgAc, 2.5 mM ethylenediaminetetraacetic acid, 50 µM myelin basic protein peptide$_{4-14}$, 20 µg of phosphatidyl serine, 0.4 µg of diolein.

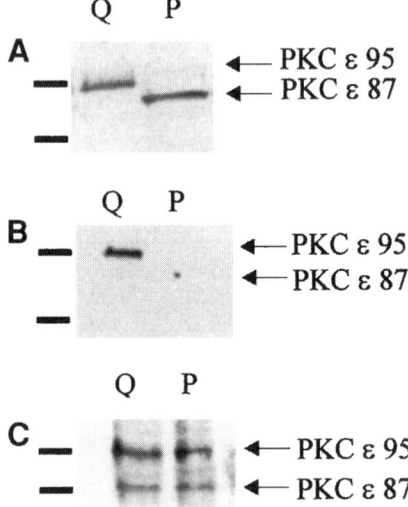

Fig. 1. Detection of PKCε dephosphorylation upon passage by Western blotting. 3T3 cells were grown to confluency and total protein harvested (Q) or were passaged into new media and flasks for 15 minutes before harvesting adherent cells (P). (A) Immunoblotting was performed for PKCε as described in **Subheading 3.1.5.** Markers shown are at 97 and 79 kDa. (B) Immunoblotting was performed for PKCε with a phospho Ser729 specific antibody as described in **Subheading 2.2.1.** Markers shown are at 97 and 79 kDa. (C) Immunoblotting was performed for PKCε with a phospho Thr566 specific antibody as described in **Subheading 2.2.1.** Markers shown are at 97 and 79 kDa.

2. [γ-³³P]ATP.
3. PKC inhibitor peptide: RFAEKGSLRQKNV.
4. Laemmli protein loading buffer.

3. Methods

3.1. Identification of Site-Specific Dephosphorylation in Vivo

Our studies of PKCε in fibroblast cells show that PKC can be dephosphorylated at one site independently of the others *(14,15)*. Upon passage, PKCε loses its phosphate at Ser729, the hydrophobic domain, whereas phosphorylation at the other two sites remains intact. This change in dephosphorylation was initially detected by Western blotting as a change in mobility of the protein from 95 to 87 kDa using a general PKCε antibody (**Fig. 1A**) (*see* **Note 1**). Using a combination of general PKCε antibodies, a phospho-Ser729-specific antibody and a phospho-Thr566 we were able to show that PKCε 95 and 87 were phosphorylated at Thr566 but that only PKCε 95 was phosphorylated

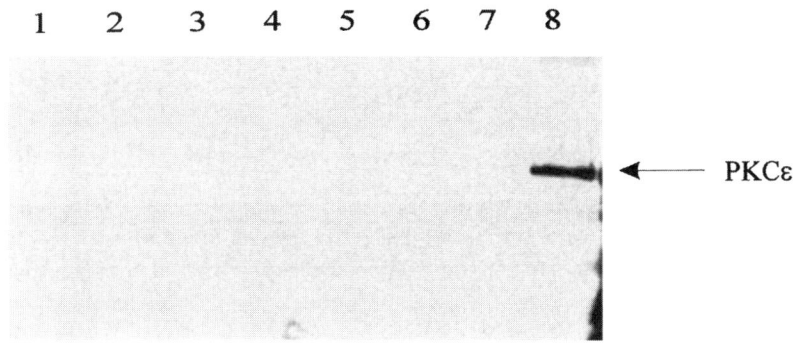

Fig. 2. PKCε is not synthesized in 3T3 cells until 72 h after passage. A flourogram of PKCε immunoprecipitated from [^{35}S]Met-labeled 3T3 cells at lane 1, 15 min after passage; lane 2, 30 mins after passage; lane 3, 1 h after passage; lane 4, 4 h after passage; lane 5, 8 h after passage; lane 6, 12 h after passage; lane 7, 24 h after passage; lane 8, 48 h after passage. PKCε is indicated.

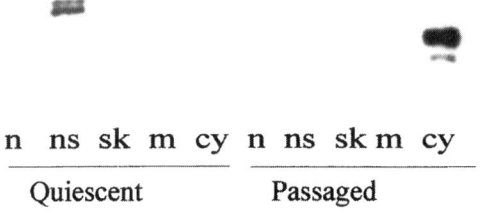

Fig. 3. PKCε87 and PKCε95 are differentially localized within the cell. Total cell protein from quiescent (Q) and passaged (P) 3T3 cells was fractionated into nuclear (n), nuclear scaffold (ns), cytoskeleton (sk), postnuclear membrane (m), and cytoplasm (c). Western blotting was performed for PKCε as described in **Subheading 3.** Markers shown are at 97 and 79 kDa.

at Ser729 (**Fig. 1B,C**). We also probed Western blots of immunoprecipitates of PKCε from quiescent and newly passaged cells with antiphospho-Ser and Phospho-Thr antibodies to show that both forms of PKC contained phosphoserine and phosphothreonine (*see* **Note 2**; **ref. *15***). MALDI-TOF m/s fingerprinting was used to confirm that PKCε 95 is phosphorylated at both Ser703 and Ser729, whereas PKCε 87 is not phosphorylated at Ser729 (*see* **Table 1**). [^{35}S]Met labeling was used to confirm that the less-phosphorylated form of PKCε is a product of dephosphorylation rather than newly synthesized protein, which has not yet been fully phosphorylated (**Fig. 2**). This change in the mobility of the protein was also associated with a change in its localization; PKCε 95, the triphosphorylated form of the protein, is localized to the perinuclear region as detected by cell fractionation (**Fig. 3**). In this perinuclear region of the cell, the appropriate PKC co-factors phosphatidyl serine and diacylglycerol are found and so it is likely that PKC could be active here. PKCε 87, however, was found in the cytoplasm.

3.1.1. Cell Passage to Induce Dephosphorylation of PKCε

1. Maintain 3T3 and 3T6 cells in DMEM supplemented with 10% fetal calf serum in a humidified incubator at 5% CO_2, 37°C. Grow cells to quiescence by allowing them to grow to confluency and leaving them for 2–3 d or by serum starving for 24 h. For most experiments, one 25-cm² tissue culture flask of confluent fibroblasts is sufficient.
2. Passage by rinsing with Tris-HCl saline, followed by release of cells with Tris-HCl Trypsin.
3. Resuspend cells in fresh medium and transfer into fresh tissue culture flasks and allowed to settle for 15 min before harvesting.

3.1.2.Whole Cell Lysate Preparation (see **Note 3**)

1. Work with the tissue culture flasks on ice to reduce any decrease in dephosphorylation. Rinse the cell monolayer three times with phosphate-buffered saline (PBS) containing a cocktail of phosphatase inhibitors.
2. Add 100 μL lysis buffer per 6-cm² cell culture dish and scrape cells into buffer; transfer to an Eppendorf tube.
3. Shear the DNA, using a probe sonicator or alternatively shear DNA through a 26-gauge needle and determine protein concentration (*see* **Note 4**).

3.1.3. Cell Fractionation (see **Note 5**)

3T3 and 3T6 fibroblasts were fractionated based on the method by Ohmori et al. (*16*).

1. Harvest cells in homogenization buffer.
2. Pass the homogenate through a 26-gauge needle several times and determine protein concentration by BCA assay (Pierce). Use equal amounts of total cell protein for fractionation.
3. Centrifuge at 800g for 10 min; the pellet is the nuclear fraction. The supernatant represents the postnuclear fraction.
4. Treat the nuclear pellet with 1% TX-100 in homogenization buffer for 30 min on ice and centrifuge at 100,000g in a benchtop ultracentrifugefor 20 min. The detergent-insoluble fraction is the nuclear scaffold, whereas the detergent-soluble fraction contains the nuclear membrane.
5. Centrifuge the postnuclear supernatant for 20 min at 100,000g as above. The supernatant is the cytosolic fraction. The resulting pellet was treated with TX-100 as above and recentrifuged as above. The detergent-insoluble fraction represents the cytoskeleton and the soluble fraction is the postnuclear membrane.

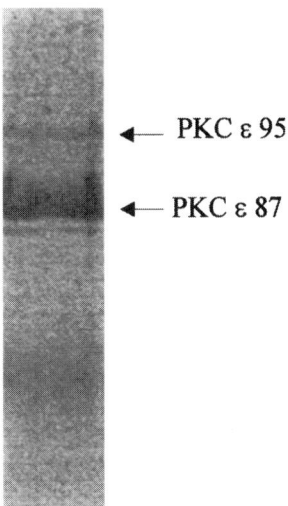

Fig. 4. Immunoprecipitation of PKCε. PKCε was immunoprecipitated from quiescent 3T3 cells and resolved on a 7% SDS-PAGE and the gel was stained with Coomassie blue. PKCε is indicated.

6. Lyophilize all samples in a freeze dryer overnight and rehydrate in 100 μL of lysis buffer by sonication.
7. Resolve 20-μL aliquots of each fraction by SDS-PAGE for Western blotting.

3.1.4. PKC Immunoprecipitation

This method can be used to obtain differentially phosphorylated forms of PKC for MALDI-TOF m/s, activity assays and analysis with anti phospho-Ser/Thr/Tyr antibodies.

1. Extract total cell protein or protein from the relevant cell fraction, in lysis buffer as above (*see* **Notes 6** and **7**).
2. Preclear the lysate with 50 μL of PAS on a rotating stirrer for 1 h at 4°C.
3. Pellet the PAS in a bench top centrifuge at full speed for 3 min and transfer the supernatant to a new tube incubate with 50 μL of PKC-specific antibody at 4°C on mixer overnight. Add 50 μL of PAS for 2 h on the stirrer.
4. Pellet the PAS as above and wash three times with lysis buffer.
5. Resuspend in Laemmli loading buffer and boil for 8 min. (If you are immunoprecipitating before assaying, do not perform this step but resuspend in assay buffer).
6. Pellet down PAS beads in a bench top centrifuge at full speed for 3 min.
7. Load on a SDS-PAGE or 2D gel (*see* **Fig. 4**).

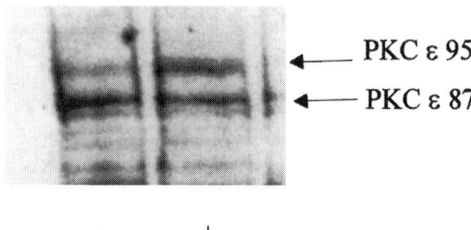

Fig. 5. In vitro dephosphorylation of PKCε. Total cell protein from quiescent 3T3 cells was incubated with alkaline phosphatase for 30 min +/– phosphatase inhibitors before analysis for PKCε by western blotting. Markers shown are at 97 and 79 kDa.

3.1.5. Western Blotting

1. Boil equal amounts of protein (30 μg) in Laemmli loading buffer for 8 min before separation on 10% SDS-PAGE gels.
2. Transfer protein to nitro-cellulose at 250 mA for 1 h 20 min in the Mini Protean II transfer unit. Monitor transfer efficiency and equal loading before antibody probing using Ponceau S. Stain the membrane by incubating in Ponceau for 5 min, then wash in TBST until the blot is clean.
3. Block membranes in 5% non-fat dried milk in TBST for 1 h on a rocking table, incubate overnight in 1/500 of the required PKCε antibody in 1% non-fat dried milk TBST in a cold room on a rocking table. Wash two times in TBST for 5 min, 20 min in 0.5 M NaCl, and two times in TBST for 5 min. Then incubate for 1 h with 1/5000 secondary antibody and detect PKC using ECL (Pierce).

3.2. In Vitro Dephosphorylation of PKCε

To confirm that the changes in mobility of PKCε on SDS-PAGE were caused by dephosphorylation, protein was extracted from quiescent cells (containing the fully phosphorylated form of the protein) and exposed to alkaline phosphatase or lambda phosphatase. The sample was resolved by SDS-PAGE and probed with anti-PKC antibodies (**Fig. 5**). This showed a decrease in PKCε 95 and an increase in PKCε 87, showing that PKCε 87 production can be caused by dephosphorylation. This method also produces dephosphorylated forms of PKC, which can be used as controls, especially for MALDI TOF m/s fingerprinting.

3.2.1. For Dephosphorylation with Alkaline Phosphatase

1. Harvest total cell protein (*see* **Note 8**) into alkaline phosphatase buffer and pass through a 26-gauge needle approximately five times.
2. Add 30 units of alkaline phosphatase (Sigma) +/–10 mM sodium pyrophosphate as a control.

3. Incubate samples at 37°C for 2 h before dissolving in Laemmli loading buffer and analyzing for PKCε by Western blotting.

3.2.2. For Lambda Phosphatase Treatment

1. Harvest total cell protein in lambda phosphatase buffer and pass through a 26-gauge needle approximately five times.
2. Add 200 units λ phosphatase (Calbiochem) +/–200 m*M* EGTA and 50 m*M* NaF as a control. Incubate at 30°C with for 30 min before dissolving in Laemmli loading buffer and analyzing for PKCε by Western blotting.

3.3. MALDI-TOF Mass Spectrometry to Identify Phosphorylation Sites

This method complimented the antibody studies and also allowed us to identify the phosphorylation status at Ser703 for which we had no phospho-specific antibody (*see* **Note 9**). It also allowed us to prove that the change in PKC mobility on the Western blot was caused by cleavage of PKC because peptides that covered the N and C terminals of the protein were detected.

3.3.1. Extraction of PKC for MALDI-TOF Mass Spectrometry

Protein digestion method is based on that of Clauser and Andrews (http:\\donatello.ucsf.edu/ingel.ht) and Rosenfeld et al. *(17)*.

1. Resolve PKCε immunoprecipitates (see **Subheading 3.1.4.**) on a 7% SDS-PAGE gel (*see* **Note 10**).
2. Stain the gel with Coomassie blue stain and excise the PKC bands of interest with a clean razor blade (*see* **Note 11**)
3. Place gel pieces in Eppendorf tubes in sufficient destain solution to cover them and either homogenise or chop finely with a razor.
4. Remove destain solution with a gel loading tip (this helps avoid removal of the gel pieces) and replace with fresh solution. Repeat until the gel pieces appear white and have shrunk in appearance, typically three washes.
5. Remove excess solution and dry down gel pellets in a Speed Vac (30 min on low works well for most samples).
6. Resuspend the gel pieces in just enough trypsin solution to cover them. Incubate overnight at 37°C. Do not allow the pellets to dry out.
7. Add 50 μL of high-performance liquid chromatography-grade water to each tube, vortex, and centrifuge briefly in a benchtop centrifuge. Use a gel-loading tip to transfer the supernatant to a fresh tube.
8. Add 50 μL of elution buffer, vortex, and centrifuge. Transfer the supernatant into the same tube as above. The same gel-loading tip can be used to reduce loss of peptide. Repeat until the gel pieces are white again.
9. Reduce the volume of the supernatant to about 10 μL in a speed vac. Add 50μL of high-performance liquid chromatography-grade water and reduce volume again.

Table 1
MALDI-TOF mg Detection of PKCε Phosphopeptides

Phosphorylation site	Peptide	Enzyme	M/z	Detected in PKCε 87	Detected in PKCε 95
Ser-703	678-724	Glu C	5737.7	Yes	Yes
	693-704	Trypsin	1604.7	Yes	Yes
	693-717	Trypsin	3157.3	Yes	Yes
Ser-729	727-737	Trypsin	1358.6	No	Yes
	719-737	Trypsin	2361.7	No	Yes

10. Further removal of salt can be achieved by passing the sample through a C18 ZipTip (Millipore, Bedford, MA).
11. Mix 1 µL of sample with 1 µL of matrix and spot onto a MALDI-TOF m/s sample plate.
12. Run sample on MALDI-TOF m/s from set conditions using a suitable calibration. Note that negative mode may be better to visualize phosphopeptides.
13. Phosphorylation of a peptide will increase its mass by 80. Therefore, by comparing the predicted peptide mass fingerprint for PKC (generated by PeptideMass available at ExPasy) with the peptide mass fingerprints obtained, any peptide of a mass 80 greater then predicted are potentially phosphopeptides. If the same peptides are not present in samples treated with phosphatase and the dephosphorylated peptide mass is detected we can be confident that the peptide is phosphorylated, see **Table 1** below (*see* **Note 12**).

3.4. [³⁵S]-Met Incorporation Studies

Multiplied phosphorylated forms of PKC within the cell may be derived in two alternative forms. They may be immature proteins that have not as yet been fully phosphorylated or they may be dephosphorylated forms of the protein. One way to distinguish between the two possibilities is to label newly synthesized protein with [³⁵S]Met and determine whether newly phosphorylated protein is contained in the partially phosphorylated form of PKC. We therefore monitored PKCε synthesis from quiescent cells, from cells 15 min after passage and from cells growing to quiescence again. We demonstrated that PKCε was synthesized only at 48 h after passage. Therefore, the change in mobility seen upon passage was caused by dephosphorylation and not new synthesis of PKC. Because PKCε 95 reappears 1 h after passage it is likely that PKCε is rephosphorylated because there is still no new protein synthesis at this time point.

1. Incubate cells in methionine-free DMEM (Gibco-BRL) for 1 h. Add 1 MBq of [³⁵S]Met/Cys (Promix, Amersham Radiochemicals, UK) for 2 h.

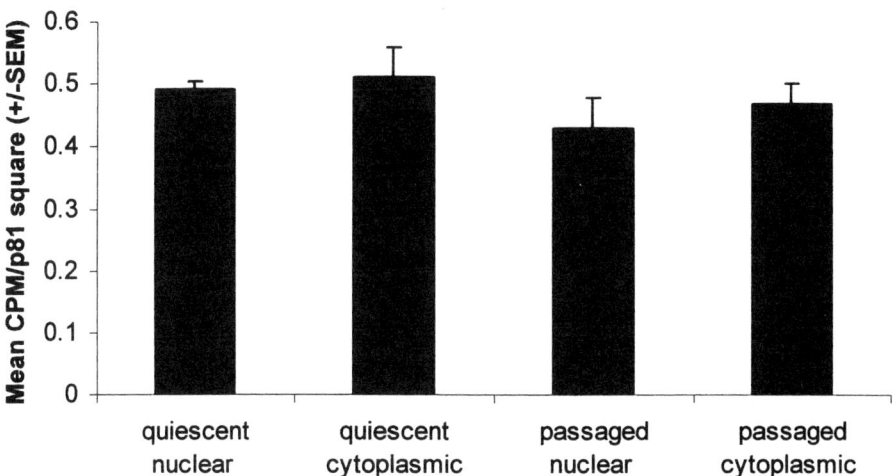

Fig. 6. Both PKCε87 and PKCε95 can phosphorylate MBP peptide$_{4-14}$ in the presence of appropriate cofactors. PKCε was immunoprecipitated from the nuclear and cytoplasmic fractions from quiescent and newly passaged 3T3 cells. The nuclear fraction contains predominantly PKCε95 and the cytoplasmic fraction predominantly PKCε87. The ability of these immunoprecipitates to phosphorylate MBP$_{4-14}$ was assessed in the presence of cofactors with and without PKC inhibitor peptide. Incorporation of [γ-^{33}P]ATP into MBP$_{4-14}$ was measured (CPM) on the y-axis as described in **Subheading 3.5.** Data are shown with values for PKC inhibitor peptides subtracted.

2. Passage cells into fresh medium and flasks and retain 1 Mbq [^{35}S]-Met/Cys in the medium (*see* **Note 13**).
3. Harvest cells at different time points—typically 15 min, 30 min, 1, 2, 4, 24, 48, and 72 h after passage.
4. Immunoprecipitate PKCε as described in **Subheading 3.1.4.**
5. Resolve samples by SDS-PAGE, fix gel in 25/20/65 (v/v/v) isopropanol/acetic acid, H$_2$O for 1 h, then incubate for 15 min in Amplify (Amersham, UK) before drying on a gel dryer and exposing to film.

3.5. PKC Activity Assays of Differentially Phosphorylated Forms of PKCε

Because the different phosphates have different effects on PKC function, it was important to look at the activity of the two differentially phosphorylated forms of PKCε detected in our cells using PKC activity assays. Our results showed that PKCε 95 and PKCε 87 were both catalytically active in the presence of the appropriate co-factors (**Fig. 6**). However, as mentioned earlier, the two forms of PKC differ in their localization and this may alter their activity in vivo through regulation of the availability of cofactors.

1. Immunoprecipitate PKCε from quiescent and newly passaged cells to give the tri- and di-phosphorylated forms of the protein, respectively. Do not resuspend the pellet in protein-loading buffer, but instead wash three times, and resuspend in 20 mM Tris-HCl pH 7.5. Divide this mix in half. Analyze half the immunoprecipitate by Western blotting as before to confirm equal PKC levels and the correct phosphorylation status.
2. Add 200 μL per tube PKC activity assay buffer containing myelin basic protein substrate.
3. Start the reaction with 10 μL of 2.5 nM ATP containing 1.86 kBq [γ-^{33}P]ATP and incubate at 37°C for 30 min (*see* **Note 14**). Perform the reactions in triplicate +/− the PKC inhibitor peptide RFAEKGSLRQKNV.
4. Spot 20 μL aliquots in triplicate onto P81 phosphocellulose paper squares. Wash four times in 80 mM phosphoric acid, then water, and finally in acetone. Allow to dry.
5. Add paper squares to Ultima Gold scintillant and count radioactivity using a suitable program on a scintillation counter.

3.6. Toward Identification of PKC Phosphatases and the Signaling Pathways That Control Them

One way to understand the regulation of PKC dephosphorylation is to investigate signaling pathways that control this in vivo. The simplest way to investigate is to start with pharmacological inhibitors of cell-signaling molecules because these are commercially available and relatively inexpensive and easy to use. Once candidates are identified, more detailed studies can be conducted using a variety of techniques, dominant-negative proteins, antisense, etc. *(13)*. The efficiency of the inhibitor may also be monitored, for example, we monitored the effectiveness of P98059 in our system by looking at its effect of mitogen-activated protein kinase (MAPK) phosphorylation using a phospho-specific MAPK antibody and Western blotting. This is especially important if no effect is seen (*see* **Note 15**).

3.6.1. Inhibitor Studies

Cells were treated with inhibitors at concentrations described in **Table 2**. Inhibitors were dissolved in dimethyl sulfoxide unless otherwise stated and effects compared with dimethyl sulfoxide control-treated cells. Solvent concentrations were never greater than 0.01%.

1. Add inhibitor to quiescent cells 30 min before passage and add fresh inhibitor after passage.
2. Monitor changes in the phosphorylation of PKCε through Western blotting as before.

Table 2
Effect of Chemical Inhibitors on PKCε[87] Formation on Cell Passage

Inhibitor	Target	Concentration	Effect on PKCε[87] Production
Okadaic acid	PP2A, PP1	10 μM	Increased
Calyculin	PP2A, PP1	10 nM	Increased
Cyclosporin A	PP2B	10 μM	Increased
Ascomycin C	PP2B	10 μM	Increased
Ro-31-8220	PKC	100 nM	Inhibited
Chelerythrine	PKC	1 μM	Inhibited
Wortmannin	PI3 kinase	1 μM	Increased
LY294002	PI3 kinase	10 μM	Increased
Cytochalasin D	Disrupts microfilaments	200 nM	Increased
Nocodazole	Disrupts microtubules	2 μM	None
PD98059	MEK	100 μM	Inhibited
Rapamycin	mTOR	500 μM	Inhibited
Herbimycin	Tyrosine kinases	2 μM	None
PKA inhibitor peptide	PKA	100 nM	None
SB203580	p38	20 μM	None

Our data showed that the dephosphorylation of Ser729 upon passage was not mediated through PP1, PP2A, or PP2B but was mediated through the mTor and MAPK pathways *(13)*. The involvement of the mTOR pathway in the regulation of PKCε and PKCδ has been documented by others *(18,19)*. PKCδ acted as a positive regulator of PKCε dephosphorylation. It is important, therefore, to consider the roles of individual PKC isoforms in the dephosphorylation of others.

4. Notes

1. When using PKC antibodies to look at different phosphorylated forms of the antibody, be aware of the region of the protein to which the antibody was generated. Both PKCε antibodies we used were directed to the C-terminal region of the antibody that included the Ser729 phosphorylation site and may be less immunoreactive with the phosphorylated form of the protein.

2. To identify if a particular PKC band or spot is phosphorylated antiphospho-Ser, phospho-Thr, and phospho-Tyr antibodies can be used. However, these antibodies do not give site-specific information and the phospho-Ser/Thr antibodies, in particular, are not as yet well used and their substrate specificity is not yet fully defined. Alternatively, ^{32}P or ^{33}P incorporation into PKC bands can be used to study their phosphorylation.

3. Dephosphorylation of PKC may not always result in a shift in the mobility of the protein that is detectable by Western blotting. It may therefore be required to use 2D gels to identify the presence of differentially phosphorylated forms of PKC within cells.

4. Be wary of centrifuging the cell lysate before sonication. Differentially phosphorylated forms of PKC may be localized to different cellular compartments and centrifuging before the cell is fully lysed may preferentially lose some of these fractions.

5. If the antibodies you are using are highly specific (i.e., do not react with any other proteins) especially the phospho-specific antibodies, they can be used for immunofluorescence staining to confirm PKC localization to different regions of the cell. Methods need to be determined for each cell line but a good starting place for fibroblast cells is as follows: grow cells on glass cover-slips as described earlier, rinse three times with warm PBS, fix for 10 min in 3% paraformaldehyde in PBS before quenching in 50 mM NH$_4$Cl for 1 h, and rinse twice in PBS. Permeablize with 0.2% TX-100 for 4 min, block by incubation with 2% BSA/5% normal goat serum in PBS, then label with an anti-PKCε antibody for 1 h in 2% BSA. Rinse several times before labeling with an appropriate secondary antibody, for example, FITC-conjugated, in 2% BSA. Invert coverslips onto mounting media (Dako, Ely, UK) and sealed with nail varnish. Examine cells with a fluorescence or confocal microscope (e.g., a Nikon Labophot fluorescence microscope).

6. For MALDI TOF m/s analysis you will need to use more initial sample to obtain enough PKC; approximately one big flask (75 cm^2) of confluent cells should give sufficient.

7. It is worth performing western blots before and after immunoprecipitation to monitor whether PKC is undergoing dephosphorylation or degradation during the immunoprecipitation process. Always include a control to which no PKC antibody has been added.

8. We tried to dephosphorylate PKCε in immunoprecipitates by incubation with various phosphatase. However, we achieved a very limited dephosphorylation. This may be because in immunoprecipitates PKC is denatured and may be in a form that is not recognized by the phosphatase. Greater success was achieved when incubation a whole cell homogenate was incubated +/– phosphatases before Western blotting. This may be because PKC is still in its natural environment, that is, in the presence of lipids and so the enzyme is in a conformation that is accessible to phosphatases.

9. MALDI TOF m/s does not require large amounts of protein or radiolabeling. However, it is hard to predict the success of this method because detection of phosphopeptides in a mix of peptides can be hard; some may not fly well in the MALDI TOF m/s and so may never be detected. Traditional methods for phosphopeptide detection, such as radiolabeling, can also be used.

10. 7% SDS-PAGE gels were used to get really good separation of the phosphorylated forms of PKC to ensure the MALDI TOF m/s samples were not cross-contaminated.

11. Samples can be frozen in Coomassie destain at this point until a suitable time. This is especially useful if you want to pool samples.
12. If using a 1D gel, be aware that other protein bands may be present in the PKC band of interest to confuse the peptide fingerprint. To get sufficient coverage of the protein, you may need to use a variety of enzymes, for example, Glu C (V8 protease) as well as trypsin.
13. Although [^{35}S] is a relatively safe isotope to work with, high levels are needed in metabolic labeling of cells and so caution should be exercised and work carried out according to local rules and regulations.
14. Although [^{33}P] is a relatively safe isotope caution should be exercised and work performed according to local rules and regulations.
15. All inhibitors are specific only until proven otherwise and many inhibitors inhibit more than one protein so care must be taken when drawing conclusions from these studies.

Acknowledgment

K. England was funded by the BBSRC Intergration of Cellular Responses Initiative.

References

1. Newton, A. C. (1997) Regulation of protein kinase C. *Curr. Opin. Cell Biol.* **9**, 161–167.
2. Dutil, E. M., Keranen, L. M., DePaoli-Roach, A. A., and Newton, A. C. (1994) In-vivo regulation of protein-kinase-C by trans-phosphorylation followed by autophosphorylation. *J. Biol. Chem.* **269**, 29,359–29,362.
3. Keranen, L. M., Dutil, E. M., and Newton, A. C. (1995) Protein kinase C is regulated in vivo by three functionally distinct phosphorylations. *Curr. Biol.* **5**, 1394–1403.
4. Garcia-Paramio, P., Cabrerizo, Y., Bornancin, F., and Parker, P. J. (1998) The broad specificity of dominant inhibitory protein kinase C mutants infers a common step in phosphorylation. *Biochem. J.* **333**, 631–636.
5. Le Good, J. A., Ziegeler, W. H., Parekh, D. B., Alessi, D. R., Cohen, P., and Parker, P. J. (1998) Protein kinase C isotypes controlled by phosphoinositide 3- kinase through the protein kinase PDK1. *Science* **281**, 2042–2045.
6. Chou, M. M., Hou, W., Johnson, J., Graham, L. K., Lee, M. L., Chen, C.-S., et al. (1998) Regulation of protein kinase C zeta by PI 3-kinase and PDK-1. *Curr. Biol.* **8**, 1069–1077.
7. Parekh, D. B., Ziegler, W., and Parker, P. J. (2000) Multiple pathways control protein kinase C phosphorylation. *EMBO J.* **19**, 496–503.
8. Chou, M. M., Hou, W., Johnson, J., Graham, L. K., Lee, M. L., Chen, C.-S., et al. (1998) Regulation of protein kinase C zeta by PI 3-kinase and PDK-1. *Curr. Biol.* **8**, 1069–1077.
9. Zhang, J., Wang, L., Schwartz, J., Bond, R. W., and Bishop, W. R. (1994) Phosphorylation of Thr (642) is an early event in the processing of newly

synthesized protein-kinase C-beta(1) and is essential for its activation. *J. Biol. Chem.* **269,** 19,578–19,584.

10. Gysin, S. and Imber, R. (1997) Phorbol-ester-activated protein kinase C-alpha lacking phosphorylation at Ser657 is down-regulated by a mechanism involving dephosphorylation. *Eur. J. Biochem.* **249,** 156–160.

11. Bornancin, F. and Parker, P. J. (1996) Phosphorylation of protein kinase C-zeta on serine 657 controls the accumulation of active enzyme and contributes to its phosphatase-resistant state. *Curr. Biol.* **6,** 1114–1123.

12. Cazaubon, S., Bornancin, F., and Parker, P. J. (1994) Threonine-497 is a critical site for permissive activation of protein-kinase C-alpha. *Biochem. J.* **301,** 443–448.

13. Orr, J. W. and Newton, A. C. (1994) Requirement for negative charge on activation loop of protein- kinase-C. *J. Biol. Chem.* **269,** 27,715–27,718.

14. England, K., Watson, J., Beale, G., Warner, M., Cross, J., and Rumsby, M. (2001) Signalling pathways regulating the dephosphorylation of Ser729 in the hydrophobic domain of Protein Kinase C ε upon cell passage. *J. Biol. Chem.* **276,** 10,437–10,442.

15. England, K. and Rumsby, M. (2000) Changes in PKCε phosphorylation status and intracellular localization as 3T3 and 3T6 fibroblasts grow to confluency and quiescence: a role for phosphorylation at Ser-729. *Biochem. J.* **352,** 19–26.

16. Ohmori, T. and Arteage, C. L. (1998) Protein kinase C epsilon translocation and phosphorylation by cis-diamminedichloroplatinum(II) (CDDP): potential role in CDDP-mediated cytotoxicity. *Cell Growth Differ.* **9,** 345–353.

17. Rosenfeld, J., Capdeveille, J., Guillemot, J. C., and Ferrara, P. (1992) In gel digestion of proteins for internal sequence analysis after 1-dimensional or 2-dimensional gel-electrophoresis. *Anal. Biochem.* **203,** 173–179.

18. Parekh, D. B., Ziegler, W., and Parker, P. J. (2000) Multiple pathways control protein kinase C phosphorylation. *EMBO J.* **19,** 496–503.

19. Parekh, D., Ziegler, W., Yonezawa, K., Hara, K., and Parker, P. J. (1999) Mammalian TOR controls one of two kinase pathways acting upon nPKC delta and nPKC epsilon. *J. Biol. Chem.* **274,** 34758–34764.

19

Phosphopeptide-Specific Antibodies to Protein Kinase C

Wayne S. Sossin

1. Introduction

Antibodies that are specific for phosphorylated residues are extremely important for research on signal transduction pathways. With these reagents one can follow a signal transduction pathway using simple immunoblotting techniques. Although many of these antibodies are commercially available, new phosphorylation sites are being found at a rapid pace and it is useful to understand what is required to make one's own phosphopeptide-specific antibody. In this chapter, I will review our laboratory's experience in producing and using these antibodies with specific reference to antibodies against sites phosphorylated in protein kinase C (PKC).

PKC is phosphorylated in distinct ways. First, PKC activity requires phosphorylation at a site in the activation loop (*1,2*). This site is phosphorylated by the enzyme phosphoinositide-dependent kinase (PDK-1; **refs. *3–5***) and there are a number of antibodies that have been raised against this site that are either isoform-specific or that recognize many isoforms of PKC (**Table 1**; **refs. *5–7***). This phosphorylation is followed by autophosphorylation of two sites in the carboxy-terminal domain of the kinase (*8–12*). The most carboxy-terminal of these two sites has been studied extensively and is well conserved among all PKCs. There are a large number of phosphopeptide antibodies that have been generated to this site and again some appear to be specific, whereas others cross-react to many PKCs (cell-signaling pan-PKC phosphopeptide antibody; **Table 1**; **ref. *7***). The more amino-terminal (or juxta-kinase) site does not appear to be the exclusive target of any phosphopeptide antibody. However, a number of antibodies do recognize this autophosphorylation site in conjunction with an

From: *Methods in Molecular Biology, vol. 233: Protein Kinase C Protocols*
Edited by: A. C. Newton © Humana Press Inc., Totowa, NJ

Table 1
Sites Used for Phosphopeptide Antibodies

Site	Location of site	PKCs recognized	Reference
FEGFS(P)YVNP-amide	Carboxy-terminal	PKC beta II (most PKCs)	Parekh et al. *(7)*
EGF[pS]YVNPQF-amide	Carboxy-terminal	PKC alpha (most PKCs)	Upstate Biotechnology
CKGF[pS]YFGEDL-amide	Carboxy-terminal	PKC epsilon (most PKCs)	Upstate Biotechnology
CTAFKGF[pS]FVNPKY-amide	Carboxy-terminal	PKC delta	Konishi et al. *(19)*
RAST(P)FCGT-amide	PDK site	PKC delta	Parekh et al. *(7)*
TTTT(P)FCGT-amide	PDK site	PKC epsilon	Parekh et al. *(7)*
DGVTTK[pT]FCGTPD-amide	PDK site	most PKCs	Chou et al. *(5)*
TENKLTQ[pT}FC-amide	PDK site	PKC Apl II	Pepio et al. *(6)*
MDKKTTR[pT]FC-amide	PDK site	PKC Apl I (PKC alpha)	Pepio et al. *(6)*
CRLQKG[pS]TKEK-amide	C2 domain	PKC Apl II	Pepio et al. *(15)*
WDRT(P)TRND-amide	C2 domain	PKC alpha	Ng et al. *(17)*
CFDREF[pT]SEAPNVT[pT]PTD-amide	Juxta-kinase	PKC Apl I	Nakhost et al. *(13)*
FDRFF[pT]RHPPVL[pT]PPD-amide	Juxta-kinase	PKC beta II	Sweatt et al. *(14)*
CNEKQPL[pS]FSDKNL-amide	Juxta-kinase	PKC delta	Konishi et al. *(19)*
CPETVGI[pY]QGFEK-amide	Hinge	PKC delta	Konishi et al. *(19)*
CIPDNNGT[pY]GKIWEG-amide	Hinge	PKC delta	Konishi et al. *(19)*
CGTPD[pY]IAPEILQG-amide	Kinase	PKC delta	Konishi et al. *(19)*

adjacent site (Table 1; **refs. *13* and *14***). PKCs are also phosphorylated at other residues during activation of the kinase that may have modulatory or feedback roles in the regulation of PKC *(13–17)*, including phosphorylation by tyrosine kinases *(18–21)*. Phosphopeptide antibodies have been used to study these modulatory sites as well (**Table 1**). Specific antibodies to PKC phosphorylation sites can be used to monitor PKC activation or to explore the role of phosphorylation by examining differences between the total kinase and the phosphorylated kinase using immunoblotting and immunocytochemistry.

2. Materials
2.1. Antigen Preparation
1. Synthesized phosphopeptide.
2. Synthesized peptide.
3. Maleimide couple carrier protein (bovine serum albumin [BSA] or keyhole limpet hemocyanin [KLH]; Pierce, Rockford, IL).
4. Dimethylsulfoxide.
5. Sulfolink (Pierce).

2.2. Affinity Purification

1. Phosphate-buffered saline (PBS).
2. Pump for controlling speed of columns.
3. Elution Buffers: 0.1 mM glycine, pH 2.3; 50 µM triethylamine, pH 12.
4. Re-equilibration buffers: 1 M Tris-HCl, pH 8.0; 1 M triethanolamine, pH 6.0.
5. Bradford Reagent for measuring protein levels.
6. Concentrators.

2.3. Testing of the Phosphopeptide Antibody

1. Antibodies to PKC of interest.
2. Source of nonphosphorylated PKC (bacterially expressed fusion protein).
3. Sodium dodecyl sulfate-polyacrylamide gel electrophoresis equipment.
4. Secondary antibodies and chemoluminescence kit.

3. Methods

3.1. Antigen Preparation

What is used as an immunogen will be critical in the success of raising a good antibody. We will discuss the important principals and choices for selecting the appropriate antigen. The epitopes that have been used for phosphopeptide antibodies to PKC are listed in **Table 1**.

3.1.1. Choice of Peptide

Unlike most peptide antibodies, where a deliberate attempt is made to pick a surface-exposed area that is fairly hydrophilic, choices are limited when making a phospho-specific antibody to the location immediately surrounding the phosphopeptide site. The peptide should be fairly short (7–15 amino acids) to increase the ease of synthesis, decrease the cost, and minimize the number of nonphosphorylated epitopes that are present. Peptides shorter than seven amino acids may be less immunogenic and may result in antibodies that are more specific for the type of phosphorylated residue (i.e., recognizing phosphoserine in any protein) and less specific for the protein of interest. For ease of coupling, we always add a cysteine residue to the N-terminus or C-terminus of the peptide if one is not naturally present at this point in the sequence chosen. The cysteine residue allows for efficient coupling to larger proteins and to columns (*see* **Subheading 3.1.3.**). If a cysteine is present in the peptide naturally, we start or end the peptide at that residue if at all possible. We have had good success with as little as one amino acid separating the phospho-amino acid from the cysteine when the cysteine is part of the natural sequence (**Table 1**). Alternatively, one can use glutaraldehyde to couple any peptide to a carrier residue, but this may reduce availability of the epitope. Another point to consider is the conservation of the sequence surrounding the phospho-amino acid. Do you want an isoform-specific antibody or one that will crossreact with other PKC isoforms, or perhaps other members of the kinase super-family? By choosing to use sequences carboxy- or amino-terminal to the phosphorylation site, one can sometimes choose between sequences conserved between all isoforms, or isoform specific sequences. A good example of this is at the PDK site, where sequences carboxy-terminal to the site are conserved in all isoforms, whereas amino-terminal residues are isoform specific (**Table 1**).

3.1.2. Synthesis of Peptide

The chemical synthesis of peptides is beyond the scope of this commentary and most investigators should consider doing this commercially. Phosphopeptides are considerably more expensive than regular peptides and we usually order in the 5–10 mg range as opposed to larger quantities for a regular peptide. Purity should be greater than 90%. Make sure the service you use has experience with generating serine and threonine phosphorylated peptides because this is more complicated than conventional solid-state peptide synthesis (*22–24*). One should also order the nonphosphorylated peptide (usually 20 mg at greater than 70% purity) for generating an affinity column to remove antibodies that recognize the nonphosphorylated epitope and to absorb these antibodies during immunoblotting (*see* **Subheadings 3.1.3.** and **3.3.2.**).

3.1.3. Coupling of Peptide

The choice of cysteine for coupling makes this step simple and reproducible but also makes it vital to prevent the formation of disulfide bonds before coupling. Peptides should be stored dry with desiccation. Before use, have all reagents ready including, dimethyl sulfoxide, dilution buffers, and other reagents required if the peptide is not highly soluble.

1. Dissolve 5 mg of peptide in 500 µL of water (*see* **Note 1**).
2. Suspend aliquot of 2 mg KLH-maleimide or BSA-maleimide with 200 µL of peptide in water. Most people working with vertebrate PKCs should use KLH-maleimide. Invertebrate organisms have tons of hemocyanin and thus should use BSA to avoid dealing with crossreaction to endogenous proteins.
3. Add 200 µL of peptide to Sulfolink Gel equilibrated in column (equilibration should be done ahead of time; follow instructions that come with kit for preparation of the column).
4. Freeze remaining 100 µL of peptide in 20-µL aliquots.
5. Use desalting column to separate coupled peptide from ethylenediamine tetraacetic acid (EDTA) (*see* **Note 2**).

6. Collect several 1-mL elutions from the column (use add 1-mL aliquots of 1×
   PBS to the column as the solution passes through; keep the column volume
   minimal, but do not let it run dry).
7. Test fractions for presence of the peptide by performing Bradford assays on a
   small sample (e.g., 10–20 µL). Keep any fractions that test positive for presence
   of the peptide (generally this will be in two to three of the fractions).
6. Aliquot into five fractions of approx 400 µL (four injections and one control).

Similarly, couple the nonphosphorylated peptide to the Sulfolink gel. This
does not have to be done at the same time, as this column is not needed until
purification of the antibody.

3.1.4 Immunization of Animals

This step is important, but beyond our expertise. One needs to choose an
appropriate adjuvant, site of injection, and animal to inject. One should consult
the animal resource facility at one's institution or there are many commercial
sites that can provide these services. Usually, antibodies are produced in rabbits,
although there are alternatives (e.g., rat, chicken, and goat). One consideration
is whether one wishes to measure phosphorylation with immunocytochemistry.
In this case, an alternative choice of animal for injection may ease double
labeling studies.

3.2. Affinity Purification of Antibodies

We have found that a critical step in the use of antibodies to phosphopeptides
is at the affinity purification step. Usually, peptides are not strong enough
antigens to produce sera that are useful without purification. Also critical in the
purification of a phosphopeptide antibody is the attempt to remove antibodies
that recognize epitopes in the peptide that do not include the phospho-amino
acid. We also purify serums from each bleed of the rabbit independently as
the quality of the antibodies can change from one injection to the next, so
the protocol is designed for small amounts of serum (5–10 mL). Usually this
does not saturate the column. With larger amounts of serum, saturation can
become a problem.

3.2.1. Affinity Purification Protocol

Our standard protocol is outlined below and pictured schematically in **Fig. 1A**.
A typical output from 10 mL of serum is shown in **Fig. 1B,C**.

1. Dilute 10 mL of serum to 50 mL with PBS.
2. Load slowly over Sulfolink affinity column of nonphosphorylated peptide
   (0.2 mL / min). Save flowthrough.

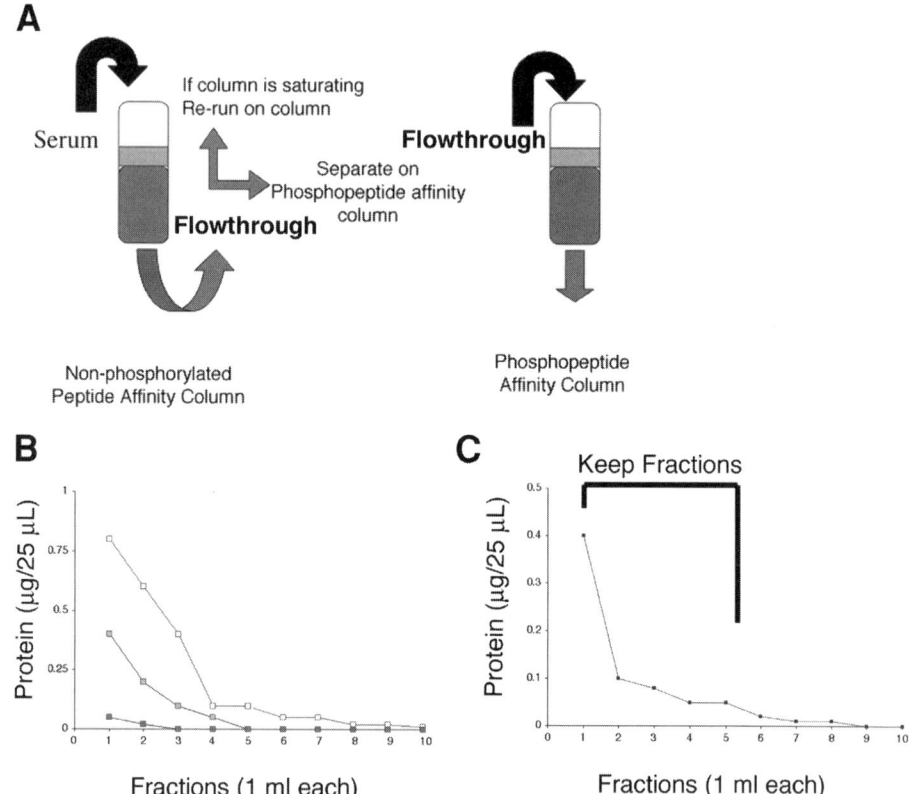

Fig. 1. **(A)** Schematic of affinity purification protocol for phosphopeptide-specific antibodies. Serum is first loaded on a column made from the nonphosphorylated peptide. The flowthrough is saved and the antibodies bound are eluted by low and/or high pH solution. Depending on the amount of antibody bound, the flowthrough is either reloaded on this column or now loaded onto an affinity column containing the phosphopeptide. **(B)** Elution profiles from the nonphosphopeptide affinity column (no shading, first load, medium shading, second load, dark shading, third load). **(C)** Elution profile from phosphopeptide affinity column. The fractions selected are shown.

3. Wash with 10 mL of PBS (all wash steps at 1 mL /min).
4. Wash with 10 mL of PBS + 0.5 *M* NaCl.
5. Wash with 10 mL of PBS.
6. Elute with 10 mL of 0.1 *M* glycine, pH 2.3 (0.2 mL/min). Collect 10 1-mL fractions using tubes preloaded with 100 µL of 1 *M* Tris-HCl, pH 8.0.
7. Wash with 20 mL of PBS (0.5 mL/min). Check that pH is re-equilibrated.
8. Elute with 10 mL of 50 µ*M* triethylamine (pH 12). Collect 10 1-mL fractions into tubes preloaded with 100 µL of 1 *M* triethanolamine, pH 6.0.

9. Wash column with 20 mL of PBS (+ Na azide (0.02%) for storage). Check that pH is re-equilibrated.
10. Assay column fractions (25 µL) for protein with Bradford Assay.
11. Using flowthrough, repeat **steps 1–10** until minimal antibody (<0.5 µg/25 µL) is eluted from column (*see* **Fig. 1B**).
12. Repeats **steps 1–10** using Sulfolink column of phosphorylated peptide.
13. Pool all elutions with concentrations greater than 0.2 µg/25 µL (*see* **Note 3**).
14. Concentrate antibodies to a final concentration of at least 1 mg/mL and aliquot. The antibody is good for approx 1 yr at 4°C and probably for 5–10 yr frozen at –70°C, although some antibody may be lost by the freeze–thaw cycle.

3.2.2. Protocol 2

Depending on the epitope, it may occur that all the antibodies that recognize the phosphopeptide also have enough affinity for the nonphosphorylated peptide to be retained on the first column. Perhaps, for example, the binding site is made up of three elements, only one of which is the phosphate group. Antibodies will still retain an affinity for the nonphosphorylated peptide that may cause them to be retained on the nonphosphopeptide affinity column. In this case, we use a protocol where only the phosphopeptide column is used to purify the antibody. Discrimination of the phospho-specific antibodies is done by competition with the nonphosphorylated peptide prior to and during the primary antibody incubation (*see* **Subheading 3.3.2.**).

3.3. Testing of the Phosphopeptide Antibody

Even after affinity purification, we still find that there can be some affinity for nonphosphorylated kinase. It is critical to be able to test this explicitly and optimize conditions to both enhance sensitivity of the antibody and to ensure phosphospecificity.

3.3.1. The Nonphosphorylated Control

It is critical to have a source of nonphosphorylated protein to determine the optimum conditions for use of the phosphopeptide antibody and as a control to run on immunoblotting experiments. Usually, a fusion protein expressed in a bacterial system is the optimal control because most eukaryotic sites are not phosphorylated during expression in bacteria. This is true for most PKC autophosphorylation sites as well. Dephosphorylation of purified PKC may be an alternative option, although it is difficult to know the completeness of the dephosphorylation. Expression of a PKC with a mutation at the phosphorylation site (i.e., a serine to alanine change) is also an appropriate control, although it is possible that the antibody could still recognize the protein with a nonphosphorylated serine, but not recognize the alanine containing protein. Finally, coupling of the nonphosphorylated peptide to an alternative fusion protein can provide a control for immunoreactivity to the nonphosphorylated protein.

3.3.2. Determining the Range of Phosphosensitivity

1. What is required for this step is samples of protein that are phosphorylated and samples that are not phosphorylated (*see* above for nonphosphorylated control). Samples of phosphoprotein could include extracts from cells where some of the protein is endogenously phosphorylated, extracts from stimulated cells, or products of in vitro kinase reactions (*see* **Note 4**). Ideally, the antibody should be optimized in the range of protein levels that will be used in the final experiments (e.g., 10 μg of tissue). A range of the nonphosphorylated control should be run using approximately equal amounts of nonphosphorylated protein to a 20-fold molar excess of phosphorylated protein (e.g., 1, 2, 5, and 20).

2. To start with, approximately four equivalent samples are needed. We usually begin with primary antibody concentrations of approx 1 μg/mL. The four conditions we use are no peptide, 3:1 molar ratio of peptide/antibody, 30:1 molar ratio of peptide, and 300:1 molar ratio of peptide.

3. The peptide is incubated with the antibody before dilution and then centrifuged for 5 min in a desktop centrifuge.

4. The supernatant is then diluted with blocking solution (*see* **Note 5**) and added to the blots in plastic bags to minimize volumes. The blots are incubated for 1 h at room temperature (*see* **Note 6**), washed three times for 5 min, and then processed with secondary antibody and developing reagent. The protocol described above was optimized using goat anti-rabbit peroxidase and an enhanced chemoluminescence protocol; different concentrations of primary antibody may be necessary with other development procedures. One is looking for the optimum amount of signal in the phosphorylated sample, and no signal in the nonphosphorylated sample. The blots should be stripped and re-probed with an antibody that is not specific for the phosphorylation state of the protein to ensure an equivalent range of protein in the controls.

If no reaction is seen with the phosphopeptide antibody, even in the absence of peptide the following steps can be tried: (1) increase antibody concentration up to 10 μg/mL; (2) incubate overnight at 4°C (this may assist antibodies with slow binding); and (3) reduce blocking concentration of milk to 1% milk or replace milk with 5% BSA. If crossreaction is observed with the nonphosphorylated sample even at the highest concentration of peptide, there is probably too much protein on the blot. Try different protein levels (all using the highest concentration of peptide) until phosphospecificity is established. Then, attempt to lower peptide levels to enhance signal while retaining phosphospecificity. It may also be useful to decrease antibody concentration.

3.4. Using the Phosphopeptide Antibody

Once conditions have been optimized for phosphospecificity, experiments can be performed to measure phosphate incorporation under physiological conditions.

3.4.1. Immunoblots

It is useful to run positive and negative controls of all gels to ensure that in the range of each particular experiment phospho-sensitivity is retained. It is also important to strip blots and reprobe with the antibody to total protein to establish whether changes in immunoreactivity with the phosphopeptide antibody are to the result of changes in phosphorylation or to changes in total protein level. Blotting with the regular antibody is also important in order to show that the phosphorylated protein seen on immunoblots corresponds to the protein of interest. A major problem with phosphopeptide antibodies is crossreaction with other proteins, especially ones with similar phosphorylation sites. The band of interest may not be the darkest band on the blot.

3.4.2. Quantitation of Immunoblots

We have found calculation of a phospho-ratio (i.e., intensity of signal with phospho-peptide antibody/intensity with regular antibody) to be useful when comparing phosphorylation between different conditions *(13,15)*. This requires both blots to be in the linear range of detection. It is also important to note that this ratio is only useful for in-gel comparisons and cannot be used to compare samples run on separate gels and developed separately.

3.4.3. Immunoprecipitations

Even if the phosphopeptide antibody is not specific enough for the protein of interest, it can still be useful to measure changes in phosphorylation if (1) it is phosphospecifc and (2) one has an antibody to the protein of interest that can successfully immunoprecipitate the protein. The immunoprecipitates are then immunoblotted with the phosphopeptide antibody and there is no worry about crossreactivity to other proteins since they are no longer on the gel.

3.4.4. Immunocytochemistry

Antibodies to phosphopeptides must recognize only a single band on the blot to be useful in standard immunocytochemistry. Even then, it is difficult to control for phospho-specificity without some known mechanism for regulating the phosphorylation in cells or double labeling with an antibody that is not phosphospecifc. However, in some cases, even with a phosphopeptide antibody that is not absolutely clean, one can visualize phosphorylated PKC using variations of fluorescence resonance energy transfer *(17)*.

4. Notes

1. Usually if you have 10 mg or less of protein, it is difficult to weigh out an appropriate amount, so simply re-suspend it in water. If after 30 s or so of vortexing, the solution is not clear, add dimethyl sulfoxide to 10% and then 20% to get the peptide into solution quickly; the longer the peptide waits in the absence of the coupling reaction, the more disulfide bonds will form. Usually a ratio of 1–3 mg peptide to 2 mg of KLH or BSA maleimide will saturate the sites on the carrier protein, even if some disulfide bonds form. One can assay the free cysteines with Ellman's reagent (available from Pierce). However, this is of limited use because in an aqueous solution, even in the absence of the maleimide carrier, the peptide is completely disulfide bonded after the 2 h of coupling. There are some reagents to disassociate the bonds once formed, but in my experience, this is quite difficult to do with high efficiency.
2. The gel filtration is required to remove the EDTA in the stabilization buffer before injection into animals. I usually pre-run about a 5-mL P-50 column with 2 mg of BSA to block any nonspecific binding sites on the column and to measure the void volume on the column. Try to collect all the conjugated peptide in at most 2 mL. Another option is to dialyze to remove the EDTA. This usually involves an overnight step at 4°C, but usually the conjugated peptide is stable.
3. It is important to remember that higher affinity antibodies may take longer to elute from the column or may require the high pH step with triethylamine. Although the quantities of these antibodies may be less, they are still quite valuable and may be the antibodies that are most useful. We therefore are quite liberal in taking extra fractions from the flowthrough, and generally keep at least one fraction beyond that where minimal antibody was detected (*see* **Fig. 1**). We do not in general treat these fractions separately, but pool all fractions and concentrate them to a final concentration of 1 mg/mL.
4. The control for in vitro kinase reactions can be a bit problematic. We found that even on ice, there is some kinase activity and thus could not easily generate a 0 time point. Instead, mock reactions in the absence of ATP serve as a useful control.
5. There are many phosphorylated proteins in milk (i.e., casein). For antibodies with weak reactivity to other phosphorylated residues, milk can absorb all of the antibody and must be replaced with a nonphosphorylated block like BSA. In contrast, the phosphorylated proteins in milk may absorb out all antibodies that recognize all phosphorylated residues, leaving behind the very specific antibodies that only recognize the phosphorylated protein of interest.
6. The time of incubation that is required is related to how long the antibody takes to bind. High-affinity antibodies often bind slowly and the signal may be improved by incubation overnight at 4°C.

Acknowledgments

I thank Jonathan Hislop and Arash Nakhost for comments on the chapter. This work was supported by a grant MT-12046 from the Canadian Institute of Health Research (CIHR). WSS is a Killiam Scholar and a CIHR investigator.

References

1. Dutil, E. M., Keranen, L. M., DePaoli, R. A., and Newton, A. C. (1994) In vivo regulation of protein kinase C by trans-phosphorylation followed by autophosphorylation. *J. Biol. Chem.* **269,** 29,359–29,362.

2. Newton, A. C. (1995) Protein kinase C: structure, function, and regulation. *J. Biol. Chem.* **270,** 28,495–28,498.

3. Le Good, J. A., Ziegler, W. H., Parekh, D. B., Alessi, D. R., Cohen, P., and Parker, P. J. (1998) Protein kinase C isotypes controlled by phosphoinositide 3-kinase through the protein kinase PDK1. *Science* **281,** 2042–2045.

4. Dutil, E. M., Toker, A., and Newton, A. C. (1998) Regulation of conventional protein kinase C isozymes by phosphoinositide-dependent kinase 1 (PDK-1). *Curr. Biol.* **8,** 1366–1375.

5. Chou, M. M., Hou, W. M., Johnson, J., Graham, L. K., Lee, M. H., Chen, C. S., et al. (1998) Regulation of protein kinase C zeta By Pi 3-kinase and Pdk-1. *Curr. Biol.* **8,** 1069–1077.

6. Pepio, A. M., Thibeault, G. L., and Sossin, W. S. (2002) Phosphoinositide dependent kinase (PDK) phosphorylation of protein kinase C (PKC) Apl II increases during intermediate facilitation in Aplysia. *J. Biol. Chem.* **277,** 37416–37122.

7. Parekh, D., Ziegler, W., Yonezawa, K., Hara, K., and Parker, P. J. (1999) Mammalian TOR controls one of two kinase pathways acting upon nPKCdelta and nPKCepsilon. *J. Biol. Chem.* **274,** 34,758–34,764.

8. Keranen, L. M., Dutil, E. M., and Newton, A. C. (1995) Protein kinase C is regulated in vivo by three functionally distinct phosphorylations. *Curr. Biol.* **5,** 1394–1403.

9. Bornancin, F. and Parker, P. J. (1996) Phosphorylation of threonine 638 critically controls the dephosphorylation and inactivation of protein kinase c-alpha. *Curr. Biol.* **6,** 1114–1123.

10. Bornancin, F. and Parker, P. J. (1997) Phosphorylation of protein kinase C-alpha on serine 657 controls the accumulation of active enzyme and contributes to its phosphatase-resistant state [published erratum appears in *J. Biol. Chem.* (1997) **272,** 13,458]. *J. Biol. Chem.* **272,** 3544–3549.

11. Behn-Krappa, A. and Newton, A. C. (1999) The hydrophobic phosphorylation motif of conventional protein kinase C is regulated by autophosphorylation. *Curr. Biol.* **9,** 728–737.

12. Tsutakawa, S. E., Medzihradszky, K. F., Flint, A. J., Burlingame, A. L., and Koshland, D. E., Jr. (1995) Determination of in vivo phosphorylation sites in protein kinase C. *J. Biol. Chem.* **270,** 26,807–26,812.

13. Nakhost, A., Dyer, J. R., Pepio, A. M., Fan, X., and Sossin, W. S. (1999) Protein kinase C phosphorylated at a conserved threonine is retained in the cytoplasm. *J. Biol. Chem.* **274,** 28,944–28,949.
14. Sweatt, J. D., Atkins, C. M., Johnson, J., English, J. D., Roberson, E. D., Chen, S. J., et al. (1998) Protected-site phosphorylation of protein kinase C in hippocampal long-term potentiation. *J. Neurochem.* **71,** 1075–1085.
15. Pepio, A. M. and Sossin, W. S. (2001) Membrane translocation of nPKCs is regulated by phosphorylation of the C2 domain. *J. Biol. Chem.* **276,** 3846–3855.
16. Flint, A. J., Paladini, R. D., and Koshland, D. E., Jr. (1990) Autophosphorylation of protein kinase C at three separated regions of its primary sequence. *Science* **249,** 408–411.
17. Ng, T., Squire, A., Hansra, G., Bornancin, F., Prevostel, C., Hanby, A., et al. (1999) Imaging protein kinase Calpha activation in cells. *Science* **283,** 2085–2089.
18. Konishi, H., Tanaka, M., Takemura, Y., Matsuzaki, H., Ono, Y., Kikkawa, U., et al. (1997) Activation of protein kinase C by tyrosine phosphorylation in response to H2O2. *Proc. Natl. Acad. Sci. USA* **94,** 11,233–11,237.
19. Konishi, H., Yamauchi, E., Taniguchi, H., Yamamoto, T., Matsuzaki, H., Takemura, Y., et al. (2001) Phosphorylation sites of protein kinase C delta in H2O2-treated cells and its activation by tyrosine kinase in vitro. *Proc. Natl. Acad. Sci. USA* **98,** 6587–6592.
20. Li, W., Chen, X. H., Kelley, C. A., Alimandi, M., Zhang, J., Chen, Q., et al. (1996) Identification of tyrosine 187 as a protein kinase C-delta phosphorylation site. *J. Biol. Chem.* **271,** 26,404–26,409.
21. Li, W., Zhang, J., Bottaro, D. P., and Pierce, J. H. (1997) Identification of serine 643 of protein kinase C-delta as an important autophosphorylation site for its enzymatic activity. *J. Biol. Chem.* **272,** 24,550–24,555.
22. Perich, J. W., Terzi, E., Carnazzi, E., Seyer, R., and Trifilieff, E. (1994) Further studies into the Boc/solid-phase synthesis of Ser(P)- and Thr(P)-containing peptides. *Int. J. Pept. Protein Res.* **44,** 305–312.
23. Lacombe, J. M., Andriamanampisoa, F., and Pavia, A. A. (1990) Solid-phase synthesis of peptides containing phosphoserine using phosphate tert.-butyl protecting group. *Int. J. Pept. Protein Res.* **36,** 275–280.
24. Otvos, L., Elekes, I., and Lee, V. M. (1989) Solid-phase synthesis of phosphopeptides. *Int. J. Pept. Protein Res.* **34,** 129–133.

VI

IDENTIFYING PROTEIN KINASE C SUBSTRATES

20

Identifying Protein Kinase C Substrates

An Introduction

Alex Toker

The identification of physiologically relevant substrates of the 10 known protein kinase C (PKC) isozymes is of obvious importance for a complete understanding of the mechanisms by which this family of Ser/Thr protein kinases relays information from extracellular stimuli to biological responses. Since its discovery, many proteins have been identified as PKC substrates both in vitro and in vivo. However, given the multitude of signaling pathways and physiological responses attributed to PKC function, it is perhaps surprising that a clear description of substrates whose phosphorylation by PKC mediates secondary signaling cascades is not available, in a manner analogous to the kinase cascades of the mitogen-activated protein kinase modules or the Akt/protein kinase B pathway. Thus, knowledge of PKC substrates is not currently matched by the near complete understanding of PKC regulation. Numerous techniques have been used in the last 20 years to identify and characterize PKC substrates in the test tube and under physiological conditions, but have been hampered by both conceptual and technical limitations. This chapter describes two recent innovations in the field that will undoubtedly accelerate the much-needed discovery of novel PKC substrates.

The problems that arise in trying to identify relevant PKC substrates are exemplified by the fact that nearly every polypeptide sequence in the proteome contains a potential Ser or Thr PKC phosphorylation site. It has been known since the mid-1980s that PKCs preferentially phosphorylate basophilic sequence motifs, such that the phosphorylatable Ser or Thr residues are flanked by Arg or Lys, typically amino-terminal to the phospho-acceptor *(1–4)*. Initial

From: *Methods in Molecular Biology, vol. 233: Protein Kinase C Protocols*
Edited by: A. C. Newton © Humana Press Inc., Totowa, NJ

in vitro experiments using synthetic peptides indicated a preferred PKC consensus phosphorylation motif: RXXS/TXRX (where X indicates any amino acid) *(4)*. The identification of this motif was also important in the definition of the PKC pseudosubstrate domain. This autoinhibitory sequence, also present in many other Ser/Thr kinases, contains an optimal PKC phosphorylation site, except that the phospho-acceptor residue is replaced by an Ala *(5)*. In the inactive enzyme, the pseudosubstrate domain binds to the substrate-binding cavity in the catalytic kinase core, preventing access to exogenous substrates. Binding of lipid activators (diacylglycerol) to the PKC regulatory domain is required for pseudosubstrate release, leading to kinase activation. Consistent with the pseudosubstrate model, synthetic peptides based on this sequence (in which the Ala is replaced with Ser) are excellent PKC substrates in vitro, with K_m values approaching 8 μM phosphate hydrolyzed per minute per mg of protein, corresponding to 10 reactions per second *(5)*. The PKC substrate motif described above has also proved useful in characterizing artificial in vitro substrates that are commonly used for the determination of PKC activity. Historically, these include histones (e.g., HIIIS) myelin basic protein, protamine, and protamine sulfate. Synthetic peptides based on the various PKC pseudosubstrate sequences are the more commonly used artificial substrates today, and are commercially available.

The finding that Ser/Thr residues flanked by basic amino acids present a good motif for PKC phosphorylation is also consistent with the identification of many in vitro and in vivo substrates. To date, numerous proteins including growth factor receptors, cytoskeletal proteins, ion channels and pumps, transcription factors, and even nuclear proteins have been shown to be bona-fide PKC substrates *(6,7)*. Perhaps the two most well-known and characterized substrates are myristoylated alanine-rich C kinase substrate (MARCKS) *(8)* and platelet and leukocyte C kinase substrate protein (Pleckstrin) *(9)*. Both of these proteins have multiple PKC phosphorylation sites that are phosphorylated upon phorbol ester stimulation of virtually all cells, although it is noteworthy that Pleckstrin is restricted in its expression, primarily found in cells of the hematopoietic lineages. Interestingly, both of these PKC substrates play fundamental roles in organization of the actin cytoskeleton, underscoring the importance of proper PKC function in mediating cell motility.

The fact that any one cell type or tissue can express up to 10 distinct isozymes has also hampered the identification of PKC substrates because genetic or pharmacological approaches to eliminate PKC protein or activity are plagued by the obvious issues of redundancy and lack of specificity. This issue is exacerbated by the fact that there are very few clear-cut cases in which a protein is a selective in vivo substrate for one PKC isozyme, but not any other. One example is the heterogeneous ribonucleo-protein A1 (hnRNP A1), which appears to be a selective PKCζ substrate *(10)*. Thus, the emerging consensus is that distinct PKC isozymes achieve substrate selectivity by being targeted to specific intracellular locations or compartments, an issue that is elegantly explained by the several targeting proteins that have been shown to modulate PKC function in the cell, such as Receptors for Activated C Kinase (RACKs) and A Kinase Anchoring Proteins (AKAPs; *see* Chapter 26).

The above discussion suggests that further advances in the field of PKC signaling will likely be made by the discovery of novel PKC substrates. Established techniques, such as expression library screening, ^{32}P labeling of cells followed by protein purification and microsequencing, which have proven useful in the discovery of substrates of many protein kinases, have provided only limited success in the PKC world. However, two recent advances have renewed enthusiasm in this area of research and are the focus of this chapter. The first, discussed by Yaffe, makes use of an oriented degenerate peptide library approach where an unbiased mixture of up to 2.5 billion peptides is presented to a highly purified protein kinase in vitro. Nonphosphorylated peptides are separated from the phosphorylated ones, which are then sequenced to determine the preferred substrate sequence motif of a given kinase. This degenerate peptide library approach was first developed by Cantley and col-leagues to determine the preferred SH2 domain-binding specificity of tyrosine phosphopeptides, and was later modified to evaluate the substrate specificity of many Ser/Thr as well as Tyr kinases *(11,12)*. Of particular relevance to this chapter is the analysis of the preferred substrate selectivity of nine human PKC isozymes, which revealed subtle, but important differences in the motifs phosphorylated by conventional (PKCα, PKCβI, PKCβII, and PKCγ), novel (PKCδ, PKCε, PKCη), and atypical (PKCζ) isozymes *(13)*. Not surprisingly, the motifs derived from this approach closely resembled (but were not identical to) the pseudosubstrate sequence in the respective PKC, and as discussed below, allowed the design and synthesis of second-generation synthetic peptides that to date display the lowest K_m values of any PKC substrate. This approach has therefore provided important new information concerning the substrate selectivity of PKC family members that can now be used to determine the phosphorylation of cellular substrates by any given isozyme.

The above methodology is likely to work well in conjunction with the distinct approach recently developed and characterized by Shokat and colleagues, which uses a chemical genetic approach to identify protein kinase substrates. As discussed below by Shokat, this methodology uses ATP derivatives in combination with a mutational strategy with the protein kinase of interest *(14)*. An amino acid mutation is introduced into the active site of a given kinase and the genetically modified kinase is now only able to phosphorylate a candidate substrate in the presence of a chemically modified ATP analogue, which is radiolabeled with ^{32}P. An alternative application of this technique is the development of chemically modified inhibitors that have been altered to selectively inhibit protein kinases with the same active site mutation *(15)*. As an example, a $[\gamma$-^{32}P]N^6-(benzyl)-ATP analogue has been successfully used to unambiguously show the direct phophorylation of novel substrates of the c-Jun amino-terminal kinase, the cyclin-dependent kinase 2, and the tyrosine kinase v-Src *(16–18)*. Because the mutation that allows binding of N^6-(benzyl)-ATP is conserved in all Ser/Thr as well as Tyr kinases, this will undoubtedly prove to be a very fruitful approach for novel substrate discovery. Although this methodology has yet to be applied to PKC isozymes, it is only a matter of time before novel substrates of this kinase family are described using this approach. The methodologies described in Chapters 21 and 22 are therefore predicted to generate much needed information in this important area of PKC signaling.

References

1. Kishimoto, A., Nishiyama, K., Nakanishi, H., Uratsuji, Y., Nomura, H., Takeyama, Y., et al. (1985) Studies on the phosphorylation of myelin basic protein by protein kinase C and adenosine 3′:5′-monophosphate-dependent protein kinase. *J. Biol. Chem.* **260,** 12,492–12,499.
2. Gould, K. L., Woodgett, J. R., Cooper, J. A., Buss, J. E., Shalloway, D., and Hunter, T. (1985) Protein kinase C phosphorylates pp60src at a novel site. *Cell* **42,** 849–857.
3. House, C., Wettenhall, R. E., and Kemp, B. E. (1987) The influence of basic residues on the substrate specificity of protein kinase C. *J. Biol. Chem.* **262,** 772–777.
4. Pearson, R. B. and Kemp, B. E. (1991) Protein kinase phosphorylation site sequences and consensus specificity motifs: tabulations. *Methods Enzymol.* **200,** 62–81.
5. House, C. and Kemp, B. E. (1987) Protein kinase C contains a pseudosubstrate prototope in its regulatory domain. *Science* **238,** 1726–1728.
6. Nishizuka, Y. (1992) Intracellular signaling by hydrolysis of phospholipids and activation of protein kinase C. *Science* **258,** 607–614.
7. Toker, A. (1998) Signaling through protein kinase C. *Front. Biosci.* **3,** D1134–D1147.
8. Hartwig, J. H., Thelen, M., Rosen, A., Janmey, P. A., Nairn, A. C., and Aderem, A. (1992) MARCKS is an actin filament crosslinking protein regulated by protein kinase C and calcium-calmodulin. *Nature* **356,** 618–622.
9. Tyers, M., Haslam, R. J., Rachubinski, R. A., and Harley, C. B. (1989) Molecular analysis of pleckstrin: the major protein kinase C substrate of platelets. *J. Cell. Biochem.* **40,** 133–145.
10. Municio, M. M., Lozano, J., Sanchez, P., Moscat, J., and Diaz-Meco, M. T. (1995) Identification of heterogeneous ribonucleoprotein A1 as a novel substrate for protein kinase C zeta. *J. Biol. Chem.* **270,** 15,884–15,891.

11. Songyang, Z., Lu, K. P., Kwon, Y. T., Tsai, L. H., Filhol, O., Cochet, C., et al. (1996) A structural basis for substrate specificities of protein Ser/Thr kinases: primary sequence preference of casein kinases I and II, NIMA, phosphorylase kinase, calmodulin-dependent kinase II, CDK5, and Erk1. *Mol. Cell Biol.* **16,** 6486–6493.

12. Songyang, Z. and Cantley, L. C. (1998) The use of peptide library for the determination of kinase peptide substrates. *Methods Mol. Biol.* **87,** 87–98.

13. Nishikawa, K., Toker, A., Johannes, F. J., Songyang, Z., and Cantley, L. C. (1997) Determination of the specific substrate sequence motifs of protein kinase C isozymes. *J. Biol. Chem.* **272,** 952–960.

14. Specht, K. M. and Shokat, K. M. (2002) The emerging power of chemical genetics. *Curr. Opin. Cell. Biol.* **14,** 155–159.

15. Bishop, A. C., Ubersax, J. A., Petsch, D. T., Matheos, D. P., Gray, N. S., Blethrow, J., et al. (2000) A chemical switch for inhibitor-sensitive alleles of any protein kinase. *Nature* **407,** 395–401.

16. Habelhah, H., Shah, K., Huang, L., Burlingame, A. L., Shokat, K. M., and Ronai, Z. (2001) Identification of new JNK substrate using ATP pocket mutant JNK and a corresponding ATP analogue. *J. Biol. Chem.* **276,** 18,090–18,095.

17. Polson, A. G., Huang, L., Lukac, D. M., Blethrow, J. D., Morgan, D. O., Burlingame, A. L., et al. (2001) Kaposi's sarcoma-associated herpesvirus K-bZIP protein is phosphorylated by cyclin-dependent kinases. *J. Virol.* **75,** 3175–3184.

18. Shah, K. and Shokat, K. M. (2002) A chemical genetic screen for direct v-Src substrates reveals ordered assembly of a retrograde signaling pathway. *Chem. Biol.* **9,** 35–47.

21

A Chemical Genetic Approach for the Identification of Direct Substrates of Protein Kinases

Kavita Shah and Kevan M. Shokat

1. Introduction

Protein kinases form one of the largest superfamily of enzymes that play pivotal roles in controlling almost every signaling pathway *(1)*. Deregulated kinase activity thus leads to multiple diseases, including various forms of cancer *(2)*, inflammatory and autoimmune diseases *(3)*, neurodegenerative diseases *(4,5)*, diabetes *(6)*, and HIV infection (*(7)*). Signaling networks regulated by kinases are complex and highly interconnected. Additionally, many kinases display overlapping substrate specificities in vitro and can functionally compensate for each other in single gene knockout experiments *(8,9)*. Therefore, unraveling of these pathways to dissect the role of any particular kinase (normal or oncogenic) has remained one of the major challenges ever since the first kinase was purified. Ideally, if the substrate of each kinase could be identified in a cell, it would provide a baseline for understanding the complex cellular functions and consequently also provide a blueprint for novel targets for drug discovery.

To date, multiple strategies for kinase substrate identification have been developed. Screening of degenerate peptide *(10)*, cDNA *(11–13)*, and phage display expression libraries *(14)* have proven to be valuable for the identification of several putative targets of multiple purified kinases. Recently, Zhu et al., *(15)* used nanowell chip technology to analyze the kinase activities of 119 yeast protein kinases with 17 different substrates, revealing novel substrate specificities. Phosphorylation of the proteome chips expressing all the yeast proteins with different purified kinases is potentially a powerful tool to spatially resolve kinase substrate specificity at the proteome level *(16)*.

From: *Methods in Molecular Biology, vol. 233: Protein Kinase C Protocols*
Edited by: A. C. Newton © Humana Press Inc., Totowa, NJ

The key determinant of kinase substrate specificity in vivo however, depends on cellular localization and protein–protein interactions, hard to recapitulate in an in vitro system *(17)*. As a result, various techniques have been used to dissect the precise substrate phosphorylation events in signaling cascades. Recently, an approach has been described for the simultaneous measurement of multiple active kinase states using polychromatic flow cytometry, allowing the identification of distinct signaling cascades in various cellular processes *(18)*. This analysis at a single cell level allows monitoring of signaling mechanisms in rare cell populations. However, the requirement of phosphospecific antibodies for detection, limits the efficacy of this method to the known substrates only in a signaling cascade. Other techniques for kinase substrates identification include immunoprecipitation of relevant kinase complexes *(19,20)*, mass spectrometry and functional proteomics *(21,22)*, and modulation of divalent metal requirements in cell lysates *(23)*. Although, the fundamental question of which kinase is responsible for a given phosphorylation event remains elusive because of the presence of many kinases in the same pathway that exhibit overlapping substrate specificities *(24)* and because all kinases use ATP as the phosphodonor substrate.

Recently, we have developed a chemical genetic approach to address this question *(25–27)*. This review will focus on advances in this chemical genetic approach for identification of direct substrates of multiple protein kinases. A kinase of interest is engineered to accept a non-natural phosphate donor substrate (A*TP) that is poorly accepted by wild-type protein kinases in the cell. The modification of the kinase's active site, so that it accepts a structurally modified γ-^{32}P-labeled nucleotide analog provides a unique handle by which the direct substrates of any particular kinase can be traced in the presence of other protein kinases for the first time (**Fig. 1**). The design strategy uses the engineering of a unique active site pocket in the ATP binding site of the target kinase and a complementary substituent on the A*TP analog. This unnatural pocket is created by the replacement of a conserved bulky residue with a glycine or an alanine in the active site and the complementing substituent on ATP is created by attaching bulky substituents at the N-6 position of ATP (ex: N^6-(benzyl) ATP, N^6-(phenethyl) ATP etc., **refs. 26,28**).

To confer highly specific sensitivity of the engineered kinase to unnatural ATP analogs, engineered amino acid residue should meet following design criteria.

- The wild-type residue should have a conserved bulky side chain such that mutation to a smaller residue creates a new pocket in the active site.
- The residue should be highly conserved across the whole kinase family in terms of similarity of side chain volume at the minimum.

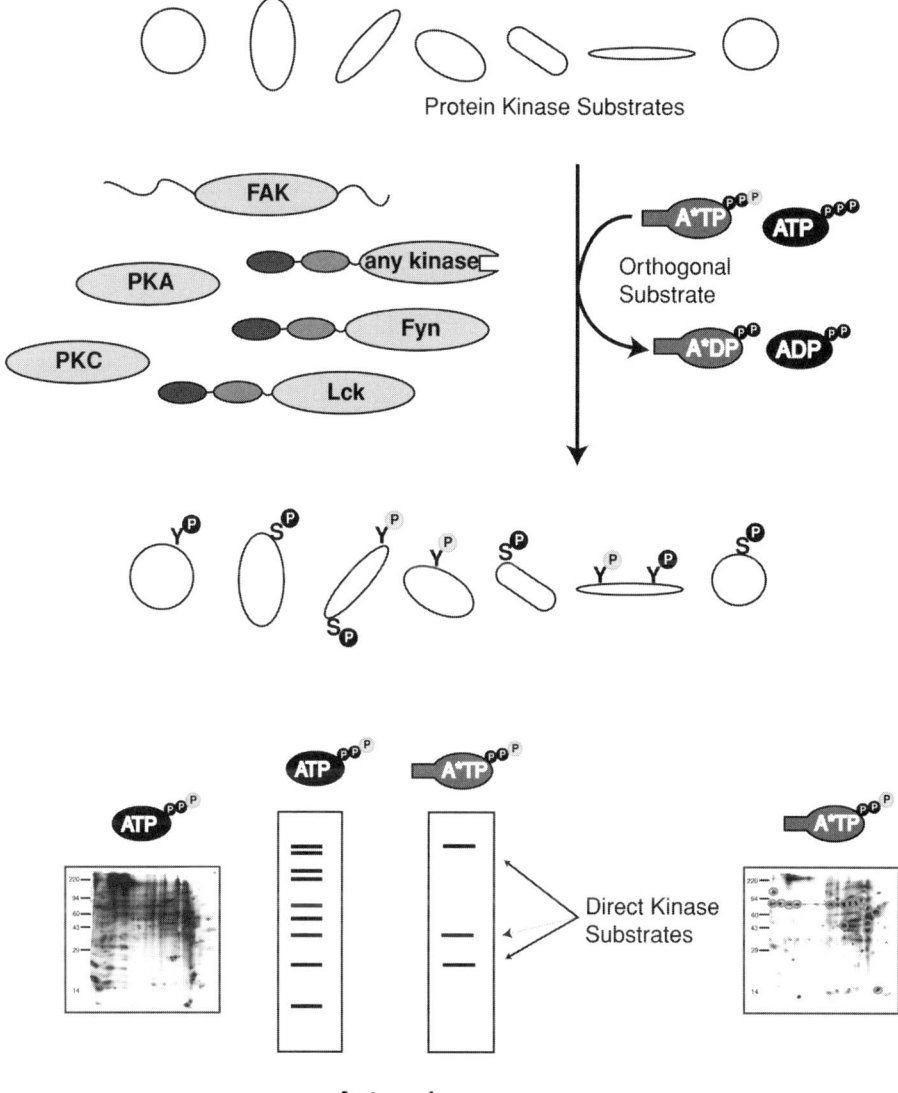

Autorad

Fig. 1. Schematic representation of the use of A*TP by an engineered kinase. The use of specific phosphodonor substrate (A*TP) for one kinase allows for its direct substrates to be uniquely radiolabeled (grey P represents ^{32}P). S indicates serine and Y indicates tyrosine. Ovals represent catalytic or regulatory domains (black = Src-homology 3 domain, dark grey = Src-homology 2 domain, light grey = catalytic domain).

```
                Subdomain IV                        338  Subdomain V
                                                      ▼
v-Src         [318]RHEKLVQLYAMVSE----------EPIYIVIEYMSK--GSLLDFLKGEMG
c-Fyn         [319]KHDKLVQLYAVVSE----------EPIYIVTEYMNK--GSLLDFLKDGEG
c-Abl         [294]KHPNLVQLLGVCTRE---------PPFYIITEFMTY--GNLLDYLRECNR
CamK IIα      [68] KHPNIVRLHDSISEE---------GHHYLIFDLVTG--GELFEDIVAREY
Cdk2          [59] NHPNIVKLLDVIHTE--------NKLYLVFEFLHQ---DLKKFMDASAL
Cdc28(Cdk1)   [66] KDDNIVRLYDIVHSDA-------HKLYLVFEFLDL---DLKRYMEGIPK
Fus3          [67] KHENIITIFNIQRPDSFENF----NEVYIIQELMQT---DLHRVISTQM
```

Fig. 2. Kinase domain sequence conservation and partial sequence alignment of the protein kinases. The highly conserved "gatekeeper" residue is shown in bold.

- Mutation of the residue should be functionally silent and should not dysregulate the kinase or alter its phosphoacceptor specificity.
- Ideally the mutation should be generalizable to all members of the protein kinase superfamily (est. 500 human kinases)

A functionally conserved residue was identified in the V subdomain of the kinase active site (I338 in v-Src) that could be mutated to an alanine or a glycine, thus creating a new pocket in the active site (**Fig. 2**). Further studies revealed that this conserved large residue in fact functions as a "gatekeeper" by blocking an existing pocket in the kinase active site. Thus, by clipping the bulky side chain of the key residue, we uncover a "path" for the bulky N^6-substituents on ATP analogs to gain access to the existing pocket rather than engineering a whole new pocket *(27)*. The distinctive importance of the gatekeeper residue has also been well exploited for designing specific inhibitors for both wild-type serine/threonine and tyrosine kinases *(29)*. Some kinases require an additional sensitizing mutation immediately amino-terminal to the conserved DFG motif in kinase subdomain VII.

This key residue makes close contact with ATP and often has a smaller side chain. However, several kinases have large amino acids at this position that interfer with the binding of ATP analog. Mutation of this residue to an alanine or a glycine effectively allows efficient catalysis with A*TP analogs *(30,31)*. The engineered kinase fulfills several necessary criteria to uniquely tag its authentic substrates in the presence of all wild-type kinases. Engineered kinase:

- Accepts an ATP analog (A*TP) that is a poor substrate for all wild-type protein kinases;
- Uses A*TP with high catalytic efficiency;
- Exhibits reduced catalytic efficiency for the natural nucleotide substrate (ATP) so that, in the presence of cellular levels of ATP (~5 m*M*), the engineered kinase preferentially uses A*TP as the phosphodonor.

And finally all of these criteria are met by creating a functionally silent mutation generalizable across the entire kinase superfamily. Structural and functional assessment of peptide specificity of mutant protein kinases with orthogonal ATP analogs showed that the creation of a unique nucleotide binding pocket does not alter the phospho-acceptor binding site of the kinase *(32)*. A panel of optimal peptide substrates of defined sequence, as well as a degenerate peptide library was utilized to assess the phospho-acceptor specificity of the engineered "traceable" kinases. The specificity profiles for the mutant kinases were found to be identical to those of their wild-type counterparts, further confirming that the engineered mutation is indeed functionally silent.

This strategy has recently been applied to identify the direct targets of two different serine/threonine kinases in cellular extracts: c-Jun amino terminal kinase (JNK; **ref. *31***) and CDK2 *(33)*. Addition of immunopurified mutant JNK (JNK analog specific-3 [JNK-as3]) and $[\gamma\text{-}^{32}\text{P}]N^6\text{-}$(phenethyl) ATP analog into whole cell extract resulted in the specific phosphorylation of hnRNP-K protein. This study further confirmed hnRNP-K as a true physiological target of JNK by in vivo ^{32}P labeling, whose phosphorylation at serine 216 and serine 353 by JNK increases its transcriptional effects. In another study, phosphorylation by baculovirus expressed mutant CDK2/cyclins (CDK2-as1) and A*TP revealed Kaposi's Sarcoma-Associated Herpesvirus K-bZIP protein as its direct substrate in BCBL-1 cell extracts. This is the first report that shows K-bZIP protein as a direct target of CDK2, suggesting its potential involvement in cell cycle.

Most recently, we applied this approach to look for the direct substrates of v-Src in fibroblasts *(34)*. v-Src-induced transformation of fibroblasts leads to the tyrosine phosphorylation of more than 50 proteins; however, it is not clear if all of these proteins are direct v-Src substrates or if they are phosphorylated as a consequence of the v-Src activation of other kinases. Therefore, a global analysis of direct v-Src substrates was carried out by the addition of bacterially expressed mutant XD4 (constitutively active form of Src, consisting mainly SH1 domain) and $[\gamma\text{-}^{32}\text{P}]N^6\text{-}$(benzyl) ATP into NIH3T3 cell lysate. Interestingly, no substrate phosphorylation was observed suggesting that may be the lack of SH2/SH3 domains in XD4 prevented it to form relevant complexes. However, the addition of immunopurified mutant v-Src (v-Src-as1) also showed minimal phosphorylation of substrates in cell lysate. Interestingly, when mutant v-Src was endogenously expressed in NIH3T3 cells, addition of $[\gamma\text{-}^{32}\text{P}]N^6\text{-}$(benzyl) ATP into the cell lysate showed phosphorylation of more than 20 proteins on a 2D gel (**Fig. 3**). As a control, addition of radiolabeled A*TP analog to wild-type v-Src expressing cell lysate showed no phosphorylation. Mass spectral analysis of several of these spots revealed novel substrates

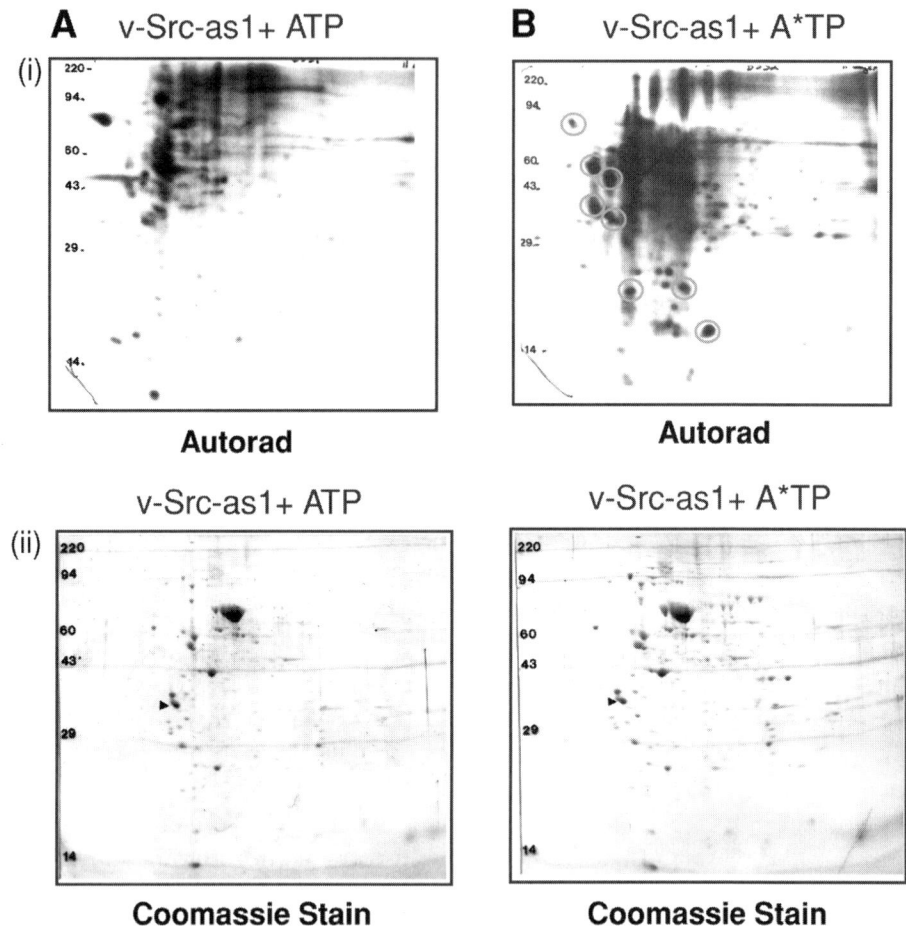

Fig. 3. Differential phosphorylation of direct v-Src substrates in *v-Src-as1* NIH3T3 cell lysate with [γ-^{32}P]N^6(benzyl)ATP. **(A)** 2D gel electrophoresis of kinase reaction of *v-Src-as1* NIH3T3 cell lysate and [γ-^{32}P]ATP in the presence of 100 μ*M* ATP for 45 min. Internal standard (tropomyosin) is shown as an arrow at MW 33 K and pI 5.2. (i) Autorad; (ii) The gels were Coomassie blue stained, dried, and scanned. **(B)** 2D gel electrophoresis of kinase reaction of v-Src-as1 NIH3T3 cell lysate and [γ-^{32}P]A*TP in the presence of 100 μM ATP for 45 min. (i) Autorad; (ii) The gels were Coomassie blue stained, dried, and scanned.

of v-Src. This result strongly suggests that kinases anchored in proper signaling scaffolds best phosphorylate their physiological substrate proteins and added kinases (especially tyrosine kinases) can't easily access the relevant complexes.

This result further prompted us to look for substrates of v-Src in different signaling complexes. Additionally, we predicted that use of v-Src immuno-precipitation to enrich for v-Src substrates would allow the identification of substrates present at low concentration and those substrates that are phosphorylated at low stoichiometry. We isolated multiple complexes (anti-v-Src, -Fak, -cortactin) in which v-Src was directly responsible for phosphorylation of all proteins in the complex as might be expected for this highly active oncoprotein. Interestingly, a very low abundance scaffolding protein Dok-1 was found to be a direct substrate of v-Src in all of these complexes. Yet, in one particular complex isolated by immunoprecipitation with an anti-RasGAP antibody, both v-Src and Dok-1 were present yet v-Src was only responsible for a subset of the phosphorylation products, and not of Dok-1. This result suggested that although v-Src was present and active in multiple complexes, it is not a promiscuous kinase that always phosphorylates a given protein (Dok-1) in all signaling complexes. Other kinases in these complexes are responsible for highly specific phosphorylation events, revealing such protein complexes to be more dynamic and specifically regulated than previously appreciated.

Further studies revealed a number of surprising features of v-Src substrates that contradict currently accepted model of tyrosine kinase specificity. For example, analysis of the individual tyrosine phosphorylation sites on the multiply phosphorylated protein Dok-1 (Dok-1 had 14 potential tyrosine phosphorylation sites) revealed that out of four phosphorylation sites, only two (Y361 and Y450) were definitively sites phosphorylated by v-Src, even though three (Y295, Y314, and Y361) of the four sites shared a similar consensus sequence (YXXP). Moreover, the site that was predicted to provide a nonoptimal v-Src phosphorylation site, was one of the sequences directly phosphorylated ($LY_{450}SQV$). One of the v-Src phosphorylation sites (Y361) on Dok-1 produces an optimal binding site for at least two different SH2 containing proteins, v-Src itself and RasGAP. Because the other v-Src phosphorylation site Y450 on Dok-1 is not an optimal consensus phosphorylation site for Src, its phosphorylation may require proper binding of v-Src to Dok-1. Thus, stable association of v-Src and Dok-1 may allow v-Src to phosphorylate the sub-optimal v-Src consensus site, which in turn creates a binding site for the SH2 domain of Csk. Such a processive phosphorylation model has been proposed by Zhou and Cantley *(24)* and by Scott and Miller *(35)*. Because both Src and rasGAP compete for binding at pY361, it is likely that bound rasGAP blocks the binding of v-Src at pY361, resulting in the observed failure of v-Src to phosphorylate Dok-1 in rasGAP complex. In summary, we found that v-Src phosphorylation of substrates can be regulated at the level of protein complex formation and that a limited subset of sites of heavily phosphorylated proteins are directly phosphorylated by a single kinase, v-Src. It further suggests that identification of direct phosphorylation sites solely based on the optimal sequence specificity of a particular kinase may be misleading.

The role of Dok-1 phosphorylation by v-Src revealed a model for the assembly of negative regulatory proteins (rasGAP and Csk) onto the scaffolding protein Dok-1. Further studies revealed that Csk negatively regulates the duration of c-Src signaling in a feedback mechanism. Because *v-Src* lacks this critical phosphorylation site, it is not responsive to this arm of the retrograde signaling apparatus, contributing to its potent transformation. Mutation of v-Src phosphorylation sites Y295 or Y361 (rasGAP binding sites) abrogated the inhibitory effect of Dok-1 on cellular transformation, indicating that these sites are necessary for the repression of transformation activity, thereby highlighting yet another arm of retrograde signaling that is controlled by v-Src itself. Identification of the direct substrates of v-Src leads to a model for the precise order of assembly of a retrograde signaling pathway in v-Src transformed cells and has provided a new insight into the balance between those signals that promote cell transformation mediated by v-Src catalyzed tyrosine phosphorylation and those that inhibit it.

This chemical genetic approach has also been used in the identification of substrates in multiple cell lines in response to different stimuli and in various tissues. Addition of any particular analog sensitive serine/threonine kinase to fractionated tissue homogenate results in specific phosphorylation of their direct substrates. This method offers the advantage of looking at the differential phosphorylation pattern of any kinase of interest in normal versus diseased tissue type. The direct radiolabeled targets thus obtained can be further confirmed by in vivo labeling or by phosphotyrosine immunoblotting (Kim et al., unpublished results).

More recently, we have designed novel ATP analogs with enhanced selectivity for mutant kinases *(36)*. These ATP analogs are more orthogonal than previously reported analogs and should allow the detection of protein substrates too scarce to be detected previously *(34)*. However, intact cells are the best context to identify bona fide kinase substrates, because subcellular localization and protein-protein interactions are powerful determinants of kinase signaling specificity *(17)*. To accomplish this goal, we recently engineered a kinase that accepts an unnatural triphosphate analog, but is insensitive to cellular milieu of ATP (~5 mM). N^4-(benzyl) AICAR triphosphate is orthogonal to wild type kinases and is a substrate of T106G p38. Thus, generation of a unique kinase/phosphodonor pair that is insensitive to ATP will provide a higher signal to noise ratio of substrate labeling, thereby allowing the identification of low abundance kinase substrates in intact cells (Ulrich et al., unpublished results).

In addition to using a chemical genetic approach for substrate identification, we have also applied our strategy to probe the role of individual kinases by using allele-specific orthogonal inhibitors *(37,38)*. By using the same analog-sensitive mutant of a particular kinase together with the chemically modified "bumped" inhibitor, we have recently shown the role of three kinases in yeast: Cdc28p *(39)*, Cla4p *(30)*, and Pho85p *(40)*. This approach is extremely powerful because it allows elucidation of the role of any kinase of interest in vivo.

2. Materials

1. 6-Chloropurine riboside, benzylamine, carbonyl diimidazole, trimethyl phosphate and $POCl_3$ (Aldrich) were stored at room temperature under anhydrous conditions.
2. 1 *M* TEAB buffer was prepared by suspending 1 mole of triethylamine (Aldrich) in water in a total volume of 1000 mL, followed by CO_2 bubbling until a pH of 7.5 was attained.
3. Diethylaminoethyl-Sephadex (DEAE-Sephadix A1-25) and protein A Sepharose (Pharmacia) were soaked in water overnight and washed several times with water before use.
4. Luria Bertani (LB) Superbroth, isopropyl β-D-thiogalactopyranoside (IPTG), ethylenediaminetetraacetic acid (EDTA), dithiothreitol (DTT), glutathione S transferase (GST) beads, and glutathione (Sigma or any other supplier).
5. Optimal peptide substrate for the kinase of interest can be purchased if commercially available or can be custom synthesized.
6. $[\gamma\text{-}^{32}P]ATP$ and $[\gamma\text{-}^{32}P]H_3PO_4$ (NEN or any other supplier).
7. p81 filter and polyvinylidene membrane (Millipore).
8. GFP protein-based kinase substrates were generated as published previously *(41)*.
9. 4G10 (Upstate) and HRP-conjugated anti-mouse antibody (VWR cat. no. 55550).
10. Supersignal kit (Pierce).
11. Kodak Biomax MS film, Kodak X-OMAT AR film and Biomax MS intensifying screen (Kodak).
12. Strong anion exchange column (Rainin).

3. Methods

The methods described below outline (1) the steps for chemical genetic dissection of a kinase signaling cascade and (2) experimental procedures for synthesis of ATP analogs, protein expression, screening of ATP analogs, and determination of direct substrates in cellular systems.

3.1. Steps for Chemical Genetic Dissection of a Kinase Signaling Cascade

3.1.1. Identify Site of Mutation

The "gatekeeper" and the other key residue immediately amino-terminal to the conserved DFG motif in kinase subdomain VII can be readily identified by kinase catalytic domain sequence alignment of a desired kinase using Kinase Sequence Database located at http://kinase.ucsf.edu/ksd. This database

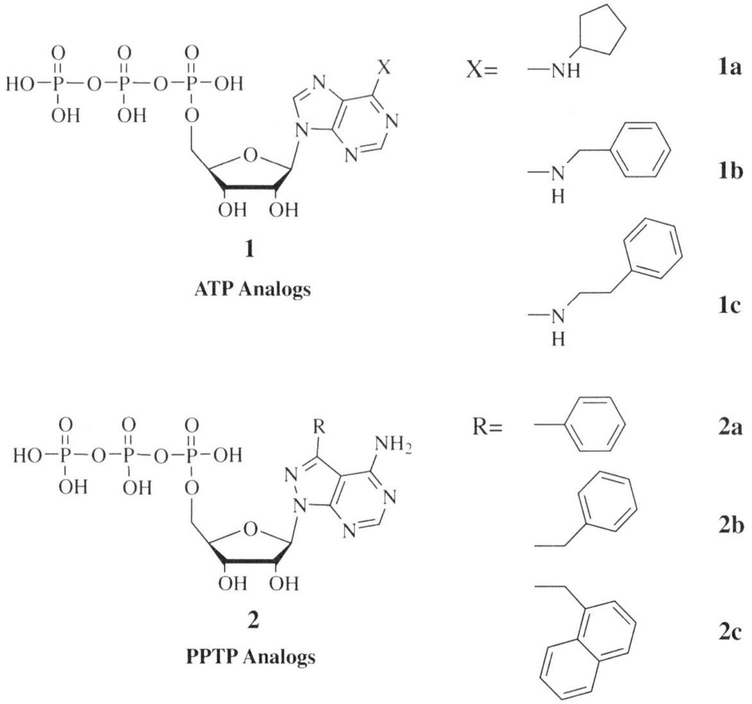

Fig. 4. Structure of ATP and PPTP (pyrazolopyrimidine triphosphate) analogs.

contains information on 290 protein kinase families for a total of 5041 protein kinases from more than 100 organisms *(42)*.

3.1.2. Construction of Analog-Sensitive Kinase Allele

1. For most kinases, mutation of the "gatekeeper" amino acid residue in the ATP binding site (equivalent to I338 in v-Src) to either a glycine (analog sensitive v-Src-1, v-Src-as1) or an alanine (v-Src-as2) renders them sensitive to orthogonal ATP analogs. Additional mutation may be required of the residue flanking the amino terminal of DFG motif to an alanine (as-3) for those kinases that have a bulky side chain on this residue.
2. Recombinant-engineered kinase is expressed as a fusion protein in *Escherichia coli*, insect cells, or mammalian cells following precedents with the corresponding wild-type kinase.

3.1.3. Determining the Optimal ATP Analog for the Engineered Kinase

1. A panel of cold ATP analogs (**Fig. 4**) are tested as inhibitors of [γ-^{32}P] ATP-dependent phosphorylation of an optimal peptide substrate by the engineered kinase in an in vitro kinase reaction.

2. For assay of catalytic activity, the engineered tyrosine kinase is tested with ATP analogs and highly efficient GFP protein based kinase substrates may be used, or alternatively any known phosphoacceptor substrate of the kinase of interest *(41)*.
3. The ATP analog with the highest catalytic efficiency paired with the desired engineered kinase are chosen.

3.1.4. Choice of Cell System

1. Recombinant-engineered kinase is added exogenously to the desired cell lysate or tissue extracts for an unbiased analysis of direct substrates of the kinase of interest.
2. Alternatively, engineered kinase is expressed endogenously in appropriate cell type. This approach is particularly useful for tyrosine kinases that require proper SH2/SH3 interactions for substrate phosphorylation (see above).

3.1.5. Optimize Kinase Reaction Conditions

1. The kinase reaction is performed with appropriate radioactive $[\gamma\text{-}^{32}P]A^*TP$ in the presence of cold ATP. Because the unnatural ATP analog is orthogonal to wild-type kinases, cold ATP is added to the kinase reaction to maintain proper stoichiometry of phosphorylation. Therefore, amount of radioactive A^*TP/cold ATP is empirically determined.
2. Reaction time (10 s–20 min) should be empirically optimized.
3. Optimal cell lysis buffer conditions are determined to maximize kinase activity and thus substrate phosphorylation *(43)*.

3.1.6. 2D Gels or Immunoprecipitation

1. Substrate phosphorylation is conducted in different immune complexes containing the engineered kinase. This method allows for the enrichment of potential substrates and thus can be resolved completely by 1D polyacrylamide gel electrophoresis (PAGE) *(34)*.
2. For the identification of substrates in total cell lysate, phosphorylated substrates are further concentrated by immunoprecipitating with phosphospecific antibodies or by other affinity resins and then separated on 1D PAGE (Shokat et al., unpublished results).
3. Alternatively, phosphorylated substrates from total cell lysate are also isolated by 2D PAGE *(34)*

3.1.7. Mass Spectral Analysis or Immunoblotting

1. Isolated radiolabeled substrates are identified by mass spectral analysis. The only limitation of this method is for the substrates that are present in low abundance.
2. Direct substrates of a kinase can also be identified by immunoblotting or by double immunoprecitation.

3.1.8. Confirmation In Vivo

1. Substrates identified by in vitro kinase assay are further confirmed in an in vivo system. For confirming the substrates of oncogenic-engineered tyrosine kinase, substrates are immunoprecitated from the transformed cells and immunoblotted with a phosphotyrosine-specific antibody *(34)*. These substrates can be further confirmed by using allele-specific orthogonal inhibitors, which should inhibit the phosphorylation event catalyzed by the engineered kinase only.
2. The engineered kinase can also be activated appropriately in the cell and the substrates are confirmed similarly.
3. For confirming the substrates of serine/threonine kinases, cells are labeled in vivo with ^{32}P phosphoric acid and substrates are isolated after stimulation and checked on by autoradiography for the incorporation of phosphate. These substrates are also further confirmed by using allele specific inhibitors (Shah et al., unpublished results).

3.1.9. Functional Studies

1. Depending on the signaling pathways controlled by the kinase of interest, the role of phosphorylation of novel direct substrates are explored.
2. On the other hand, identification of novel substrates also leads to the identification of novel pathways.

3.2. Experimental Procedures

3.2.1. Synthesis of [γ-^{32}P]N^6(benzyl) ATP ([γ-^{32}P]A*TP)

1. N^6(benzyl)adenosine was synthesized by refluxing 6-chloropurine riboside (1 mmol) (Aldrich) with benzylamine (5 mmol) in ethanol (10 mL) overnight. Ethanol was removed in vacuo and the resulting oily residue obtained was crystallized from ethanol (yield 90%). N^6 (benzyl) ADP was synthesized by sequential phosphorylation according the method of Hecht and Kozarich *(44)*. To an ice-cooled suspension of N^6(benzyl) adenosine (68 mg, 0.2 mmol) in trimethyl phosphate (0.5 mmol), $POCl_3$ (0.025 mL) was added and the reaction mixture was stirred at 0°C for 1 h, after which the reaction was quenched with 5 mL of 1 *M* triethylammonium bicarbonate (TEAB buffer, pH 7.5). Solvent was removed in vacuo at <40°C by rotary evaporation. The resulting slurry was purified on DEAE (A-25) Sephadex (Pharmacia) column using TEAB pH 7.5 (0.1–0.5 *M* gradient). The purified N^6(benzyl) AMP shows a retention time of 7.5 min on a strong anion exchange high-performance liquid chromatography column (SAX, cat. no. 83-E03-ETI, Varian) using a gradient of 5–750 m*M* ammonium phosphate, pH 3.9, in 10 min at a flow rate of 0.5 mL/min.
2. In the second step, a solution of N^6(benzyl) AMP (44 mg, 0.1 mmol) and carbonyl diimidazole (81 mg, 0.5 mmol) in dimethylformamide (DMF) (5 mL) was stirred at room temperature for 20 h, after which methanol (35 µL) was added. After 1 h, a solution of tributyl ammonium phosphate (1 mmol) was added in DMF (1 mL).

The reaction was stirred for additional 24 h. After quenching the reaction mixture with 2 mL of TEAB buffer (pH 7.5), solvent was removed in vacuo at <40°C and the residue was purified as described above (retention time 9.7 min).

3. N^6(benzyl) ADP (2.5 µmol, molar absorbancy (ε_{max}) 15.4 × 10^3 at 265 nm at pH 7.00) was dissolved in DMF (200 µL) and carbonyl diimidazole (8 mg, 10 µmol) was added to it. The reaction mixture was stirred for 24 h at room temperature after which methanol (4 µL) was added and the reaction was stirred for an additional 1 h. ^{32}P orthophosphoric acid (5 mCi, 8500 Ci/mmol) was dried in vacuo, dissolved in DMF (100 µL), and was added to the reaction mixture. After stirring for 24 h, DMF was removed in vacuo and the radiolabeled analog was purified by ion-exchange chromatography using DEAE (1.5- × 7-cm packed volume) and a gradient of 0.1–1 M TEAB buffer pH 7.5 at a flow rate of 1.5 mL/min. The purified product was concentrated in vacuo at <40°C by rotary evaporation. The concentrated triphosphate was redissolved in 200 µL of water and the concentration was determined by scintillation counting (yield 20%). The [γ-^{32}P] N^6(benzyl) ATP was characterized by co-injection of the radiolabeled material with an authentic sample of N^6(benzyl) ATP *(25)* on a strong anion exchange-high-performance liquid chromatography column (retention time 11.2 min).

3.2.2. Expression of Wild-Type and Mutant Kinase in E. coli (GST-XD4 as a Model System)

XD4 is a truncated form of v-Src, that contains an intact SH1 catalytic domain, but lacks SH2 and SH3 domains and exhibits higher specific activity than full length v-Src *(45)*.

1. Expression of wild-type and mutant XD4 was carried out in DH5α cells. A single colony was inoculated into 25 mL of LB superbroth liquid media with 100 µg/mL ampicillin. The culture was grown at 37°C overnight. This culture was added to 250 mL of LB superbroth. After 2 h at 37°C (OD$_{600}$ = 0.5), IPTG was added to a final concentration of 1 mM. The culture was shaken for 5 to 6 h at 37°C and then was centrifuged at 2500g for 15 min.
2. The pellet was resuspended in 5 mL of 25 mM Tris-HCl, pH 8.0, 1 mM EDTA, 1 mM DTT buffer, and lysed twice at 8000 psi in a French press at 4°C.
3. The lysate was centrifuged again at 2500g for 10 min. The supernatant was added to 1 mL of reconstituted glutathione beads and was gently shaken for 30 min on ice.
4. The slurry was added to a column (Fisher Scientific cat. no. 11-387-50) and the beads were washed with 10 mL of 25 mM Tris-HCl, pH 8.0, 10 mM EDTA buffer followed by 50 mM Tris-HCl pH 8.0 (5 mL). The glutathione fusion protein was eluted with 4 mL of 10 mM free glutathione, 50 mM Tris-HCl, pH 8.0, 150 mM NaCl solution.

3.2.3. Kinase Assays with GST-XD4 to Determine Optimal ATP Analog

3.2.3.1. INHIBITION ASSAY

1. Purified XD4 and XD4-as1 proteins were mixed with kinase buffer, 250 μM Src peptide substrate (EIYGEFKKK), [γ-^{32}P]ATP (5 μCi) and 100 μM cold A*TP analog in a total volume of 30 μL.
2. After 20 min incubation at room temperature, the reaction was terminated by the addition of 50 μL of 10% phosphoric acid.
3. After a brief centrifugation, 50 μL of each supernatant was spotted on p81 filters, washed four times with 1% phosphoric acid, rinsed with acetone, and finally radioactivity remaining on the filters was determined.

3.2.3.2. CATALYTIC ACTIVITY ASSAY

1. Kinase assays were performed using GFP protein-based kinase substrates *(41)* with both wild-type and mutant XD4 in the presence of cold ATP analogs for 20 min at room temperature.
2. The reaction mixtures were boiled in the sample buffer for 5 min. Proteins were resolved on 12% SDS-PAGE, electrophoretically transferred to nitrocellulose membrane and were probed with anti-phosphotyrosine (4G10), followed by horseradish peroxidase conjugated anti-mouse antibody. Enhanced chemiluminescence was used for antibody detection as described in the manufacturer's instructions and visualized by Supersignal kit.

3.2.4. Transfection and Retroviral Infection of v-Src and v-Src-as1 into NIH3T3 Cells

1. Both *v-Src* and *v-Src-as1* (cloned into pBabe puro plasmids) were transiently transfected into Bosc 23 cells using the calcium phosphate transfection method *(46)*.
2. Culture medium containing the retroviruses were harvested from 72 h to 5 d after transfection. Viral supernatants were stored at –80°C.
3. Negative control retroviruses lacking *v-Src* were prepared by using pBabe puro vector alone.
4. 5×10^5 NIH3T3 cells were seeded on a 10-cm culture dish and were allowed to attach to the plate for several hours. Infection was performed by adding the retrovirus supernatant (3 mL) in the presence of polybrene (4 μg/mL).
5. After 3 h the viral supernatant was removed and replaced with fresh media (10% bovine calf serum/Dulbecco's modified eagles medium). After 48 h, the infected cells were selected in the presence of 2.5 μg/mL of puromycin. After another 24 h greater than 95% cells were viable by trypan blue staining.
6. The expression of wild-type and *v-Src-as1* in NIH3T3 cells was confirmed by immunoprecipitation and immunoblotting. Both wild-type and v-Src-as1 proteins were present in equal amounts.

3.2.5. Kinase Assays and 2D Gel Electrophoresis of v-Src and v-Src-as1 NIH3T3 Cells

1. Cells were lysed at 4°C in modified RIPA buffer (1% NP 40, 50 mM Tris-HCl, pH 7.5, 150 mM NaCl, 0.25% sodium deoxycholate, 2 mM ethylenebis-(oxyethylenenitrilo)tetraacetic acid, 1 mM phenylmethylsulfonyl fluoride, 10 μg/mL leupeptin, and 10 μg/mL aprotinin) and cleared by centrifugation at 20,000g for 15 min at 4°C.
2. Cleared lysates were mixed with 10 MgCl$_2$, 10 mM MnCl$_2$, 1 mM sodium orthovanadate, 100 μM ATP, 5 μCi of [γ-^{32}P] ATP (6000 Ci/mmol) or [γ-^{32}P]N^6 (benzyl) ATP (8500 Ci/mmol) and were incubated for 45 min at room temperature.
3. Cell lysates were boiled for 5 min in sample buffer (62.5 mM Tris-HCl, pH 6.8, 2.5% SDS, 10% glycerol, 2.5 mg/mL DTT, 2.5% β-mercaptoethanol) and separated on 10% SDS-PAGE gels. Separated proteins were electrophoretically transferred either to immobilon polyvinylidene membranes with a semidry apparatus (Owl Scientific) for 1 h at 10V. For autoradiography, the membranes were exposed to Kodak Biomax MS film (Kodak) with a Biomax MS intensifying screen at −70°C.
4. 2D gel electrophoresis of radiolabeled samples was conducted by Kendrick Labs (Madison, WI) using 2% pH 3.5–10 ampholines (Pharmacia) for 9600 volt h. One mg of an IEF internal standard, tropomyosin, was added to each sample. 10% SDS slab electrophoresis was performed for about 4 h at 12.5 mA/gel. The gels were Coomasassie blue stained, dried, and exposed to Kodak X-OMAT AR film.

3.2.6. Immunoprecipitation and Kinase Assays of v-Src and v-Src-as1 NIH3T3 Cells

1. Cells were lysed at 4°C in modified RIPA buffer with 0.1% SDS and 2 mM sodium orthovanadate.
2. After centrifugation, the cleared lysate was incubated with 1–5 μg of the desired antibody at 4°C for 2 h with gentle rotation, followed by incubation with 50 μL of 50% protein A Sepharose slurry.
3. The immune complexes were washed three times (500 μL each) in modified RIPA buffer without SDS and twice in kinase buffer (30 mM Tris-HCl, pH 7.5, 10 mM MgCl$_2$, and 10 mM MnCl$_2$).
4. Kinase reactions were initiated by adding 0.5 μCi of [γ-^{32}P]ATP alone or 1 μCi of [γ-^{32}P]N^6(benzyl) ATP in the presence of 10 μM ATP and incubated for 10–15 min (as indicated in figure legends) at room temperature.
5. After boiling for 5 min in sample buffer immune complex proteins were separated on 10% SDS-PAGE gels, electrophoretically transferred and exposed as described before.

3.2.6. Mass Spectral Analysis

1. Gel spots were digested by adding 0.05 μg of modified trypsin in the minimum amount of 0.025 *M* Tris-HCl, pH 8.5, and leaving the gel overnight at 32°C.
2. Peptides are extracted with 2 × 50 μL of 50% acetonitrile/2% TFA and the combined extracts are dried and resuspended in matrix solution prepared by making a 10 mg/mL solution of 4-hydroxy-α-cyanocinnamic acid in 50% acetonitrile/0.1% TFA and adding two internal standards, angiotensin and bovine insulin to the matrix solution.
3. Matrix-assisted laser desorption/ionisation mass spectrometric analysis was performed on the digest using a PerSeptive Voyager DE-RP mass spectrometer in the linear mode. These peptides were analyzed by peptide mass fingerprinting and data base searching using Protein Prospector (Columbia University/HHMI, Protein Core Facility, New York, NY 10032).

4. Notes

1. A*TP are unstable at room temperature and susceptible to freeze/thaw. Therefore, it is best to make small aliquots of A*TP solution in water and freeze them at –20°C.
2. Optimal storage conditions (temperature and buffers) for the kinase of interest should be determined to retain maximal activity over a period of time.
3. 1 *M* TEAB buffer is best when prepared fresh, however, it can be stored at 4°C for couple of days.

References

1. Hunter, T. (2000) Signaling-2000 and Beyond. *Cell* **100,** 113–127.
2. Sawyers, C. L. (2002) Rational therapeutic intervention in cancer: kinases as drug targets. *Curr. Opin. Genet. Dev.* **12,** 111–115.
3. Lewis, A. J. and Manning, A. M. (1999) New targets for anti-inflammatory drugs. *Curr. Opin. Chem. Biol.* **3,** 489–494.
4. Maccioni, R. B., Munoz, J. P., and Barbeito, L. (2001) The molecular bases of Alzheimer's disease and other neurodegenerative disorders. *Arch. Med. Res.* **32,** 367–381.
5. Wagey, R. T. and Krieger, C. (1998) Abnormalities of protein kinases in neurodegenerative diseases. *Prog. Drug Res.* **51,** 133–183
6. Jiang, G. and Zhang, B. B. (2002) Pi 3-kinase and its up- and down-stream modulators as potential targets for the treatment of type II diabetes. *Front. Biosci.* **7,** d903–d907.
7. Arasteh, K. and Hannah, A. (2000) The role of vascular endothelial growth factor (VEGF) in AIDS-related Kaposi's sarcoma. *Oncologist* **5(Suppl. 1),** 28–31.
8. Ghaffari, S., Wu, H., Gerlach, M., Han, Y., Lodish, H. F., and Daley, G. Q. (1999) BCR-ABL and v-SRC tyrosine kinase oncoproteins support normal erythroid development in erythropoietin receptor-deficient progenitor cells. *Proc. Natl. Acad. Sci. USA* **96,** 13,186–13,190.

9. Ihle, J. N. (2000) The challenges of translating knockout phenotypes into gene function. *Cell* **102**, 131–134.

10. Songyang, Z. and Cantley, L. C. (1998) The use of peptide library for the determination of kinase peptide substrates. *Methods Mol. Biol.* **87**, 87–98.

11. Fukunaga, R. and Hunter, T. (1997) MNK1, a new MAP kinase-activated protein kinase, isolated by a novel expression screening method for identifying protein kinase substrates. *EMBO J.* **16**, 1921–1933.

12. Lock, P., Abram, C. L., Gibson, T., and Courtneidge, S. A. (1998) A new method for isolating tyrosine kinase substrates used to identify fish, an SH3 and PX domain-containing protein, and Src substrate. *EMBO J.* **17**, 4346–4357.

13. Stukenberg, P. T., Lustig, K. D., McGarry, T. J., King, R. W., Kuang, J., and Kirschner, M. W. (1997) Systematic identification of mitotic phosphoproteins. *Curr. Biol.* **7**, 338–348.

14. Deng, S. J., Liu, W., Simmons, C. A., Moore, J. T., and Tian, G. (2001) Identifying substrates for endothelium-specific Tie-2 receptor tyrosine kinase from phage-displayed peptide libraries for high throughput screening. *Comb. Chem. High Throughput Screen* **4**, 525–533.

15. Zhu, H., Klemic, J. F., Chang, S., Bertone, P., Casamayor, A., Klemic, K. G., et al. (2000) Analysis of yeast protein kinases using protein chips. *Nat. Genet.* **26**, 283–289.

16. Zhu, H., Bilgin, M., Bangham, R., Hall, D., Casamayor, A., Bertone, P., et al. (2001) Global analysis of protein activities using proteome chips. *Science* **293**, 2101–2105.

17. Pawson, T. and Nash, P. (2000) Protein-protein interactions define specificity in signal transduction. *Genes Dev.* **14**, 1027–1047.

18. Perez, O. D. and Nolan, G. P. (2002) Simultaneous measurement of multiple active kinase states using polychromatic flow cytometry. *Nat. Biotechnol.* **20**, 155–162.

19. Neet, K. and Hunter, T. (1995) The nonreceptor protein-tyrosine kinase CSK complexes directly with the GTPase-activating protein-associated p62 protein in cells expressing v-Src or activated c-Src. *Mol. Cell Biol.* **15**, 4908–4920.

20. Wu, L. W., Mayo, L. D., Dunbar, J. D., Kessler, K. M., Baerwald, M. R., Jaffe, E. A., et al. (2000) Utilization of distinct signaling pathways by receptors for vascular endothelial cell growth factor and other mitogens in the induction of endothelial cell proliferation. *J. Biol. Chem.* **275**, 5096–5103.

21. Lewis, T. S., Hunt, J. B., Aveline, L. D., Jonscher, K. R., Louie, D. F., Yeh, J. M., et al. (2000) Identification of novel MAP kinase pathway signaling targets by functional proteomics and mass spectrometry. *Mol. Cell* **6**, 1343–1354.

22. Yoshimura, Y., Shinkawa, T., Taoka, M., Kobayashi, K., Isobe, T., and Yamauchi, T. (2002) Identification of protein substrates of Ca(2+)/calmodulin-dependent protein kinase II in the postsynaptic density by protein sequencing and mass spectrometry. *Biochem. Biophys. Res. Commun.* **290**, 948–954.

23. Knebel, A., Morrice, N., and Cohen, P. (2001) A novel method to identify protein kinase substrates: eEF2 kinase is phosphorylated and inhibited by SAPK4/p38delta. *EMBO. J.* **20**, 4360–4369.

24. Zhou, S. and Cantley, L. C. (1995) Recognition and specificity in protein tyrosine kinase-mediated signalling. *Trends Biochem. Sci.* **20,** 470–475.

25. Shah, K., Liu, Y., Deirmengian, C., and Shokat, K. M. (1997) Engineering unnatural nucleotide specificity for Rous sarcoma virus tyrosine kinase to uniquely label its direct substrates. *Proc. Natl. Acad. Sci. USA* **94,** 3565–3570.

26. Liu, Y., Shah, K., Yang, F., Witucki, L., and Shokat, K. M. (1998) Engineering Src family protein kinases with unnatural nucleotide specificity. *Chem. Biol.* **5,** 91–101.

27. Liu, Y., Shah, K., Yang, F., Witucki, L., and Shokat, K. M. (1998) A molecular gate which controls unnatural ATP analogue recognition by the tyrosine kinase v-Src. *Bioorg. Med. Chem.* **6,** 1219–1226.

28. Liu, Y., Bishop, A., Witucki, L., Kraybill, B., Shimizu, E., Tsien, J., et al. (1999) Structural basis for selective inhibition of Src family kinases by PP1. *Chem. Biol.* **6,** 671–678.

29. Adams, J., Huang, P., and Patrick, D. (2002) A strategy for the design of multiplex inhibitors for kinase-mediated signalling in angiogenesis. *Curr. Opin. Chem. Biol.* **6,** 486–492.

30. Weiss, E. L., Bishop, A. C., Shokat, K. M., and Drubin, D. G. (2000) Chemical genetic analysis of the budding-yeast p21-activated kinase Cla4p. *Nat. Cell Biol.* **2,** 677–685.

31. Habelhah, H., Shah, K., Huang, L., Burlingame, A. L., Shokat, K. M., and Ronai, Z. (2001) Identification of new JNK substrate using ATP pocket mutant JNK and a corresponding ATP analogue. *J. Biol. Chem.* **276,** 18,090–18,095.

32. Witucki, L. A., Huang, X., Shah, K., Liu, Y., Kyin, S., Eck, M. J., et al. (2002) Mutant tyrosine kinases with unnatural nucleotide specificity retain the structure and phospho-acceptor specificity of the wild-type enzyme. *Chem. Biol.* **9,** 25–33.

33. Polson, A. G., Huang, L., Lukac, D. M., Blethrow, J. D., Morgan, D. O., Burlingame, A. L., et al. (2001) Kaposi's sarcoma-associated herpesvirus K-bZIP protein is phosphorylated by cyclin-dependent kinases. *J. Virol.* **75,** 3175–3184.

34. Shah, K. and Shokat, K. M. (2002) A chemical genetic screen for direct v-Src substrates reveals ordered assembly of a retrograde signaling pathway. *Chem. Biol.* **9,** 35–47.

35. Scott, M. P. and Miller, W. T. (2000) A peptide model system for processive phosphorylation by Src family kinases. *Biochemistry* **39,** 14,531–14,537.

36. Kraybill, B. C., Elkin, L. L., Blethrow, J. D., Morgan, D. O., and Shokat, K. M. (2002) Inhibitor Scaffolds as New Allele Specific Kinase Substrates. *J. Am. Chem. Soc.* **124,** 12,118–12,128.

37. Bishop, A. C., Shah, K., Liu, Y., Witucki, L., Kung, C., and Shokat, K. M. (1998) Design of allele-specific inhibitors to probe protein kinase signaling. *Curr. Biol.* **8,** 257–266.

38. Bishop, A. C., Buzko, O., and Shokat, K. M. (2001) Magic bullets for protein kinases. *Trends Cell Biol.* **11,** 167–172.

39. Bishop, A. C., Ubersax, J. A., Petsch, D. T., Matheos, D. P., Gray, N. S., Blethrow, J., et al. (2000) A chemical switch for inhibitor-sensitive alleles of any protein kinase. *Nature* **407,** 395–401.

40. Carroll, A. S., Bishop, A. C., DeRisi, J. L., Shokat, K. M., and O'Shea, E. K. (2001) Chemical inhibition of the Pho85 cyclin-dependent kinase reveals a role in the environmental stress response. *Proc. Natl. Acad. Sci. USA* **98,** 12,578–12,583.

41. Yang, F., Liu, Y., Bixby, S. D., Friedman, J. D., and Shokat, K. M. (1999) Highly efficient green fluorescent protein-based kinase substrates. *Anal. Biochem.* **266,** 167–173.

42. Buzko, O. V. and Shokat, K. M. (2002) A Kinase Sequence database:sequence alignments and family assignment. *Bioinformatics* **18,** 1274–1275.

43. Polte, T. R. and Hanks, S. K. (1997) Complexes of focal adhesion kinase (FAK) and Crk-associated substrate (p130(Cas)) are elevated in cytoskeleton-associated fractions following adhesion and Src transformation. Requirements for Src kinase activity and FAK proline-rich motifs. *J. Biol. Chem.* **272,** 5501–5509.

44. Hecht, S. M. and Kozarich, J. W. (1973) A chemical synthesis of adenosine 5′-(gamma-32P)triphosphate. *Biochim. Biophys. Acta* **331,** 307–309.

45. DeClue, J. E. and Martin, G. S. (1989) Linker insertion-deletion mutagenesis of the v-src gene: isolation of host- and temperature-dependent mutants. *J. Virol.* **63,** 542–554.

46. Pear, W. S., Nolan, G. P., Scott, M. L., and Baltimore, D. (1993) Production of high-titer helper-free retroviruses by transient transfection. *Proc. Natl. Acad. Sci. USA* **90,** 8392–8396.

22

Studying the Optimal Peptide Substrate Motifs of Protein Kinase C Using Oriented Peptide Libraries

Michael B. Yaffe

1. Introduction

Determining the in vivo substrates of different protein kinase C (PKC) family members is critical to understanding how these enzymes regulate cell division, differentiation, and growth *(1–3)*. The evolution of new technologies, particularly mass spectroscopy, has greatly facilitated this analysis *(4–6)*, although in this case the investigator must still perform large-scale biochemical purification of the presumed substrate to determine its identity. The use of oriented peptide library screening combined with bioinformatics offers an alternative approach to predicting the likely substrates of protein kinases through motif identification and database searching *(7–12)*. In this approach, peptide libraries containing an orienting Ser (or Thr) residue flanked by a series of degenerate positions containing a mixture of all possible amino acids are phosphorylated in vitro by the PKC (or other kinase) of interest, and the subset of phosphorylated peptides separated from the bulk of nonphosphory-lated peptides by immobilized metal affinity chromatography *(13–15)*. The recovered phosphopeptides are then sequenced by Edman degradation in bulk, and selection for each amino acid in positions flanking the fixed Ser or Thr is determined. The end result of this process is a matrix of selection values that describe the optimal phosphorylation motif for the various PKC isotype that was investigated. These matrices can then be used to search protein sequence databases and identify potential substrate proteins that contain the best matches to the optimal PKC phosphorylation motif. The putative substrates can then be examined in vitro and in vivo to determine whether they are, in fact, PKC substrates under physiological conditions. A basic outline of the technique is presented in **Fig. 1**.

From: *Methods in Molecular Biology, vol. 233: Protein Kinase C Protocols*
Edited by: A. C. Newton © Humana Press Inc., Totowa, NJ

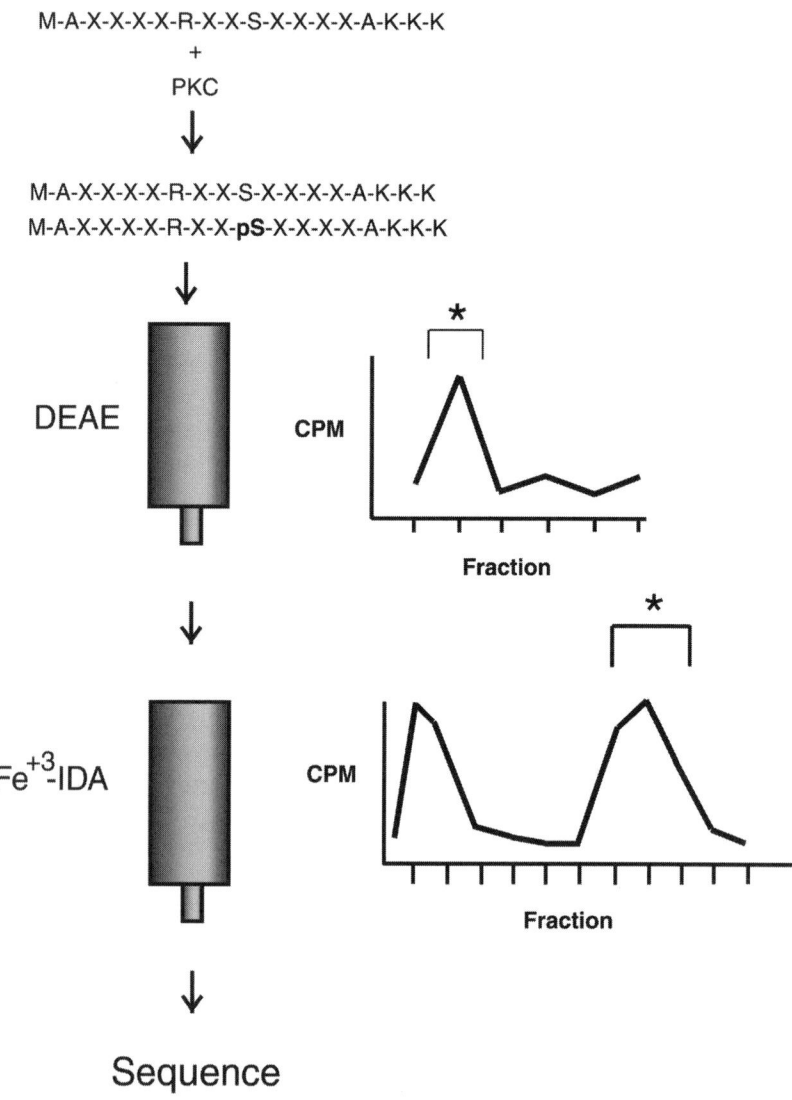

Fig. 1. Schematic of oriented peptide library screening procedure for PKCs.

2. Materials

2.1. Reagents

1. Standard reagents for peptide synthesis and protein sequencing.
2. DEAE Sephacel (Sigma cat. no. I6505).
3. 30% acetic acid in H_2O.

4. 0.1 *M* acetic acid in H$_2$O.
5. 85% phosphoric acid.
6. 10 m*M* ATP, pH 7.0 (store at –20°C).
7. γ-^{32}P-ATP (specific activity 3000 Ci/mmol, 10 mCi/mL).
8. 100 m*M* FeCl$_3$ in 0.1 *M* acetic acid (prepare fresh).
9. 0.1 *M* ethylenediaminetetraacetic acid, pH 8.0.
10. 0.1% ammonium acetate, pH 8.0, freshly prepared and stored at 4°C.
11. 0.1% ammonium acetate, pH 9.5, freshly prepared and stored at 4°C.
12. 0.1% ammonium acetate, pH 11.5, freshly prepared and stored at 4°C.
13. Degenerate Ser or Thr-containing peptide libraries (see below).
14. Immobilized Iminodiacetic Acid Agarose (IDA Agarose) with 1,4 butanediol diglycidyl ether spacer (Pierce cat. no. 20277).
15. Enzymatically active PKC of interest to be screened (*see* **Note 1**).
16. Phosphatidylserine (1 mg/mL) and diacyl glycerol (500 µ*M*) solution in 25 m*M* Tris-HCl, pH 7.5 (store at –20°C under nitrogen; *see* **Note 2***)*.

2.2. Equipment

1. Peptide Synthesizer (Applied Biosystems 431A) and protein sequencer (Applied Biosystems Procise, single- or multicartridge instrument, or Applied Biosystems 477A).
2. Polypropylene columns (12 mL and 3 mL capacity) with tip plugs (Bio-Rad Poly-Prep).
3. Chromatography columns.
4. P81 paper.
5. Scintillation counter.
6. Ring stand with clamps.
7. Speed-Vac.
8. Microcentrifuge.
9. Microcentrifuge tubes.
10. Disposable 50- and 15-mL polyproylene centrifuge tubes (Corning cat. nos. 430291 and 430052 or equivalent).

3. Methods
3.1. Design and Synthesis of Degenerate Peptide Libraries

1. The selection of the one or more residues around which to orient a degenerate peptide library is critical and determines the success or failure of the technique. In general, all peptide libraries used in the initial screening of a protein kinases should contain a single fixed Ser residue, flanked by additional fixed and degenerate residues as necessary. It has been our experience that substitution of Thr for Ser in the orienting position has no effect on the optimal phosphorylation motif selected by the protein kinase, but tends to perform somewhat less robustly as a general kinase substrate. Consequently, we do not recommend that Thr-oriented libraries be used in the initial determination of kinase motifs, although

they can, and perhaps should, be used during motif refinement experiments using secondary libraries.

Frequently, libraries containing only a single fixed Ser residue are insufficiently phosphorylated by the kinase of interest to provide enough phosphopeptides for accurate analysis. In these cases, the inclusion of additional fixed amino acids besides Ser that known to be important for phosphorylation usually suffices to provide an adequate kinase substrate for motif determination. For basophilic kinases such as PKC family members, this is accomplished by constructing libraries containing a fixed Arg residue 3 positions amino-terminal to the fixed Ser position (i.e., an Arg-Xxx-Xxx-Ser sequence, see below). For most PKC family members, this fixed Arg residue is required for efficient phosphorylation of peptide substrates.

Libraries containing eight or fewer degenerate positions work best for initial screening. These libraries already have more than 25 billion degenerate peptide combinations, and one can position the fixed amino acid residue within the center of the degenerate stretch, or along either end. Incorporating more than 10 degenerate positions in the starting library will increase the initial complexity of the peptide mixture; however, it can simultaneously decrease the relative fraction of peptides within the starting mixture that are phosphorylated, yielding inadequate amounts for sequencing. Longer stretches of degenerate positions are possible, however, in secondary libraries. Libraries are routinely constructed beginning with the sequence Met-Ala, and terminated after the last degenerate position with the sequence Ala-Lys-Lys-Lys. A typical library used to deduce the phosphorylation motif for PKC family members, for example, contains the sequence Met-Ala-Xxx-Xxx-Xxx-Xxx-Arg-Xxx-Xxx-Ser-Xxx-Xxx-Xxx-Xxx-Ala-Lys-Lys-Lys, where Xxx indicates all amino acids except Cys, Ser, Thr, and Tyr. The Met-Ala sequence at the N-terminus of the peptide libraries allows verification that peptides from this mixture are being sequenced, and provides a quantitative estimate of peptide binding. The N- and C-terminal Ala residues bracketing the degenerate positions allow estimation of how much peptide has been lost during sequencing (i.e., repetitive yield) and gives an indication of cleavage lag and carryover between sequencing cycles. The polylysine tail insures solubility of the libraries and aids in retention of the peptides on the Biobrene-coated filters during the sequencing steps. The omission of Ser, Thr, and Tyr from the degenerate positions forces phosphorylation of the library by the kinase of interest to occur only at the position of the orienting Ser residue within the degenerate sequence motif.

Once an initial binding motif has been determined, it is often helpful to construct additional secondary libraries to further refine and expand individual amino acid preferences. For example, many PKC family members select for Phe in the Ser+1 position (9). One could then construct a secondary library containing the general sequence Met-Ala-Xxx Xxx- Xxx- Xxx-Arg- Xxx- Xxx-Ser-Phe- Xxx- Xxx- Xxx- Xxx-Ala-Lys-Lys-Lys. Because the general affinity of the PKC for any random peptide within this Arg-Xxx-Xxx-Ser-Phe library is higher than that for a peptide from the initial Arg-Xxx-Xxx-Ser-Xxx library, weaker interactions from residues within and bracketing the Arg-Xxx-Xxx-Ser-Phe core will now dominate the selectivity. In the initial screening performed with the Arg-Xxx-Xxx-Ser-only library, information about these interactions was reduced, because the binding was partially dominated by selection for Phe in the Ser+1 position. Furthermore, once the secondary amino acid selection has been determined for the degenerate positions by this approach, a retrospective analysis can be used to determine the relative importance of the various residues that were fixed in the initial screening. For example, an analysis performed on PKCs using a library of general sequence Met-Ala-Xxx Xxx- Xxx- Xxx- Xxx-Xxx-Ser-Phe- Xxx- Xxx-Ala-Lys-Lys-Lys would reveal the relative selection for the Arg residue in the Ser–3 position, which was fixed in the previous library screening experiments.

2. Solid-phase synthesis of the degenerate peptide libraries is performed using a Rink amide resin matrix and an automated peptide synthesizer (Applied Biosystems 431A) according to standard BOP/HOBt coupling protocols with commercially available Fmoc-protected amino acids. At the degenerate positions, equal amounts of Fmoc-blocked amino acids (except Cys, Ser, Thr, and Tyr) are weighed out and loaded into the synthesizer cartridges so that the mixture is in fourfold molar excess to the coupling resin. The ratio of Fmoc-blocked amino acids needs to be adjusted slightly on different synthesizers in order to obtain a roughly even distribution of degenerate amino acids. Synthesis of oriented degenerate peptide libraries can also be performed by an outside peptide synthesis facility for a reasonable cost. One such supplier is the Tufts University core facility (http://www.tucf.com).

3. Once the synthesis is complete, the degenerate peptide libraries are deprotected and cleaved from the resin for 3 h at room temperature using 4 mL of a cleavage reagent prepared by dissolving 0.75 g of phenol in 10 mL of trifluoroacetic acid and adding 0.5 mL of H_2O, 0.5 mL of thioanisole, and 0.25 mL of ethane dithiol. The crude library mixture is precipitated by slow addition into 50 mL of diethyl ether followed by cooling to –20°C for 1 h. The precipitate is filtered using a fretted funnel, washed six times with chilled diethyl ether, dissolved in H_2O, and lyophilized. The final peptide libraries are stored as dry powders at –20°C.

4. Working stocks of the libraries are prepared at 10–30 mg/mL in either phosphate-buffered saline or in the kinase buffer of interest and adjusted by addition of dry *N*-hydroxyethylpiperazine-*N′*-2-ethanesulfonate or Tris-HCl to pH of approx 7.0. A small amount of the library is sequenced to insure that all amino acids are present at similar amounts (within a factor of 3) in each degenerate position.

3.2. Phosphorylating the Oriented Degenerate Peptide Library by PKCs

The PKC of interest should be incubated with 1 mg of the degenerate peptide library mixture until 0.2–1% of the library has been phosphorylated.

For a typical peptide library with an average Mw of approx 2000 Da, this corresponds to approx 1–5 nmol of phosphorylated product that is further purified as described in **Subheadings 3.3.** and **3.4.**

1. In a 1.5-mL microfuge tube add the following:

   30 µl of 10X kinase buffer. (Use whatever buffer the kinase prefers for phosphory-lation of a known substrate. For many PKCs, a good 10× buffer is 500 m*M* Tris-HCl, pH 7.5, 100 m*M* MgCl$_2$, 10 m*M* dithiothreitol.)

   1 mg of the degenerate peptide library mixture (100 µL of a 10 mg/mL solution or 33 µL of a 30 mg/mL solution.

   3 µL of 10 m*M* ATP.

   6 µL of a phosphatidylserine (1 mg/mL) and diacyl glycerol (500 µ*M*) solution in 25 m*M* Tris-HCl, pH 7.5.

   x µL of purified active PKC (typically 5–25 µL).

   y µL of double distilled deionized H$_2$O to bring the final volume to 300 µL.

   For conventional PKC assays, the final reaction should also contain 200 µ*M* CaCl$_2$, whereas assays with novel and atypical PKCs should include 0.5 m*M* ethylenebis(oxyethylenenitrilo)tetraacetic acid.

2. Incubate the kinase reaction at 30°C. Determine the extent of library phosphoryla-tion by removing a 5-µL sample of the reaction mixture at 0, 1, and 2 h and spotting it onto labeled 2- × 2-cm squares of P81 paper. Allow the squares to air dry, and then wash the squares four times in a beaker containing 50 mL of 0.45% phosphoric acid for 2 min/wash. Place the squares in scintillation vials, cover with 5–10 mL of scintillation fluid, and count.

3. Based on the P81-bound counts and the specific activity of the γ-^{32}P-ATP on the day the assay was performed, it is a straightforward matter to determine the amount of phosphorylated peptide. Each 300-µL reaction contains a total of 30 nmol of ATP into which the trace amount of radioactive ATP is distributed. Sub-tract the cpm incorporated from the 1- and 2-h sample from the 0 h background control and multiply by 60 (=300 µL total/5 µL counted) to get the total cpm incor-porated into the peptide library. Calculate *cpm_incorporated* ×30 to determine the *total_^{32}P_cpm_in_reaction* to determine the total nmol of phosphorylated peptide library. For example, assuming a specific activity of 10 µCi/mL and a counting efficiency of 100%, 1 nmol of ATP incorporation into the peptide library mixture would give a total of 7400 cpm incorporated. It may be necessary to allow the kinase reaction to proceed overnight to obtain a sufficient amount of phosphorylated product.

4. Stop the reaction by adding 30% acetic acid to a final concentration of 15%.

3.3. Removal of Unincorporated ATP and Separation of Phosphorylated from Nonphosphorylated Peptides

The peptide library mixture is partially purified from the unincorporated ATP using DEAE-Sephacel, and the phosphorylated peptides separated from their nonphosphorylated counterparts using an immobilized Fe^{+3} column.

1. Prepare the DEAE-Sephacel resin by placing 3 mL of the slurry and 10 mL of 30% acetic acid in a 50-mL disposable polypropylene centrifuge tube (Corning cat. no. 430291 or equivalent). Mix gently, and then pellet the resin by low-speed centrifugation (250g for 2 min). Discard the supernatant and resuspend the washed resin in 1.5 mL of 30% acetic acid.

2. Clamp a disposable 12-mL chromatography column in a ring stand, keeping the tip plugged. Fill the column with 8 mL of 30% acetic acid and add 1.5 mL of the washed resuspended DEAE-Sephacel slurry. Allow resin to settle for 20 min by gravity, giving a packed bed volume of approx 1 mL.

3. Number seven microfuge tubes 1 through 7. Using a marker, indicate the position of 600 μL on tube 1, 1 mL on tube 2, and 500 μL on tubes 3–7 on the sides of the tubes.

4. Remove the plug from the column and allow the column to run until the top of the resin is just exposed. Apply the phosphorylated library mixture to the top of the resin. Allow the column to run by gravity, collecting the first 600 μL of eluent (the dead volume) into tube 1. Switch to tube 2 and add an additional 600 μL of 30% acetic acid to the top of the DEAE column. Collect the effluent in tube 2. Finally, add 3 mL of 30% acetic acid to the top of the column, and collect the next 400 μL into tube 2 so that the final volume in this tube is 1 mL. Collect 500-μL fractions into tubes 3–7.

5. Count 5 μL of each fraction on a scintillation counter, and then calculate the total cpm contained in each fraction. Plot the cpm as a function of fraction number. The phosphorylated peptide library should elute in fraction 2, whereas the unincorporated ATP generally elutes in fractions 5–7 and later (**Fig. 1**).

6. Dry down the fraction containing the peptide library (fraction 2) overnight on a Speed-Vac apparatus. At this point the sample can be stored at –20°C before proceeding to the next step.

7. Clamp a disposable 3-mL chromatography column in a ring stand, keeping the tip plugged. Fill the column with 2–3 mL of H_2O, add 900 μL of a 1:1 slurry of the immobilized iminodiacetic acid Agarose, and allow the resin to settle by gravity (*see* **Note 3**).

8. Remove the tip plug from the column and allow the covering H_2O to drain until the surface of the resin is exposed. Strip the resin by washing the column sequentially with 4 mL of H_2O, 2 mL of 0.1 M ethylenediaminetetraacetic acid, 4 mL of H_2O, and 2 mL of 0.1 M acetic acid.

9. Charge the resin by adding 2 mL of 100 mM $FeCl_3$ in 0.1 M acetic acid, followed by a single wash using 2 mL of 0.1 M acetic acid. The column should be yellow in color.

10. Prerun the column to elute any contaminants prior to sample loading. Wash the column sequentially with 2 mL of 0.1 M acetic acid, 1 mL of H_2O, 1 mL of 0.1% ammonium acetate, pH 8.0, 2 mL of 0.1% ammonium acetate, pH 9.5, 10 mL of H_2O, and 6 mL of 0.1 M acetic acid. The color of the column will change progressively from yellow to brown, and then back to light yellow again.

11. During the pre-run step, resuspend the dried peptide library sample in 50 µL of 50 mM MES (pH 5.5), 1 M NaCl (Buffer A) by aggressive vortexing.

12. Apply the sample to the Fe^{3+} column. Rinse the sample tube with another 50 µL of Buffer A. A hand-held scintillation counter is useful to insure that most of the radioactivity, corresponding to the phosphorylated peptide library, has been adequately transferred to the column.

13. Number microcentrifuge tubes 1–10. Develop the column, collecting 0.5-mL fractions, by eluting with 1 mL of 0.1 M acetic acid (fractions 1, 2), 1 mL of H_2O (fractions 3, 4), 1 mL of 0.1% ammonium acetate pH 8.0 (fractions 5, 6), and 2 mL of 0.1% ammonium acetate pH 11.5 (fractions 7–10). Count 5 µL of each fraction in a scintillation counter and plot the cpm as a function of fraction number. Most of the radioactivity should elute in fractions 1–2 and 7–10. Fractions 1–2 correspond to nonpeptide-containing ^{32}P, whereas fractions 7–10 correspond to the PKC-phosphorylated peptides (**Fig. 1**).

14. Pool the tubes containing the phosphorylated peptide library (generally fractions 7–9) into a single tube and dry down in a Speed-Vac apparatus. Dissolve the pellet in 250–500 µL of H_2O and relyophilize several times to remove as much volatile salt as possible. It may be necessary to add small amounts of 30% acetic acid to neutralize the ammonium acetate and resuspend the sample, as evidenced by release of radioactivity into the liquid phase. Resuspend the final pellet in 80 µL of H_2O and sequence 40 µL by automated Edman degradation.

3.4. Peptide Sequencing and Data Analysis

Samples can be sequenced by any commercial sequencing facility, using any sequencing protocol sufficient to detect individual amino acids at the 5–10 pmol range. It is necessary to sequence the initial peptide library mixture as well. In our lab, the peptide libraries are spotted onto Biobrene-coated glass-fiber filter discs and loaded into the cartridges of an Applied Biosystems Procise sequencer. A pulsed-liquid solvent delivery cycle is used with standard times for phenylisothiocyanate (PITC) coupling, extraction, cleavage, and transfer recommended by the manufacturer, along with a 9-min phenylthiohydantain (PTH) conversion at 64°C.

The PTH-derivitized amino acids are then detected by high-performance liquid chromatography with a gradient optimized for separation between amino acids, particularly Phe, Ile, Lys, and Leu, which elute late. The quantity of each amino acid in pmol, within each sequencing cycle, is reported. For best accuracy, amino acid standards should be prepared weekly, and the calibration run preceding each sequencing reaction performed using amounts of each PTH amino acid reasonably close to what is expected for the actual peptide library being sequenced.

To determine the optimal peptide motif phosphorylated by the PKC, the relative abundance of each amino acid at a given sequencing cycle of the phosphorylated peptide mixture is divided by the relative abundance of the same amino acid in that cycle from the starting library mixture. This ratio corrects for variations in the mol percentages of particular amino acids in the starting library as well as variations in yield of amino acid recoveries during sequencing. The calculations can easily be performed in spreadsheet fashion.

1. Let A_{ij} represent the amount, in pmol, of amino acid i reported for the PKC-phosphorylated library sample in sequencing cycle j. For example, A_{34} might represent the amount of Glu (i.e., letting i = 3 denote Glu) in sequencing cycle 4 (j = 4). Begin by calculating the mol percentage of each amino acids in one sequencing cycle (fixed j) that contains the degenerate mixture of amino acids. This value, denoted MP_{ij} (for Sample Mol Percentile), is given by the following:

$$MP_{ij} = \frac{A_{ij}}{\sum_{i=1}^{i=16} A_{ij}}$$

where the sum of A_{ij} in the denominator is performed over all 16 amino acids in that sequencing cycle j. Perform this same calculation for all remaining sequencing cycles j+1, j+2, etc., which contain degenerate amino positions.

2. Perform the identical calculations for the mol percentages of each amino acid present in the degenerate positions in the initial unphosphorylated peptide library mixture. Let B_{ij} represent the amount, in pmol, of amino acid i reported for the starting library mixture in sequencing cycle j. The mol percentage of amino acid i in cycle j for the control, denoted by CP_{ij} (Control Percentile), is similarly given by the following:

$$CP_{ij} = \frac{B_{ij}}{\sum_{i=1}^{i=16} B_{ij}}$$

3. Calculate the raw selectivity value for a given amino acid i in a particular sequencing cycle j, denoted S_{ij}^{raw}, by $S_{ij}^{raw} = MP_{ij} / CP_{ij}$.

4. Normalize the raw selectivity values so that the sum of all selectivity values in a given sequencing cycle j is equal to the total number of possible amino acids. If the library has been synthesized without Cys, Ser, Thr, or Tyr in the degenerate positions, then the sum of all selectivity values should be normalized to 16, by defining:

$$S_{ij}^{norm} = \left[\frac{S_{ij}^{raw}}{\sum_i S_{ij}^{raw}} \right] \times 16$$

where the calculation is performed for individual fixed values of j.

Graphic plots of normalized preference values vs amino acid for each sequencing cycle are very useful for revealing the optimal peptide motif. If the kinase has no specificity at a particular residue position within the motif, then the mol-percentage of all amino acids in this position will be the same as that present in the initial mixture, and the preference values for all amino acids will be approx 1. In identifying amino acids that are selected at particular positions within a motif, one should pay special attention to those amino acids that change from cycle to cycle, rather than those that remain persistently slightly elevated or depressed in relative abundance. This latter effect usually results from small systematic errors in sequencing for a particular residue between the sample and the starting library mixture (*see* **Note 4**).

It is strongly recommended that individual peptides containing the optimal motif sequence, or containing the motif with amino acid substitutions at key selected positions be synthesized and evaluated for phosphorylation by the kinase used in peptide library screening *(9,11)*. Measurements of k_m and V_{max} values for these peptides by the PKC is important to validate the peptide library motif selection.

3.5. Using the Optimal Domain Motif to Elucidate Protein Interactions and Cell-Signaling Pathways

Once the phosphorylation motif for a particular PKC has been determined, it can then be used to prospectively identify likely phosphorylation sites on known substrates. Alternatively, one can use the matrix of selectivity values to identify new potential PKC substrates via searching in protein or translated nucleic acid databases. In database searching it is important to allow for partial matches to the motif in scoring query sequences to avoid missing potential substrates because many proteins will match the motif at some but not all residue positions. A variety of software available over the internet can accept the matrix of selection values obtained from peptide library screening and search for proteins containing high-scoring matches to the optimal motif. Relevant programs include PatScan (www.unix.mcs.anl.gov/compbio/PatScan,) and Scansite (http://scansite.mit.edu) *(12)*, the latter of which is specifically designed to accept user-input matrices for searching individual protein or protein database queries, and reports a rank-ordered list of matching sites and putative substrates. One can also perform a restricted search for the motif in a subset of proteins having homology to a known target using Pattern Hit Initiated Blast, or PHI-BLAST *(16)*.

4. Notes

1. The PKC of interest can be obtained by purification from brain or other tissue or by overexpression of the recombinant protein in either baculovirus-infected insect cells or in mammalian cell culture. It may be best to express the kinase fused to an affinity tag such as glutathione-S-transferase or hexahistidine to

facilitate purification. In some cases, the introduction of point mutations that convert phosphorylated Ser or Thr residues within the kinase activation loop to Asp are necessary to increase the intrinsic phosphotransferase activity of the enzyme. Alternatively, the recombinant protein can be phosphorylated in vitro by recombinant phosphoinositide-dependent protein kinase-1 in the presence of phosphatidylinositol(3,4,5)P$_3$ *(17)*. It is critically important to purify the PKC away from co-associating protein kinases before peptide library screening to insure that the activity being measured corresponds to that of the PKC. Performing an identical peptide library screen using a catalytically inactive mutant form of the PKC can serve as a useful control in cases where the kinase is affinity-purified from cells in culture. The amount of PKC required for the assay depends entirely on its specific activity. It has been our experience that microgram quantities are usually sufficient.

2. We typically prepare stock solutions of phosphatidylserine (2 mg/mL) and diacylglycerol 10 mg/mL in chloroform, and store these under nitrogen at –20°C. To prepare the working solution, aliquot 50 µL of phosphatidyl serine stock and 1.7 µL of the diacylglycerol stock in a 1.5-mL microfuge tube and dry down under nitrogen until the lipids form a dry layer on the tube wall and bottom. Add 100 µL of a 25 mM Tris-HCl, pH 7.5 solution, and sonicate the microfuge tube for 5–10 min in a cup horn sonicator on a medium power setting. The working solution will be 1 mg/mL phosphatidylserine and 500 µM diacylglycerol and should be diluted ×50 in the final PKC reaction.

3. Alternatively, one can use nitrilotriacetic acid agarose, metal free (Sigma cat. no. N9153). In this case, the sample after DEAE chromatography should be resuspended in 50 µL of 0.1 M acetic acid. Adjust the pH to approx 3.5 using 30% acetic acid and resuspend by aggressive vortexing

4. Although the Fe^{3+} column separates phosphorylated from nonphosphorylated peptides, there is a background from nonphosphorylated peptides that contain large amounts of Asp and Glu in the degenerate positions, and consequently adhere to the Fe^{3+} column. Consequently, great care should be used in assigning selectivity values to these amino acids. The amount of nonphosphorylated peptide background can be estimated based on the amount of Ser remaining in the fixed position of the phosphorylated library sample. (Phosphoserine undergoes β-elimination to form dithiothreitol adducts during sequencing, and consequently chromatographs at different positions than the unphosphorylated Ser residue.) This background estimate of nonphosphorylated peptides can be subtracted from the final sequenced peptide data to correct for this, although we do not routinely do this unless the background signal exceeds approx 30% of the sample signal.

Acknowledgments

I am grateful to Drs. Kiyotaka Nishikawa and Lewis C. Cantley for sharing their expertise on the application of this technique to analysis of PKC family members. This work was supported by NIH grants GM59281 and GM-60594.

References

1. Mellor, H. and Parker, P.J. (1998) The extended protein kinase C superfamily. *Biochem. J.* **332**, 281–292.
2. Dempsey, E. C., Newton, A. C., Mochly-Rosen, D., Fields, A. P., Reyland, M. E., Insel, P. A., et al. (2000) Protein kinase C isozymes and the regulation of diverse cell responses. *Am. J. Physiol.* **279**, L429–L438.
3. Newton, A. C. (2001) Protein kinase C: structural and spatial regulation by phosphorylation, cofactors, and macromolecular interactions. *Chem. Rev.* **101**, 2353–2364.
4. Aitken, A. and Learmonth, M. (1997) Analysis of sites of protein phosphorylation. *Methods Mol. Biol.* **64**, 293–306.
5. Resing, K. A. and Ahn, N. G. (1997) Protein phosphorylation analysis by electrospray ionization-mass spectrometry. *Methods Enzymol.* **283**, 29–44.
6. Zhang, W., Czernik, A. J., Yungwirth, T., Aebersold, R., and Chait, B. T. (1994) Matrix-assisted laser desorption mass spectrometric peptide mapping of proteins separated by two-dimensional gel electrophoresis: determination of phosphorylation in synapsin I. *Protein Sci.* **3**, 677–686.
7. Wu, J., Ma, Q. N., and Lam, K. S. (1994) Identifying substrate motifs of protein kinases by a random library approach. *Biochemistry* **33**, 14,825–14,833.
8. Songyang, Z., Blechner, S., Hoagland, N., Hoekstra, M. F., Piwnica-Worms, H., and Cantley, L. C.. (1994) Use of an oriented peptide library to determine the optimal substrates of protein kinases. *Curr. Biol.* **4**, 973–982.
9. Nishikawa, K., Toker, A., Johannes, F. J., Songyang, Z., and Cantley, L. C. (1997) Determination of the specific substrate sequence motifs of protein kinase C isozymes. *J. Biol. Chem.* **272**, 952–960.
10. Songyang, Z. and Cantley, L. C. (1998) The use of peptide library for the determination of kinase peptide substrates. *Methods Mol. Biol.* **87**, 87–98.
11. Obata, T., Yaffe, M. B., Leparc, G. G., Piro, E. T., Maegawa, H., Kashiwagi. A., et al. (2000) Peptide and protein library screening defines optimal substrate motifs for AKT/PKB. *J. Biol. Chem.* **275**, 36,108–36,115.
12. Yaffe, M. B., Leparc, G. G., Lai, J., Obata, T., Volinia, S., and Cantley, L. C. (2001) A motif-based profile scanning approach for genome-wide prediction of signaling pathways. *Nat. Biotechnol.* **19**, 348–353.
13. Muszynska, G., Andersson, L., and Porath, J. (1986) Selective adsorption of phosphoproteins on gel-immobilized ferric chelate. *Biochemistry* **25**, 6850–6853.
14. Holmes, L. D. and Schiller, M. R. (1997) Immobilized Iron(III) metal affinity chromatography for the separation of phosphorylated macromolecules: ligands and applications. *J. Liquid Chrom. Rel. Technol.* **20**, 123–142.
15. Neville, D. C., Rozanas, C. R., Price, E. M., Gruis, D. B., Verkman, A. S., and Townsend. R. R. (1997) Evidence for phosphorylation of serine 753 in CFTR using a novel metal-ion affinity resin and matrix-assisted laser desorption mass spectrometry. *Protein Sci.* **6**, 2436–2445.

16. Zhang, Z., Schaffer, A. A., Miller, W., Madden, T. L., Lipman, D. J., Koonin, E. V., et al. (1998) Protein sequence similarity searches using patterns as seeds. *Nucleic Acids Res.* **26,** 3986–3990.
17. Le Good, J. A., Ziegler, W. H., Parekh, D. B., Alessi, D. R., Cohen, P., and Parker, P. J. (1998) Protein kinase C isotypes controlled by phosphoinositide 3-kinase through the protein kinase PDK1. *Science* **281,** 2042–2045.

VII

STRUCTURAL ANALYSIS OF PROTEIN KINASE C

23

Structural Analysis of Protein Kinase C

An Introduction

James H. Hurley

Understanding the three-dimensional structure of protein kinase C (PKC) has been a fundamental goal in the PKC field. The determination of the structure is a prerequisite for a complete understanding of the complex activation mechanisms of PKC. Considerable effort has been spent by many laboratories in efforts to obtain crystals of intact PKC isozymes suitable for high-resolution structure determination. Previous efforts have fallen short of this goal but have led to lower resolution analyses by extended X-ray absorption fine structure spectroscopy *(1)*, electron microscopy *(2)*, and electron diffraction *(3)*.

Given the difficulty in obtaining diffraction quality crystals of an intact PKC, a piecemeal approach has been taken by a number of laboratories in which the structures of individual PKC domains are solved one at a time. The X-ray structure of the PKCδ C1B domain has been solved in complex with phorbol 13-acetate *(4)*. Several nuclear magnetic resonance structures of PKC C1 domains have been solved *(5–7)*, including one in the presence of short-chain lipid micelles *(7)*. Four structures of PKC C2 domains have been solved. The C2 domains of the conventional isozymes PKCα *(8)* and β *(9)* have been solved, the former in complex with Ca^{2+} and a short-chain phosphatidylserine, and the latter in complex with Ca^{2+}. The C2 domains of the novel isozymes PKCδ *(10)* and ε *(11)* have also been determined. No structure of a PKC catalytic domain has been published as determined by experimental methods, but the sequence identity to protein kinase A, for which the structure has been solved, is sufficiently high that a credible model has been constructed based on homology *(12)*.

From: *Methods in Molecular Biology, vol. 233: Protein Kinase C Protocols*
Edited by: A. C. Newton © Humana Press Inc., Totowa, NJ

Current understanding of PKC structure rests primarily on combining the known high-resolution three-dimensional structures of the individual domains with indirect structural information obtained from biochemical approaches such as limited proteolysis. This section of the book will deal with the protocols involved in these approaches.

References

1. Hubbard, S. R., Bishop, W. R., Kirschmeier, P., George, S. J., Cramer, S. P., and Hendrickson, W. A. (1991) Identification and characterization of zinc-binding sites in protein kinase-C. *Science* **254,** 1776–1779.
2. Newman, R. H., Carpenter, E., Freemont, P. S., Blundell, T. L., and Parker, P. J. (1994) Microcrystals of the β (1) isoenzyme of protein kinase-C—An electron microscopic study. *Biochem. J.* **298,** 391–393.
3. Owens, J. M., Kretsinger, R. H., Sando, J. J., and Chertihin, O. I. (1998) Two-dimensional crystals of protein kinase C. *J. Struct. Biol.* **121,** 61–67.
4. Zhang, G., Kazanietz, M. G., Blumberg, P. M., and Hurley, J. H. (1995) Crystal structure of the Cys2 activator-binding domain of protein kinase Cδ in complex with phorbol ester. *Cell* **81,** 917–924.
5. Hommel, U., Zurini, M., and Luyten, M. (1994) Solution structure of a cysteine-rich domain of rat protein kinase C. *Nat. Struct. Biol.* **1,** 383–387.
6. Ichikawa, S., Hatanaka, H., Takeuchi, Y., Ohno, S., and Inagaki, F. (1995) Solution structure of cysteine-rich domain of protein kinase C. *J. Biochem.* **117,** 566–574.
7. Xu, R. X., Pawelczyk, T., Xia, T.-H., and Brown, S. C. (1997) NMR structure of a protein kinase C-phorbol-binding domain and study of protein-lipid micelle interactions. *Biochemistry* **36,** 10,709–10,717.
8. Verdaguer, N. Corbalan-Garcia, S., Ochoa, W.F., Fita, I., and Gomez-Fernandez, J. C. (1999) Ca²⁺ bridges the C2 membrane-binding domain of protein kinase Cα directly to phosphatidylserine. *EMBO J.* **18,** 6329–6338.
9. Sutton, R. B. and Sprang, S. R. (1998) Structure of the protein kinase C-β phospholipid-binding C2 domain complexed with Ca²⁺. *Struct. Fold. Des.* **6,** 1395–1405.
10. Pappa, H., Murray-Rust, J., Dekker, L. V., Parker, P. J., and McDonald, N. Q. (1998) Crystal structure of the C2 domain from protein kinase C-δ. *Struct. Fold. Des.* **6,** 885–894.
11. Ochoa, W.F., Garcia-Garcia, J., Corbalan-Garcia, I. F. S., Verdaguer, N., and Gomez-Fernandez, J. C. (2001) Structure of the C2 domain from novel protein kinase C-ε. A membrane binding model for Ca²⁺-independent C2 domains. *J. Mol. Biol.* **311,** 837–849.
12. Orr, J. W. and Newton, A. C. (1994) Intrapeptide regulation of protein kinase C. *J. Biol. Chem.* **269,** 8383–8367.

24

Bacterial Expression and Purification of C1 and C2 Domains of Protein Kinase C Isoforms

Wonhwa Cho, Michelle Digman, Bharath Ananthanarayanan, and Robert V. Stahelin

1. Introduction

Protein kinase C (PKC) contains two types of membrane-targeting domains, C1 and C2 domains, in its regulatory region *(1–5)*. It has been shown that C1 and C2 domains of conventional PKCs (α, β_I, β_{II}, and γ) and C1 domains of novel (δ, ε, θ, and η) play pivotal roles in their subcellular targeting and regulation *(1–5)*. C1 domains are small (~50 amino acids) cysteine-rich structures that contain two structurally important zinc ions *(6,7)*. In conventional and novel PKCs, C1 domains occur in a tandem repeat (C1A and C1B) and serve as an interaction site for diacylglycerol and its structural analog, phorbol ester *(1)*. C1 domains found in atypical PKCs (ζ and ι/λ), however, do not bind diacylglycerol/phorbol ester because of minor sequence variations and might be involved in protein–protein interactions *(4)*. C2 domains (~130 residues) of PKCs have a common structure in which eight antiparallel β strands are connected by variable loops *(5,8,9)*. For C2 domains of conventional PKCs that bind the membrane in a calcium-dependent manner, three loops located at one side of the domain serve as Ca^{2+}-binding sites. C2 domains of novel PKCs do not bind calcium as a result of the lack of calcium ligands in these loops and might be involved in calcium-independent membrane interactions *(10)* and/or protein–protein interactions *(8,9)*. To elucidate the exact roles of C1 and C2 domains in cellular regulation of PKCs and identify critical residues involved in PKC regulation, it is often necessary to perform in vitro structure–function studies of isolated domains as well as the full-length proteins. Thus, functional expression of isolated C1 and C2 domains and their respective mutant proteins

From: *Methods in Molecular Biology, vol. 233: Protein Kinase C Protocols*
Edited by: A. C. Newton © Humana Press Inc., Totowa, NJ

in quantities sufficient for biophysical characterization is an important step toward understanding of cellular regulation of PKC activities. C1 and C2 domains of PKCs have been expressed in *Escherichia coli*, mainly as gluta-thione S-transferase (GST)-fusion proteins *(7,11–13)* or as hexahistidine (His$_6$)-tagged proteins *(14,15)*. In general, both bacterial expression methods give good protein expression yield and allow simple affinity chromatography separation of expressed proteins. A main disadvantage of the former method is the removal of GST-fusion that entails extra steps in protein purification. Furthermore, in the case that a GST-fusion protein forms the inclusion body in bacteria, it is extremely difficult to refold the detergent-solubilized inclusion body. On the other hand, the latter method has a potential problem that the His$_6$-tag might interfere with functions of a domain. We have mainly used the latter method for the expression and purification of C1 and C2 domains of PKCs because we have found that the carboxy-terminal His$_6$-tags do not affect the membrane binding properties of these domains in neutral pH and that some His$_6$-tagged domains that form inclusion bodies in *E. coli* can be readily refolded. This chapter describes the bacterial expression and purification of His$_6$-tagged C1 and C2 domains of conventional and novel PKCs. All C2 domains are expressed as soluble proteins in *E. coli* and purified to near homogeneity by two-step chromatographic separation involving Ni affinity and gel filtration chromatography. Most C1 domains are also expressed as soluble proteins, which are purified by the same protocol. Some C1 domains (e.g., PKC-α C1 domains) have a high tendency to form inclusion bodies in bacteria. These proteins are solubilized in urea, refolded, and purified to near homogeneity by Ni affinity chromatography.

2. Materials
2.1. Expression and Purification of C1 Domains

1. Expression vectors for isolated C1a and C1b domains. Expression vectors were constructed by subcloning C1 domain sequences (*see* **Fig. 1**) of rat PKCs into the pET21a vector (Novagen, Madison, WI). These vectors contain sequences for a Leu-Glu linker and a His$_6$ tag just before a stop codon at the 3′ end.
2. *E. coli* strain BL21(DE3) as a host for protein expression (Novagen).
3. Luria broth (LB): 10 g of bactotrypton (Fisher), 5 g of yeast extract (Fisher), and 10 g of NaCl in 1 L of deionized water, pH 7.4.
4. Ampicillin (Fisher).
5. Isopropyl β-D-thiogalactopyranoside (IPTG) (Research Products, Mount Prospect, IL).
6. 0.01 mM ZnSO$_4$ solution (Fisher).
7. Cell lysis buffers. For soluble C1 (and C2) domains: 50 mM Tris-HCl buffer, pH 8.0, containing 300 mM NaCl, 10 mM imidazole, 0.4% (v/v) Triton X-100

		37	86
C1A	PKC-α	HKFIARFFKQPTFCSHCTDFIWGFGKQGFQCQVCCFVVHKRCHEFVTFSC	
C1B	PKC-α	HKFKIHTYGSPTFCDHCGSLLYGLIHQGMKCDTCDMNVHKQCVINVPSLC	
C1A	PKC-γ	HKFTARFFKQPTFCSHCTDFIWGIGKQGLQCQVCSFVVHRRCHEFVTFEC	
C1B	PKC-γ	HKFRLHSYSSPTFCDHCGSLLYGLVHQGMKCSCCEMNVHRRCVRSVPSLC	
C1A	PKC-δ	HEFIATFFGQPTFCSVCKEFVWGLNKQGYKCRQCNAAIHKKCIDKIIGRC	
C1B	PKC-δ	HRFKVYNYMSPTFCDHCGTLLWGLVKQGLKCEDCGMNVHHKCREKVANLC	

154

C2 PKC-α DHTEKRGRIYLKAEV-TDEKLHVTVRDAKNLIPMDPNGLSDPYVKLKLIPDPKNESKQKTKTIRSTLNPQWNESF

C2 PKC-γ DHTERRGRLQLEIRAPTSDEIHITVGEARNLIPMDPNGLSDPYVKLKLIPDPRNLTKQKTKTVKATLNPVWNETF

C2 PKC-δ ----------MAPFLRISFNSYELGSLQAEDDASQPFCAVKMKEALTTDRGKTLVQKKPTMYPEWKSTF

228   289

C2 PKC-α TFKLKPSDKDRRLSVEIWDWDRTTRNDFMGSLSFGVSELMKMP-----VDGWFKLLSQEEGEYFNVP

C2 PKC-γ VFNLKPGDVERRLSVEVWDWDRTSRNDFMGAMSFGVSELLKAP-----ASGWYKLLNQEEGEYYNVP

C2 PKC-δ DAHIYEG--RVIQIVLMRAAEDPMSEVTVGVSVLAERCKKNNGK---AEFWLDLQP--QAKVLMCV

Fig. 1. Amino acid sequences of C1 and C2 domains of PKCs bacterially expressed using the protocls described herein. All expressed domains have an LE linker and a His$_6$ tag in their carboxy termini. Numbers indicate the amino acid sequence numbers for PKC-α.

(Pierce, Rockford, IL), and 1 mM phenylmethylsulfonyl fluoride (Sigma). For insoluble C1 domains (PKC-α C1a and C1b): 50 mM Tris-HCl buffer, pH 8.0, containing 50 mM NaCl, 5.6 mM cetyltrimethylammonium bromide, and 1 mM phenylmethylsulfonyl fluoride.

8. Refolding buffers for insoluble C1 domains. Buffer A: 50 mM Tris-HCl, pH 8.0, 50 mM NaCl, 8 M urea, 5 mM dithiothreitol. Buffer B: 50 mM Tris-HCl pH 8.0, 50 mM NaCl, 1.5 M urea, 50 μM ZnSO$_4$, 0.5 mM dithiothreitol. Buffer C: 50 mM Tris-HCl, pH 8.0.

9. Filters (0.2-μm; Fisher).

10. Affinity chromatography system. Ni-NTA agarose resins (Qiagen, Valencia, CA). Wash buffer: 50 mM Tris-HCl buffer, pH 8.0, 300 mM NaCl, and 15 mM imidazole. Elution buffer: 50 mM Tris-HCl buffer, pH 8.0, 300 mM NaCl, and 300 mM imidazole.

11. Gel filtration chromatography system. A Superdex100 Hi-Load column (Amersham-Pharmacia) attached to an Äkta FPLC system (Amersham-Pharmacia). Elution buffer: 20 mM Tris-HCl buffer, pH 7.4, 0.16 M KCl.

12. Sodium dodecylsulfate (SDS) polyacrylamide gel electrophoresis system. 20% SDS polyacrylamide gels, a Novex (San Diego, CA) Xcell SureLock Minicell protein electrophoresis unit, a power supply (Fisher), and protein molecular weight standards (Bio-Rad).

13. Protein concentration unit. An Amicon Centriplus centrifugal filter devise with YM-3 membranes (3000 molecular weight cut-off; Millipore, Bedford, MA).

14. Glycerol (Sigma).

15. 20-[^{3}H]Phorbol-12,13-dibutyrate (American Radiolabeled Chemicals, St. Louis, MO) for C1 domain assays (*see* **Note 1**).

2.2. Expression and Purification of C2 Domains

1. Expression vectors for isolated C2 domains. Expression vectors were constructed by subcloning C2 domain sequences (*see* **Fig. 1**) of rat PKCs into the pET21a vector. These vectors contain sequences for a Leu-Glu linker and a His$_6$ tag just before a stop codon at the 3′ end.

2. Other materials are essentially identical with those used for the expression and purification of soluble C1 domains, except that 15% SDS polyacrylamide gels are used for electrophoresis and that ZnSO$_4$ solution is not necessary.

3. Methods

3.1. Bacterial Expression and Purification of C1 Domains

3.1.1. Expression and Purification of Soluble C1 Domains

1. Plate *E. coli* BL21(DE3) harboring each C1 domain plasmid on an agar plate supplemented with LB and ampicillin.

2. Select one colony and grow overnight in 5 mL of LB containing 0.1 mg/mL of ampicillin at 37°C with constant agitation.

3. Inoculate 1 L of LB supplemented with 100 µg/mL ampicillin with 5 mL of the overnight culture. Grow cells at 37°C until their absorbance at 600 nm reached 0.8 to 1.4 (*see* **Note 2**).

4. Induce the protein expression with 0.5 mM IPTG (final concentration) for 4 h. To ensure proper folding of the C1 domains, inoculate the cells also with 0.01 mM ZnSO$_4$ (final concentration).

5. Harvest the cells by centrifugation at 5000g and at 4°C for 10 min.

6. Resuspend the cells in 20 mL of lysis buffer and sonicate the suspension on ice for 15 s followed by a 45-s incubation on ice (×10).

7. Collect the supernatant by centrifugation at 50,000g for 25 min at 4°C and filter the supernatant through a 0.2-µm syringe filter into a 50-mL tube. Add 1 mL of Ni-NTA solution (Qiagen) and incubate the mixture on ice for 1 h with constant agitation.

8. Pour the mixture onto a gravity column. After the entire volume flows through, wash the column with 10 mL of wash buffer. Elute the C1 domain in 0.5-mL fractions using elution buffer (×6). Analyze all fractions by SDS-polyacrylamide gel electrophoresis using an 18% SDS-polyacrylamide gel.

9. Pool the C1 domain fractions and concentrate the solution to 1 mL using a Centriplus filter devise with a YM3 membrane. To remove high molecular weight bacterial impurities, load the concentrated C1 domain solution onto a Superdex100 gel filtration column equilibrated and eluted with 20 mM Tris-HCl buffer, pH 7.4, 0.1 M NaCl. Pool the C1 domain fractions and concentrate the solution to 1 mL as described above.

10. Check the protein purity by SDS polyacrylamide electrophoresis (*see* **Note 3**), determine the protein concentration, and add 30% glycerol (final concentration, v/v) to 0.5-mL aliquots of protein. Store all aliquots of purified protein at −20°C.

3.1.2. Refolding and Purification of C1 Domains Expressed as Inclusion Bodies

1. Grow and harvest cells as described in **Subheading 3.1.1.**

2. Resuspend cells in 25 mL of lysis buffer, sonicate the suspension and collect the inclusion body pellet by centrifugation at 100,000g for 15 min at 4°C.

3. Resuspend the pellet in 25 mL of lysis buffer and centrifuge it again.

4. Resuspend the washed inclusion body in 10 mL of refolding buffer A, stir the mixture at room temperature for 15 min, and centrifuge at 100,000g for 15 min at 4°C.

5. Remove the insoluble matter and dialyze the supernatant against refolding buffer B for 12 h and then against refolding buffer C for 12 h (×2).

6. Purify the refolded C1 domain using a Ni-NTA agarose resin as described above.

7. Dialyze the resulting protein (4–5 mL) against 50 mM Tris-HCl, pH 8.0 for 4–5 h to remove the imidazole (*see* **Note 4**).

8. Check the protein purity (*see* **Note 3**), determine the protein concentration, and add 40% glycerol (final concentration; v/v) to 0.5-mL aliquots of protein. Store aliquots of purified proteins at –20°C *(4)*.

3.2. Expression and Purification of C2 Domains

The protocol for the expression and purification of His_6-tagged C2 domains are essentially the same as that for soluble His_6-tagged C1 domains. All C2 domains including C2-like domains of novel PKCs are expressed as soluble proteins in *E. coli* in high yield.

1. Inoculate 1 L of LB supplemented with 100 µg/mL ampicillin with 1 mL of overnight grown culture of *E. coli* BL21(DE3) harboring each C2 domain.
2. Grow cells at 37°C until their absorbance at 600 nm reached 0.8 and induce protein expression by the addition of 0.5 m*M* of IPTG for 4 h.
3. Harvest the cells by centrifugation (5000*g* for 10 min at 4°C).
4. Resuspend cells in 20 mL of lysis buffer and sonicate the suspension on ice for 15 s followed by a 45-s incubation on ice (×12).
5. Collect the supernatant by centrifugation at 50,000*g* for 25 min at 4°C and filter the supernatant through a 0.2-µm syringe filter into a 50-mL tube.
6. Purify the C2 domain using a Ni-NTA agarose resin as described above.
7. Pool the C2 domain fractions and concentrate the solution to 1 mL using a Centriplus filter devise with a YM3 membrane.
8. To remove high molecular weight bacterial impurities, load the concentrated C2 domain solution onto a Superdex100 gel filtration column equilibrated and eluted with 20 m*M* Tris-HCl, pH 8.0, containing 0.1 *M* NaCl. Pool the C2 domain fractions and concentrate the solution to 1 mL as described above.
9. Check the purity of the protein *(3)*, determine the protein concentration, and add 30% glycerol (final concentration; v/v) to 0.5-mL aliquots of protein. Store all aliquots of purified protein at –20°C.

4. Notes

1. Some C1 domains are difficult to express, purify, and store as a result of their hydrophobic nature. Thus, it is recommended that their affinities for radiolabeled phorbol-12,13-dibutyrate be regularly checked according to the published protocol to ensure that they are properly and functionally folded *(16)*. C2 domains are generally more stable and easier to handle than C1 domains and thus the regular monitoring of structural integrity is not as essential for C2 domains.
2. Final protein yield varies from 0.5 to 1.5 mg per liter of growth medium for soluble C1 domains. For C1 domains expressed as inclusion bodies, this value ranges from 1 to 8 mg per liter. Finally, the yield for C2 domains is in the range of 2 to 5 mg per liter. The difference is mainly the result of the different levels of protein expression for soluble C1 and C2 domains, whereas the yield for inclusion body-forming domains largely depends on the efficiency of refolding.

For those proteins with lower expression yield (i.e., low protein copies per cell), cells are grown to near saturation before IPTG induction. For instance, C1a domains of PKC-γ and -δ are induced when the absorbance of culture at 600 nm reaches 1.4.

3. Soluble C1 and C2 domains are typically greater than 90% electrophoretically pure after two-step chromatographic separation. Because the refolded proteins are relatively pure, a single chromatographic separation using Ni-NTA yields more than 95% electrophoretically pure proteins.

4. During the refolding of the inclusion body-forming C1 domains dissolved in urea, the protein concentration should be kept as low as possible to prevent the protein aggregation. Therefore, any step that entails protein concentration should be avoided. Also, a higher concentration (i.e., 40%) of glycerol was necessary for a long-term storage.

Acknowledgments

This work was supported by NIH grants GM52598 and GM53987.

References

1. Nishizuka, Y. (1988) The molecular heterogeneity of protein kinase C and its implications for cellular regulation. *Nature* **334**, 661–665.
2. Newton, A. C. (1995) Protein kinase C: structure, function, and regulation. *J. Biol. Chem.* **270**, 28,495–28,498.
3. Newton, A. C. (1995) Protein kinase C. Seeing two domains. *Curr. Biol.* **5**, 973–976.
4. Ron, D. and Kazanietz, M. G. (1999) New insights into the regulation of protein kinase C and novel phorbol ester receptors. *FASEB J.* **13**, 1658–1676.
5. Cho, W. (2001) Membrane targeting by c1 and c2 domains. *J. Biol. Chem.* **276**, 32,407–32,410.
6. Hommel, U., Zurini, M., and Luyten, M. (1994) Solution structure of a cysteine rich domain of rat protein kinase C. *Struct. Biol.* **1**, 383–387.
7. Zhang, G., Kazanietz, M. G., Blumberg, P. M., and Hurley, J. H. (1995) Crystal Structure of the Cys2 Activator-Binding Domain of Protein Kinase Cδ in Complex with Phorbol Ester. *Cell* **81**, 917–924.
8. Nalefski, E. A., Slazas, M. M., and Falke, J. J. (1997) Ca2+-signaling cycle of a membrane-docking C2 domain. *Biochemistry* **36**, 12,011–12,018.
9. Rizo, J. and Sudhof, T. C. (1998) C2-domains, structure and function of a universal Ca2+-binding domain. *J. Biol. Chem.* **273**, 15,879–15,882.
10. Pepio, A. M. and Sossin, W. S. (2001) Membrane translocation of novel protein kinase Cs is regulated by phosphorylation of the C2 domain. *J. Biol. Chem.* **276**, 3846–3855.
11. Xu, R. X., Pawelczyk, T., Xia, T.-H., and Brown, S. C. (1997) NMR structure of a protein kinase c-gamma phorbol-binding domain and study of protein-lipid micelle interactions. *Biochemistry* **36**, 10,709–10,717.

12. Sutton, R. B. and Sprang, S. R. (1998) Structure of the protein kinase Cbeta phospholipid-binding C2 domain complexed with Ca2+. *Structure* **6,** 1395–1405.
13. Johnson, J. E., Giorgione, J., and Newton, A. C. (2000) The C1 and C2 domains of protein kinase C are independent membrane targeting modules, with specificity for phosphatidylserine conferred by the C1 domain. *Biochemistry* **39,** 11,360–11,369.
14. Medkova, M. and Cho, W. (1999) Interplay of C1 and C2 domains of protein kinase C-alpha in its membrane binding and activation. *J. Biol. Chem.* **274,** 19,852–19,861.
15. Garcia-Garcia, J., Corbalan-Garcia, S., and Gomez-Fernandez, J. C. (1999) Effect of calcium and phosphatidic acid binding on the C2 domain of PKC alpha as studied by Fourier transform infrared spectroscopy. *Biochemistry* **38,** 9667–9675.
16. Sharkey, N. A. and Blumberg, P. M. (1985) Highly lipophilic phorbol esters as inhibitors of specific [^{3}H]phorbol 12,13-dibutyrate binding. *Cancer Res.* **45,** 1924.

25

Crystallization of the Protein Kinase Cδ C1B Domain

Gongyi Zhang and James H. Hurley

1. Introduction

The activation of conventional and novel protein kinase C (PKC) isozymes by diacylglycerol (DAG) and phorbol esters is mediated by their C1 domains (*1–4*. Several structures of PKC C1 domains have been determined using nuclear magnetic resonance (*5–7*) and X-ray crystallography (*8*).

C1 domains contain approx 50 amino acids that are organized into two β-sheets and a short C-terminal α-helix (**Fig. 1**). The secondary structural elements are arranged around two Zn^{2+} ion-binding motifs, each containing three cysteines and one histidine. These buried Zn^{2+} clusters are integral to the domain structure. At the tip of the domain, two β-strands have been "unzipped" or pulled apart, leaving a pocket where main chain hydrogen bonding groups are exposed to solvent.

Most PKC isozymes contain two C1 domains within their regulatory region. The crystal structure of the second C1 domain (C1B) from PKCδ bound to a phorbol acetate revealed the "unzipped" pocket to be the binding site for phorbol esters. Key Pro and Gln residues help maintain the unzipped conformation. Conserved C1 residues make hydrogen bonds to oxygen on the multiringed phorbol group, which are believed to be analogous in part to the interactions made with DAG. In particular, the main chain polar groups of residues Thr242, Leu251, and Gly253 provide binding sites for the 3-, 4-, and 20-oxygens of the phorbol and for the 3-hydroxyl and one of the acyl oxygens of a modeled DAG. The conserved Pro, Gly, and Gln residues together provide a fingerprint for DAG binding functionality that allows typical and atypical C1 domains to be classified and their DAG-binding properties to be predicted based on their sequence.

From: *Methods in Molecular Biology, vol. 233: Protein Kinase C Protocols*
Edited by: A. C. Newton © Humana Press Inc., Totowa, NJ

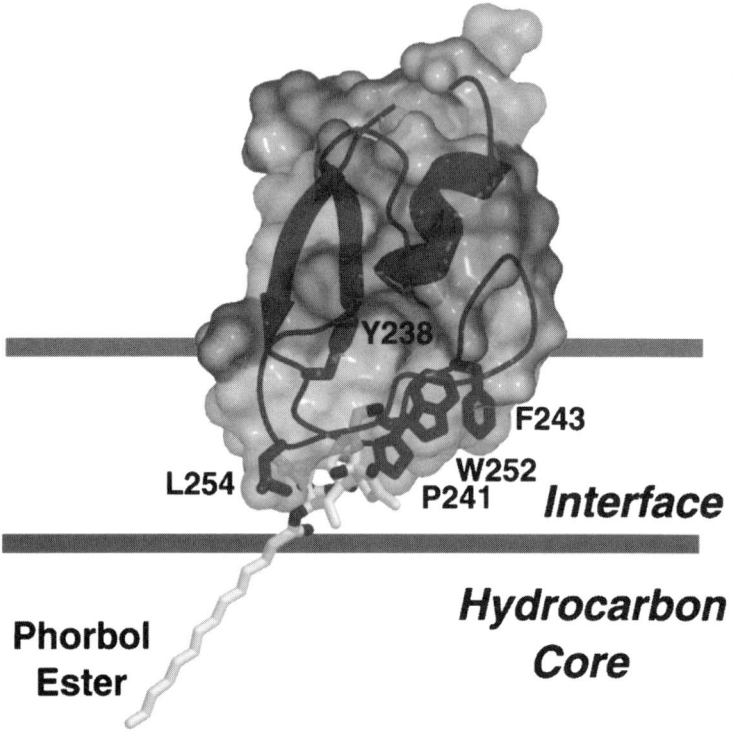

Fig. 1. Structure of the C1B domain of PKCδ complexed with phorbol ester and docked onto a membrane surface (reprinted from **ref. 5**). The myristate tail has been modeled onto the phorbol ester. Selected phorbol-binding and membrane-interacting residues are highlighted.

The structural conclusions described above could only be obtained after the C1 domain was crystallized bound to an activating ligand. This chapter describes the methods used in the crystallization of the PKCδ-C1B domain in the presence and absence of one of its ligands, phorbol 13-acetate.

2. Materials

1. PGEX-2TK plasmid with DNA coding for residues 231–280 of mouse PKCδ C1B domain subcloned into the *Bam*HI and *Eco*RI sites.
2. *Escherichia coli* strain XL-1 blue.
3. Isopropyl β-D-thiogalactopyranoside.
4. Phosphate-buffered saline: 150 mM NaCl, 5 mM Na$_2$HPO$_4$, 1.7 mM KH$_2$PO$_4$, pH 7.2.
5. Fermenter (8 L).
6. Triton X-100.

Fig. 2. A crystal of PKCδ-C1B domain. The crystal is 0.4 mm along the longest dimension.

7. Lysozyme.
8. Phenylmethylsulfonyl fluoride (PMSF).
9. Leupeptin.
10. Thrombin.
11. Superdex S-200.
12. Glutathione Sepharose-4B (Pharmacia).
13. Phosphate buffer, pH 6.8.
14. Phorbol 13-acetate (LC laboratories).
15. Linbro plates.
16. Vacuum grease.
17. Siliconized glass cover slips.
18. Microbridges (Hampton).
19. Stereo dissecting microscope.
20. Capillary micropipet (e.g., Drummond).

3. Methods

3.1. Protein Expression

1. Transform competent *E. coli* XL1-Blue with pGEX-2TK-PKCδ(231-280) (*see* **Note 1**) using standard molecular biology procedures and plate on Luria broth (LB)/ampicillin *(9)*.
2. Grow 1 L of transformed cells overnight in LB/ampicillin.
3. Harvest the overnight culture and inoculate a fermenter according to the manufacturer's directions. Grow cells to an OD (600 nm) = 10.0 and induce with 0.3 mM isopropyl β-D-thiogalactopyranoside. Note that the procedure can be

scaled down and executed in Fernbach flasks in a shaker. The yield is relatively low (~1 mg C1B/liter culture); therefore, a large number of flasks are required to obtain enough material for crystallization trials. The fermenter procedure described produces a cell mass equivalent to approx 80 L of shaker-grown culture. Solutions described below should be scaled down proportionately.

4. Harvest cells by centrifugation after 6 h.
5. Lyse cells by sonication in 1 L of ice-cold phosphate-buffered saline, 1 mg/mL lysozyme, 1% Triton X-100, 1 µg/mL leupeptin, and 1 mM PMSF.
6. Centrifuge lysate at 30,000g for 30 min.
7. Equilibrate with 20 mL of glutathione-sepharose and wash according to the manufacturer's instructions.
8. Cleave the resin-bound GST-fusion protein overnight (16 h) with 20 U thrombin.
9. Load 10 mL of the supernatant on a 2.6 × 60 cm Superdex S-200 gel filtration column equilibrated in 0.5 M NaCl, 0.1 M triethanolamine HCl (pH 7.0). Run at 1 mL/min using a FPLC system.
10. Pool peak fractions, adjust to 10% glycerol, 1 mM PMSF, 1 mM DTT, and store at –80°C.

3.2. Protein Crystallization

1. Concentrate the C1B protein to 8 mg/mL by precipitating it with 55% saturated ammonium sulfate. Resuspend the precipitate in 0.1 M phosphate buffer (pH 6.8), 1 mM dithiothreitol.
2. Prepare a well solution of 25% isopropanol, 0.2 M ammonium acetate, and 0.1 M sodium phosphate (pH 6.8).
3. For each crystallization trial, mix 6 µL of protein with 6 µL of well solution and place on a glass cover slip following standard procedures for hanging drop vapor diffusion crystallization experiments *(10)*.
4. Incubate at 4°C until small crystals appear. Harvest crystals **(Fig. 2)** and wash crystals in well solution.
5. Set up new sitting drop crystallization trials on Hampton microbridges using the same well solution as in **Subheading 3.2.3.**, with the following modification: the drops now should consist of 8–10 µL of sample, 6–8 µL of well solution (same as in **Subheading 3.2.3.**), and 2–4 µL of glycerol. Using a capillary micropipet and a stereoscopic dissecting microscope, inoculate the sitting drops with crystals harvested in the previous step. Incubate the crystallization trials at 4°C as in **Subheading 3.2.4.**.
6. To obtain crystals with phorbol 13-acetate, prepare a stock solution of 20 mM phorbol 13-acetate in methanol. Add 0.6 µL of this solution to the 12 µL crystallization trial described above by using native C1 microcrystals as seeds (*see* **Notes 2–5**).

4. Notes

1. The protein sequences (N terminus: GSRRASVGS, C terminus: EFIVTD) derived from the pGEX-2TK vector are required for crystal formation.

2. Protein crystallization is not a robust technique even under ideal circumstances. The growth of wild-type C1B domain crystals is unpredictable and requires a large number of seeding attempts to produce a small number of acceptable crystals. In our past attempts, single-site mutants of the C1B that were expected to have modest effects on the overall structure failed to crystallize. The odds of success with mutants and complexes with alternative ligands may be increased by using uncomplexed wild-type C1B domain microcrystals as seeds.

3. The solubility of the C1B domain is affected by the ligand to which it is bound. Some ligands decrease the solubility so markedly, even in the presence of organic solvent, that it is not possible to crystallize it. For example, the C1B domain complexed with high-affinity ligands, such as bryostatin, was not soluble enough to initiate crystallization trials.

4. The affinity of ligands for the C1B domain appears to be markedly lowered by crystal packing constraints and by the organic solvent used in crystallization. Although it was possible to obtain a crystalline complex with the relatively high-affinity ligand phorbol 13-acetate, lower affinity ligands such as short-chain diacylglycerols and conformationally constrained diacylglycerols failed to bind in the crystal. The lack of binding was judged by determining the crystal structures of the putative complexes and computing difference electron density maps. These maps showed no significant differences, indicating that the complex was not actually formed.

5. The phorbol/DAG binding site is not directly blocked by crystal contacts, but much of the remainder of the protein surface is blocked. C1 domain ligands (e.g., phosphatidylserine, free fatty acids) that do not compete for the DAG binding site presumably bind elsewhere on the surface. It seems unlikely that the crystals we described here will support binding of these ligands, although we have not tested this.

Acknowledgment

R. Grisshammer is thanked for comments on the manuscript.

References

1. Hurley, J. H., Newton, A. C., Parker, P. J., Blumberg, P. M., and Nishizuka, Y. (1997) Taxonomy and function of C1 protein kinase C homology domains. *Protein Sci.* **6,** 477–480.
2. Newton, A. C. and Johnson, J. E. (1998) Protein kinase C: a paradigm for regulation of protein function by two membrane-targeting modules. *Biochim. Biophys. Acta Rev. Biomembr.* **1376,** 155–172.
3. Ron, D. and Kazanietz, M. G. (1999) New insights into the regulation of protein kinase C and novel phorbol ester receptors. *FASEB J.* **13,** 1658–1676.
4. Hurley, J. H. and Misra, S. (2000) Signaling and subcellular targeting by membrane-binding domains. *Annu. Rev. Biophys. Biomol. Struct.* **29,** 49–79.
5. Hommel, U., Zurini, M., and Luyten, M. (1994) Solution structure of a cysteine-rich domain of rat protein kinase C. *Nat. Struct. Biol.* **1,** 383–387.

6. Xu, R. X., Pawelczyk, T., Xia, T.-H., and Brown, S. C. (1997) NMR structure of a protein kinase C-phorbol-binding domain and study of protein-lipid micelle interactions. *Biochemistry* **36,** 10,709–10,717.

7. Ichikawa, S., Hatanaka, H., Takeuchi, Y., Ohno, S., and Inagaki, F. (1995) Solution structure of cysteine-rich domain of protein kinase C. *J. Biochem.* **117,** 566–574.

8. Zhang, G., Kazanietz, M. G., Blumberg, P. M., and Hurley, J. H. (1995) Crystal structure of the Cys2 activator-binding domain of protein kinase Cδ in complex with phorbol ester. *Cell* **81,** 917–924.

9. Sambrook, J., Fritsch, E. F., and Maniatis, T. (1989) *Molecular Cloning, A Laboratory Manual,* Second ed. Cold Spring Harbor Laboratory Press, Cold Spring Harbor, New York.

10. McPherson, A. (1989) *Preparation and Analysis of Protein Crystals.* Robert E. Krieger Publishing Co., Malabar, FL.

VIII

METHODS FOR DETECTING BINDING PROTEINS

26

Methods for Detecting Binding Proteins

An Introduction

Gerda Endemann and Daria Mochly-Rosen

1. Introduction

The identification of proteins that bind protein kinase C (PKC) is essential to understanding the contributions of PKCs to signaling pathways. A number of the methods are uniquely applicable to this family of serine/threonine kinases. The PKC family contains at least 10 closely related members with overlapping substrate specificities *(1,2)*. Therefore, isozyme-specific phosphorylation of substrates must be caused in part by isozyme-specific subcellular localization, isozyme-specific proximity to substrates, and/or to regulation of activity in an isozyme-specific manner *(3)*. There is some selectivity in the regulation of catalytic activity of PKC classes rather than individual isozymes. The classic (cPKC) enzymes α, βI, βII, and γPKC are activated by phosphatidylserine (PS), diacylglycerol (DG), and calcium, the novel (nPKC) enzymes δ, ε, θ, and ηPKC require PS and DG only, and the atypical (aPKC) enzymes ζ and λ/ιPKC are activated by PS and possibly phosphatidylinositol-3,4,5-trishosphate, and ceramide *(1,4,5)*.

Regulation of the subcellular localization of PKCs is complex, given that the localization of these enzymes is cell-type specific and isozyme specific, and further, depends on the activation state of the enzyme. Thus, cell-specific mechanisms must exist to localize individual PKC isozymes in their inactive states, to translocate these isozymes upon activation, and to localize them in sites associated with active enzyme. The translocation process is poorly characterized but could involve diffusion toward diacylglycerol-enriched membranes containing PS *(6)*, or could involve an escort or transport mecha-

From: *Methods in Molecular Biology, vol. 233: Protein Kinase C Protocols*
Edited by: A. C. Newton © Humana Press Inc., Totowa, NJ

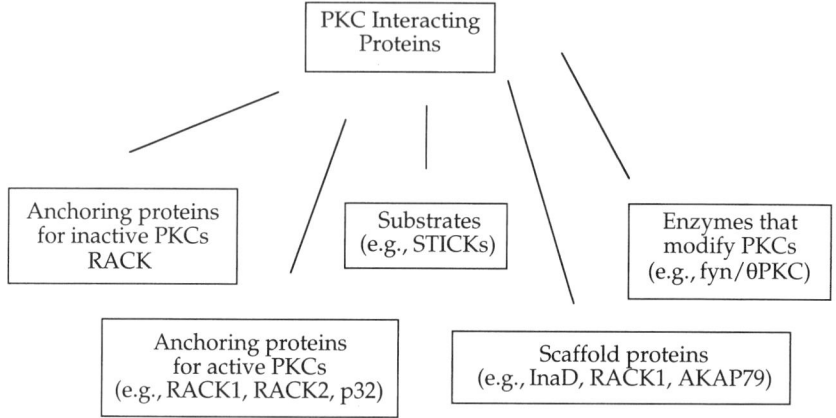

Fig. 1. Classes of PKC interacting proteins.

nism. Retention at specific membrane sites is likely to involve anchoring proteins that may also function as scaffolds maintaining PKCs in proximity to substrates and signaling partners. Proteins that are not substrates for PKCs, and that function as anchoring proteins or scaffolds for activated PKCs, have been termed RACKs, for receptors for activated C-kinase. RACKs for βIIPKC and for εPKC have been identified as described below. The classes of proteins that are expected to interact with PKCs are summarized in **Fig. 1**.

PKCs have been shown to interact with several domains known to mediate protein–protein interactions, including LIM domains, PH domains, PDZ domains, SH2 and SH3 domains, and WD40 domains (**Table 1**).

Surprisingly few known protein–protein interaction motifs have been identified in PKCs. A caveolin-binding motif has been found in the kinase domains of α, ε, and ζPKCs *(7)*, an actin-binding motif in the C1 domain of εPKC *(8)*, and a PDZ-binding sequence at the C-terminus of αPKC *(9)*. The classic and novel PKCs bind DG and phorbol esters in a phospholipid-dependent manner via a tandemly duplicated cysteine-rich, zinc finger motif in the C1domain, $HX_{12}CX_2CX_nCX_2CX_4HX_2CX_7C$ (n is 13 or 14, C is cysteine, H is histidine, and X is any amino acid). Although in some proteins, such as n-chimaerin and Unc-13, one of these zinc finger motifs appears to be sufficient for phorbol ester binding, the single zinc finger motif of atypical PKCs is not able to bind phorbol ester *(10)*. Rather than function to bind diacylglycerol, the zinc finger motif of the atypical PKCs may mediate protein–protein interactions. As discussed below, lambda interacting protein (LIP) binds the zinc finger domain of λ/ιPKC, but not other PKCs, and the zinc finger motif in ζPKC mediates binding of PAR-4. The zinc finger domain in other PKC isozymes may also

Table 1
Protein–Protein Interaction Domains in PKC Interacting Proteins

Protein–protein interaction domain and PKC-binding partner the domain is found in.[a]	Motif recognized by domain	Interacting PKCs	Domain of PKC implicated
LIM domain [$CX_2CX_{16-23}HX_2CX_2CX_{16-21}CX_2(C/H/D)$]			
Enigma homologue (ENH) 3 LIM domains		$\beta IPKC$, γPKC, ϵPKC (**17**)	N-terminal 9 aa
Enigma (3 LIM domains)		α, βI, ζPKC (**17**)	N-terminal 133 aa
LIM-1 kinase-1 (2 LIM domains)		α, βI, γ, δ, ϵ, $\zeta PKCs$ (**17**)	
Cypher1		α, βI, γ, δ, ϵ, $\zeta PKCs$ (**18**)	
PDZ domain (~90 amino acids, with six β strands and two α helices, with the sequence GLGF between the first two β strands)	C-terminal: XS/TXϕ XϕXϕ XD/EXϕ		
PICK1		αPKC (**9**)	C-terminal QSAV
InaD (PDZ 2 or 4)		InaC (eyePKC) (**19**)	
WD40 (~40 amino acids with four β strands, usually containing WD at end)			
RACK1	DpSBXXpS; pSXXXpS	$\beta IIPKC$ (**20**)	C2 and V5
Pleckstrin homology (PH)	Lipids, proteins		
Dynamin, Spectrin		βPKC (**21**)	
Bruton tyrosine kinase		$\beta IPKC$ (**22**)	
RAC/Akt kinase α, β, γ		ζPKC (**23**)	Regulatory domain

(continued)

Table 1 (continued)

Protein–protein interaction domain and PKC-binding partner the domain is found in.[a]	Motif recognized by domain	Interacting PKCs	Domain of PKC implicated
SH2	Py		
SH3	PXXP		
src		ϵPKC (24)	Regulatory domain
LIP (Lamda interacting protein)		λ/ιPKC (25)	Zinc finger domain
PAR-4		ζPKC (26)	Zinc finger domain
ZIP (ζPKC interacting protein)	YXDEDX$_5$SDEE/D	ζPKC (27)	Pseudosubstrate sequence in N-terminus of regulatory domain
Caveolin	ψXψXXXXψ ψXXXXψXXψ	α, ϵ, ζPKCs (7)	Kinase domain
F-actin		ϵPKC (8)	Amino acids 222-230 in C1 domain, LKXXEX

[a]pS, pT, and pY are phosphoserine, phosphothreonine, and phosphotyrosine, respectively; X is any amino acid, ϕ is a hydrophobic amino acid, ψ is the aromatic amino acids W (tryptophan), F (phenylalanine), or Y (tyrosine).

be involved in protein–protein interactions. For example, in NIH 3T3 cells, the zinc finger domain of εPKC appears to localize this isozyme to the Golgi network and to regulate the secretion of sulfated glycosaminoglycans *(11)*, although the binding partner for this interaction was not identified.

Information on the roles in protein–protein interactions of the C2/V1 domains in the classic and novel PKC isozymes, respectively, has been obtained in the past few years. Our laboratory showed that anchoring of βPKC to its anchoring protein, RACK, is mediated by specific sequences within the C2 domain of this isozyme *(12,13)*. The V1 domain of the nPKC isozymes δ and εPKC was also found to be required for localization of the corresponding isozymes in vivo; introduction of the δV1 or εV1 domain into cells caused a selective inhibition of anchorage and function of δ or εPKC, respectively *(14)*.

Finally, the V5 domain of classic PKC isozymes determines the subcellular localization of these enzymes after activation. Gokmen-Plar and Fields *(15)* showed that nuclear localization of βIIPKC is determined by the last 13 amino acids in the V5 region. We showed that isozyme-specific RACK binding sites in βI and βII PKCs are in unique sequences within the V5 region (a domain of ~50 amino acids) *(16)*.

Many PKC interacting proteins have been identified using a variety of methods that are summarized below. Details on many of these methods are available in previously published reviews *(28,29)*, the references cited below, and in chapters 27–31 of this volume. Although examples of binding proteins are given, this introduction is not intended to comprehensively review all known PKC-binding proteins. Establishing a direct interaction of a PKC and a putative binding partner, as well as the physiological relevance of this interaction, has been carried out for some, but not all, of the binding partners described below. At a minimum, this requires demonstrating (1) the direct interaction of purified/recombinant proteins and (2) their interaction in vivo in a relevant cell type, for by example, co-immunoprecipitation.

2. Methods for Detecting Binding Proteins
2.1. Expression Cloning from Phage Libraries: Plaque Lift and Detection with PKC

Several laboratories have successfully identified proteins that bind PKCs through expression cloning in phage libraries (the protocol is described in detail in chapter 29 in this book and in *[28]*). Proteins expressed by phage-infected bacteria are immobilized on nitrocellulose, which is then incubated with PKCs to allow interaction. Antibodies to PKC are used to detect bound PKC (as well as PKCs expressed in the library) and to identify positive plaques. This protocol has been used by several laboratories to identify scaffold proteins that are not PKC substrates, as well as scaffolds and other proteins that are PKC substrates.

The plaque lift method was used to clone the PKC anchoring protein RACK1 from a rat brain cDNA expression library *(20)*. Nitrocellulose lifts were made from phage plaques expressing the library and incubated with a mixture of PKCs in the presence of the PKC activators PS, DG, and calcium. Binding of PKC to expressed proteins was detected with anti-PKC antibodies. Specificity of RACK1 for βIIPKC has since been demonstrated *(16,30)*. The two proteins are colocalized by confocal microscopy and are co-immunoprecipitated after PKC activation *(30)*. RACK1 is not a substrate for phosphorylation by PKC, and the interaction of these two proteins depends on activation of PKC; thus, RACK1 appears to function as an anchoring protein for activated βIIPKC.

RACK1 consists largely of seven repeats of the WD40 motif, a protein-binding motif also found in the β subunit of heterotrimeric G proteins, and indirect evidence suggests that the sequence SIKIWDL in the sixth WD40 repeat mediates binding to βIIPKC *(31)*. Specific sequences within both the C2 and V5 domains of βIIPKC have been identified that contribute to RACK1 binding in vitro *(13,16)*. Disruption of this interaction with peptides from the V5 domain (isozyme-specific sequences) and C2 domain (common to α, βI, βII, and γPKCs) of βIIPKC prevents PKC localization and blocks PKC function in vivo in an isozyme-specific or family-specific manner, respectively *(13,16)*.

A scaffold function for RACK1 is supported by the interaction of RACK1 with other proteins. RACK1 has been shown to interact with several proteins via their PH domains, including GTPase activating protein (GAP), dynamin-1, and spectrin *(21)*. RACK1 also binds Src via its SH2 domain and inhibits Src kinase activity *(32)*. In addition, RACK1 interacts with the receptor tyrosine phosphatase, PTPμ, the integrin β subunit, cAMP phosphodiesterase 4D5, the influenza A virus M1 protein, the Epstein-Barr virus activator protein, BZLF1, and Grb2-related protein with insert domain (GRID) *(33,34)*.

A second RACK was cloned with the same plaque lift methodology from a rat heart cDNA expression library as a protein that binds recombinant εPKC-V1 domain *(35)*. However, because the V1 domain does not contain the cysteine-rich domains mediating lipid interactions, binding of εPKC-V1 fragment to nitrocellulose lifts was performed in the absence of PKC activators. εRACK and εPKC are colocalized in cardiac myocytes after activation and co-immunoprecipitate from cardiac myocyte lysates *(35)*. Recently, both in vitro and in vivo data have shown that nitration of εPKC increases its interaction with εRACK *(36)*. εRACK is also known as β'-COP, which is found on the cytoplasmic face of Golgi membranes, exocytic transport vesicles, and in the cytosol *(37)*. β'-COP is part of a multiprotein complex known as the Golgi coatomer complex, that makes up the coat of nonclathrin-coated vesicles, and thus likely plays a role in vesicle formation and/or exocytosis *(37)*.

εRACK, like RACK1, contains seven repeats of the WD40 motif; however, six of the WD40 repeats are found in the N-terminal third of εRACK, whereas the εPKC-V1 domain binds to the C-terminal half of εRACK *(35)*. Although binding of εPKC-V1 domain to εRACK does not require PKC activators, the binding of full length εPKC to immobilized recombinant εRACK in vitro is dependent on PS, enhanced by DG, and inhibited by calcium *(35)*. Indirect evidence suggests that the sequence NNVALGYD just C-terminal of the WD40 repeats in εRACK mediates an interaction with the sequence EAVSLKPT in the V1 domain of εPKC *(29)*. Disruption of this interaction with the EAVSLKPT peptide has been shown to prevent both the normal subcellular localization of activated εPKC and its signaling functions *(14,38)*.

With the same expression cloning technique, calsequestrin was found to bind the V1 domain of εPKC *(39)*. Calsequestrin is also a substrate for phosphorylation by εPKC, thus it appears to interact with εPKC in two ways, via the interaction with the isozyme specific V1-domain, and with the catalytic domain *(39)*.

A third protein that does not appear to be a PKC substrate, but instead to function as a PKC anchor, was identified with a slight variation on this plaque lift method. Atypical PKC-specific interacting protein (ASIP) was cloned from an NIH 3T3 cell expression library as a protein that binds mouse ζPKC (produced in Sf21 cells) in the presence of PS *(40)*. The ζPKC was labeled with ^{32}P by autophosphorylation to allow direct detection of positive plaques. ASIP shows sequence similarity to *C. elegans* PAR-3, which is required for the establishment of cell polarity in early embryos, and both bind atypical PKCs in vitro *(41)*. λPKC and ASIP also co-immunoprecipitate from NIH 3T3 cells and from MDCKII cells, and further, they are colocalized to the cell junctional complex in cultured epithelial cells *(40)*. λPKC also co-immunoprecipitates with a mammalian homologue of *C. elegans* PAR-6 and mPAR-6 *(41)*. λPKC appears to mediate formation of a ternary complex of λPKC, PAR-3, and PAR-6 localizing to the apical junctional region of MDCK cells *(41)*. Epithelial cell polarity and tight junction structure are impaired upon expression of a kinase negative λPKC *(41)*. N-terminal to the PKC-binding domains of ASIP and PAR-3 are three PDZ domains, and thus both proteins are proposed to be scaffolds mediating the formation of multiprotein complexes regulating epithelial polarity and junctional structures *(41)*.

Autophosphorylated ^{32}P-labeled δPKC was similarly used to clone SRBC (sdr-related gene product that binds to c-kinase) from an NIH 3T3 cell expression library. SRBC is induced by serum starvation or retinoic acid, and is likely to be a δPKC substrate *(42)*.

The plaque lift and PKC overlay method has also been used to identify several PKC substrates collectively termed STICKS, for substrates that interact with C-kinase, from a rat kidney expression library *(43,44)*. The technique used was extremely similar to the technique used to identify RACK1 except for the absence of DG and the presence of high salt in the overlay buffer. In addition, formaldehyde was used at the end of this procedure to fix the PKC probe to the filter before detection with an antibody. Substrates identified in this way include MARCKS and MacMARCKS, vinculin, talin, annexins I and II, α-adducin, STICK72/SSeCKS, gravin, STICK34/sdr, GAP43, and AKAP79 *(44)*. These substrates are phospholipid-binding proteins and in general phosphorylation by PKC decreases phospholipid binding and modifies protein function *(44)*. PKC binding to STICKS is dependent on phosphatidyl-serine in overlay assays but not in solution assays *(44)*. Co-immunoprecipitation of endogenous STICKs with PKCs is usually not observed.

Some of these substrates also appear to serve as anchors or scaffolds for PKCs. αPKC may be localized to caveolae through an interaction with STICK34/sdr *(45)*. The regulatory domain of αPKC binds STICK34/sdr residues 145–250 in the presence of phosphatidylserine, and the 145–250 peptide from STICK34/sdr blocks the binding of αPKC to caveolae membranes *(45)*. Gravin is a 250-kDa scaffold protein that interacts with PKA and actin as well as PKC. The gravin complex regulates recycling of the β2-adrenergic receptor after agonist-induced desensitization *(46)*. SSeCKS is aPKC substrate that binds to PKA, PKC, and actin, and mediates actin-remodeling events *(46)*. AKAPs are A-kinase anchoring proteins that form multivalent signaling complexes and are thought to integrate signaling pathways regulating protein phosphorylation *(46)*. PKC binds to the T1 targeting domain of AKAP79, and AKAP79 is a substrate for βIIPKC in vitro *(47)*.

2.2. Expression Cloning by Yeast Two-Hybrid Screening Using PKC Domains as Bait

Two important types of PKC interactions involving zinc fingers have been identified by yeast two-hybrid screening. First, by screening with the regulatory domains of atypical PKCs as bait, proteins that interact with the zinc finger motif in atypical PKCs have been identified as described below. Second, using the regulatory domain of βIPKC as bait, the LIM protein, Enigma homologue (ENH), was identified as a binding partner. LIM domains, which contain a type of zinc finger motif, have since been shown to interact with most PKC isozymes.

LIP was identified by yeast two-hybrid analysis using the regulatory domain of λ/ι PKC as bait *(25)*. LIP binds the zinc finger domain of λ/ιPKC (but not to α, ε, or ζPKC), and activates the catalytic activity of λ/ιPKC comparable to activation by PS. LIP and λ/ιPKC also co-immunoprecipitate from serum-treated, but not quiescent, NIH 3T3 cells, and therefore LIP may be functioning as a RACK for λ/ιPKC. The zinc finger domains of the atypical PKCs do not bind diacylglycerol or phorbol esters, and in this case seem to have taken on an alternate function in mediating activation via a protein–protein interaction. LIP and λ/ιPKC appear to synergize in the transactivation of a κB-dependent promoter *(25)*.

Inhibition of PKC activity through the interaction of a binding partner with the zinc finger motifs of ζ and λ/ιPKCs has also been demonstrated. The regulatory domain of ζPKC was used as bait in the yeast two-hybrid system to identify the product of the par-4 (prostate apoptosis response) gene as a ζPKC-binding partner *(26)*. The zinc finger region of ζPKC is sufficient for this interaction, and the zinc finger region of λ/ιPKC binds PAR-4 in the yeast two-hybrid assay as well. This interaction appears to occur in mammalian cells as well, as endogenous PAR-4 and ζPKC co-immunoprecipitate from ultraviolet irradiated NIH-3T3 cells. PAR-4 binds ζPKC in vitro and this interaction requires the zinc finger motif in the regulatory domain of ζPKC. The kinase activity of both atypical PKCs is strongly inhibited by PAR-4 in vitro *(26)*, and this appears to be important in the induction of apoptosis by PAR-4. Further regulation of this pathway may occur via formation of a ternary complex of p62 with PAR-4 and ζPKC, and reactivation of PAR-4 bound, inactive ζPKC by p62 *(48)*.

RBCC protein interacting with PKC (RBCK1) was identified by yeast two-hybrid screening using the regulatory domain of βPKC as bait *(49)*. RBCK1 contains two putative coiled-coil regions, a RING finger, which may be a DNA binding motif and/or may mediate multiprotein complex assembly, a B-box, and a B-box-like motif. RBCK1 can bind DNA and has been shown to possess transcriptional activity *(49)*. The interacting domains of βIPKC and RBCK1 have not yet been determined.

The regulatory domain of βPKC was also used as bait in a yeast two-hybrid assay to identify the LIM protein ENH, as a PKC-binding protein, and thus the LIM domain as a PKC-binding motif *(17)*. ENH contains three LIM domains, which are 50–60 amino acid cysteine-rich domains containing two zinc coordinated fingers, with the consensus sequence of $CXXCX_{16-23}$ $HXXCXXCXXCX_{16-21}CXXH/D/C$ *(17)*. Unlike other types of zinc fingers, LIM domains do not bind DNA; rather, LIM domains appear to interact with other proteins. Each of the three LIM domains of ENH can associate with the regulatory domain of βIPKC. The ENH LIM domains interact with βI, γ, and εPKCs, but not with α, δ, and ζPKCs. The 19 amino acid N-terminal V1 domain of βIPKC and the 133 N-terminal V1 domain of εPKC are sufficient for LIM binding activity *(17)*. This interaction may be physiologically relevant, as overexpressed βIPKC and ENH co-immunoprecipitate from cells, and ENH is a substrate for phosphorylation by rat brain PKC in vitro *(17)*.

Human Enigma contains LIM domains highly related to those in ENH. When these domains are expressed as a GST-LIM1-3 fusion protein they can bind α, βI, and ζPKCs, but not γ, δ, and εPKCs, in cell lysates *(17)*. Furthermore, LIM-kinase-1 contains two less closely related LIM domains, and this kinase co-immunoprecipitates γ and ζPKCs when these proteins are over expressed in COS-7 cells *(17)*. In addition, Cypher-1 was identified as a novel LIM domain-containing gene from the mouse EST database and shown to bind to α, βI, ζ, γ, δ, and εPKCs in vitro via its LIM domains *(18)*. LIM domains may be responsible for selective binding of proteins to specific subsets of PKC isozymes.

Proteins containing LIM domains play many roles, including cytoskeletal organization, organ development, and oncogenesis, and protein–protein interactions mediated by the LIM domains have been shown to be crucial for at least some of these functions *(50)*. LIM proteins can contain one, two, or multiple LIM domains as well as other protein-binding motifs, and multiple binding partners for LIM proteins have been identified, leading to the proposal that these proteins are scaffolds mediating assembly of multiprotein complexes *(50)*. For example, Enigma binds not only selective PKC isozymes, but also the insulin receptor via its third LIM domain and another receptor tyrosine kinase, Ret, via its second LIM domain *(50)*. Both LIM domains bind tyrosine-containing motifs in these receptors, however, tyrosine phosphorylation of the motifs is not required for binding. LIM proteins may provide a rich source of information on PKC-signaling pathways. For example, disruption of the LIM protein, MLP, in mice results in dilated cardiomyopathy with hypertrophy and heart failure after birth *(51)*, and it will be interesting to determine whether this is mediated by signaling via a particular PKC.

Yeast two-hybrid screening has also resulted in the identification of substrates for PKC. For example, the catalytic domain of αPKC was used as bait to identify PICK1 (perinuclear PKC substrate), as a PKC-binding protein. The PDZ domain in PICK1 has been shown to interact with a PDZ-binding sequence (QSAV) at the C-terminus of αPKC *(9)*. Zeta-interacting protein (ZIP) was cloned from a rat brain cDNA library by a yeast two-hybrid screen for ζPKC-binding partners *(27)*. ZIP binds to the pseudosubstrate region in the regulatory domain of ζPKC, and is a substrate for ζPKC phosphorylation. ζPKC binding was mapped to a region of ZIP containing the possible protein-binding motif YXDEDX$_5$SDEE/D. ZIP also interacts with the SH2 domain of p56^{lck}, associates with the cytokine receptor EBI-3, and is induced by oxidative stress in macrophages, and may function as a scaffolding protein *(27)*.

2.3. Identification of PKC-Binding Proteins by Immunoprecipitation

A large number of PKC-binding partners have been identified because of their presence in immunoprecipitates of PKCs from cell lysates. Both endogenously expressed PKCs and binding partners, and overexpressed PKCs and binding partners have been useful for this purpose. Although co-immunoprecipitation is indicative of a physiologically relevant interaction in vivo, it is not indicative of a direct interaction of the two proteins, which must be demonstrated in vitro with purified proteins. Typically, PKC (or the regulatory or catalytic domain of PKC) and the proposed binding partner are expressed recombinantly with tags allowing immobilization and purification. The proposed binding partner is immobilized, for example, on glutathione S-transferase beads, followed by co-incubation with the purified PKC in the presence or absence of PKC activators, and the beads are washed and analyzed for retention of PKC.

Many proteins in complexes with εPKC were identified by immunopre-cipitation of εPKC from cardiac tissue of mice expressing low levels of constitutively active εPKC, followed by two-dimensional gel electrophoresis of the immunoprecipitates, and mass spectrometry to identify proteins (Chapter 31 in this volume and **ref. 52**). These included structural and skeletal proteins (α-actin, troponin T2, α-troposyosin, prohibitin, BM130, villin, lap 2, desmin, caveolin-3, myosin light chain), signaling proteins (PI3kinase, connexin-43, pyk2 kinase, src and lck tyrosine kinases, BMX, Akt, p38 MAPK, MAP-KAPK2, JNK1, JNK2, MEK1, ERK1 and 2, ras), and stress-activated proteins (iNOS, eNOS, COX-2, Hif-1α, heme oxygenase-1, aldose reductase, HSP 70, HSP 27, αβ-crystallin). When samples from control mice were compared with tissue from animals expressing low levels of constitutively active εPKC, many of these proteins were shown to be expressed at higher levels or differently posttranslationally modified in the transgenics. The interaction of Src with εPKC is mediated by the SH2 and SH3 domains of Src and by the regulatory domain of εPKC *(24)*. Src binds preferentially to εPKC that has been activated with PS or is constitutively active because of a mutation that prevents the binding of the pseudosubstrate domain to the catalytic domain *(24)*.

Immunoprecipitation experiments in various cell types has demonstrated associations (not known to be direct) of δPKC with phosphatidylinositol-3-kinase *(53)*, RAFT1 (the mammalian homologue of mTOR) *(54)*, and the insulin receptor *(55)*, and interactions with the δPKC substrates Stat3 *(56)* and the protein tyrosine phosphatase SHPTP1 *(57)*. δPKC and c-src also co-immunoprecipitate from MCF-7 cell lysates, and src can phosphorylate δPKC in vitro, in support of the hypothesis that the c-src is responsible for the tyrosine phosphorylation of δPKC *(58)*. Immunoprecipitation has also demonstrated associations, subsequently shown to be direct, of αPKC with syndecan-4 *(59)*, and λPKC with p70 S6 kinase *(60)*.

The stringency of immunoprecipitation conditions can be increased after treatment of cells with a cross-linking agent. The association of endogenous cPKCs with members of the transmembrane-4 superfamily (TM4SF) is maintained even in the presence of 0.2% sodium dodecyl sulfate if intact cells are pretreated with dithiobis(succinimidyl propionate), a membrane permeant homobifunctional cross-linking agent with a span of 12 Å *(61)*. In K562 cells activated with PMA, αPKC interacts with four TM4SF proteins, CD9, CD53, CD81, and CD82, and βIIPKC interacts with CD9 *(61)*. Integrins are also found in complexes with PKC and TM4SF, and it has been proposed that complex formation is mediated by simultaneous binding of the extracellular domains of TM4SF to integrin α3 subunit, and the intracellular/transmembrane domains of TM4SF to αPKC *(61)*. Treatment of cells with dithiobis succinimidyl propionate has also been used to allow detection of a transient interaction of αPKC with vinculin during the adherence of HeLa cells *(62)*. Although vinculin is a substrate for phosphorylation by αPKC, vinculin binding to αPKC in vitro does not require a functional kinase domain, and does require calcium and an acidic phospholipid (phosphatidylinositol-4,5-bisphosphate), suggesting an interaction of vinculin with the regulatory domain of αPKC *(62)*.

The presence of PKCs in complexes with several anchoring or scaffolding proteins has also been detected with immunoprecipitation assays. Centrosome and Golgi-localized PKN-associated protein (CG-NAP)/AKAP450 is a scaffold for the protein kinase A regulatory subunit, protein phosphatase 2A, and protein phosphatase 1, and in immunoprecipitates from rat brain, εPKC also associates with CG-NAP *(63)*. Direct binding studies showed that hypo or nonphosphorylated εPKC binds to CG-NAP via its catalytic domain, and then disassociates after phosphorylation *(63)*. FRS2 is a docking protein that mediates signaling from the FGF receptor via downstream signaling molecules such as Grb-2 and SHP2. λPKC was found to associate with FRS2 in Swiss 3T3 cell lysates following treatment with bFGF, and direct interaction of the catalytic domain of λPKC with FRS2 was demonstrated by yeast two-hybrid analysis *(64)*.

InaD is a multivalent scaffolding protein that appears to organize photoreceptor complexes in *Drosophila (65)*. Mutants of inaC (the eye PKC) and inaD were originally classified in *Drosophila* as defective in inactivation of the light response. Eye PKC is found in immunoprecipitates of the InaD protein from the eye of the fly Calliphora *(19)*. InaD contains five distinct PDZ domains, and expression of individual PDZ domains as glutathione S-transferase fusions and incubation with membrane extracts from fly heads demonstrated specific interactions of the TRP ion channel with PDZ-3, PKC with PDZ-4, and phospholipase C with PDZ-5 *(65)*. However, yeast two-hybrid analysis and protein overlay demonstrated an interaction between PKC and PDZ-2 *(66)*. In an inaD null mutant PKC, phospholipase C, and TRP are no longer localized to the rhabdomere, a specialized subcellular compartment involved in phototransduction, instead eye PKC is found in the cytoplasm *(65)*.

2.4. Standard Protein Purification Techniques

Several variations of standard protein purification techniques have been used to identify PKC-interacting proteins, including the use of immobilized PKC affinity reagents. Binding partners have also been identified through purification by standard fractionation techniques, with the assay for positive fractions consisting of analysis by far Western blot probed with PKC.

A RACK for δPKC has been identified by purification from the Triton-X-100 insoluble fraction of hepatocytes *(67)*. Proteins were purified on the basis of a far Western assay (a type of overlay assay) in which proteins were separated by sodium dodecyl sulfate polyacrylamide gel electrophoresis, transferred to nitrocellulose, incubated with partially purified PKCs in the presence of DG, PS, and calcium, and PKC-binding proteins detected with antibodies to PKC. One of the proteins identified in this way is p32/gC1qBP, a receptor for the globular head of complement C1q. p32 is not a PKC substrate, and its binding increases PKC activity twofold. δPKC binds p32 only in the presence of PKC activators; therefore, p32 fits the definition of a δRACK. However, p32 may play a complex role in signaling by other PKC isozymes because the other isozymes tested bind to p32 in both the presence and absence of activators *(67)*. Another hepatocyte PKC-binding protein that was identified with the far Western assay is calreticulin, a calcium-binding protein involved in integrin function *(68)*. Calreticulin is phosphorylated by PKC, and it interacts with several PKCs in vitro and in vivo *(68)*.

An affinity reagent consisting of the V1 domain of θPKC was immobilized and used to purify PKC interacting proteins from T cells *(69)*. Recombinant θPKC-V1 domain was immobilized via a histidine tag on a nickel-agarose resin, incubated with cell lysates, and eluates were analyzed by Western blot for known T cell-signaling proteins and for phosphotyrosine content. The tyrosine kinase, p59fyn, was a major phosphotyrosine-containing protein in the eluate, and θPKC and p59fyn were also shown to co-immunoprecipitate from T-cell lysates. The interaction of p59fyn and the V1 domain of θPKC was further demonstrated in vitro with recombinant proteins. Finally, p59fyn phosphorylates θPKC on tyrosine residues in vitro, resulting in increased catalytic activity of θPKC *(69)*.

3. Conclusions

A variety of methodologies have been used to identify PKC-interacting proteins that function as anchoring proteins, scaffolds of multiple signaling proteins, and/or substrates. Expression cloning by plaque lift and binding of PKC or a PKC fragment resulted in the identification of scaffold and anchoring proteins RACK1 and εRACK as well as a number of the substrates termed STICKs. Protein purification and assay by far Western blot have also been used to identify both a RACK (p32/gC1qBP) and a PKC substrate (calreticulin). Expression cloning by yeast two-hybrid screening has also led to the identification of PKC scaffolds (e.g., LIP, possibly a λ/ιPKC RACK) as well as PKC substrates (e.g., PICK1). However, yeast two-hybrid screening with PKC regulatory domains has been the only technique so far resulting in the identification of protein–protein interactions dependent on zinc finger motifs, both in aPKCs and in binding partners (the LIM domain).

The discovery that the zinc finger motif in aPKCs and the C2/V1 domain of c and nPKCs mediate protein–protein interactions, suggests that in the future, recombinant protein fragments encompassing this portion of the regulatory domain may be useful probes or bait in the identification of other PKC-interacting proteins. The converse is also likely to be true, that proteins containing LIM, PDZ, PH, and WD40 domains are good candidates for PKC-binding partners, and recombinant protein fragments encompassing these domains will be useful probes or bait in the identification of relevant PKCs.

References

1. Newton, A. C. (1995) Protein kinase C: Structure, function, and regulation. *J. Biol. Chem.* **270,** 28,495–28,498.
2. Nishikawa, K., Toler, A., Johannes, F.-J., Zhou, S., and Cantley, L. C. (1997) Determination of the specific substrate sequence motifs of protein kinase C isozymes. *J. Biol. Chem.* **272,** 952–960.
3. Mochly-Rosen, D. (1995) Localization of protein kinases by anchoring proteins: a theme in signal transduction. *Science* **268,** 247–51.
4. Nishizuka, Y. (1992) Intracellular signaling by hydrolysis of phospholipids and activation of protein kinase C. *Science* **258,** 607–614.
5. Toker, A. (1998) Signaling through protein kinase C. *Fronti. Biosci.* **3,** d1134–d1147.
6. Teruel, M. N. and Meyer, T. (2000) Translocation and reversible localization of signaling proteins: a dynamic future for signal transduction. *Cell* **103,** 181–184.

7. Oka, N., Yamamoto, M., Schwencke, C., Kawabe, J. Ebina, T., Ohno, S., et al. (1997) Caveolin interaction with protein kinase C. *J. Biol. Chem.* **272,** 33,416–33,421.

8. Prekeris, T., Hernandez, R. M., Mayhew, M. W., White, M. K., and Terrian, D. M. (1998) Molecular analysis of the interactions between protein kinase C-epsilon and filamentous actin. *J. Biol. Chem.* **273,** 26,790–26,798.

9. Staudinger, J., Lu, J., and Olson, E. N. (1997) Specific interaction of the PDZ domain protein PICK1 with the COOH terminus of protein kinase C-alpha. *J. Biol. Chem.* **272,** 32,019–32,024.

10. Ohno, Y., Fuji, T., Ogita, K., Kikkawa, U., Igarashi, K., and Nishizuka, Y. (1989) Protein kinase C zetz subspecies from rat brain: its structure, expression, and properties. *Proc. Natl. Acad. Sci. USA* **86,** 3099–3103.

11. Lehel, C., Olah, Z., Jakab, G., and Anderson, W. B. (1995)Protein kinase C epsilon is localized to the Golge via its zinc-finger domain and modulates Golgi function. *Proc. Natl. Acad. Sci. USA* **92,** 1406–1410.

12. Mochly-Rosen, D., Miller, K. G., Scheller, R. H., Khaner, H., Lopez, J., and Smith, B. L. (1992) p65 fragments, homologous to the C2 region of protein kinase C, bind to intracellular receptors for protein kinase C. *Biochemistry* **31,** 8120–8124.

13. Ron, D., Luo, J., and Mochly-Rosen, D. (1995) C2 region-derived peptides inhibit translocation and function of beta protein kinase C in vivo. *J. Biol. Chem.* **270,** 24,180–24,187.

14. Johnson, J. A., Gray, M. O., Chen, C. H., and Mochly-Rosen, D. (1996) A protein kinase C translocation inhibitor as an isozymeselective antagonist of cardiac function. *J. Biol. Chem.* **271,** 24,962–24,966.

15. Gokmen-Polar, Y. and Fields, A. P. (1998) Mapping of a molecular determinant for protein kinase C betaII isozyme function. *J. Biol. Chem.* **273,** 20,261–20,266.

16. Stebbins, E. G. and Mochly-Rosen, D. (2001) Binding specificity for RACK1 resides in the V5 region of betaII protein kinase C. *J. Biol. Chem.* **276,** 29,644–29,650.

17. Kuroda, S., Tokunaga, C. Kiyohara, Y., Higuchi, O., Konishi, H., Mizuno, K., et al. (1996) Protein-protein interaction of zinc finger LIM domains with protein kinase C. *J. Biol. Chem.* **271,** 31,029–31,032.

18. Zhou, Q., RuizLozano, P., Martone, M. E., and Chen, J. (1999) Cypher, a striated muscle-restricted PDZ and LIM domain-containing protein, binds to alpha-actinin-2 and protein kinase C. *J. Biol. Chem.* **274,** 19,807–19,813.

19. Huber, A., Sander, P., and Paulsen, R. (1996) Phosphorylation of the InaD gene product, a photoreceptor membrane protein required for recovery of visual excitation. *J. Biol. Chem.* **271,** pp. 11,710–11,717.

20. Ron, D., Chen, C. H., Caldwell, J., Jamieson, L., Orr, E., and Mochly-Rosen, D. (1994) Cloning of an intracellular receptor for protein kinase C: a homolog of the beta subunit of G proteins (published erratum appears in *Proc. Natl. Acad. Sci. USA* 1995 Feb 28;92(5):2016). *Proc. Natl. Acad. Sci. USA* **91,** 839–843.

21. Rodriguez, M. M., Ron, D., Touhara, K., Chen, C. H., and Mochly-Rosen, D. (1999) RACK1, a protein kinase C anchoring protein, coordinates the binding

of activated protein kinase C and select pleckstrin homology domains in vitro. *Biochemistry* **38,** 13,787–13,794.

22. Yao, L., Kawakami, Y., and Kawakami, T. (1994) The pleckstrin homology domain of Bruton tyrosine kinase interacts with protein kinase C. *Proc. Natl. Acad. Sci. USA* **91,** 9175–9179.

23. Konishi, H., Shinomura, T., Kuroda, S., Ono, Y., and Kikkawa, U. (1994) Molecular cloning of rat RAC protein kinase alpha and beta and their association with protein kinase C zeta. *Biochem. Biophys. Res. Commun.* **205,** 817–825.

24. Song, C., Vondriska, T. M., Wang, G. W., Klein, J. B., Cao, X., Zhang, J., et al. (2002) Molecular conformation dictates signaling module formation: example of PKCepsilon and Src tyrosine kinase. *Am. J. Physiol. Heart Circ. Physiol.* **282,** H1166–H1171.

25. Diaz-Meco, M. T., Municio, M. M., Sanchez, P., Lozano, J., and Moscat, J. (1996) Lambda–interacting protein, a novel protein that specifically interacts with the zinc finger domain of the atypical protein kinase C isotype lambda/iota and stimulates its kinase activity in vitro and in vivo. *Mol. Cell Biol.* **16,** 105–114.

26. Diaz-Meco, M. T., Municio, M. M., Frutos, S., Sanchez, P., Lozano, J., Sanz, L., et al. (1996) The product of par-4, a gene induced during apoptosis, interacts selectively with the atypical isoforms of protein kinase C. *Cell* **86,** 777–786.

27. Puls, A., Schmidt, S., Grawe, F., and Stabel, S. (1997) Interaction of protein kinase C zeta with ZIP, a novel protein kinase C-binding protein. *Proc. Natl. Acad. Sci. USA* **94,** 6191–6196.

28. Csukai, M. and Mochly-Rosen, D. (1998) Molecular genetic approaches II: expression interaction cloning, in *Methods in Molecular Biology* (R. A. Clegg, ed.), Humana Press Inc., Totowa, NJ, pp. 133–139.

29. Schechtman, D. and Mochly-Rosen, D. (2002) Isozyme-specific inhibitors and activators of protein kinase C. *Methods Enzymol.* **345,** 470–489.

30. Ron, D., Jiang, Z., Yao, L., Vagts, A., Diamond, I., and Gordon, A. (1999) Coordinated movement of RACK1 with activated betaII PKC. *J. Biol. Chem.* **274,** 27,039–27,046.

31. Ron, D. and Mochly-Rosen, D. (1995) An autoregulatory region in protein kinase C: the pseudoanchoring site. *Proc. Natl. Acad. Sci. USA* **92,** 492–496.

32. Chang, B. Y., Conroy, K. B., Machleder, E. M., and Cartwright, C. A. (1998) RACK1, a receptor for activated C Kinase and a homologue of the beta subunit of G proteins, inhibits activity of src tyrosine kinases and growth of NIH 3T3 cells. *Mol. Cell. Biol.* **18,** 3245–3256.

33. Schechtman, D. and Mochly-Rosen, D. (2001) Adaptor proteins in protein kinase C-mediated signal transduction. *Oncogene* **20,** 6339–6347.

34. Ellis, J. H., Ashman, C., Burden, M. N., Kilpatrick, K. E., Morse, M. A., and Hamblin, P. A. (2000) GRID: a novel Grb-2-related adapter protein that interacts with the activated T cell costimulatory receptor CD28. *J. Immunol.* **164,** 5805–5814.

35. Csukai, M., Chen, C. H., De Matteis, M. A., and Mochly-Rosen, D. (1997) The coatomer protein beta'-COP, a selective binding protein (RACK) for protein kinase Cepsilon. *J. Biol. Chem.* **272,** 29,200–29,206.

36. Balafanova, Z., Bolli, R., Zhang, J., Zheng, Y., Pass, J. M., Bhatnagar, A., et al. (2002) Nitric oxide induces nitration of PKC epsilon, facilitating PKC epsilon translocation via inhanced PKC epsilon-RACK2 interactions: a novel mechanism of NO-triggered activation of PKCepsilon. *J. Biol. Chem.* **277,** 15,021–15,027.

37. Harrison-Lavoie, K. J., Lewis, V. A., Hynes, G. M., Collison, K. S., Nutland, E., and Willison, K. R. (1993) A 102 kDa subunit of a Golgi-associated particle has homology to beta subunits of trimeric G proteins. *EMBO J.* **12,** 2847–2853.

38. Chen, C. H., Gray, M. O., and Mochly-Rosen, D. (1999) Cardioprotection from ischemia by a brief exposure to physiological levels of ethanol: role of epsilon protein kinase C. *Proc. Natl. Acad. Sci. USA* **96,** 12,784–12,789.

39. Rodriguez, M. M., Chen, C. H., Smith, B. L., and Mochly-Rosen, D. (1999) Characterization of the binding and phosphorylation of cardiac calsequestrin by epsilon protein kinase C. *FEBS Lett.* **454,** 240–246.

40. Izumi, Y., Hirose, T., Tamai, Y., Hirai, S., Nagashima, Y., Fujimoto, T., et al. (1998) An atypical PKC directly associates and colocalizes at the epithelial tight junction with ASIP, a mammalian homoloque of *Caenorhabditis elegans* polarity protein PAR–3. *J. Cell. Biol.* **143,** 95–106.

41. Suzuki, A., Yamanaka, T., Hirose, T., Manabe, N., Mizuno, K., Shimizu, M., et al. (2001) Atypical protein kinase C is involved in the evolutionarily conserved PAR protein complex and plays a critical role in establishing epithelia-specific junctional structures. *J. Cell. Biol.* **152,** 1183–1196.

42. Izumi, Y., Hirai, S., Tamai, Y., Fujise-Matsuoka, A., Nishimura, Y., and Ohno, S. (1997) A protein kinase C delta-binding protein SRBC whose expression is induced by serum starvation. *J. Biol. Chem.* **272,** 7381–7389.

43. Chapline, C., Ramsay, K., Klauck, T., and Jaken, S. (1993) Interaction cloning of protein kinase C substrates. *J. Biol. Chem.* **268,** 6858–6861.

44. Jaken, S. and Parker, P. J. (2000) Protein kinase C binding partners. *Bioessays* **22,** 245–254.

45. Mineo, C., Ying, Y. S., Chapline, C., Jaken, S., and Anderson, R. G. (1998) Targeting of protein kinase Calpha to caveolae. *J. Cell. Biol.* **141,** 601–610.

46. Diviani, D. and Scott, J. D. (2001) AKAP signaling complexes at the cytoskeleton. *J. Cell Sci.* **114,** 1431–1437.

47. Feliciello, A., Gottesman, M. E., and Avvedimento, E. V. (2001) The biological functions of A-kinase anchor proteins. *J. Mol. Biol.* **308,** 99–114.

48. Chang, S., Kim, J. H., and Shin, J. (2002) p62 forms a ternary complex with PKCzeta and PAR-4 and antagonizes PAR-4-induced PKCzeta inhibition. *FEBS Lett.* **510,** 57–61.

49. Tokunaga, C., Kuroda, S., Tatematsu, K., Nakagawa, N., Ono, Y., and Kikkawa, U. (1998) Molecular cloning and characterization of a novel protein kinase

C-interacting protein with structural motifs related to RBCC family proteins. *Biochem. Biophys. Res. Commun.* **244,** 353–359.

50. Bach, I. (2000) The LIM domain: regulation by association. *Mech. Dev.* **91,** 5–17.

51. Arber, S., Hunter, J. J., Ross, J., Jr., Hongo, M., Sansig, G., Borg, J., et al. (1997) MLP-deficient mice exhibit a disruption of cardiac cytoarchitectural organization, dilated cardiomyopathy, and heart failure. *Cell* **88,** 393–403.

52. Ping, P., Zhang, J., Pierce, W. M., Jr., and Bolli, R. (2001) Functional proteomic analysis of protein kinase C epsilon signaling complexes in the normal heart and during cardioprotection. *Circ. Res.* **88,** 59–62.

53. Ettinger, S. L., Lauener, R. W., and Duronio, V. (1996) Protein kinase C delta specifically associates with phosphatidylinositol 3-kinase following cytokine stimulation. *J. Biol. Chem.* **271,** 14,514–14,518.

54. Kumar, V., Pandey, P., Sabatini, D., Kumar, M., Majumder, P. K., Bharti, A., et al. (2000) Functional interaction between RAFT1/FRAP/mTOR and protein kinase cdelta in the regulation of cap-dependent initiation of translation. *EMBO J.* **19,** 1087–1097.-

55. Formisano, P., Oriente, F., Miele, C., Caruso, M., Auricchio, R., Vigliotta, G., et al. (1998) In NIH-3T3 fibroblasts, insulin receptor interaction with specific protein kinase C isoforms controls receptor intracellular routing. *J. Biol. Chem.* **273,** 13,197–13,202.

56. Jain, N., Zhang, T., Kee, W. H., Li, W., and Cao, X. (1999) Protein kinase C delta associates with and phosphorylates Stat3 in an interleukin-6-dependent manner. *J. Biol. Chem.* **274,** 24,392–24,400.

57. Yoshida, K. and Kufe, D. (2001) Negative regulation of the SHPTP1 protein tyrosine phosphatase by protein kinase C delta in response to DNA damage. *Mol. Pharmacol.* **60,** 1431–1438.

58. Shanmugam, M., Krett, N. L., Peters, C. A., Maizels, E. T., Murad, F. M., Kawakatsu, H., et al. (1998) Association of PKC delta and active Src in PMA-treated MCF-7 human breast cancer cells. *Oncogene* **16,** 1649–1654.

59. Oh, E. S., Woods, A., and Couchman, J. R. (1997) Syndecan-4 proteoglycan regulates the distribution and activity of protein kinase C. *J. Biol. Chem.* **272,** 8133–8136.

60. Akimoto, K., Nakaya, M., Yamanaka, T., Tanaka, J., Matsuda, S., Weng, Q. P., et al. (1998) Atypical protein kinase Clambda binds and regulates p70 S6 kinase. *Biochem. J.* **335,** 417–424.

61. Zhang, X. A., Bontrager, A. L., and Hemler, M. E. (2001) Transmembrane-4 superfamily proteins associate with activated protein kinase C (PKC) and link PKC to specific beta(1) integrins. *J. Biol. Chem.* **276,** 25,005–25,013.

62. Ziegler, W. H., Tigges, U., Zieseniss, A., and Jockusch, B. M. (2002) A lipid-regulated docking site on vinculin for protein kinase C. *J. Biol. Chem.* **277,** 7396–7404.

63. Takahashi, M., Mukai, H., Oishi, K., Isagawa, T., and Ono, Y. (2000) Association of immature hypophosphorylated protein kinase cepsilon with an anchoring protein CG-NAP. *J. Biol. Chem.* **275,** 34,592–34,596.

64. Lim, Y. P., Low, B. C., Lim, J. Wong, E. S., and Guy, G. R. (1999) Association of atypical protein kinase C isotypes with the docker protein FRS2 in fibroblast growth factor signaling. *J. Biol. Chem.* **274,** 19,025–19,034.

65. Tsunoda, S., Sierralta, J., Sun, Y., Bodner, R., Suzuki, E., Becker, A., et al. (1997) A multivalent PDZ-domain protein assembles signalling complexes in a G-protein-coupled cascade. *Nature* **388,** 243–249.

66. Kumar, R. and Shieh, B. H. (2001)The second PDZ domain of INAD is a type I domain involved in binding to eye protein kinase C. Mutational analysis and naturally occurring variants. *J. Biol. Chem.* **276,** 24,971–24,977.

67. Robles-Flores, M., Rendon-Huerta, E., Gonzalez-Aguilar, H., Mendoza-Hernandez, G., Islas, S., Mendoza, V., Ponce-Castaneda, M. V., et al. (2002) p32 (gC1qBP) is a general protein kinase C (PKC)-binding protein; interaction and cellular localization of P32-PKC complexes in ray hepatocytes. *J. Biol. Chem.* **277,** 5247–5255.

68. Rendon-Huerta, E., Mendoza-Hernandez, G., and Robles-Flores, M. (1999) Characterization of calreticulin as a protein interacting with protein kinase C. *Biochem. J.* **344,** 469–475.

69. Ron, D., Napolitano, E. W., Voronova, A., Vasquez, N. J., Roberts, D. N., Calio, B. L., et al. (1999) Direct interaction in T-cells between thetaPKC and the tyrosine kinase p59fyn. *J. Biol. Chem.* **274,** 19,003–19,010.

27

Detection of Protein Kinase-Binding Partners by the Yeast Two-Hybrid Analysis

Betty Y. Chang and Christine A. Cartwright

1. Introduction

One mechanism by which protein kinases are regulated and function is via their interaction with other cellular proteins (reviewed in **ref. 1**). Thus, identifying proteins that interact with kinases will reveal important mechanisms of kinase action. The two-hybrid method is a genetic assay for detecting protein–protein interactions in vivo in yeast *(2)*. This approach has advantages over other available methods for studying protein interactions. The assay can be used to define interactions between two known proteins or to search genomic or cDNA libraries for proteins that interact with a bait protein. For the latter approach, the method not only identifies the interacting protein but also provides the gene for the protein. Or, the method can be used to identify specific domains or specific amino acids involved in protein–protein interactions. The assay is performed in vivo; thus, the interacting proteins are more likely to be in their native conformations. The method can detect transient and weak interactions, which is probably how many interactions occur in signaling complexes in vivo. This approach avoids the need to purify interacting proteins or to generate antibodies in order to identify the proteins. The assay is sensitive, efficient, and rapid. A disadvantage of the method is generation of false positives. Thus, care must be taken to eliminate false positives and to verify protein interactions that are detected in the screen. Nonetheless, the yeast two-hybrid analysis has been used successfully to identify many important protein interactions. These include interaction of the yeast SNF1 protein kinase with a substrate, SIP1 *(3)*, mammalian Ras with the serine/threonine kinase Raf *(4,5)*, the retinoblastoma protein with the protein phosphatase type 1 catalytic

From: *Methods in Molecular Biology, vol. 233: Protein Kinase C Protocols*
Edited by: A. C. Newton © Humana Press Inc., Totowa, NJ

subunit *(6)*, cyclin-dependent kinase Cdk2 with Cip1, an inhibitor of G1 cyclin-dependent kinases *(7)*, the human breast cancer susceptibility protein BRCA2 with the RAD51 DNA repair protein *(8,9)*, and the Src tyrosine kinase with RACK1 (receptor for activated C kinase) *(10)*. The identification of RACK1 as an Src-binding protein will be used as an example to illustrate the feasibility of the yeast two-hybrid method for detecting protein-kinase binding partners.

2. Materials
2.1. Preparation for the Two-Hybrid Analysis
2.1.1. Construction of Hybrid Genes
2.1.1.1. GAL4 DNA-BINDING DOMAIN VECTOR

1. pAS2 is an 8.4-kb plasmid that we used to fuse the bait protein (Src-UD/SH3/SH2 domain) with the GAL4 DNA-BD (GenBank Accession: #U30496; Clontech, Palo Alto, CA; **refs. *6,7,11–13***). pAS2 contains the GAL4 DNA-BD, the *CYH^{S}2* gene that confers sensitivity to cycloheximide, the *TRP1* gene for selection in Trp⁻ auxotrophic yeast strains, and the HA epitope tag.
2. Alternatively, one can use pAS2-1, which is derived from pAS2 and contains a neutral, short peptide instead of an HA epitope tag. Removing the HA epitope tag and converting a.a. 149 from Glu to Val completely eliminates the autonomous activation activity of pAS2 *(11,12)*.
3. pAS2 (or pAS2-1) plasmid, lacking the bait protein, is used as a negative control when verifying bait protein expression in yeast transformants.

2.1.2. Transformation of Hybrid Genes into Yeast

1. YPD medium: 20 g/L Difco peptone, 10 g/L yeast extract, pH 5.8, autoclave. Add dextrose (glucose) to 2% by adding 40 mL of a 50% stock solution (separately filter-sterilized or autoclaved) to each liter of media.
2. Yeast strain *MATa Saccharomyces cerevisiae* Y190 *(7)* in YPD medium/25% glycerol (Clontech). Use when screening AD fusion libraries for proteins that interact with a bait protein.
3. 10X TE : 0.1 *M* Tris-HCl, 10 m*M* ethylenediaminetetraacetic acid (EDTA), pH 7.5, autoclave.
4. 10X LiAc: 1 *M* lithium acetate (Sigma). Adjust to pH 7.5 with dilute acetic acid and autoclave.
5. Polyethylene glycol/lithium acetate (PEG/LiAc) solution: 40% PEG 3350 (Sigma) in 1X TE/LiAc.
6. Herring testes carrier DNA (10 mg/mL; Sigma); denature carrier DNA by boiling for 30 min and immediately cool it on ice.
7. Appropriate SD agar plates.

Prepare the selection media (SD powder form; Qbiogene, Carlsbad, CA) and pour the agar plates in advance. Allow the plates to dry for 2–3 d at room temperature or for 3 h at 30°C before plating. Moisture on the agar surface may cause uneven plating.

2.1.3. Verification of Fusion Protein Expression in Yeast

1. Src-specific monoclonal antibody 327 (MAb 327) *(14)*.
2. MAb 12CA5, which recognizes the HA epitope (Boehringer Mannheim Biochemicals, Indianapolis, IN).

2.1.4. Verification that Hybrid Genes do not Autonomously Activate Reporter Genes

1. pLAM 5′ is a 6.0-kb plasmid (pGBT9) that encodes a fusion protein of human lamin C (a.a. 66–230) and GAL4 DNA-BD *(11–13)* GenBank Accession #M13451; Clontech). pLAM 5′ is used to detect false-positive clones.
2. pCL1 is a 15.3-kb plasmid that encodes the full-length, wild-type GAL4 protein *(2)* (Clontech). pCL1 is used as a positive control for the β-galactosidase transcription assay.

2.2. Screening Libraries for Interacting Proteins

2.2.1. Activation Domain Fusion Library

1. A cDNA library of interest cloned into a pGAL4 AD vector.
2. The AD fusion library that we used contained a WI-38 human lung fibroblast cell line cDNA library fused to the GAL4 activation domain in the pGAD GL vector (Clontech). pGAD-GL is a 6.9-kb plasmid containing the GAL4 AD, the *Amp^r* gene that confers resistance to ampicillin, and the LEU2 gene for selection in Leu⁻ auxotrophic yeast strains *(4,15)*.

2.2.2. Transformation of the cDNA Library into Yeast

1. 3-aminotriazole (3-AT; Sigma); 1 *M*, filter sterilized.

2.2.3. Assay for Activation of the LacZ Reporter Gene (β-Galactosidase Filter Assay)

1. Whatman no. 5 filters: 125-mm diameter filters for 150-mm diameter plates, sterile.
2. Z-buffer: $Na_2HPO_4 \cdot 7H_2O$ (16.1 g/L), $NaH_2PO_4 \cdot 7H_2O$ (5.5 g/L), KCl (0.75 g/L), $MgSO_4 \cdot 7H_2O$, (0.246 g/L), pH 7.0, autoclave. Can be stored at room temperature for up to 1 yr.
3. X-gal stock solution: dissolve 5-bromo-4-chloro-3-indolyl β-D-galactoside (X-gal) in *N,N*-dimethylformamide at a concentration of 20 mg/mL. Store in dark at –20°C.
4. Z-buffer/X-gal solution: 100 mL Z buffer, 0.27 mL β-mercaptoethanol, 1.67 mL X-gal stock solution.

2.3. Analysis and Verification of Putative Positive Clones

1. *Escherichia coli* HB101 (which carried a *leuB* mutation that can be complemented by *LEU2* from yeast; Invitrogen, Carlsbad, CA).
2. M9 minimal medium (Sigma).

2.4. Confirmation of Protein Interactions

1. pGEX plasmids for generating glutathione-S-transferase (GST) fusion proteins; in our case: pGEX*src*, pGEX*src*-UD/SH3/SH2 and pGEX-RACK1.
2. Plasmids for generating in vitro translated proteins; in our case: pGEM*src* and pcDNA3-RACK1.
3. In vitro translation system (TNT-coupled rabbit reticulocyte lysate system; Promega Inc, Madison, WI).

RACK1-specific monoclonal antibody (Transduction Laboratories, Lexington, KY).

3. Methods

The two-hybrid system exploits the two-domain structure of many eukaryotic transcription factors to detect interactions between two different proteins (reviewed in **refs. *2,11–13***). The transcription factor consists of a DNA-binding domain, which is distinct from a transcriptional activation domain. A hybrid containing a DNA-binding domain fused to protein A is unable to activate transcription of a reporter gene because it lacks a transcriptional activation domain. A second hybrid containing an activation domain fused to protein B is unable to localize to the reporter gene. If both hybrids are expressed in the same cell and proteins A and B interact, then the activation domain is anchored to the DNA-binding site and the reporter gene is expressed *(2,11–13)*.

The methods described below outline the (1) preparation for the two-hybrid analysis: construction of hybrid genes and introduction of the genes into yeast; (2) screening of libraries for interacting proteins; (3) analysis and verification of putative positive clones; and (4) confirmation of protein interactions.

3.1. Preparation for the Two-Hybrid Analysis

3.1.1. Construction of Hybrid Genes

The 5′ coding region of murine c-src cDNA, encoding the unique, SH3 and SH2 domains (UD/SH3/SH2), was inserted into pAS2 to generate pAS2-Src-UD/SH3/SH2 *(10)*. pAS2 contains the GAL4 DNA-BD, the *CYH^S2* gene (which confers sensitivity to cycloheximide), the *TRP1* gene for selection in Trp⁻ auxotrophic yeast strains, and the HA epitope tag *(7)*.

1. Generate a fusion gene using compatible restriction sites in the bait gene and the DNA-BD/vector, or by polymerase chain reaction amplification of the gene with compatible restriction sites incorporated into the primers *(16)*.

2. Purify a small amount of plasmid (about 100 μg) using either CsCl gradients *(16)* or Maxi-prep columns (Qiagen, Valencia, CA).

3.1.2. Transformation of Hybrid Genes into Yeast

The cycloheximide-resistant (Chx^r), *MATa Saccharomyces cerevisiae* strain Y190 was transformed with pAS2-Src-UD/SH3/SH2 or pAS2 *(10)*, using the PEG/LiAc method for yeast transformation (*see* **Note 1**; **refs.** *17,18*).

1. Inoculate cells from a single Y190 yeast colony into 5 mL of YPD medium and incubate overnight (16–18 h) at 30°C with shaking (250 rpm) to stationary phase (OD$_{600}$ >1.4).
2. Transfer 0.1 mL of the overnight culture into 50 mL of YPD medium and incubate for 3–6 h at 30°C with shaking (250 rpm) until the OD$_{600}$ is 0.3–0.4.
3. Centrifuge the cells at 1000*g* for 5 min at room temperature. Discard the supernatant and resuspend the cell pellet by vortexing in sterile TE buffer. Centrifuge again, decant the supernatant and resuspend in 0.5–1.0 mL of freshly prepared, 1X TE/LiAc. This will produce competent yeast cells.
4. Transform 0.1 mL of the competent cells by adding 0.1 μg of pAS2-Src-UD/SH3/SH2 or pAS2 and 0.1 mg of carrier DNA; mix well by vortexing.
5. Add 0.5 mL of sterile PEG/LiAc solution and vortex at high speed to mix.
6. Incubate at 30°C for 30 min with shaking (at 150 rpm).
7. Add 50 μL of dimethyl sulfoxide, and gently tap to mix.
8. Heat shock this mixture in a 42°C water bath for 15 min.
9. Chill cells on ice for 5 min. Centrifuge cells at 11,000*g* in a microfuge for 5 s. Remove the supernatant and resuspend cells in 0.25 mL of 1X TE.
10. Plate cells on SD/-Trp plates. Under these conditions, only cells that harbor the DNA-BD/bait plasmid (which contains the *TRP1* gene) will grow.
11. Pick yeast colonies and culture in SD/-Trp liquid media.
12. Transfer 1 mL of yeast culture into a 1.5 mL Eppendorf and centrifuge at high speed for 5 min. Decant the supernatant, resuspend the yeast pellet in 1X sodium dodecyl sulfate (SDS)-sample buffer and boil for 20 min. Centrifuge briefly and load the supernatant onto an SDS gel for immunoblot analysis to verify expression of the bait protein in yeast (as described in **Subheading 3.1.3.**).
13. Alternatively, yeast colonies can be picked from agar plates (**step 10**), transferred directly to Eppendorf tubes, and boiled in SDS-sample buffer prior to loading onto an SDS gel.

3.1.3. Verification of Fusion Protein Expression in Yeast

Verify expression of the bait protein in Trp$^+$ transformants by performing immunoblot analysis with an antibody specific for the protein or to the HA tag. Expression of the DNA-BD/Src-UDSH3/SH2 fusion protein was confirmed by immunoblot analysis of yeast cell lysates with MAb 327, which recognizes the SH3 domain of Src *(14)*, and with MAb 12CA5, which recognizes the

HA epitope. Use proteins from yeast cells transformed with pAS2 and from untransformed yeast cells as negative controls.

3.1.4. Verification that Hybrid Genes Do Not Autonomously Activate Reporter Genes

In Trp⁺ transformants that appropriately express the bait protein, verify that the DNA-BD/bait fusion does not autonomously active a reporter gene (*see* **Note 2**). To do so, assay for activation of the *lacZ* reporter gene using the β-galactosidase assay, as described in **Subheading 3.2.3.** Include positive (pCL1) and negative (pAS2 and pLAM5′) controls. If autonomous activation is not observed for the bait plasmid, prepare stock plates and liquid cultures (use SD/-TP) of Trp⁺ transformants for freezing.

3.2. Screening Libraries for Interacting Proteins

Trp⁺ transformants that expressed the DNA-BD/ Src-UD/SH3/SH2 fusion and did not autonomously activate the *lacZ* reporter gene, were sequentially transformed with the human lung fibroblast cell line WI-38 cDNA library fused to the GAL4 activation domain *(10)*. 2×10^6 transformants were screened for interacting proteins (*see* **Note 1**).

3.2.1. Activation Domain Fusion Library

Obtain a cDNA library cloned into a pGAD-AD vector or construct your own AD fusion library using recombinant DNA techniques *(16)*. We used a WI-38 human lung fibroblast cell line cDNA library fused to the GAL4-AD in the pGAD-GL vector (Clontech). In addition to the GAL4 AD, pGAD-GL contains the *Amp^r* gene that confers resistance to ampicillin and the *LEU2* gene for selection in Leu⁻ auxotrophic yeast strains *(4,15)*.

3.2.2. Transformation of the cDNA Library into Yeast

Trp⁺ transformants that expressed the DNA-BD/Src-UD/SH3/SH2 fusion were sequentially transformed with the AD/library fusion using the method described in **Subheading 3.1.2.** except that all the solutions and reagents were scaled up 50 times. Yeast transformants were plated onto SD/-Leu/-Trp/-His plates containing 25–100 m*M* of 3-AT for medium-stringency screening (*see* **Note 3**).

3.2.3. Assay for Activation of the LacZ Reporter Gene (β-galactosidase Filter Assay)

1. Plate yeast transformants on SD/-Leu/-Trp/-His plates containing 3-AT (150-mm diameter plates) and incubate at 30°C for 2–3 d until colonies reach approx 2 mm

in diameter. These conditions will select for Leu$^+$ Trp$^+$, His$^+$ transformants, that is, cells that harbor an AD/library plasmid (which contains the *LEU2* gene), a DNA-BD/bait plasmid (which contains the *TRP1* gene), and have interacting hybrid proteins (which results in activation of the *HIS3* reporter gene).

2. Prepare Z buffer/X-gal solution *(16)*.

3. Add 5 mL of Z buffer/X-gal solution to a clean plate (150-mm diameter) and soak a sterile Whatman no. 5 filter (125-mm diameter) in the liquid. Prepare one filter for each plate of yeast transformants to be assayed.

4. Carefully place a dry Whatman no. 5 filter over the surface of the agar plate containing the yeast transformants. Gently use a gloved finger to smooth out and remove any air bubbles. Make sure that the filter and yeast colonies are well attached but not smeared.

5. Orient the filter on the agar by poking holes (with a needle) through the filter into the agar in at least three asymmetric places at the periphery of the filter. Mark the holes with a pen to facilitate orienting the filter back on the plate after the assay, for colony picking. Also, number the filters and corresponding plates to facilitate matching them correctly later on.

6. After the filter is completely wet by the agar plate, carefully lift the filter off the agar plate with forceps. Be careful not to smear the colonies as you lift the filter. Smeared colonies result in unidentifiable colonies.

7. Immediately freeze the filter in a pool of liquid nitrogen for 10–20 s (colonies facing up), and allowing it to thaw at room temperature. The freeze/thaw step permeabilizes the cells.

8. Carefully place the filter, colony side up, onto the filter that has been presoaked in Z buffer/X-gal solution (**step 3**). Be careful not to trap air bubbles between the filters. Incubate filters at 30°C for several hours to overnight, until blue colonies appear.

9. Locate the β-galactosidase-producing colonies on the agar plate by aligning the filter to the plate (align the holes in the filter with those in the agar). Pick the agar colonies that correspond to the blue colonies on the filter. If most of the agar colonies stuck to the filter during the lift, then allow the original colonies to regrow on the original plate for another 2 d.

10. Often, more than one AD/library plasmid will be present in each β-galactosidase-positive colony. Thus, the β-galactosidase-positive clones should be purified by resteaking on SD/-Leu/-Trp/-His plates. The additional reculturing will allow segregation of the AD/library plasmids to occur. Well-isolated colonies should be reassayed for β-galactosidase activity to verify that they maintain the correct phenotype. After two to three rounds of restreaking and verification, retrieve positive single colonies and save them as glycerol stocks. These single colonies include the pAS2-GAL4 DNA-BD-bait plasmid and at least one copy of GAL4AD-library plasmid. Also, streak all positive colonies onto one to two plates with positive and negative controls to establish the master plates.

11. This will generate Leu$^+$Trp$^+$His$^+$LacZ$^+$ transformants, that is, cells that harbor a AD/library plasmid (which contains the *LEU2* gene), a DNA-BD/bait plasmid

(which contains the *TRP1* gene), and have interacting hybrid proteins (with resulting activation of the *HIS3* and *LacZ* reporter genes).

3.2.4. Calculation of Transformation Efficiency and Number of Clones Screened

We screened 2×10^6 transformants for interacting proteins *(10)*, which was about twice the size of the fibroblast cDNA library.

1. To calculate the co-transformation efficiency, count the colonies (cfu) growing on the SD/-Leu/-Trp/-His dilution plate that has 30–300 cfu. The ratio of cfu × total suspension volume (μL) to volume plated (μL) × dilution factor × amount of DNA used (μg)* = cfu/μg DNA *(11,12)*. (*For simultaneous co-transformations, this is the amount for the limiting plasmid, i.e., the lesser of the two plasmids, not the total amount of DNA.)
2. To estimate the number of clones screened: cfu/μg × amount of library plasmid used (μg) = number of clones screened *(11,12)*. If you screened $<10^6$ clones, you may want to repeat the transformation using more DNA.

3.3. Analysis and Verification of Putative Positive Clones

Of 2×10^6 colonies transformed with pAS2-SrcUD/SH3/SH2 and pGAD GL/fibroblast library, 48 were Leu⁺Trp⁺His⁺LacZ⁺ *(10)*. After eliminating false positives by cycloheximide selection in SD/-Leu/+Trp/+His medium and recloning the plasmids in *Escherichia coli* that carry a *leuB* mutation, 12 clones were identified that interacted with the Src fusion protein and not a negative control fusion protein (lamin). Redundant clones were eliminated by analyzing insert sizes, and the eight remaining clones were sequenced. One clone had 100% homology at the nucleotide sequence level with human H12.3, a homolog of the heterotrimeric G-protein β subunit, and 100% homology at the amino acid level with rat brain RACK1, a known PKC-binding partner (reviewed in **refs. *19–21***).

3.3.1. Verification that Activation Domain/Library Hybrids Do Not Activate Reporter Genes

False positives are transformants that are His⁺ or turn blue in a β-galactosidase assay but contain plasmids that do not encode hybrid proteins that directly interact. False positives can occur if either the DNA-BD/bait or the AD/library plasmid alone can activate transcription of the reporter gene (*see* **Note 2**). The DNA-BD/bait plasmid should already have been tested for autonomous activation of the reporter genes before the cells were transformed with the AD/library plasmid (*see* **Subheading 3.1.4.**). Thus, at this point it is necessary to test only the AD/library plasmid for autonomous activation of the reporter genes. To do so, the following must be performed:

1. Culture individual Leu⁺Trp⁺His⁺LacZ⁺ transformant colonies in liquid SD/ -Leu/+Trp/+His medium containing cycloheximide (10 μg/mL) at 30°C for 2–3 d. Under these conditions, the AD/library plasmid will be maintained (because pGAD GL contains the *LEU2* gene) and the DNA-BD/bait plasmid will be lost (because pAS2 does not contain a *LEU* gene, and it does contain the *CYH^S2* gene, which confers sensitivity to cycloheximide).
2. Assay the surviving Leu⁺Chx^r transformant colonies for β-galactosidase activity. Colonies that are negative for β-galactosidase activity should be saved because they contain AD/library plasmids that do not autonomously activate the reporter gene. Colonies that are positive for the assay should be discarded because they probably contain an AD/library that encodes a transcriptional activator capable of recognizing the *GAL1* promoter.
3. Isolate plasmid DNA from Leu⁺Chx^r transformants that are negative for β-galactosidase activity *(16)*.
4. Transform *E. coli* HB101 (which carries a *leuB* mutation that can be complemented by *LEU2* from yeast) with the yeast plasmid DNA by electroporation (*see* **Note 4**; **ref. 16**).
5. After electroporation, allow the cells to recover by growing them in 1 mL of rich medium (Luria broth) without antibiotics at 37°C for 60 min with shaking (250 rpm).
6. Wash cells twice with M9 minimal medium.
7. Plate cells on M9 agar minimal medium supplemented with 50 μg/mL ampicillin, 40 μg/mL proline, 1 m*M* thiamine-HC, and a 1× mixture of amino acids (lacking Leu). The same (–Leu) DO solution used to supplement the yeast minimal medium can be used to supplement the bacterial minimal medium.
8. Isolate plasmid DNA from Leu⁺Chx^r HB101 transformants using a standard plasmid mini-prep procedure *(16)*.
9. Verify plasmid construction by restriction enzyme digestion.

3.3.2. Elimination of False-Positive Clones

1. Retransform the AD/library hybrid plasmids (Leu⁺Trp⁻ transformants) that do not autonomously active the reporter genes, back into the original yeast host strain, Y190, harboring: 1) pAS2-Src-UD/SH3/SH2, 2) pAS2, 3) pLAM5′, or 4) no plasmid.
2. Reassay these transformants for growth on Trp⁻, Leu⁻, and His⁻ medium containing 50 m*M* 3-AT (except for transformation 4, which should be selected on SD medium lacking Leu only), and for β-galactosidase activity. The expected β-galactosidase assay results for true positives are shown in **Table 1**. The AD/library fusions that result in β-galactosidase activity when transformed into yeast cells harboring pAS2-Src-UD/SH3/SH2, but not when transformed into

Table 1
Transformations to Eliminate False-Positive Protein Interactions

#	Plasmid 1	Plasmid 2	Expected β-galactosidase activity for true positives
1	pAS2-Src-UD/SH3/SH2 (DNA-BD/protein A)	AD/library (AD/protein B)	Positive
2	pAS2 (DNA-BD only)	AD/library (AD/protein B)	Negative
3	pLAM5′ (DNA-BD/Lamin C hybrid)	AD/library (AD/protein B)	Negative
4	No plasmid	AD/library (AD/protein B)	Negative

cells harboring pAS2, pLAM5′, or no plasmid, are likely to be true positives and should be saved for further testing. The AD/library fusions that result in β-galactosidase activity when transformed into cells harboring pAS2-Src-UD/SH3/SH2, and also when transformed into cells harboring pAS2, pLAM5′, and/or no plasmid, are false positives and should be discarded (*see* **Note 5**).

3.3.3. Sequence Activation Domain/Library Inserts of Positive Clones

1. Isolate plasmid DNA from yeast transformants that contain AD/library inserts that are likely to be true positives (*see* **Subheading 3.3.2.**).
2. Eliminate redundant clones by performing restriction enzyme digestion and analyzing plasmid insert sizes.
3. Sequence nonredundant cDNA clones using the dideoxynucleotide chain termination method *(16)*. We used Sequenase 2.0 (United States Biochemicals, Cleveland, OH).
4. Search databases for homologous sequences. We searched GenBank and EMBL nucleotide sequence databases using FASTDB analysis (Intelligenetics, Santa Clara, CA).

3.4. Confirmation of Protein Interactions

3.4.1. In Vitro Analysis of Protein Interactions

To confirm the results obtained with the yeast two-hybrid assay, we first assessed for binding of Src and RACK1 in vitro *(10)*. To do so, full-length Src or the UD/SH3/SH2 domain of Src were expressed as a bacterial fusion protein with GST, purified to homogeneity on glutathione agarose beads, and tested for the ability to bind [^{35}S]methionine-labeled, in vitro-translated RACK1 (**Fig. 1A**). One-tenth of the unbound translation reaction product (FT for flowthrough) was loaded on the gel as a measure of the amount of protein trans-

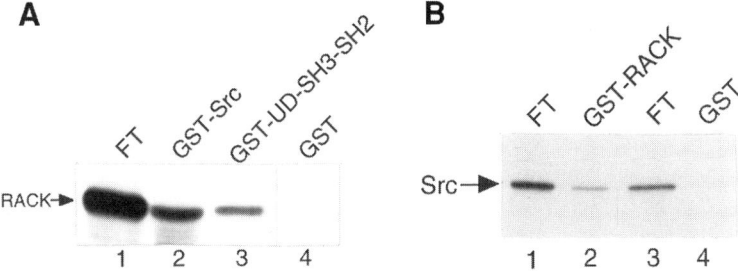

Fig. 1. Association of RACK1 and Src in vitro. (**A**) Binding of in vitro-translated RACK1 to GST-Src fusion proteins. [^{35}S]methionine/cysteine-labeled RACK1 was synthesized in rabbit reticulocyte lysates and incubated with 1 μg of purified GST-Src (lane 2), GST-SrcUD/SH3/SH2 (lane 3), or GST (lane 4) for 3 h at 4°C. Protein complexes were collected on glutathione-agarose beads, washed and boiled in SDS sample buffer. Proteins were resolved by SDS-PAGE. ^{35}S-labeled proteins were visualized by autoradiography. Lane 1: For each reaction, 1/10 of the unbound translation reaction product (FT for flowthrough) was loaded directly on the gel as a measure of the amount of protein translated. (**B**) Binding of in vitro-translated Src to GST-RACK1. In vitro-translated Src was assessed for binding to GST-RACK1 (lane 2) or GST alone (lane 4) as described in (A). Lanes 1 and 3: 1/20 of the FT was loaded directly on the gel. Adapted from Chang et al. *(10)*.

lated (lane 1). We observed that GST-Src (lane 2) and GST-Src-UD/SH3/SH2 (lane 3) bound RACK1, whereas the GST control (lane 4) did not.

Conversely, GST-RACK1 was tested for its ability to bind [^{35}S]methionine-labeled, in vitro-translated Src (**Fig. 1B**). Here, 1/20 of the FT was loaded on the gel (lanes 1 and 3). We observed that GST-RACK1 bound Src (lane 2), whereas GST alone (lane 4) and GST fused to a protein unrelated to RACK1 (Elongation Factor-Tu; not shown) did not.

3.4.2. In Vitro Translation and GST Fusion Protein-Binding Assays

1. Circular plasmid DNAs (2 μg of pGEM*src* or pcDNA3-RACK1) were transcribed and translated in vitro using a T$_N$T-coupled rabbit reticulocyte lysate system, as instructed by the manufacturer (Promega) *(10,22)*.
2. Cultures of *E. coli* DH5α containing pGEX-RACK1 or various pGEX*src* plasmids were induced with 0.1 m*M* isopropyl-β-D-thiogalactopyranoside (United States Biochemicals) for 3 h at 30°C. Bacteria were harvested, resuspended in Tris-buffered saline (TBS) containing 1% Triton X-100 and 100 m*M* EDTA and sonicated. After centrifugation at 12,000*g* for 10 min to remove debris, the supernatant was incubated with glutathione-agarose beads (Sigma) for 2 h at 4°C with agitation. Beads were washed three times with TBS. GST fusion proteins were eluted by the addition of 100 m*M* Tris-pH 8.0, 120 m*M* NaCl, and 20 m*M* glutathione and dialyzed four times against TBS *(10,22)*.

3. PRO-MIX[^{35}S] (70% L-[^{35}S]methionine and 30% L-[^{35}S]cysteine; >1000 Ci/mmol; Amersham, Arlington Heights, IL)-labeled in vitro translation products were diluted in binding buffer (50 mM Tris-pH 7.5, 150 mM NaCl, and 0.2% Nonidet P-40) and incubated with 1 µg of purified GST fusion protein for 3 h at 4°C. Protein complexes were collected with the addition of 30 µL of glutathione beads, washed four times in buffer containing 0.5% NP-40, 20 mM Tris pH-8.0, 100 mM NaCl, and 1 mM EDTA, and boiled in SDS-sample buffer. Proteins were resolved by SDS-polyacrylamide gel electrophoresis (PAGE). One-tenth or 1/20 of the unbound translation reaction product was loaded directly on the gel as a measure of the amount of protein translated. Gels were stained with Coumassie brilliant blue G-250 and treated with Fluoro-Hance (Research Products International Corp., Mount Prospect, IL). Radiolabeled proteins were detected by fluorography *(10,22)*.

3.4.3. In Vivo Analysis of Protein Interactions

To determine whether RACK1 associates with Src in mammalian cells (*see* **Note 6**), proteins from NP-40 lysates of NIH 3T3 cells or proteins precipitated from the lysates with pre-immune serum or excess Src polyclonal antibody R7 were resolved by SDS-PAGE, transferred to polyvinylidene diflouride (PVDF) membranes, and immunoblotted with a MAb specific for RACK1 (**Fig. 2A**; **ref. *10***). RACK1 protein was detected in lysate (lane 3) and in Src immunoprecipitates (lane 2) but not in control immunoprecipitates (lane 1).

Conversely, proteins from NP-40 lysates of NIH 3T3 overexpressing Src (3T3/c-Src cells) were immunoprecipitated with MAbs specific for RACK1 or Src (MAb 327), or rabbit anti-mouse IgG and subjected to immunoblot analysis with MAb 327 (**Fig. 2B**; **ref. *10***). Src protein (≈60 kDa) was detected in Src (lane 2, upper band) and RACK1 (lane 3, upper band) immunoprecipitates, but not in control immunoprecipitates (lane 1). The ≈55-kDa band running below Src in lanes 2 and 3 is mouse IgG heavy chain.

3.4.3.1. PROTEIN EXTRACTION AND IMMUNOPRECIPITATION

Cells were washed three times with ice-cold TBS and lysed in ice-cold NP-40 buffer (0.5% Nonidet P-40; 20 mM Tris-pH 8.0, 100 mM NaCl, 1 mM EDTA, 100 µM sodium vanadate, 50 mM sodium fluoride, 50 µM leupeptin, 1% aprotinin, and 1 mM dithiothreitol; *see* **Note 7**, **refs. *10,22–27***). Lysates were centrifuged at 14,000g for 1 h at 4°C. Protein concentrations were measured by the BCA protein assay (Pierce, Rockford, IL) and samples were standardized to equal amounts of total cellular protein. Lysates were incubated for 3 h at 4°C with 1 µg of Src MAb 327, Src anti-peptide R7, or MAb RACK1 and protein complexes were collected with the addition of 30 µL of protein A/G Sepharose beads (Pharmacia, Biotech, Piscataway, NJ) for MAb 327 or

A RACK blot **B** Src blot

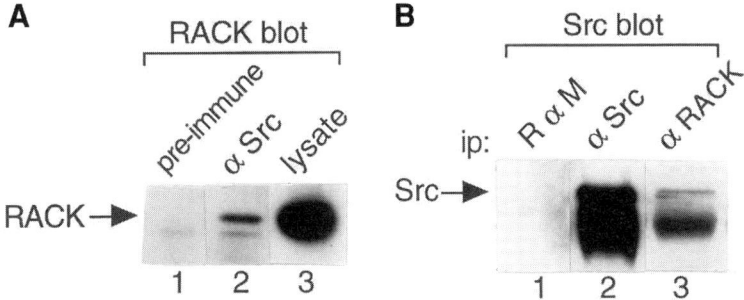

Fig. 2. Association of RACK1 and Src in mammalian cells. (**A**) RACK1 is present in Src immunoprecipitates. Proteins were precipitated with pre-immune serum (lane 1) or excess Src polyclonal antibody R7 (lane 2) from NP-40 lysates of NIH 3T3 cells containing 800 µg of total cellular protein, resolved by SDS-PAGE, transferred to PVDF membranes and immunoblotted with RACK1 antibody. Proteins were detected by enhanced chemilumin escence (ECL; Amersham). The band below Src is mouse IgG heavy chain. Lane 3: Lysate containing 20 µg of total cellular protein was loaded directly on the gel prior to transfer and immunoblot analysis with anti-RACK1. (**B**) Src is present in RACK1 immunoprecipitates. Proteins were precipitated with rabbit anti-mouse IgG (R α M, lane 1), Src MAb 327 (lane 2), or RACK1 antibody (lane 3) from NP-40 lysates of 3T3/c-Src cells containing 500 µg of total cellular protein, and subjected to immunoblot analysis with MAb 327. Adapted from Chang et al. (*10*).

R7 immunoprecipitates, or with the addition of 30 µL of goat anti-mouse IgM conjugated with agarose beads (Sigma) for MAb RACK1 immunopre-cipitates (*10,22–27*). Proteins were resolved on 7% SDS-polyacrylamide gels (acrylamide-bisacrylamide 20:1) to achieve maximum separation between 60 kDa Src and 55 kDa IgG.

3.4.3.2. IMMUNOBLOT ANALYSIS

Src or RACK1 immunoprecipitates were resolved on 10% SDS-poly-acrylamide gels (acrylamide-bisacrylamide, 29:0.8). Proteins were transferred to PVDF membranes (Immobilon-P™; Millipore, Bedford, MA) in transfer buffer (25 m*M* Tris-HCl-pH 7.4, 192 m*M* glycine, and 15% methanol) using a Trans-Blot apparatus (Bio-Rad, Hercules, CA) for 2 h at 60 V (*10,22–27*). Protein-binding sites on the membranes were blocked by incubating membranes overnight in TNT buffer (10 m*M* Tris-HCl-pH 7.5, 100 m*M* sodium chloride, 0.1% v/v Tween-20; Sigma) containing 3% nonfat, powdered milk (blocking buffer). Membranes were incubated with MAb RACK1 (0.08 µg/mL) or affinity-purified MAb 327 ascites (2 µg/mL) for 1 h, washed in TNT buffer with changes every 5 min for 30 min, and incubated with horseradish peroxidase-conjugated donkey anti-mouse IgM (Zymed, San Francisco, CA) for RACK1 blots or goat anti-mouse IgG (Bio-Rad) for 327 blots *(10,22–27)*. Proteins were detected by ECL according to the manufacturer's protocol.

The association of Src and RACK1 following PKC activation was also demonstrated by confocal immunofluorescence microscopy (*see* **Note 6**; **ref.** *22*).

4. Notes

1. A useful modification of the method is to sequentially transform the DNA-BD/bait fusion (first) and the AD/library fusion (second) into the yeast strain, rather than to simultaneously co-transform them. Advantages to this approach are as follows: (1) the efficiency of transformation is an order of magnitude higher with sequential than with simultaneous co-transformation; (2) the DNA-BD/bait fusion needs to be tested for autonomous activation of reporter genes before it is used in a library screening (*see* **Note 2**), so the first step of the sequential transformation will be performed anyway, even if one plans to do simultaneous co-transformations; (3) transformation of the DNA-BD/bait fusion first allows for verification that the bait protein is appropriately expressed in the fusion, before it is used in a library screening; and (4) with sequential transformation, the original yeast clones expressing the DNA-BD/bait fusion can later be used again when eliminating false positives, that is, for retransformation with select AD/library fusions that have been shown not to autonomously activate the reporter genes. The sequential approach is a very efficient, effective way to perform the screen, unless expression of the DNA-BD fusion is toxic to the cells.

2. Some proteins may have intrinsic DNA-binding and/or transcriptional activating capabilities that could result in false positives in the screen. For example, a DNA-BD/bait protein could activate reporter gene expression without an AD/library protein if the bait protein had a transcriptional activation domain. Thus, it is important to check the DNA-BD/bait fusion for autonomous activation of reporter genes before using it in a library screen (the sequential transformation approach facilitates this; *see* **Note 1**). It may be possible to delete an activating domain of a bait protein, but doing so might also delete the interacting domain. After the screen, it is also important to check the AD/library fusion for autonomous activation of reporter genes. This could occur if the library protein had a DNA-binding domain.

3. The stringency of the screen can be adjusted using medium (SD/-Leu/-Trp/-His) or low (SD/-Leu/-Trp)-stringency media. Less stringent conditions will result in higher numbers of false positives. More stringent conditions may not detect weak or transient protein interactions, or those that are enhanced by posttranslational modifications. The Y190 yeast strain is very leaky for HIS3 expression and therefore can grow normally on SD/-His plates. Thus, the addition of 3-amino-1,2,4-triazole, which is a competitive inhibitor of the His3 protein, is necessary

to suppress background growth in low-stringency screens. It is important to optimize the 3-AT concentration because too much can kill cells, whereas too little will result in high background growth. To optimize the 3-AT concentration, plate DNA-BD/bait transformants onto a series of SD/-Trp/-His plates containing various concentrations of 3-AT (25 to 100 m*M*). Use the lowest concentration of 3-AT that allows only small (<1 mm) colonies to grow after 1 wk. By adjusting both the stringency of the media and the concentration of 3-AT, one should obtain the desired protein–protein interactions.

4. Electroporation or standard chemical methods can be used to transform *E. coli* with plasmid DNA isolated from yeast. Electroporation is recommended because it achieves higher efficiency of transformation, which is important because plasmid DNA isolated from yeast generally contains large amounts of genomic DNA.

5. Several variations on the original GAL4-based two-hybrid assay have been developed in recent years and may offer advantages for detecting some protein interactions. One variation utilizes the prokaryotic lexA protein for the DNA-binding domain and an acidic peptide for the activation domain, and activates reporter genes via the LexA operator *(28)*. The inducible yeast *GAL1* promoter is used to induce expression of the fusion proteins only at the time of the screen, which limits toxicity to the cells.

6. The interaction of mammalian proteins in the heterologous yeast environment does not necessarily mean that the interaction occurs when the proteins are in their native environment. Thus, it is important to demonstrate the protein interactions in mammalian cells. Conversely, the environment in yeast cells may not provide the proper folding or posttranslation modifications (such as phosphorylation) required for interaction of some mammalian proteins. One approach that has been used successfully to detect phosphotyrosine-dependent interactions (when the bait protein is not a tyrosine kinase), is to co-transform a tyrosine kinase-catalytic domain together with the bait protein *(29,30)*.

7. The stringency for detecting protein interactions can be adjusted by changing the type or concentration of detergent in the cell lysis buffer. For higher stringency, use SDS-containing lysis buffers (e.g., RIPA buffer), which will decrease the number of nonspecific interactions, but may preserve only strong interactions. For lower stringency, use NP-40-containing lysis buffers, which will preserve low affinity or transient interactions, or those that are enhanced by posttranslational modifications, but may increase the number of nonspecific interactions.

Acknowledgments

We thank Tony Hunter and Martin Broome for providing pGEM*src*. We thank Joosang Park and Annette Walter for generating pGEX*src* (full-length) and pGEX*src*-UD/SH3/SH2. This work was supported by grants from the National Institutes of Health to C. A. C. (R01 DK43743) and to B. Y. C. (National Research Service Award CA69810).

References

1. Thomas, S. M. and Brugge, J. A. (1997) Cellular functions regulated by Src family kinases. *Annu. Rev. Cell Dev. Biol.* **13,** 513–609.
2. Fields, S. and Song, O. (1989) A novel genetic system to detect protein–protein interactions. *Nature* **340,** 245–246.
3. Yang, X., Hubbard, J. A., and Carlson, M. (1992) A protein kinase substrate identified by the two-hybrid system. *Science* **257,** 680–682.
4. van Aelst, L., Barr, M., Marcus, S., Polverino, A., and Wigler, M. (1993) Complex formation between RAS and RAF and other protein kinases. *Proc. Natl. Acad. Sci. USA* **90,** 6213–6217.
5. Vojtek, A. B., Hollenberg, S. M., and Cooper, J. A. (1993) Mammalian Ras interacts directly with the serine/threonine kinase Raf. *Cell* **74,** 205–214.
6. Durfee, T., Becherer, K., Chen, P. L., Yeh, S. H., Yang, Y., Kibburn, A. E., Lee, W. H., and Elledge, S. J. (1993) The retinoblastoma protein associates with the protein phosphatase type I catalytic subunit. *Genes Devel.* **7,** 555–569.
7. Harper, J. W., Adami, G. R., Wei, N., Keyomarsi, K., and Elledge, S. J. (1993) The p21 Cdk-interacting protein Cip-1 is a potent inhibitor of G1 cyclin-dependent kinases. *Cell* **75,** 805–816.
8. Sharan, S. K., Morimatsu, M., Albrecht, U., Lim, D. S., Regel, E., Dinh, C., et al. (1997) Embryonic lethality and radiation hypersensitivity mediated by Rad51 mice lacking Brca2. *Nature* **386,** 804–810.
9. Mizuta, R., LaSalle, J. M., Cheng, H. L., Shinohara, A., Ogawa, H., Copeland, N., et al. (1997) RAB22 and RAB163/mouse BRCA2: proteins that specifically interact with the RAD51 protein. *Proc. Natl. Acad. Sci. USA* **94,** 6927–6932.
10. Chang, B. Y., Conroy, K. B., Machleder, E. M., and Cartwright, C. A. (1998) RACK1, a receptor for activated C kinase and a homology of the β subunit of G proteins, inhibits activity of Src tyrosine kinases and growth of NIH 3T3 cells. *Mol. Cell. Biol.* **18,** 3245–3256.
11. *Clontech Matchmaker GAL4 Two-hybrid System and Libraries User Manual* (PT 3061-1) *and Vectors Handbook* (PT3062-1) (1998). Clontech Laboratories Inc., Palo Alto, CA.
12. *Clontech Matchmaker GAL4 Two-hybrid System 3 and Libraries User Manual* (PT3247-1/PR94575) (1999). Clontech Laboratories Inc., Palo Alto, CA.
13. Bartel, P. L., Chien, C. T., Sternglanz, R., and Fields, S. (1993) Using the two-hybrid system to detect protein–protein interactions. In *Cellular Interactions in Development: A Practical Approach* (Hartley, D. A., ed.), Oxford University Press, Oxford, pp. 153–179.
14. Lipsich, L. A., Lewis, A. J., and Brugge, J. S. (1983) Isolation of monoclonal antibodies that recognize the transforming proteins of avian sarcoma viruses. *J. Virol.* **48,** 352–360.
15. Chien, C. T., Bartel, P. L., Sternglanz, R., and Fields, S. (1991) The two-hybrid system: a method to identify and clone genes for proteins that interact with a protein of interest. *Proc. Nat. Acad. Sci. USA* **88,** 9578–9582.

16. Sambrook, J., Fritsch, E. F., and Maniatis, T. (eds.) (1989) *Molecular Cloning: A Laboratory Manual*, Cold Spring Harbor Laboratory, Cold Spring Harbor, NY.

17. Ito, H., Fukuda, Y., Murata, K., and Kimura, A. (1983) Transformation of intact yeast cells treated with alkali cations. *J. Bacteriol.* **153**, 163–168.

18. Gietz, D., St. Jean, A., Woods, R. A., and Schietstl, R. H. (1992) Improved method for high efficiency transformation of intact yeast cells. *Nucleic Acids Res.* **20**, 1425–1431.

19. Mochly-Rosen, D. (1995) Localization of protein kinases by anchoring proteins: a theme in signal transduction. *Science* **268**, 247–251.

20. Mochly-Rosen, D. and Kauvar, L. M. (1998) Modulating protein kinase C signal transduction. *Adv. Pharm.* **44**, 91–145.

21. Schechtman, D. and Mochly-Rosen, D. (2001) Adaptor proteins in protein kinase C-mediated signal transduction. *Oncogene* **20**, 6339–6347.

22. Chang, B. Y., Chiang M., and Cartwright, C. A. (2001) The interaction of Src and RACK1 is enhanced by activation of protein kinase C and tyrosine phosphorylation of RACK1. *J. Biol. Chem.* **276**, 20,346–20,356.

23. Cartwright, C. A., Hutchinson, M., and Eckhart, W. (1985) Structural and functional modifications of pp60$^{c\text{-}src}$ associated with polyoma middle tumor antigen from infected or transformed cells. *Mol. Cell. Biol.* **5**, 2647–2652.

24. Cartwright, C. A., Kaplan, P. L., Cooper, J. A., Hunter, T., and Eckhart, W. (1986) Altered sites of tyrosine phosphorylation pp60$^{c\text{-}src}$ associated with polyomavirus middle tumor antigen. *Mol. Cell. Biol.* **6**, 1562–1570.

25. Cartwright, C. A., Eckhart, W., Simon S., and Kaplan, P. L. (1987) Cell transformation by pp60$^{c\text{-}src}$ mutated in the carboxyl-terminal regulatory domain. *Cell* **49**, 83–91.

26. Cartwright, C. A., Meisler, M. A., and Eckhart, W. (1990) Activation of the pp60$^{c\text{-}src}$ protein kinase is an early event in colonic carcinogenesis. *Proc. Natl. Acad. Sci. USA* **87**, 558–562.

27. Park, J. and Cartwright, C. A. (1995) Src activity increases and Yes activity decreases during mitosis of human colon carcinoma cells. *Mol. Cell. Biol.* **15**, 2374–2382.

28. Fashena, S. J., Serebriiskii, I. G., and Golemis, E. A. (2000) LexA-based two-hybrid systems. *Methods Enzymol.* **328**, 14–26.

29. Wang, B., Lemay, S., Tsai, S., and Veillete, A. (2001) SH2 domain-mediated interaction of inhibitory protein tyrosine kinase Csk with protein tyrosine phosphatase-HSCF. *Mol. Cell. Biol.* **21**, 1077–1088.

30. Yamada, M., Suzuki, K., Mizutani, M., Asada, A., Matozaki, T., Ikechi, T., Koizumi, S., and Hatanaka, H. (2001) Analysis of tyrosine phosphorylation-dependent protein–protein interactions in TrkB-mediated intracellular signaling using modified yeast two-hybrid system. *J. Biochem. (Tokyo)* **130**, 157–165.

28

Glutathione S-Transferase Pull-Down Assay

Deborah Schechtman, Daria Mochly-Rosen, and Dorit Ron

1. Introduction

The glutathione S-transferase (GST) pull-down assay is a relatively easy, straightforward method to identify potential protein kinase C (PKC)-binding partners. The method is also extensively used to confirm known interactions and to map interaction sites. The pull-down method relies on the immobilization of a GST fusion protein on glutathione sepharose beads that serve as a solid phase. The first step requires the expression of the PKC domain of interest as a fusion protein with the GST moiety. After binding of the GST fusion protein to the glutathione sepharose matrix, the mixture is incubated with whole-cell homogenate or a purified protein. Nonbound material is washed off the column, and subsequently the binding complex is eluted. Upon elution, the mixture is resolved by sodium dodecyl sulfate polyacrylamide gel electrophoresis (SDS-PAGE) and analyzed by Coomassie staining, silver staining, or Western blot. This chapter focuses on the GST tag; however, other tags, such as hexa-Histidine (6xHis) and maltose-binding protein (MBP), are also commonly used and will be mentioned throughout the chapter (*see* **Note 1**).

Numerous studies have used this method to identify PKC-binding proteins. For example, β-tubulin was identified as a PKCζ-binding protein by expressing the regulatory domain of PKCζ as a GST fusion protein and incubating the immobilized complex with rat brain extract *(1)*. Fyn kinase was identified as a PKCθ-binding protein by incubation of Jurkat T-cell lymphoma extract with an immobilized V1 region of PKCθ *(2)*.

From: *Methods in Molecular Biology, vol. 233: Protein Kinase C Protocols*
Edited by: A. C. Newton © Humana Press Inc., Totowa, NJ

This chapter concentrates mainly on the use of the GST pull-down assay to identify new PKC-binding proteins. However, the in vitro pull-down assay is also useful to confirm interactions identified by other methods and to determine domains that are essential for a specific protein–protein interaction. Listed below are a few examples. The GST pull-down assay was used to identify the C2 and C2-like regions of PKCα and PKCε, respectively, as the domains that mediate the interaction between PKC and the cytoskeletal protein calponin (3). By expressing different regions from the regulatory domain of PKCδ as GST fusion proteins, Dekker and Parker (4) identified the C2-like domain as the binding site of PKCδ with GAP43. Using the pull-down assay, we found that the C2 and the V5 domains of PKCβII bind the receptor for activated C kinase-1 (RACK1) and inhibit the interaction between the two proteins (5,6). We have also used this assay to identify several pleckstrin homology domains that bind to both PKCβ and RACK1 (7). In addition, the assay was used to determine which fragments of RACK2 (β'COP) bind to the V1 region of PKCε (8). Finally, Meller et al. (9) elegantly combined the transfection of tag-fusion proteins together with the pull-down assay to identify an interaction between PKCθ and 14-3-3 in cells and to confirm the interaction in vitro.

2. Materials
2.1. GST Fusion Protein Cloning and Expression

1. pGEX gene fusion vector (Amersham Pharmacia Biotech).
2. *Escherichia coli* BL21 (Novagene).
3. Isopropyl-β-D-thiogalactopyranoside (Roche Diagnostic Corporation). Store powder at 4°C. Prepare 1M solution in ddH$_2$O, sterilize by filtration, aliquot, and store at –20°C.
4. Ampicillin (Roche Diagnostic Corporation). Store powder at 4°C. Prepare 100 mg/mL solution in ddH$_2$O, sterilize by filtration, aliquot, and store at –20°C.
5. Luria broth (LB; Difco): 25 g in 1 L of dd H$_2$O, autoclave at 120°C for 15 min.
6. Lysis buffer: 1 M Tris-HCl, pH 7.5, 5 M NaCl, 0.5 M ethylenediamine tetraacetic acid (EDTA), and protease inhibitor cocktail (Roche Diagnostic Corporation; 1 tablet/10 mL buffer).
7. Triton X-100 (Sigma) stored at room temperature.

2.2. Immobilizing of GST Fusion Protein on Beads and Overlay with Tissue or Cell Lysate

1. Glutathione sepharose 4B bead column (Amersham Pharmacia Biotech).
2. Wash buffer: 200 mM NaCl, 50 mM Tris-HCl, pH 7.5.
3. PKC cofactors; Phospholipids phosphatidylserine (PS) in CHCl$_3$ (Avanti), sn-1,2 diacylglycerol (DAG) in CHCl$_3$ (Avanti), and 1 mM CaCl$_2$ (*see* **Note 2**).

4. Overlay buffer: 200 m*M* NaCl, 50 m*M* Tris-HCl (pH 7.5), 0.1% w/v PEG 8000, 12 m*M* β-mercaptoethanol, with 0.1% bovine serum albumin (BSA) and protease inhibitor cocktail (Roche Diagnostic Corporation; 1 tablet/10 mL buffer).
5. Overlay wash buffer: 200 m*M* NaCl, 50 m*M* Tris-HCl (pH 7.5), 12 m*M* β-mercaptoethanol, 0.1% w/v PEG 10,000-15,000.
6. Igepal (NP-40) (Sigma), store at room temperature.
7. Triton X-100 (Sigma), store at room temperature.

2.3. SDS-PAGE Gel and Analysis

1. Glutathione (Roche Diagnostic Corporation) 5 m*M* in 50 m*M* Tris-HCl, pH 8. Make a fresh solution for each experiment. Keep on ice before using.
2. Silver staining kit: Gelcode silver SNAP (Pierce).
3. Coomassie Blue: 0.6% brilliant blue R-250 (Bio-Rad), 40% methanol, 10% acetic acid, H$_2$O to 100%.

3. Methods
3.1. GST Fusion Protein Cloning and Expression

1. Clone the cDNA of the protein of interest into the GST expression vector pGEX.
2. Transform constructs into the *E. coli* strain BL21 (*see* **Note 3**)
3. Transfer a single colony into 5 mL of LB and 50 µg/mL ampicillin. Grow overnight at 37°C while shaking.
4. Transfer the 5 mL into 500 mL of LB, add 50 µg/mL ampicillin, and grow at 37°C while shaking at 250 rpm in an incubating shaker until the OD (595 nm) reaches the absorbance reading of approx 0.6 (approx 3 h).
5. Transfer flask to 30°C, add 1 m*M* isopropyl-β-D-thiogalactopyranoside and incubate while shaking at 250 rpm in an incubating shaker for 5 h.
6. Centrifuge at 7700*g* for 10 min at 4°C and discard the supernatant.
7. Resuspend the pellet on ice in lysis buffer; 500 mL of culture is usually suspended in 10 mL of buffer. Snap-freeze the suspension in dry ice/ethanol slurry.
8. Thaw the suspension on ice water.
9. When the suspension is completely thawed, sonicate five times (pulse 1 s, duty 4), 1 min each time, with a 1 min interval on ice in between sonication steps.
10. Add 20% triton X-100 to final concentration of 1%. Leave on ice for 30 min mixing gently by tapping tube periodically.
11. Centrifuge at 12,000*g* for 10 min at 4°C. Transfer the supernatant to new tube and conduct an SDS-PAGE analysis on the supernatant and the pellet (*see* **Note 4**).
12. Bacterial lysates can be stored at –80°C.

3.2. Immobilizing of GST Fusion Protein on Beads and Overlay with Tissue or Cell Lysate

1. Pack a glutathione sepharose 4B column as follows; transfer 25 µL of beads into 1.5-mL Eppendorf tube. Centrifuge beads at 1000*g* for 1 min at 4°C, remove

the supernatant, and wash the beads with 3 × 1 volume of lysis buffer or PBS to equilibrate the column. Allow the beads to mix for 10 min at room temperature between each rinse. Load the column with the GST fusion protein lysate (10 mL of lysate per 2 mL of glutathione sepharose suspension) and allow the mixture to rotate for 60 min at 4°C (*see* **Note 5**).

2. Centrifuge tubes at 1000*g* for 1 min at 4°C and collect flow through to determine efficiency of loading by SDS-PAGE analysis. Be careful not to disturb the beads layer.
3. Wash beads three times with wash buffer.
4. Prepare tissue or cell homogenate as we recently described elsewhere *(10)*.
5. Add cell/tissue homogenate (0.1–1 mg protein) in overlay buffer. Final binding assay volume should be between 0.2 to 1 mL (*see* **Note 6**). To determine the contributions of PKC co-factors to the binding, $CaCl_2$ and/or PS/DAG can be added to the overlay buffer (*see* **Note 7**). Allow the tube to rotate for 1 h at 4°C or for 15–30 min at room temperature.
6. Centrifuge tubes and aspirate the solution.
7. Wash beads with equal volumes of overlay buffer in wash buffer with 0.1% Igepal detergent (NP-40). Monitor the OD_{280} of the flow through until it is less than 0.5 OD_{280} (usually three washes).
8. Resuspend the final pellet in a small volume of PBS (50 µL; *see* **Note 8**).

3.3. SDS-PAGE Gel and Analysis

1. Add 20% sample buffer to the tube, boil sample for 5 min at 100°C. Alternatively, elute the protein mixture with 5 × 1 mL 5 m*M* glutathione. Collect 1-mL fractions and resolve the samples on one-dimensional (1D) or two-dimensional (2D) gel electrophoresis.
3. Stain SDS-PAGE gel with Coomassie or silver staining (*see* **Note 9**).
4. Identification of PKC-binding protein(s) can be determined by excising the band of interest and micro-sequencing after trypsinization. Alternatively, the trypsinized sample is subjected to mass spectroscopy analysis (*see* **Note 10**).
5. Western blot analysis is used if the interaction is between GST-PKC fragment and a purified protein.

4. Notes

1. Protein of interest can also be expressed in bacteria as a fusion protein with tags, such as 6xHis (Qiagen) or MBP (New England Biolab). 6xHis fusion proteins are immobilized on a Ni-NTA agarose column, and eluted with increasing concentrations of imidazole (Qiagen). MBP fusion proteins are immobilized on an amylose agarose column and eluted with maltose (New England Biolab).
2. Preparation of DAG/PS micelles: (1) Use the 25- and 500-µL Hamilton syringes and rinse with ethanol. In a chloroform-resistant test tube add: 480 µL (60 µg/mL) PS and 16 µL (2 µg/mL) DAG. (2) Gently dry off all the chloroform

with Argon or Nitrogen. Stop when you see a few drops of an oily solution left at the bottom of the test tube. (3) Add 4 mL of 20 mM Tris-HCl, pH 7.4. Solution should turn cloudy. (4). Sonicate on ice 3× for 1 min with 30-s intervals on ice. Solution should go from cloudy to clear. (5) Store in a dark container at 4°C; the phospholipids solution lasts approx 2 wk.

3. The *E. coli* strain XL-1 blue is also commonly used for the expression of GST fusion proteins.

4. A common problem when expressing GST fusion protein is the localization of the recombinant protein in inclusion bodies. When that happens, the fusion protein will be detected only in the pellet fraction due to its insolubility. Please refer to Amersham Pharmacia Biotech data sheet for suggestions.

   Frequently, the pellet will still contain a significant amount of recombinant protein due to incomplete lysis of bacteria. For complete solubilization, repeat the freeze-thaw and sonication steps several times.

5. If lysates are stored in glycerol, add lysis buffer to dilute glycerol to less than 10%. Note that the capacity of the beads depends on the level of expression of the GST fusion protein. Therefore, the ratio between glutathione sepharose suspension and the GST fusion protein may need to be adjusted.

6. Difficulties that may occur with this type of assay include nonspecific binding. This can be alleviated by adding BSA (3% BSA weight/volume). In addition, the GST moiety may be sticky. Therefore, nonspecific binding can be reduced by preblocking the beads with excess of purified GST. After the beads are loaded with the GST fusion protein lysates, the beads are extensively washed and incubated with GST (from 1 to 5 µg) for 1 h at 4°C temperature, and the excess GST is removed by centrifugation. Whole-cell homogenate is then added to the beads without a washing step. Adding 50 µM NaCl to the overlay buffer during the washing steps may also decrease nonspecific binding.

7. Please note that the PKC cofactors may be necessary to enhance the binding between the two proteins. For example, we have found that RACK1 interacts preferably with activated PKC *(11)*.

8. The pull-down assay can also be used to differentiate between PKC-binding proteins in unstimulated and stimulated conditions. For example, homogenates can be made from cells treated with vehicle and phorbol ester (PMA), agonists of Gq-coupled receptors and growth factors.

9. If the GST fusion is masking the band of the interacting protein(s) in the 1D gel electrophoresis, two alternative methods can be taken: (1) the sample can be resolved on a 2D gel electrophoresis that is likely to yield a better separation of bands or (2) the binding protein(s) are eluted from the column with high salt concentration that disrupts protein–protein interactions.

10. It is possible that the protein–protein interaction seen in vitro does not occur in vivo. Therefore, it is necessary to confirm that the interaction identified using the pull-down method, by other approaches, such as immunohistochemistry and co-immunoprecipitation.

References

1. Garcia-Rocha, M., Avila, J., and Lozano, J. (1997) The zeta isozyme of protein kinase C binds to tubulin through the pseudosubstrate domain. *Exp. Cell Res.* **230,** 1–8.
2. Ron, D., Napolitano, E. W., Voronova A., Vasquez, N. J., Roberts, D.N., Calio B. L., et al. (1999) Direct interaction in T-cells between theta KC and the tyrosine kinase p59fyn. *J. Biol. Chem.* **274,** 19,003–19,010.
3. Leinweber, B., Parissenti, A. M., Gallant, C., Gangopadhyay, S. S., Kirwan-Rhude, A., Leavis, P. C., et al. (2000) Regulation of protein kinase C by the cytoskeletal protein calponin. *J. Biol. Chem.* **275,** 40,329–40,336.
4. Dekker, L.V. and Parker, P. J. (1997) Regulated binding of the protein kinase C substrate GAP-43 to the V0/C2 region of protein kinase C-delta. *J. Biol. Chem.* **272,** 12,747–12,753.
5. Ron, D., Luo, J., and Mochly-Rosen, D. (1995) C2 region-derived peptides inhibit translocation and function of beta protein kinase C in vivo. *J. Biol. Chem.* **270,** 24,180–24,187.
6. Stebbins, E.G. and Mochly-Rosen, D. (2001) Binding specificity for RACK1 resides in the V5 region of beta II protein kinase C. *J. Biol. Chem.* **276,** 29,644–29,650.
7. Rodriguez, M. M., Ron, D., Touhara, K., Chen, C. H., and Mochly-Rosen, D. (1999) RACK1, a protein kinase C anchoring protein, coordinates the binding of activated protein kinase C and select pleckstrin homology domains in vitro. *Biochemistry* **38,** 13,787–13,794.
8. Csukai, M., Chen, C. H., De Matteis, M. A., and Mochly-Rosen, D. (1997) The coatomer protein beta'-COP, a selective binding protein (RACK) for protein kinase Cepsilon. *J. Biol. Chem.* **272,** 29,200–29,206.
9. Meller, N., Liu, Y. C., Collins, T. L., Bonnefoy-Berard, N., Baier, G. Isakov, N., et al. (1996) Direct interaction between protein kinase C theta (PKC theta) and 14-3-3 tau in T cells: 14-3-3 overexpression results in inhibition of PKC theta translocation and function. *Mol. Cell Biol.* **16,** 5782–5791.
10. Schechtman, D. and Mochly-Rosen, D. (2002) Isozyme-specific inhibitors and activators of protein kinase C. *Methods Enzymol.* **345,** 470–489.
11. Ron, D., Chen, C. H., Caldwell, J., Jamieson, L., Orr, E., and Mochly-Rosen, D. (1994) Cloning of an intracellular receptor for protein kinase C: a homolog of the beta subunit of G proteins. *Proc. Natl. Acad. Sci. USA* **91,** 839–843.

29

Overlay Method for Detecting Protein–Protein Interactions

Deborah Schechtman, Christopher Murriel, Rachel Bright, and Daria Mochly-Rosen

1. Introduction

A simple method to identify protein kinase C (PKC)-interacting proteins is the overlay or far Western assay. This technique uses proteins and/or their putative binding domains as probes to detect binding to immobilized proteins, which are resolved by sodium dodecyl sulfate polyacrylamide gel electrophoresis (SDS-PAGE). This is a simple and fast assay that may be used to complement other assays used to detect protein–protein interactions.

The overlay method was initially used by Wolf and Sahyoun (*1*) to detect the interaction of PKC with proteins in the cellular cytoskeletal fraction. Mochly-Rosen and collaborators (*2*) adapted Wolf and Sahyoun's method to identify intracellular PKC-binding proteins characterized as receptors for activated C kinase (RACKs). Because each PKC isozyme binds to a specific RACK, binding of PKC to RACK confers PKC signaling specificity (*3*). In addition, RACKs are adaptor proteins (reviewed in **ref. 4**). For example, RACK1, the RACK for βIIPKC has been shown to bind many proteins, including phosphodiesterase (*5*), Src kinase (*6*), and dynamin I, a βIIPKC substrate (*7*). Overlays may be used to identify new proteins in tissue lysates that bind to PKC or PKC domains. The overlay assay can also be useful in characterizing the binding properties of PKC to known proteins as well as for screening peptides and pharmacologic agents that enhance or inhibit a protein–protein interaction. We tested rationally designed peptides for their inhibition of βIIPKC binding to RACK1 using the overlay assay (*8*). A peptide (peptide I), which was able to inhibit PKC binding to RACK, was shown to have a

From: *Methods in Molecular Biology, vol. 233: Protein Kinase C Protocols*
Edited by: A. C. Newton © Humana Press Inc., Totowa, NJ

complementary effect when tested in vivo *(8,9)*. Microinjection of peptide I into *Xenopus* oocytes inhibited insulin-induced translocation of βPKC and insulin-induced oocyte maturation *(9)*.

Jaken and collaborators also identified PKC-binding proteins using a similar overlay method (*see* Chapter 30). However, their method included a glutaraldehyde cross-linking step that covalently fixed PKC protein complexes to the nitrocellulose membrane *(10)*. Many PKC substrates were detected by the Jaken's overlay method, including a well-characterized PKC substrate, MARCKS *(10)*. Crosslinking reagents are therefore useful for detecting PKC substrates and lower affinity PKC-binding proteins and to map the interacting domains. For example, Fujise et al. *(11)* showed by the overlay assay that MARCKS binds both the regulatory and catalytic domains of δPKC *(11)*.

Other PKC-binding proteins identified using overlay assays include p32 (gC1q-R), a pan PKC-binding protein detected by Robles-Flores and collaborators. By means of the overlay assay, different PKC isozymes were shown to have unique binding affinities and cofactor requirements for binding to p32 *(12)*. For example, δPKC binds to p32 only in the presence of PKC activators whereas β, ε, and θ PKCs bind to p32 equally well regardless of PKC activators *(12)*.

Pawelczyk and collaborators *(13)* showed that the Cys2 domain of γPKC binds different proteins in fractionated rat brain extract depending on the presence or absence of phospholipids and calcium *(13)*. Caruso and collaborators *(14)* investigated the role of δPKC in a pyruvate dehydrogenase complex and found that δPKC bound to pyruvate dehydrogenase phosphatase 1/2 both by co-imunoprecipitation and by overlay assay *(14)*. Leinweber and collaborators *(15)* showed that calponin binds to the regulatory domain of εPKC and αPKCs particularly to the C2 and possibly the C1b domains. Interestingly, they performed their overlay assay by running PKC domains fused to glutathione S-transferase on an SDS-PAGE gel and overlayed them with calponin *(15)*.

Overlays may be used to clone PKC-binding protein. RACK1 (specific for βII PKC) *(16)* and RACK 2 (specific for εPKC) *(17)* have been successfully cloned by screening λgt11 expression libraries using the overlay method (for more details on this method, *see* Csukai and Mochly-Rosen, **ref.** *18*). Chapline and collaborators *(19)* have also successfully cloned PKC substrates with their overlay method, which includes a cross-linking step. Here, we describe the protocol we have used to detect RACKs that are high-affinity PKC-binding proteins.

2. Materials

2.1. Probe

PKC or PKC fragments (probe) should be used at a minimal concentration of 20 µg/mL. The protein probe may be expressed and purified from bacteria, insect cells, or mammalian cells.

2.2. Tissue or Cell Homogenate

1. Homogenization buffer (final concentrations): 20 m*M* Tris-HCl, pH 7.5, 2 m*M* ethylenediamine tetraacetic acid, 10 m*M* ethylenebis(oxyethylenenitrilo) tetra-acetic acid, 12 m*M* β-mercapthoethanol, 0.25 *M* sucrose, and protease (10 μg/ml) inhibitors diluted 1 : 1000 (added just before use).
2. Stock solutions of protease inhibitors.
   a. Phenylmethylsulfonyl fluoride (17 mg/mL) in isopropanol.
   b. Soybean trypsin inhibitor (20 mg/mL) in 20 m*M* Tris-HCl, pH 7.5.
   c. Leupeptin (25 mg/mL), in 20 m*M* Tris-HCl, pH 7.5.
   d. Aprotinin (25 mg/mL) in 20 m*M* Tris-HCl, pH 7.5.

2.3. Overlay Assay

1. Overlay blocking buffer: 50 m*M* Tris-HCl, pH 7.5, 200 m*M* NaCl, 3% (w/v) bovine serum albumin, 0.1% polyethylene glycol (MW 15,000–20,000).
2. Overlay buffer: 50 m*M* Tris-HCl, pH 7.5, containing 200 m*M* NaCl, 12 m*M* β-mercaptoethanol, 1% (w/v) polyethylene glycol (MW 15,000–20,000), and 10 μg/mL of protease inhibitors (aprotinin, leupeptin, soybean trypsin inhibitor, and phenylmethylsulfonyl fluoride, prepared as described above).
3. Overlay wash buffer: 50 m*M* Tris-HCl, pH 7.5, containing 200 m*M* NaCl, 12 m*M* β-mercaptoethanol, and 0.1% (w/v) polyethylene glycol (MW 15,000–20,000).

2.4. Detection of Binding of the PKC to Protein in Tissue or Cell Homogenate

1. TBS-Tween: 20 m*M* Tris-HCl, pH 7.5, 100 m*M* NaCl, and 0.05% Tween 20.
2. Primary antibody against the PKC probe or the tag on the PKC probe.
3. Secondary antibody: The secondary antibody may be labeled with ^{125}I or conjugated to either alkaline phosphatase or horseradish peroxidase.
4. Detection method appropriate to the secondary antibody.

3. Methods
3.1. Preparation of the PKC Probe

The PKC probe may be expressed in bacteria, insect cells, or be native to or overexpressed in mammalian cells. To facilitate probe purification, the PKC probe may be expressed as a fusion protein with for example, glutathione S-transferase *(20)*, or maltose-binding protein *(21)*. In some cases, the fusion protein may interfere with the binding between PKC or the PKC fragment and its binding protein. Many fusion proteins are constructed to contain a factor Xa or thrombin proteolysis site. Treatment with the corresponding protease can release the PKC probe from the fusion protein. A tag may also be added to the PKC probe (such as a FLAG or Myc epitope). Detection with antibodies against the tag is particularly useful when the antigenic determinant for the anti-PKC antibody is masked when PKC associates with its binding protein. Finally, using various PKC probes that carry the same tag enables a simple comparison of relative affinity of each probe to the binding protein.

3.2. Preparation of Tissue Homogenate

Homogenize the tissue or cell sample of interest in homogenization buffer. If total cell lysates are desired, add 1% Triton X-100 to the buffer. Homogenization of the cells or tissue may be obtained by dounce homogenization, sonication, or by passing the resuspended cells or tissue several times through a 25-gauge needle.

If separation of the soluble and Triton-soluble fractions is desired, first solubilize the sample with homogenization buffer without Triton. After centrifugation at $100,000g$, remove the supernatant (soluble fraction), and resuspend the pellet in homogenization buffer containing 1% Triton X-100. After incubation at 4°C for 30 min, the homogenate is centrifuged for 30 min at $100,000g$. The supernatant from the last centrifugation step contains the Triton-soluble fraction.

3.3. Overlay Assay

3.3.1. Overlaying Cell or Tissue Lysate with PKC Probe

1. Resolve 50–100 µg of proteins from whole tissue or cell homegenates by SDS-PAGE. Alternatively, when characterizing binding to a pure protein, 0.5–2 µg of the pure protein should be resolved by SDS-PAGE.
2. Immobilize proteins by transferring them to either nitrocellulose (Schleicher & Schuell Protran pure nitrocellulose) or polyvinylidene diflouride membranes (Millipore).
3. After transferring the proteins to a nitrocellulose or polyvinylidene diflouride membrane, briefly wash the blots with double distilled water for 5 min, and block the membrane with overlay blocking buffer for 1 h.
4. Incubate blots with overlay buffer containing the PKC probe at approx 20 µg/mL for 1 h at room temperature. This incubation may be carried out in the presence or absence of activators (phospholipids) or other cofactors (*see* **Note 1**). For example, RACK1 only binds activated PKC (*16*). Therefore, phosphatidylserine, diacylglycerol, and calcium are necessary for detecting PKC binding to RACK1. In some cases, it may be necessary to increase the temperature to 34–37°C to facilitate binding.
5. Wash blots with overlay wash buffer four times for 5 min at room temperature.

3.3.2. Detection of PKC Binding

1. Block the overlayed blots (see above) with 5% milk in TBS-Tween-20 for 30 min.

2. Incubate blots with an appropriate primary antibody generated against the overlay protein probe or its epitope tag. This step may be performed for three hours at room temperature or alternatively, overnight at 4°C.
3. Wash blots with TBS-Tween-20 three times for 10 min each.
4. Incubate blots with a secondary antibody, specific for the primary antibody species, in TBS-Tween-20 for 1 h at room temperature.
5. Developed the blots using the appropriate detection method according to the secondary antibody (*see* **Notes 2–4**).

4. Notes

1. Some protein–protein interactions occur only in the presence of co-factors as is the case for PKC, which interacts with RACKs only in the presence of phosphatidylserine and diacylglycerol *(17)*.
2. In this overlay assay, one of the proteins is denatured by SDS-PAGE; to detect binding, the denatured protein must retain its binding capability to the PKC probe. Therefore, a negative result does not mean that there is no protein–protein interaction in a physiological setting. For example, we found that denatured RACK1 binds βII PKC *(16)*, but denatured RACK2 (εRACK) lost its PKC-binding activity *(17)*.
3. False negatives can also occur when posttranslational modifications are necessary for a protein–protein interaction. For example, 14-3-3 interacts with phosphorylated serine residues *(22)*. Therefore, if the expressed protein is not posttranslationally modified properly (i.e., due to expression in bacteria), binding in the overlay will appear negative.
4. A positive overlay assay alone provides insufficient proof that a certain protein–protein interaction occurs in cells. Positive assays using at least two other methods including in vitro pull-down assays (*see* Chapters 27–31 in this volume), co-localization studies in cells (e.g., **ref.** *17*), and in vivo functional studies (e.g., refs. *9,23,24*) are necessary to conclude that a particular protein–protein interaction occurs in cells.

Acknowledgments

This work was supported in part by NIH 52141 and AA11147 grants to Daria Mochly-Rosen.

References

1. Wolf, M. and Sahyoun, N. (1986) Protein kinase C and phosphatidylserine bind to Mr 110,000/115,000 polypeptides enriched in cytoskeletal and postsynaptic density preparations. *J. Biol. Chem.* **261,** 13,327–13,332.
2. Mochly-Rosen, D., Khaner H., and Lopez, J. (1991) Identification of intracellular receptor proteins for activated protein kinase C. *Proc. Natl. Acad. Sci. USA* **88,** 3997–4000.

3. Mochly-Rosen, D. and Gordon, A. S. (1998) Anchoring proteins for protein kinase C: a means for isozyme selectivity. *FASEB J.* **12,** 35–42.

4. Schechtman, D. and Mochly-Rosen, D. (2001) Adaptor proteins in protein kinase C-mediated signal transduction. *Oncogene* **20,** 6339–6347.

5. Yarwood, S. J., Steele, M. R., Scotland, G., Houslay, M. D., and Bolger, G. B. (1999) The RACK1 signaling scaffold protein selectively interacts with the cAMP-specific phosphodiesterase PDE4D5 isoform. *J. Biol. Chem.* **274,** 14,909–14,917.

6. Chang, B. Y., Chiang, M., and Cartwright, C. A. (2001) The interaction of Src and RACK1 is mediated by the SH2 domain of Src and by phosphotyrosines in the sixth WD repeat of RACK1, and is enhanced by activation of PKC and tyrosine phosphorylation of RACK1. *J. Biol. Chem.* **276,** 20,346–20,356.

7. Rodriguez, M. M., Ron, D., Touhara K., Chen, C. H., and Mochly-Rosen, D. (1999) RACK1, a protein kinase C anchoring protein, coordinates the binding of activated protein kinase C and select pleckstrin homology domains in vitro. *Biochemistry* **38,** 13,787–13,794.

8. Mochly-Rosen, D., Khaner, H., Lopez, J., and Smith, B. L. (1991) Intracellular receptors for activated protein kinase C. Identification of a binding site for the enzyme. *J. Biol. Chem.* **266,** 14,866–14,868.

9. Ron, D. and Mochly-Rosen, D. (1994) Agonists and antagonists of protein kinase C function, derived from its binding proteins. *J. Biol. Chem.* **269,** 21,395–21,398.

10. Hyatt, S. L., Liao, L., Aderem, A., Nairn, A. C., and Jaken, S. (1994) Correlation between protein kinase C binding proteins and substrates in REF52 cells. *Cell Growth Differ.* **5,** 495–502.

11. Fujise, A., Mizuno, K., Ueda Y., Osada, S., Hirai, S., Takayanagi, A., et al. (1994) Specificity of the high affinity interaction of protein kinase C with a physiological substrate, myristoylated alanine-rich protein kinase C substrate. *J. Biol. Chem.* **269,** 31,642–31,648.

12. Robles-Flores, M., Rendon-Huerta, E., Gonzalez-Aguilar, H., Mendoza-Hernandez, G., Islas, S., Mendoza, V., et al. (2002) p32 (gC1qBP) is a general protein kinase C (PKC)-binding protein; interaction and cellular localization of P32-PKC complexes in ray hepatocytes. *J. Biol. Chem.* **277,** 5247–5255.

13. Pawelczyk, T., Kowara, R., and Matecki, A. (2000) Protein kinase C-gamma phorbol-binding domain involved in protein- protein interaction. *Mol. Cell. Biochem.* **209,** 69–77.

14. Caruso, M., Maitan, M. A., Bifulco, G., Miele, C., Vigliotta, G., Oriente, F., et al. (2001) Activation and mitochondrial translocation of protein kinase Cdelta are necessary for insulin stimulation of pyruvate dehydrogenase complex activity in muscle and liver cells. *J. Biol. Chem.* **276,** 45,088–45,097.

15. Leinweber, B., Parissenti, A. M., Gallant, C., Gangopadhyay, S. S., Kirwan-Rhude, A., Leavis, P. C., et al. (2000) Regulation of protein kinase C by the cytoskeletal protein calponin. *J. Biol. Chem.* **275,** 40,329–40,336.

16. Ron, D. Chen, C. H., Caldwell, J., Jamieson, L., Orr E., and Mochly-Rosen, D. (1994) Cloning of an intracellular receptor for protein kinase C: a homolog of the beta subunit of G proteins. *Proc. Natl. Acad. Sci. USA* **91,** 839–843.

17. Csukai, M., Chen, C. H., De Matteis, M. A., and Mochly-Rosen, D. (1997) The coatomer protein beta'-COP, a selective binding protein (RACK) for protein kinase C epsilon. *J. Biol. Chem.* **272,** 29,200–29,206.
18. Csukai, M. and Mochly-Rosen, D. (1998) Molecular genetic approaches. II. Expression-interaction cloning. *Methods Mol. Biol.* **88,** 133–139.
19. Chapline, C., Ramsay, K., Klauck, T., and Jaken, S. (1993) Interaction cloning of protein kinase C substrates. *J. Biol. Chem.* **268,** 6858–6861.
20. Smith, D. B. and Johnson, K. S. (1988) Single-step purification of polypeptides expressed in Escherichia coli as fusions with glutathione S-transferase. *Gene* **67,** 31–40.
21. Maina, C. V., Riggs, P. D., Grandea, A. G., III, Slatko, B. E., Moran, L. S., Tagliamonte, J. A., et al. (1988) An Escherichia coli vector to express and purify foreign proteins by fusion to and separation from maltose-binding protein. *Gene* **74,** 365–373.
22. Van Der Hoeven, P. C. J., Van Der Wal, J. C. M., Ruurs, P., Van Dijk, M. C. M., and Van Blitterswijk, W. J. (2000) 14-3-3 Isotypes facilitate coupling of protein kinase C-ζ to Raf-1: negative regulation by 14-3-3 phosphorylation. *Biochem. J.* **345,** 297–306.
23. Dorn, G. W., II, Souroujon, M. C., Liron, T., Chen, C. H., Gray, M. O., Zhou, H. Z., et al. (1999) Sustained in vivo cardiac protection by a rationally designed peptide that causes epsilon protein kinase C translocation. *Proc. Natl. Acad. Sci. USA* **96,** 12,798–12,803.
24. Chen, L., Hahn, H., Wu, G., Chen, C. H., Liron, T., Schechtman, D., et al. (2001) Opposing cardioprotective actions and parallel hypertrophic effects of delta PKC and epsilon PKC. *Proc. Natl. Acad. Sci. USA* **98,** 11,114–11,119.

30

An Overlay Assay for Detecting Protein Kinase C-Binding Proteins and Substrates

Susan Jaken

1. Introduction

Several different approaches have been used to identify protein kinase C (PKC) protein-binding partners. It should be noted that each method may be biased toward identifying a certain type of binding protein, possibly because the methods used emphasize different aspects of PKC-binding protein interactions. This chapter describes features of the overlay assay for identifying PKC-binding partners and briefly compares and contrasts this method with others. Because of the abundance of PKC-binding proteins *(1)*, those cases in which functional relevance has been established are emphasized.

1.1. Overlay Assay

The first method to identify PKC-binding proteins was a modified Western blot approach in which cell lysate proteins were separated by sodium dodecyl sulfate polyacrylamide gel electrophoresis, immobilized on nitrocellulose, and then "overlayed" with purified PKC. After stringent washing, proteins to which PKC remains bound were identified by probing blots with PKC antibodies *(2–4)*. To clone PKC-binding proteins, we modified the assay to screen lifts of λgt11 expression libraries for expressed sequences that directly bind PKC *(5)*. Of more than 10^6 colonies screened, 10 distinct clones were isolated (several clones were isolated multiple times). All were found to be in vitro PKC substrates; many of these have now been shown to be in vivo substrates as well (discussed below).

Because these proteins bind to PKC and are also phosphorylated by PKC, we named them Substrates that Interact with C-kinase (STICKs). A separate

From: *Methods in Molecular Biology, vol. 233: Protein Kinase C Protocols*
Edited by: A. C. Newton © Humana Press Inc., Totowa, NJ

class of PKC-binding proteins that were identified by the overlay assay are known as RACKs. These are discussed in Chapter 26.

1.2. Affinity Purification

A second method for identifying PKC-binding partners is co-immuno-precipitation and/or copurification. In some cases, the PKC interaction sites were mapped to specific protein interaction domains that are widely expressed, suggesting that PKCs have the potential to interact with other proteins that also contain these motifs. For example, PKC regulatory domains directly interact with pleckstrin homology domains in Bruton's tyrosine kinase, LIM domains in ser/thr kinases *(6,7)*, and PDZ domains in INAD and other proteins. It is not yet known what features dictate binding of PKCs to some but perhaps not all proteins containing these protein-binding motifs. Direct interactions between certain PKCs and actin have also been noted *(8–10)*. In many cases, specific binding activity was defined in biochemical experiments; however, further definition of the functional significance of these interactions has not yet been described. No proteins with these interaction domains were found in our expression cloning experiments under the conditions described in methods. These results suggest that the two types of screens are complementary and emphasize different aspects of PKCs interactions with its binding partners.

1.3. Yeast Two-Hybrid Screens

A third method for identifying PKC interacting proteins is through yeast two-hybrid screens. The first PKC interacting protein identified by this approach was named proteins that interact with C-kinase (PICK1). The interaction appears to be mediated through the C-terminus of PKCα and PDZ domains within PICK1 *(11)*. PKCs interact with PDZ domains in other proteins, for example, Par-3 and atypical PKC-specific interacting protein *(12)*, indicating that PKC–PDZ domain interactions may be a broadly applied regulatory strategy (see below).

In each of these examples, the PKCs appear to bind through a type I PDZ ligand located in the PKC C-terminus. These types of proteins were not identified in our overlay assay cloning experiments. Furthermore, we could not demonstrate PKC interactions with MARCKS and other STICKs in the yeast two-hybrid system. Thus, it appears that these assay systems also emphasize different aspects of PKC–protein interactions.

1.4. Genetic Screens

Several *Drosophila* mutants with impaired vision and visual signaling dysfunctions have been noted. In particular, INAD describes a group of loci that regulate deactivation of the visual response that were identified in a screen

Table 1
PKC-Binding Proteins and Substrates Identified by the Overlay Assay

Binding protein/substrate	Location	Known properties
Vinculin/talin (3,18)	Focal contacts	Substratum adhesion
Annexins I and II (3)	Vesicles, plasma	Vesicle trafficking, secretion,
MARCKS (2)	membrane	adhesion and spreading
MARCKS-related protein (5)		
Kinesin light chain (5)		
Desmoyokin, AHNAK	Desmosomes/nucleus	Cell-cell contact and communication
α-Adducin (29–32)	Membrane skeleton	Cell polarity; permissive
γ-Adducin		environment for assembly of other CSK structures
STICK72 and gravin (22,33)	Plasma membrane and membrane ruffles	Transformation-sensitive expression
Serum deprivation response protein (34)	Cytoskeleton, caveolae	Transformation-sensitive expression
GAP43 (35)		Neurite outgrowth and neuotransmitter release
AKAP79 (25,26)	Synaptic densities	Neurotransmitter release

Many of the PKC-binding proteins identified by the overlay assay are also phospholipid binding proteins and PKC substrates associated with the cytoskeleton.

for vision mutations in *Drosophila*. INAD is an acronym for mutants displaying Inactivation No-After Potential.

Subsequent studies demonstrated that INAD is actually a scaffolding protein that organizes PLCβ, eye-PKC, and the ion channel Trp into a functional "signalplex." The combined genetic and biochemical approaches have provided compelling evidence for the requirement of eye–PKC interactions with PDZ domains in INAD for proper localization and function of eye–PKC and *Drosophila* vision in general (13). To date, there is no compelling genetic evidence for the importance of an individual STICK for PKC localization and/or function. This may be because of the fact that there are several STICKs within most cells and the impact of the loss of a particular STICK may not be noticeable. On the other hand, several STICKs are coordinately down-regulated in transformed cells (*see* **Table 1**). Thus, decreased expression of a panel of STICKs may begin to impact on cell morphology and regulation of cell cycle progression.

1.5. The Role of Phospholipids

PKC–STICK interactions in the overlay assay are facilitated by phosphatidylserine (PS). STICKs are all phospholipid-binding proteins and in general are localized to interfaces between membranes and cytoskeletal structures such as the cortical skeleton (**Table 1**). Because PKC regulatory domains also bind PS, it seems likely that the phospholipid-binding motifs in the PKC regulatory domain are important for interactions with STICKs. In fact, experimental evidence indicates that PKC regulatory domain fragments bind to STICKs with comparable efficiency as the holoenzyme (Jaken, unpublished results). Several lines of evidence indicate that the PS dependence is not simply due to PS "bridging" PKC to STICKs:

1. N-chimaerin and PKCζ (which also contain C1 domains) do not bind to STICKs (personal communication, M. Kazanietz, Philadelphia).
2. PKC does not bind to nonphysiological PS-binding proteins, such as histone.
3. Not all PS-binding proteins bind PKC *(4)*.
4. PS bridging does not support binding of STICKs to each other *(4)*.
5. We have demonstrated that STICKs are in vivo PKC substrates.
6. Phosphorylated STICKs do not bind PKC *(2,5)*.
7. Nonsubstrate basic peptides that bind PS only through electrostatic interactions (e.g., poly-lysine) do not inhibit binding; however, the MARCKS phosphorylation domain peptide does inhibit binding.
8. Binding is not inhibited by increasing the ratio of PS:PKC, indicating that PKC preferentially associates with PS-STICK complexes rather than unbound PS *(4)*.

These results indicate that the common affinity of PKC and other proteins for PS is not in itself sufficient to produce high-affinity binding. The role of PS and other anionic phospholipids is not well understood; however, recent studies demonstrate that phospholipids can substantially alter the conformation of certain proteins. For example, the phosphorylation site domain peptide of MARCKS reorganizes randomly dispersed phospholipids in vesicles into PS-rich domains *(14,15)*. PKC colocalizes with these PS-rich domains and domain organization is necessary for productive phosphorylation. The structure of the MARCKS peptide in the absence of PS is mostly random coil as determined by EPR with spin-labeled peptide *(16)*.

However, in the presence of PS:PC vesicles, the peptide lies along the membrane interface with an extended structure. In contrast with peptides that interact with anionic lipids exclusively through ionic interactions (e.g., pentalysine), MARCKS peptide contains several hydrophobic residues that penetrate the membrane interface. The effect of lipids on MARCKS peptide structure suggests that membrane binding may function to form the proper conformation for PKC phosphorylation.

Based on the colocalization of PKCα with focal adhesions in normal fibroblasts, we proposed that PKC may directly interact with certain focal adhesion proteins *(17)*. Subsequently, both vinculin and talin were identified as

PKC-binding proteins in the blot overlay assay *(3)*. In a recent more detailed study, a novel variation on the overlay assay was used to demonstrate direct binding of PKCα to the tail domain of vinculin *(18)*. These studies showed that binding was calcium and phospholipid dependent and indicated that interactions were mediated through the PKC regulatory domain. Importantly, a phospholipid-binding domain in the vinculin tail was shown to be necessary not only for PKCα binding but also for phosphorylation. Overall, the structure-function studies indicate that cytosolic vinculin is in a closed conformation in which the binding sites for talin and α-actinin (in the head domain) and F-actin (in the tail domain) are masked. Phosphorylation and/or PIP2 binding to the tail trigger a conformational change that exposes sites for protein binding partners. This phosphorylation and phospholipid binding mediated regulation of vinculin could significantly influence the formation and remodeling of adhesion complexes.

1.6. Substrates as Binding Partners

What is the function of direct PKC–substrate interactions? PKCs are relatively indiscriminate kinases in vitro. Many proteins that are not physiological substrates can be phosphorylated by PKC in vitro. Given this lack of fidelity, substrate binding may be an important mechanism for restricting substrate accessibility. There are now several examples of direct binding of kinases to their physiological substrates, including mitogen-activated protein kinases *(19)* and phosphoinositide-dependent kinase 1 *(20)* and src *(21)*. Thus, like other kinases, PKC may rely on direct, high-affinity interactions with its substrates to limit PKC action to only appropriate substrates in vivo. Such interactions may also enhance the efficiency of phosphorylation.

The localization of STICKs to the membrane–skeleton interfaces indicates that phosphorylation of STICKs is an initial event that provides a mechanistic link between PKC activation and rapid effects on cell morphology, adhesion, and cell spreading. For example, STICK72 and α-adducin phosphorylated by PKC localize to the leading lamellipodia or to peripheral focal adhesions, respectively, in migrating cells *(22)*. PKC phosphorylation functionally modifies STICK properties, especially with respect to decreased binding of anionic phospholipids, calmodulin, and actin *(23–25)*. STICKs also bind other signaling molecules and are substrates for other kinases at non-PKC sites *(26,27)*. Thus, STICKs have the appropriate properties to play a significant role in signal integration.

Interestingly, all of the PKC phosphorylation domains that we have identified in STICKS are also basic motifs containing hydrophobic residues adjacent to the phosphorylated residues (**Table 2**). Based on structural studies with MARCKS peptides *(28)*, it is reasonable to assume that hydrophobic residues

Table 2
PKC Phosphorylation Site Domains in STICKs

MARCKS:	KKKKKRfSfKKSfKLSGfSfKKNKK
ADDUCIN:	(551) KKKKKFRTPSflKKNKKK
CLONE 72A:	(278) SETTSSfKKFFTHGWAGWRKKTSfKKSKE
CLONE 72B:	(502)LKKlfSSSGLKKlSGKKQK
CLONE 35F	(121)GTGSLKRSGSfSKLRASIRR

Phosphorylatable residues in STICKs were identified by LC/MS and/or MS/MS phosphopeptide mapping and confirmed to be in vivo PKC phosphorylation sites with phosphorylation state selective antisera and mutational analysis. **S,** phosphorylatable ser; "l" and "f," adjacent hydrophobic residues.

are important for PS interactions with these peptides and that PS binding influences their structure. Mutational analysis also indicated the importance of hydrophobic residues as determinants of PS binding in the phosphorylation site domains of Clone 72 *(22)*. These results demonstrate that overlay assay screens preferentially selected for PKC substrates with phosphorylatable residues in basic environments adjacent to hydrophobic residues. Based on analogy with MARCKS, these may represent a family of membrane targeting domains.

2. Materials

1. Tris-buffered saline (TBS): 50 mM Tris-HCl (pH 7.4) containing 0.5 M sodium chloride.
2. Blocking buffer: 5% instant milk in TBS.
3. PKC overlay solution: PKC (10 µg/mL) diluted in TBS containing 10 mg/mL bovine serum albumin, 20 µg/mL phosphatidylserine, 1 mM ethylenebis(oxyethylenenitrilo) tetraacetic acid, 1.2 mM calcium, 10 µg/mL leupeptin, and 1 µg/mL aprotinin (*see* **Notes 1–3**).
4. Fixing solution: 0.5% formaldehyde in phosphate-buffered saline (PBS, 10 mM phosphate buffer [pH 7.4] containing 0.15 M sodium chloride). Formaldehyde is prepared from paraformaldehyde dissolved (10 to 60% w/v) with constant stirring at 60° in a well-ventilated fume hood.
5. Amine blocking solution: 2% glycine (w/v) in water.

3. Methods

1. The overlay assay procedure can be used on any set of proteins adsorbed or immobilized onto nitrocellulose. Proteins electrophoresed onto nitrocellulose as for an immunoblot or simply spotted onto nitrocellulose can be used. Standard procedures for immunoblotting can be used. For cloning, expression libraries are plated out at a suitable density and protein expression is induced with IPTG. Premoistened nitrocellulose sheets are placed over the plates and proteins from the bacterial colonies are adsorbed for 15 to 30 min.

2. Block the nitrocellulose with 5% instant milk in TBS for 30 min at room temperature.
3. Wash three times with TBS for 10 min each time.
4. Incubate ("overlay") with PKC solution for 1 h.
5. Wash three times with PBS for 10 min each time.
6. Incubate with fixing solution for 15 min at room temperature (*see* **Note 4**).
7. Remove fixing solution and inactivate formaldehyde by incubating with 2% glycine for 15 min at room temperature.
8. Wash three times with PBS for 10 min each time.
9. Incubate with antibodies to PKC or other appropriate antibody diluted in TBS containing 10 mg/mL bovine serum albumin for 1 h at room T.
10. Wash three times with TBS.
11. Incubate with enzyme-conjugated second antibody of the appropriate species. Alkaline-phosphatase-conjugated second antibodies are convenient for the bacterial colony lifts, whereas horseradish peroxidase-conjugated second antibody and chemiluminescence development are usually preferred for proteins electrophoretically transferred or spotted onto nitrocellulose.
12. For interaction cloning, positive colonies are identified by aligning the nitrocellulose sheet with the plate containing the bacterial colonies. Positive clones are picked and purified through two to three rounds of limited dilution. λ DNA is then isolated from the clones using standard techniques.

4. Notes

1. In our original studies, the PKCs used in the overlay assay were purified from frozen rabbit brains. The final preparation was mainly a mixture of PKCα and PKCβ. At the time, these studies were done, the importance of PKC phosphorylation was not fully appreciated. It is likely that the PKC used was heterogeneous with respect to phosphorylation at the activation loop, turn motif, and hydrophobic sites. The role of PKC phosphorylation in regulating STICK binding has not been fully evaluated.
2. Recombinant PKCs also work well in this assay, provided they are properly folded, active preparations. In principle, antibodies to tags on recombinant PKC fusion proteins could be used to screen in place of the PKC antibodies.
3. In a variation on this theme, radiolabeled PKCα was used in the overlay assay to directly monitor binding (i.e., obviating the need for indirect antibody detection of bound PKC). In this example, ^{35}S-met-labeled PKCα was produced by in vitro translation *(18)*.
4. Fixing is not essential; however, is convenient for stabilizing the interactions throughout the lengthy incubation and washing procedures.

5. References

1. Jaken, S. and Parker, P. J. (2000) Protein kinase C binding partners. *Bioessays* **22,** 245–254.

2. Hyatt, S. L., Liao, L., Aderem, A., Nairn, A., and Jaken, S. (1994) Correlation between protein kinase C binding proteins and substrates in REF52 cells. *Cell Growth Differ.* **5,** 495–502.

3. Hyatt, S. L., Liao, L., Chapline, C., and Jaken, S. (1994) Identification and characterization of α-protein kinase C binding proteins in normal and transformed REF52 cells. *Biochemistry* **33,** 1223–1228.

4. Liao, L., Hyatt, S. L., Chapline, C., and Jaken, S. (1994) Protein kinase C domains involved in interactions with other proteins. *Biochemistry* **33,** 1229–1233.

5. Chapline, C., Ramsay, K., Klauck, T., and Jaken, S. (1993) Interaction cloning of protein kinase C substrates. *J. Biol. Chem.* **268,** 6858–6861.

6. Yao, L., Kawakami, Y., and Kawakami, T. (1994) The pleckstrin homology domain of Bruton tyrosine kinase interacts with protein kinase C. *Proc. Natl. Acad. Sci. USA* **91,** 9175–9179.

7. Kuroda, S., Tokunaga, C., Kiyohara, Y., Higuchi, O., Konishi, H., Mizuno, K., et al. (1996) Protein-protein interaction of zinc finger LIM domains with protein kinase C. J. Biol. Chem. **271,** 31,029–31,032.

8. Prekeris, R., Hernandez, R. M., Mayhew, M. W., White, M. K., and Terrian, D. M. (1998) Molecular analysis of the interactions between protein kinase C-ε and filamentous actin. *J. Biol. Chem.* **273,** 26,790–26,798.

9. Blobe, G. C., Stribling, D. S., Fabbro, D., Stabel, S., and Hannun, Y. A. (1996) Protein kinase C βII specifically binds to and is activated by F-actin. *J. Biol. Chem.* **271,** 15,823–15,830.

10. Nakhost, A., Forscher, P., and Sossin, W. S. (1998) Binding of protein kinase C isoforms to actin in Aplysia. *J. Neurochem.* **71,** 1221–1231.

11. Staudinger, J., Lu, J., and Olson, E. N. (1997) Specific interaction of the PDZ domain protein PICK1 with the COOH terminus of protein kinase C-α. *J. Biol. Chem.* **272,** 32019–32024.

12. Izumi, Y., Hirose, T., Tamai, Y., Hirai, S., Nagashima, Y., Fujimoto, T., et al. (1998) An atypical PKC directly associates and colocalizes at the epithelial tight junction with ASIP, a mammalian homologue of *Caenorhabditis elegans* polarity protein PAR-3. *J. Cell Biol.* **143,** 95–106.

13. Tsunoda, S., Sun, Y., Suzuki, E., and Zuker, C. (2001) Independent anchoring and assembly mechanisms of INAD signaling complexes in Drosophila photoreceptors. *J. Neurosci.* **21,** 150–158.

14. Yang, L. and Glaser, M. (1996) Formation of membrane domains during the activation of protein kinase C. *Biochemistry* **35,** 13,966–13,974.

15. Yang, L. and Glaser, M. (1995) Membrane domains containing phosphatidylserine and substrate can be important for the activation of protein kinase C. *Biochemistry* **34,** 1500–1506.

16. Qin, Z. H. and Cafiso, D. S. (1996) Membrane structure of protein kinase C and calmodulin binding domain of myristoylated alanine rich C kinase substrate determined by site-directed spin labeling. *Biochemistry* **35,** 2917–2925.

17. Jaken, S., Leach, K., and Klauck, T. (1989) Association of type 3 protein kinase C with focal contacts in rat embryo fibroblasts. *J. Cell Biol.* **109,** 697–704.

18. Ziegler, W. H., Tigges, U., Zieseniss, A., and Jockusch, B. M. (2002) A lipid-regulated docking site on vinculin for protein kinase C. *J. Biol. Chem.* **277,** 7396–7404.
19. Sharrocks, A. D., Yang, S. H., and Galanis, A. (2000) Docking domains and substrate-specificity determination for MAP kinases. *Trends Biochem. Sci.* **25,** 448–453.
20. Gao, T., Toker, A., and Newton, A. C. (2001) The carboxyl terminus of protein kinase c provides a switch to regulate its interaction with the phosphoinositide-dependent kinase, PDK-1. *J. Biol. Chem.* **276,** 19,588–19,596.
21. Pellicena, P. and Miller, W. T. (2001) Processive phosphorylation of p130Cas by Src depends on SH3-polyproline interactions. *J. Biol. Chem.* **276,** 28,190–28,196.
22. Chapline, C., Cottom, J., Tobin, H., Hulmes, J., Crabb, J., and Jaken, S. (1998) A major, transformation-sensitive PKC binding protein is also a PKC substrate involved in cytoskeletal remodeling. *J. Biol. Chem.* **273,** 19,482–19,489.
23. McLaughlin, S. and Aderem, A. (1995) The myristoyl-electrostatic switch: a modulator of reversible protein-membrane interactions. *Trends Biochem. Sci.* **20,** 272–276.
24. Faux, M. C. and Scott, J. D. (1997) Regulation of the AKAP79-protein kinase C interaction by Ca2+/Calmodulin. *J. Biol. Chem.* **272,** 17,038–17,044.
25. Dell'Aquila, M. L., Faux, M. C., Thorburn, J., and Scott, J. D. (1998) Membrane-targeting sequences on AKAP79 bind phosphatidylinositol-4,5-bisphosphate. *EMBO J.* **17,** 2246–2260.
26. Klauck, T. M., Faux, M. C., Labudda, K., Langeberg, L. K., Jaken, S., and Scott, J. D. (1996) Coordination of three signaling enzymes by AKAP79, a mammalian scaffold protein. *Science* **271,** 1589–1592.
27. Nauert, J. B., Klauck, T. M., Langeberg, L. K., and Scott, J. D. (1997) Gravin, an autoantigen recognized by serum from myasthenia gravis patients, is a kinase scaffold protein. *Curr. Biol.* **7,** 52–62.
28. Qin, Z. H. and Cafiso, D. S. (1996) Membrane structure of protein kinase C and calmodulin binding domain of myristoylated alanine rich C kinase substrate determined by site-directed spin labeling. *Biochemistry* **35,** 2917–2925.
29. Chapline, C., Cottom, J., Fowler, L., Dong, L., Taniguchi, H., Tobin, H., et al. (1999) Direct binding of PKC to its substrate adducin. submitted.
30. Dong, L., Chapline, C., Mousseau, B., Fowler, L., Ramsay, K., Stevens, J. L., et al. (1995) 35H, a sequence isolated as a protein kinase C binding protein, is a novel member of the adducin family. *J. Biol. Chem.* **270,** 25,534–25,540.
31. Fowler, L., Dong, L. Q., Van de Water, B., Bowes, R., Stevens, J. L., and Jaken, S. (1998) Transformation-sensitive changes in expression, localization and phosphorylation of adducins in renal proximal tubule epithelial cells. *Cell Growth Differ.* **9,** 177–184.
32. Fowler, L., Everitt, J. L., Stevens, J. L., and Jaken, S. (1998) Localization, expression levels and phosphorylation state of alpha- and gamma-adducins in progressive herediatary renal cell carcinoma. *Cell Growth Differ.* **9,** 405–413.
33. Chapline, C., Mousseau, B., Ramsay, K., Duddy, S., Li, Y., Kiley, S. C., and Jaken, S. (1996) Identification of a major protein kinase C-binding protein and substrate

in rat embryo fibroblasts—Decreased expression in transformed cells. *J. Biol. Chem.* **271,** 6417–6422.

34. Mineo, C., Ying, Y. S., Chapline, C., Jaken, S., and Anderson, R. G. (1998) Targeting of protein kinase Cα to caveolae. *J. Cell Biol.* **141,** 601–610.

35. Dekker, L. V. and Parker, P. J. (1997) Regulated binding of the protein kinase C substrate GAP-43 to the VO/C2 region of protein kinase C-δ. *J. Biol. Chem.* **272,** 12,747–12,753.

31

Functional Proteomic Analysis
of the Protein Kinase C Signaling System

Jason M. Pass, Jun Zhang, Thomas M. Vondriska, and Peipei Ping

1. Introduction

Investigations designed to unravel the role of protein kinase C (PKC) isoforms in the regulation of various intracellular signaling events have been traditionally based on a linear paradigm. Activation of a given molecule (such as a PKC isoform) is modeled to sequentially activate or inhibit another kinase, which in turn affects the activity state of another "downstream" element, and the summation of many protein kinases in series constitutes a signaling pathway. This linear paradigm has served as a valuable framework for the investigation of one or a few target kinases within a particular PKC signaling pathway for several decades.

In contrast to this conventional model of signaling, several recent studies have demonstrated that signaling molecules aggregate to form multiprotein complexes, a phenomenon that appears to hold a number of various types of molecules in close vicinity. Furthermore, investigations from a number of laboratories suggest that the formation of complexes may facilitate signal transduction *(1,2)*. Through the integrated actions of multiple proteins, the modulation of biological functions within the cell is accomplished. It appears in many instances that PKC isozymes do not function in isolation; rather, they form alliances with an array of other signaling molecules. In response to this, our laboratory has submitted the signaling module hypothesis *(3)* in which we propose that the cell assembles multiprotein complexes as a means for signal transduction. Furthermore, we hypothesize that a basic functional unit of these complexes are modules, and that the formation of stimulus and subcellular

From: *Methods in Molecular Biology, vol. 233: Protein Kinase C Protocols*
Edited by: A. C. Newton © Humana Press Inc., Totowa, NJ

location-specific modules within a complex is an essential signaling mechanism whereby the cell performs a variety of distinct tasks.

In recent years, the development of proteomic technologies has made possible the examination of multiple proteins and their interactions on a large scale. The coupling of two-dimensional gel electrophoresis with advanced mass spectrometry allows for the high throughput and systematic display of multiple proteins within a cell or organ. Rather than examining a single molecule in isolation, functional proteomic strategies allow for the simultaneous collection of information regarding multiple proteins and protein interactions within a signaling system, thereby providing a highly detailed blueprint for signal transduction.

Our laboratory has used a functional proteomic approach for the analysis of the subproteome defined by the cardiac protective PKCε signaling system *(1)*. Downey and colleagues *(4)* first hypothesized that PKC was involved in myocardial resistance to ischemic cell death. They found that inhibition of PKC was sufficient to block ischemic preconditioning, a cardiac protective phenomenon whereby transient ischemic episodes (i.e., periods of low blood flow) render the heart cells dramatically resistant to a subsequent prolonged ischemic insult *(4)*. Furthermore, activation of PKC mimicked preconditioning's infarct-sparing effects. Since then, mounting evidence has demonstrated that the ε isoform of PKC is a critical mediator of cardiac protection. Importantly, multiple ancillary studies have characterized a number of other molecules that participate in preconditioning in a PKC-dependent fashion *(5)*. These findings suggest that PKCε orchestrates a series of signaling events that involve recruitment of many other factors (e.g., kinases, structural proteins, transcription factors, and stress-activated proteins) as a mechanism for the development of cardiac protection. Thus, the identification of the molecules that participate in the PKCε signaling system, as well as the description of the nature of their participation, were hypothesized to provide important insights into the molecular basis of the heart's resistance to ischemia. This challenge provided an ideal opportunity for the implementation of functional proteomics to characterize a cardiac protective subproteome. This chapter presents the proteomic experimental techniques that have been used to characterize PKCε signaling complexes (*see* **Fig. 1**) and can be readily adapted to numerous investigations for the end of defining PKC isoform signaling systems in a variety of cell types.

2. Materials

2.1. Strains of PKCε Transgenic Animals

1. Male and female mice. Insofar as the effect of PKCε activation in the heart was of interest in the authors' work, transgenic mice expressing cardiac-specific

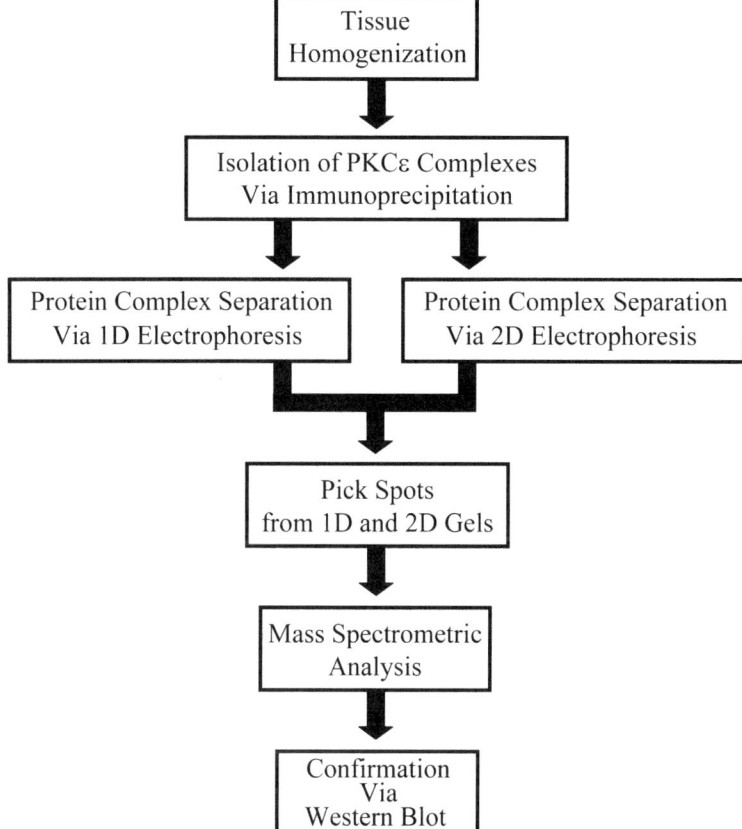

Fig. 1. Overview of the functional proteomic approach for PKC signaling complex analysis. Homogenized tissue is subjected to immunoprecipitation using PKC antibodies (here PKCε). The PKC-immunocomplex is then separated via 1-D or 2-D electrophoresis and visualized by Brilliant Blue G-Colloidal or silver staining. Gel immobilized protein spots are then chosen either manually or by a spot-picking machine and are prepared for mass spectrometric analysis. Following identification of proteins within the PKC immunocomplex via mass spectrometry, presence of these proteins within the PKC complexes is then confirmed via Western immunoblotting.

active PKCε were generated. It should be noted that this approach could easily be modified to address the role of other PKC isoforms in the heart and other organs.
2. PKCε cDNA constructs.
3. Standard tools for cDNA mutagenesis and transgenesis (*6*).

2.2. Sample Preparation

1. Glass-glass homogenizers.
2. Homogenization Buffer: 50 mM Tris-HCl (pH 7.5), 5 mM ethylenediamine tetraacetic acid, and 10 mM ethylenebis(oxyethylenenitrilo)tetraacetic acid. Store at 4°C.
3. Protease inhibitors (1X; cat. no. 1873580, Roche Diagnostics, Mannheim, Germany).

2.3. Immunoprecipitation

1. Monoclonal antibodies to PKCε (cat. no. P14820, BD Transduction Laboratories, San Diego, CA) and control IgG (cat. no. 554126, BD Transduction Laboratories).
2. All buffers and reagents are stored at 4°C.
3. Bradford protein assay reagent.
4. Immunoprecipitation (IP) buffer: 20 mM Tris-HCl, 150 mM NaCl, 10 mM ethylenediamine tetraacetic acid, 1 mM sodium orthovanadate, and 1% NP-40. Add protease inhibitors (Roche) before use.
5. Protein-G Sepharose beads (cat. no. sc-2003, Santa Cruz Biotechnologies, Santa Cruz, CA or cat. no. 15920, GIBCO BRL, Grand Island, NY) re-suspended in IP buffer.
6. 10% sucrose solution.
7. Thiourea buffer (cat. no. 704019, Genomic Solutions, Ann Arbor, MI; *see* **Note 1**).
8. Sample rocker.
9. Centrifuge.
10. Optional: Immunoprecipitation Kit (cat. no. 298142, Pierce Chemical Co., Rockford, IL) (*see* **Note 2**).

2.4. One-Dimensional (1-D) Sodium Dodecyl Sulfate Polyacrylamide Gel Electrophoresis (SDS-PAGE)

1. 5× Laemmli sample buffer (LSB): 0.3 M Tris-HCl (pH 6.8), 50% (v/v) glycerol, 5% (w/v) SDS (*see* **Note 3**), 5% (v/v) 2-mercaptoethanol, and 2.5% (w/v) bromophenol blue. LSB is stored at room temperature.
2. Gel casting and running hardware: casting stand, combs, gel running apparatus, power supply, glass plates, etc. (e.g., cat. no. 31071-020, GIBCO BRL) *(7,8)*.
3. SDS-PAGE stock reagents: 4X Tris/SDS, pH 8.8 (1.5 M Tris-HCl containing 0.4% (w/v) SDS); 4X Tris/SDS pH 6.8 (0.5 M Tris-HCl containing 0.4% [w/v] SDS); 30% acrylamide/0.8% bisacrylamide (29:1 monomer to cross-linker ratio; *see* **Note 4**); 10% (w/v) ammonium persulfate (APS); N,N,N',N'-tetramethylethylenediamine (TEMED). All SDS-PAGE reagents except for TEMED are stored at 4°C.
4. Electrophoretic running buffer: 0.05 M Tris-HCl base, 0.384 M glycine, and 3.5 mM SDS. Store at room temperature.
5. Molecular weight markers (e.g., cat. no. 10748-010, GIBCO BRL).

2.5. Two-Dimensional (2-D) PAGE

1. Thiourea buffer (Genomic Solutions).
2. 18 cm Imoboline™ Dry Imobolized pH Gradient Strips (IPG strips) (cat. no. 17-1235-01, AP Biotech, Piscataway, NJ).
3. Gel chiller/running apparatus (cat. no. 80-005, Genomic Solutions).
4. IPG strip rehydration tray (*see* **Note 5**).
5. pHaser™ Isoelectric Focusing System (cat. no. 70-3537, Genomic Solutions; *see* **Note 6**).
6. IPG strip wicks (cat. no. 70-3978, Genomic Solutions).
7. Nonconducting oil for IPG strips (cat. no. 70-3977, Genomic Solutions).
8. Power source (cat. no. 80-0004, Genomic Solutions).
9. IPG equilibration tray (cat. no. 80-0045, Genomic Solutions).
10. Equilibration buffer I (cat. no. 70-3984, Genomic Solutions): 6 M urea, 2% dithiothreitol, 30% glycerol, and 1X ESA premixed Tris/acetate equilibration buffer.
11. Equilibration buffer II (cat. no. 70-3982, Genomic Solutions): 6 M urea, 2.5% iodoacetamide, 30% glycerol, and 1X ESA pre-mixed Tris/acetate equilibration buffer.
12. 22 cm × 22 cm × 1 mm 2-D precast gel slabs (cat. no. 80-0112, Genomic Solutions).
13. Investigator 2-D Running rack (cat. no. 80-0009, Genomic Solutions).
14. 2-D running buffer: Tris/acetate lower tank buffer (1X; cat. no. 80-0114 [25×], Genomic Solutions); Tris/Tricine/SDS upper tank buffer (cat. no. 80-0115, Genomic Solutions).
15. Brilliant Blue G-colloidal concentrate (cat. no. B2025, Sigma, St. Louis, MO).
16. Silver stain reagents (cat. no. 161-0449, Bio-Rad, Hercules, CA).

2.6. Matrix-Assisted Laser Desorption/Ionization Time-of-Flight (MALDI-TOF) Mass Spectrometry

2.6.1. Protein Digestion

The following reagents are used for protein digestion from Brilliant Blue G-colloidal stained gels. Protein from silver stain gels can also be digested for MALDI-TOF as described elsewhere (*9*) or by using Bio-Rad Silver Stain Plus Kit (cat. no. 161-0449, Bio-Rad).

1. 50 mM NH$_4$HCO$_3$.
2. 50% acetonitrile (v/v).
3. 20 mM dithiothreitol with 0.1 M NH$_4$HCO$_3$ (i.e., dissolve enough dithiothreitol in 0.1 M NH$_4$HCO$_3$ to make a 20 mM dithiothreitol solution); 55 mM iodoacetamide with 0.1 M NH$_4$HCO$_3$ (ibid).
4. 0.1% trifluoroacetic acid (TFA).
5. Disgesion buffer: 20 ng/μL sequencing grade modified trypsin (cat. no. V5113, Promega, Madison, WI) in 50 mM NH$_4$HCO$_3$.

2.6.2. Matrix Preparation Reagents for MALDI-TOF

1. α-Cyano-4-hydroxycinnamic acid (α-CN; cat. no. 47687-0, Aldrich, Milwaukee, WI;*see* **Note 7**).
2. Nitrocellulose (cat. no. RPN303C, AP Biotech, Piscataway, NJ).
3. Solvent 1: ethanol and acetonitrile (1:1) with 0.1% TFA.
3. Solvent 2: 50% acetonitrile (1:1) with 0.1% TFA.
4. 5% formic acid (w/v).
5. MALDI-TOF mass spectrometer.

2.7. Western Immunoblotting

1. Gel transfer apparatus and transfer assembly cassettes (e.g., Trans-Blot Cells, cat. no. 170-3946, Bio-Rad) *(7,8)*.
2. Power source *(7,8)*.
3. All buffers are stored at room temperature.
4. Gel blotting buffer: 0.025 *M* Tris-HCl base, 0.192 *M* glycine, and 16% (v/v) methanol.
5. Electrophoresis-grade nonfat dry milk.
6. Tris-buffered saline (10X TBS): 0.1 *M* Tris-HCl, pH 7.5, and 1 *M* NaCl. For TBS with Tween-20 detergent, add 0.1% (v/v) Tween-20 to 1× TBS before use.
7. Nitrocellulose membrane.
8. Filter paper for gel-/nitrocellulose-transfer cassette interface.
9. Primary and secondary antibodies to protein of interest *(7,8)*.
10. Secondary antibody detection reagents (e.g., for a horseradish peroxidase-linked secondary antibody, electrochemiluminescence Western Blotting Detection Reagents, cat. no. RPN 2106, AP Biotech) *(7,8)*.
11. Film and film processing reagents/developer *(7,8)*.

3. Methods

3.1. Transgenic Animals

1. A PKCε cDNA construct containing an A to E point mutation at amino acid 159 in the pseudosubstrate domain was created using standard molecular biological techniques *(6)*. This mutation facilitates activation of the PKCε protein *(10)*.
2. An HA tag was inserted into the 5′ end of the construct. This tag allows for differentiation of transgene expression from that of endogenous PKCε *(10)*.
3. To achieve cardiac specific expression of the PKCε transgene, the mutated PKCε cDNA construct was inserted downstream from the α-myosin heavy chain (α*MyHC*) promoter *(6)*.The α*MyHC*-PKCε cDNA construct was then used to create transgenic mice as described elsewhere *(6)*.

3.2. Tissue Sample Preparation for Proteomic Analyses

1. For 1-D analysis of PKCε complexes, homogenize eight mouse hearts (~100 mg/heart) in 2 mL of cold (4°C) homogenization buffer per heart. Approximately 30 strokes with the glass-glass homogenizer are sufficient to

homogenize one mouse heart. Combine tissue homogenates for a total of approx 0.8 g of tissue in 16 mL of buffer.

2. For 2-D analysis of PKCε complexes, homogenize 10 mouse hearts in 2 mL of cold (4°C) homogenization buffer per heart. Combine the tissue homogenates for a total of approx 1.0 g of tissue.

3.3. Immunoprecipitation to Isolate PKCε Signaling Complexes

The immunoprecipitation protocols described herein use PKCε-specific antibodies (*see* **Note 8**) *(1,11)*. However, glutathione S-transferase affinity pull-down assays can also be used to isolate PKCε signaling complexes *(1,12)*.

3.3.1. Immunoprecipitation for 1-D Analysis

1. Using IP buffer, adjust combined heart tissue homogenate (**Subheading 3.2.**, **step 1**) to a total volume of 20 mL.
2. To preclear nonspecific aggregates from the homogenate, combine 0.8 mL of protein G Sepharose beads (*see* **Note 9**) with the 20 mL of total tissue homogenate. Incubate the beads/homogenate for 1 h while rocking at 4°C. After incubation, centrifuge the bead/homogenate slurry at 1000*g* for 10 min to pellet nonspecific aggregates. Retain the supernatant as the precleared homogenate.
3. Determine protein concentration of precleared homogenate (e.g., Bradford assay).
4. Separate the precleared homogenate into four equal portions (~5 mL precleared homogenate/portion).
5. Combine the anti-PKCε antibody and protein-G Sepharose beads with two of the precleared homogenate portions as follows: for each 1 mg of precleared protein homogenate, add 0.4 µg of PKCε antibody (*see* **Note 8**) and 50 µL of protein-G Sepharose beads.
6. To the other two precleared homogenate portions, add protein-G Sepharose beads as described above. However, IgG is substituted for anti-PKCε antibodies as a control (*see* **Fig. 2**).
7. Incubate the immunoprecipitation preparations for 12 h at 4°C while gently rocking.
8. After the incubation period, centrifuge the immunoprecipitation preparations at 1000*g* for 5 min at 4°C. Retain each pellet as the immunocomplex.
9. Resuspend each pelleted immunocomplex in 2 mL of cold IP buffer. Gently rock the immunocomplex slurry two to three times to assure even resuspension. Centrifuge the immunocomplex slurry at 1000*g* for 5 min. Discard the supernatant and retain the pellet as the washed immunocomplex. Repeat this wash step until three wash cycles have been completed.
10. After the third wash cycle, resuspend each immunocomplex in 100 µL of IP buffer and 25 µL of 5X LSB (*see* **Note 10**). Boil samples for 5 min after the addition of LSB. Each immunoprecipitation preparation is now ready for 1-D SDS-PAGE analysis (*see* **Subheading 3.4.**).

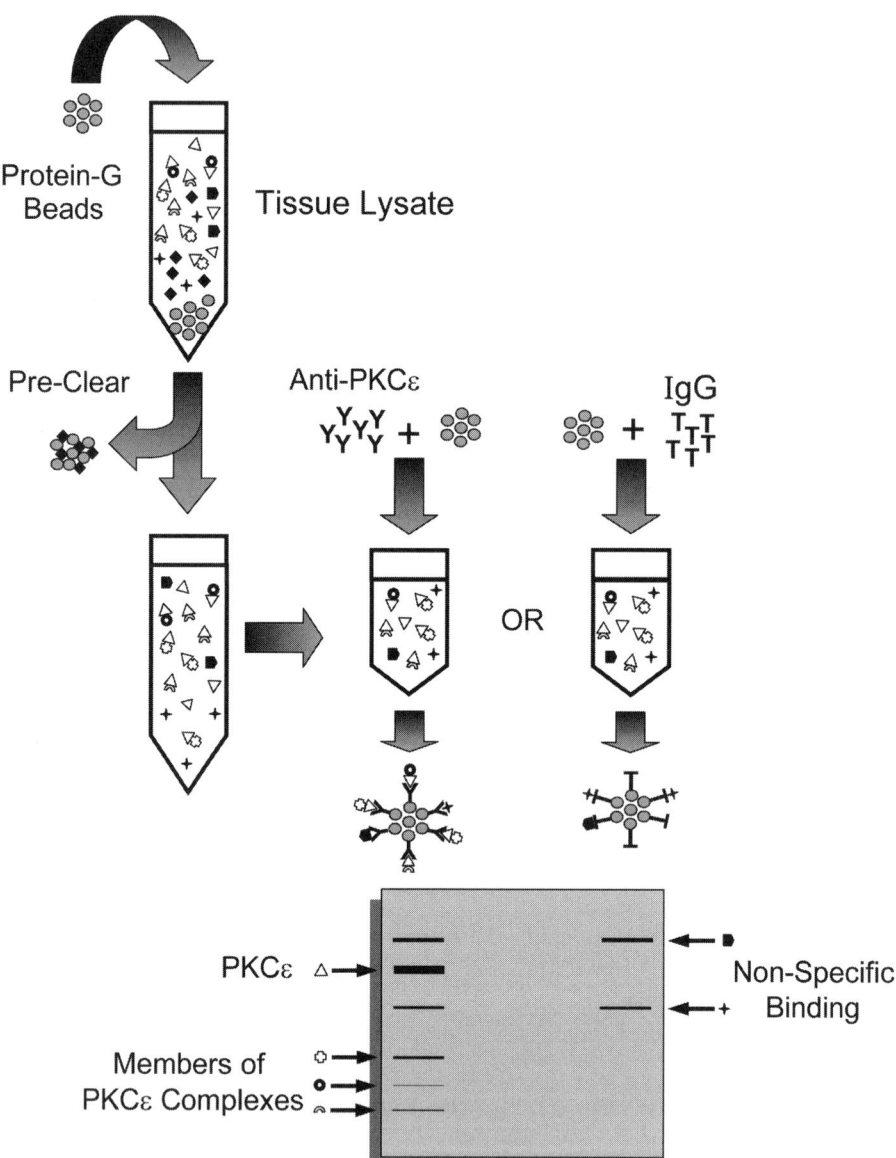

Fig. 2. Immunoprecipitation of PKC signaling complexes. Nonspecific tissue aggregates are initially precleared from the tissue homogenate by the addition of protein-G beads. After the preclearing step, the sample is split into equal portions (*see* **Subheading 3.3.1., step 4**). For immunoprecipitation of the PKCε complex, PKCε antibodies (or IgG, control) and protein-G beads are added to the precleared homogenate. After washing, the resultant immunoprecipitates are subjected to 1-D or 2-D gel electrophoresis, and nonspecific binding of proteins to the IgG is identified.

3.3.2. Immunoprecipitation for 2-D Analysis

1. Using IP buffer, adjust combined tissue homogenate (*see* **Subheading 3.2.,** **step 2**) to a total volume of 50 mL.
2. To preclear nonspecific aggregates from homogenate, combine 2.0 mL of protein-G Sepharose beads (*see* **Note 9**) with the 50 mL of total tissue homogenate. Incubate the beads/homogenate for 1 h while rocking at 4°C. After incubation, centrifuge the bead/homogenate slurry at 1000*g* for 10 min to pellet nonspecific aggregates. Retain the supernatant as the precleared homogenate.
3. Determine protein concentration of precleared homogenate (e.g., Bradford assay).
4. Separate the 50 mL of precleared homogenate into two equal portions.
5. For one of the portions, set up the immunoprecipitation reaction as follows: for every 1 mg of precleared protein homogenate, add 0.4 µg of PKCε antibody (*see* **Note 8**) and 50 µL of protein-G Sepharose beads.
6. To the other portion, add protein-G Sepharose beads and substitute IgG for anti-PKC antibodies as a control (*see* **Fig. 2**).
7. Incubate the immunoprecipitation preparations for 12 h at 4°C while gently rocking.
8. After incubation period, centrifuge the immunoprecipitation preparations at 5000 rpm for 5 min at 4°C. Retain each pellet as the immunocomplex.
9. Resuspend each immunocomplex in 50 mL of cold IP buffer. Gently rock each immunocomplex slurry two to three times to assure even resuspension. Centrifuge each immunocomplex slurry at 1000*g* for 5 min. Discard the supernatant and retain each pellet as a washed immunocomplex. Repeat this wash step until three wash cycles have been completed.
10. After the third wash step, resuspend each immunocomplex in 50 mL of cold 10% sucrose solution.
11. Centrifuge the sucrose/immunocomplex preparations at 1000*g* for 5 min. Discard the supernatant and retain the pellet.
12. Resuspend each pellet in 300 µL of thiourea buffer and incubate for 20 min at room temperature. Centrifuge each preparation at 10,000 rpm for 5 min. Retain both the supernatant and the pellet.
13. Repeat **step 12** two more times. After each centrifugation step, collect the supernatant, which contains the dissociated immunocomplex proteins. Combine the supernatants resulting from the anti-PKCε IP to form one sample and combine the supernatants from IgG IP to form another sample. These samples are now ready for 2-D analysis (*see* **Subheading 3.5.**).

3.4. Separation of Protein Bands Via 1-D SDS-PAGE

1. Assemble glass plates, gel casting stand, and gel running apparatus as detailed by manufacturer.
2. For approx 40 mL resolving gel (a volume sufficient to cast a single 12 cm × 17 cm × 1.5 mm gel), combine 13.3 mL of 30% acrylamide/0.8% bisacrylamide,

10 mL 4X Tris/SDS, pH 8.8, and 16.1 mL H_2O. Add 600 µL of 10% APS and 40 µL of TEMED to initiate gel polymerization.

3. Once the gel is poured into the casting stand, cover the top of the gel with a thin layer of H_2O. Allow the gel to polymerize completely (usually ~15–20 min). After polymerization, pour off any excess H_2O from the top of the gel. Filter paper can be used to wick away excess H_2O but care should be taken not to touch the gel surface with the filter paper.

4. For approx 10 mL of stacking gel (a volume sufficient to precede a ~40 mL resolving gel), combine 1.7 mL of 30% acrylamide/0.8% bisacrylamide, 2.5 mL 4X Tris/SDS, pH 6.8, and 5 mL H_2O. Add 200 µL of 10% APS and 20 µL of TEMED to initiate stacking gel polymerization and immediately pour stacking gel solution into the gel apparatus, on top of the polymerized resolving gel.

5. Immediately insert comb into the stacking gel before polymerization. Ensure that no bubbles are present around the comb. Once the comb is inserted, allow the gel to polymerize completely before removing the comb (usually ~15 min).

6. Remove comb and fill wells with gel running buffer. Attach cast gel to running apparatus and fill the bottom of the gel running apparatus with running buffer (*see* **Note 11**).

7. If samples have been frozen, boil for 5 min before loading.

8. Load 1-D IP samples (from **Subheading 3.3., step 10**) as shown in **Fig. 3**. If desired, molecular weight markers can also be loaded. Add 1X LSB to empty lanes.

9. Connect gel running apparatus to power supply as described by the manufacturer and run gels for 4–7 h at 70 mA or until dye front runs off of the bottom of the gel.

3.5. Display of Protein Spots Via 2-D Analysis

3.5.1. IPG Strip Rehydration

1. For a single 2-D gel, combine up to 200 µL of 2-D immunoprecipitated sample (from **Subheading 3.3., step 13**) with 200 µL of thiourea buffer for a final volume of 400 µL.

2. Pipet the 400-µL sample/thiourea buffer mixture along the groove of the rehydration tray.

3. Carefully remove protective film from 18-cm Pharmacia nonlinear IPG strip and place the IPG strip in rehydration tray face down onto the sample/rehydration buffer mixture. It is imperative that the IPG strip-sample buffer interface is contiguous along the entire length of the IPG strip and that no bubbles are trapped under the strip.

4. Cover rehydration tray and allow strips to rehydrate overnight at room temperature.

3.5.2. Isoelectric Focusing

1. Insert pHaser™ tray into pHaser housing. The pHaser tray plugs should be fully engaged into the housing and the entire housing should be placed on a chiller and maintained at 18–20°C (more details regarding pHaser assembly are supplied by manufacturer).

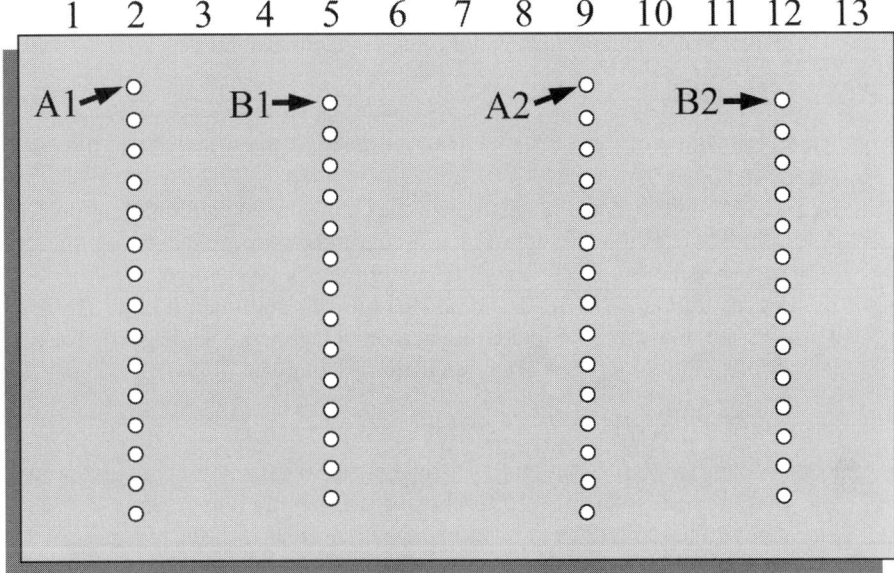

Fig. 3. 1-D gel electrophoresis loading and spot-picking pattern for mass spectrometric analysis. Protein samples that are to be subjected to mass spectrometric analysis should be loaded onto a 1-D gel as illustrated, with lanes 1, 3, 4, 6, 7, 8, 10, 11, and 13 serving as empty lanes that are to be loaded with 1X LSB only. The immunoprecipitated PKCε complexes should be loaded in lanes 2 and 5, whereas the IgG control complexes should be loaded in lanes 9 and 12. This loading pattern assures no overlap of protein between lanes and optimizes spot picking conditions. For mass spectrometric analysis, protein spots should be excised in the pattern illustrated, as this pattern allows for spot excision along the continuum of the 1-D gel. With this method, the PKCε complex member(s) from the spot labeled A1 is compared with the associated IgG complex protein spot labeled A2. Likewise, the PKCε complex member labeled B1 is compared to the associated IgG complex protein spot labeled B2, and so forth until spots from the entire length of the 1-D gel have been selected.

2. Slightly moisten two wicks with 18 MΩ H_2O and place wicks onto the cathodic and anodic ends of pHaser IPG strip groove.
3. Remove rehydrated IPG strip and place face down into the pHaser IPG strip groove of the pHaser tray on top of the wicks. Make sure that the cathodic end of the IPG strip (tapered end on AP Biotech nonlinear IPG strip) is placed on the cathode of the pHaser IPG strip groove and that the anodic end is placed on the anode.
4. Pour a small amount of mineral oil onto the strip. Ensure that the entire length of the strip is covered but do not overfill the groove with oil.
5. Close the pHaser cover and connect to power source.

6. Focus the strips using the following parameters to achieve at least 100,000 volt-hours: max voltage, 5000 V; duration, 24 h; current, 80 μA.

3.5.3. Equilibration

1. After completion of isoelectric focusing, disconnect the pHaser from the power source and open the cover.
2. Remove the IPG strip and place the strip face down in an equilibration tray.
3. Add 10 mL of equilibration buffer I to the equilibration tray and incubate the strip at room temperature for 10 min while gently rocking (*see* **Note 12**).
4. Remove equilibration buffer I and add 10 mL of equilibration buffer II to the equilibration tray. Incubate for 10 min at room temperature while gently rocking. Pour off equilibration buffer after incubation is complete.

3.5.4. Loading and Running the 2-D Gel

1. Remove the Genomic Solutions 2-D slab gels from their packaging and prepare gels as described by the manufacturer using 2-D gel running buffer.
2. Place the equilibrated IPG strip within the groove created by the two glass plates. Be sure that the strip rests face down on the gel surface and that no bubbles are between the strip-gel interface.
3. Load the slab gel into the Investigator 2-D Running Tank as described by the manufacturer.
4. Fill the upper compartment of the Investigator 2-D Running tank with 3 L of 4°C upper tank running buffer. Again, make sure that no bubbles are present in the strip-gel interface. Fill the lower tank with 3 L of 4°C lower tank running buffer.
5. Cover the Investigator 2-D Running tank and connect it to the power supply. Run the 2-D gel at a maximum voltage of 500 V until the blue dye front is 1 cm from the bottom of the gel (~4 h).

3.5.5. Protein Detection

1. For Brilliant Blue G-Colloidal staining of slab gel, follow protocol provided by manufacturer.
2. For silver staining of slab gel, follow protocol delineated in Jensen et al. *(9)* or use Bio-Rad Silver Stain Plus Kit.
3. Analyze 2-D gels using one of the many 2-D analysis software packages available. However, it should be noted that visualization of an individual protein on the gel is not necessary for its identification via MS analysis. Accordingly, the utility of image analysis software must be evaluated by the individual investigator on the basis of his or her specific goals.

3.6. Identification of Proteins by MALDI-TOF MS

The following protocol for in-gel protein digestion and MALDI-TOF preparation is adapted from Jensen and colleagues *(9)*. The method presented herein is for preparation of Brilliant Blue G-Colloidal stained gels. However, samples from silver stained gels can also be prepared for MALDI-TOF analysis as described elsewhere (*9*) or with the Bio-Rad Silver Stain Plus Kit.

3.6.1. In-Gel Protein Digestion

1. After 1-D electrophoresis of samples (*see* **Subheading 3.4.**), proteins can be visualized via Brilliant Blue G-Colloidal staining. Spots can removed either manually or via an automated "spot-picking" machine in the manner illustrated in **Fig. 3**.
2. Following 2-D electrophoresis and Brilliant Blue G-Colloidal staining, wash the 2-D slab gel with water. 2-D spots can be excised manually with a clean scalpel or by an automated "spot-picking" machine. Sections of the gel that do not contain protein should be excised to serve as controls.
3. Cut the excised pieces of gel into cubes of approx 1 mm^3 in size. Place the gel pieces into a clean (*see* **Note 13**) microcentrifuge tube and estimate the volume of total wet gel pieces. This volume is usually around 5–25 µL.
4. For a 10 µL volume of wet gel pieces (*see* **Note 14**), add 20 µL of 50 m*M* NH$_4$HCO$_3$ and incubate at room temperature for 15 min.
5. Add 20 µL of acetonitrile to the NH$_4$HCO$_3$/gel pieces (acetonitrile to NH$_4$HCO$_3$, 1:1) and incubate for 15 min (*see* **Note 15**).
6. Remove NH$_4$HCO$_3$/acetonitrile solution and dry the gel pieces with a vacuum.
7. Reduce the gel-immobilized protein by swelling the gel pieces with approx 20 µL of 20 m*M* dithiotreitol/0.1 *M* NH$_4$HCO$_3$. Incubate for 45 min at 56°C.
8. Place the vial containing the gel pieces on ice. Allow the gel pieces to reach room temperature and remove any excess liquid. Replace the liquid quickly with approx 20 µL of 55 m*M* iodoacetamide/0.1 *M* NH$_4$HCO$_3$. Incubate in the dark and at room temperature for 30 min.
9. Remove the iodoacetamide/NH$_4$HCO$_3$ solution and wash the gel pieces with 20 µL of 50 m*M* NH$_4$HCO$_3$ for 15 min.
10. Add 20 µL of acetonitrile to the NH$_4$HCO$_3$/gel pieces (acetonitrile to NH$_4$HCO$_3$, 1:1) and incubate for 15 min.
11. Remove the acetonitril/NH$_4$HCO$_3$ solution and vacuum dry the gel particles.
12. Incubate dried gel pieces in sufficient digestion buffer to cover the gel pieces, allowing the pieces to swell (for an initial 10 µL of wet gel pieces, this is approx 12–15 µL). Incubate overnight at 37°C. The sample is then ready for matrix preparation.

3.6.2. Matrix Preparation

1. Wash 5–10 mg α-CN for 2–3 sec with approx 50 µL of acetone. Discard acetone phase.
2. Dissolve washed α-CN in acetone at 10 mg/mL.
3. Dissolve nitrocellulose in solvent 1 at 10 mg/mL.
4. Combine the nitrocellulose and α-CN solutions (1:4).

5. Pipet 1 μL of the nitrocellulose/α-CN solution onto the target plate. A thin film should develop on the plate within a few seconds.
6. Wash another 5–10 mg of α-CN for 2–3 s with approx 50 μL of acetone and discard acetone phase.
7. Dissolve the washed α-CN in solvent 2 at 10 mg/mL.
8. Combine 2 μL of the α-CN/solvent 2 (**step 7**) with 2 μL of sample.
9. Pipet 1 μL of the sample-α-CN/solvent 2 solution onto the target plate (*see* **Note 16**).
10. Let the samples dry on the target plate at room temperature.
11. To each dried spot on the target plate, add 1.5 μL of 5% formic acid.
12. After 1 min, remove the formic acid solution with a Kimwipe™.
13. Repeat **steps 11** and **12** once more and let the samples dry at room temperature.
14. Subject the samples to MALDI-TOF analysis.

3.7. Confirmation of PKC Signaling Complex Proteins Via Western Immunoblotting

To confirm the proteins identified by MALDI-TOF, immunoprecipitation of heart tissue is conducted (as described in **Subheading 3.3.**) followed by Western immunoblotting. However, significantly less protein is needed for an IP that is to be followed by Western immunoblotting vs that which will be subjected to MS analysis (~150 μg vs ~2000 μg). Thus, the volumes of IP buffer and wash buffer should be reduced proportionately. Samples can be loaded on the 1-D gel as described in **Subheading 3.4.** or in desired order (*see* **Note 17**).

1. After **Subheading 3.3.9.**, remove the gel from the gel running system and place in the transfer cassette with the nitrocellulose membrane as described by the manufacturer.
2. Run the transfer overnight at 50–60 V.
3. To verify successful transfer of protein to the nitrocellulose membrane, stain the membrane with Ponceau S for 10 min. Densitometric scanning of the stained membrane can be used to determine relative levels of protein loading between the wells.
4. Rinse the membrane in 1X TBS for 5 min to remove Ponceau S stain.
5. Pour off TBS. Block the membrane for 1 h at room temperature using 5% non-fat dried milk in 1X TBS.
6. Remove the 5% milk/TBS. Add primary antibodies to 5% milk/TBS according to manufacture's specifications and incubate with the membrane for 1 h at 37°C.
7. Remove antibody-5% milk/TBS and wash the membrane using five 5-min washes with 1X TBS.
8. After the fifth wash, combine the secondary antibodies with 5% milk/TBS according to the manufacture's specifications and incubate with the membrane for 1 h at room temperature (*see* **Note 18**).

9. Remove the secondary antobody-5% milk/TBS and wash the membrane using five 5-min washes with 1X TBS (*see* **Note 19**).

10. After the fifth wash, use detection reagent and expose the membrane to X-ray film as described elsewhere *(7,8)*.

4. Notes

1. Although the thiourea buffer is supplied by Genomic Solutions (cat. no. 704019), buffer consisting of 7 *M* Urea, 2 *M* thiourea, 0.1 *M* dithiothreitol, 4% CHAPS, and 2% ampholytes is also sufficient.

2. Immunoprecipitation can also be performed using the reagents and methods described in the PIERCE Immunoprecipitation Kit (298142).

3. SDS can cause damage to skin, eyes, and lungs. Gloves, eye protection, and a mask should be worn when handling SDS.

4. Acrylamide is a neurotoxin and should only be handled with gloves.

5. Any vessel that is able to hold IPG strip and associated rehydration solution (**Subheading 2.5., step 4**) will suffice for rehydration tray. Alternatively, the pHaser Equilibration Tray (cat. no. 80-0045, Genomic Solutions) can be used for rehydration as directed by the manufacturer.

6. The complete pHaser system can be purchased from Genomic Solutions (cat. no. 70-3537). Other IPG analysis systems can be obtained from various manufacturers.

7. α-CN can be stored in 0.5- or 1.5-mL microcentrifuge tubes (5–10 mg α-CN/tube) at –20°C. α-CN solutions should always be maintained at 4°C and away from light. With these storage conditions, α-CN solutions have a shelf life of up to 3 d.

8. PKCε is used in **Subheading 3.1., step 5** as an example. Any monoclonal PKC antibody can be used for to isolate that respective isoform's complex via IP. However, it is important that the PKC antibody is in excess of the protein to be immunoprecipitated. In addition, protocols also exist (PIERCE) for the covalent cross-linkage of antibodies to protein beads. This consideration drastically reduces the nonspecific interactions that one may encounter with conventional IP procedures.

9. It is important to evenly suspend the protein-G bead slurry before pipetting. This is best achieved by rocking, as vigorous vortexing of the bead slurry may damage the beads. To further prevent damage to the agarose beads, it is beneficial to widen the orifice of the pipet tip prior to pipetting the beads. This is done by cutting the tip approx 5 mm from the pointed end.

10. When loading the solubilized immunoprecipitate into the 1-D gel, it is important to NOT load any of the beads with sample. One helpful way to remove beads prior to loading is to centrifuge the LSB-solubilized immunoprecipitate through ULTRAFREE-MC Millipore tubes (Millipore UFC30DV00) at 8000 rpm for 5 min. The filtrate is then loaded into the 1-D gel.

11. It is important that no bubbles are present at the gel/running buffer interface.

12. While IPG strip is incubating in Buffers I and II, set up the 2-D gel apparatus as described by the manufacturer.

13. Vials for in-gel protein digestion should be washed with 50% methanol in 0.1% TFA before use.

14. The procedure described herein for in-gel protein digestion is based on a starting volume of 10 μL of wet gel. The volumes of in-gel digestion solutions can be modified proportionally based the amount of starting wet gel.

15. If blue color is still present after initial washing (**Subheading 3.6.1., steps 4** and **5**), repeat **steps 4** and **5** until blue color is gone before moving on to the drying step (**Subheading 3.6.1., step 6**).

16. A high concentration of organic solvent within the sample can damage the plate. If the samples contain a high concentration of organic solvent, 0.5–1.0 μL of water can be deposited on the plate before the addition of the samples.

17. It should be noted that while MS analysis may provide for the detection of some nonspecific interactions, only Western immunoblotting offers unequivocal evidence for the presence or absence of a protein within a complex, be it the complex of one's PKC isoform of interest, or the IgG. Therefore, the Western immunoblot confirmation step considerably strengthens a positive result in the experimental complex (PKC) if this same protein does not appear in the control complex (IgG).

18. For antibodies that produce significant background, add Tween-20 (0.1%) to the 5% milk/TBS (i.e., make TTBS). This will help reduce nonspecific binding of the antibody to the membrane and to membrane bound proteins other than your protein of interest.

19. Again, for antibodies that produce significant background, wash with TTBS for the first three washes. However, be sure to switch to TBS (sans Tween-20) for the last two washes, as Tween-20 can interfere with some ECL reagents.

Acknowledgments

The authors would like to thank Drs. Michael Pisano, Genomic Solutions, Ann Arbor, MI, and Jian Cai, University of Louisville, Department of Pharmacology and Toxicology, Louisville, KY, for their invaluable technical assistance. Peipei Ping is supported by the National Heart, Lung, and Blood Institute.

References

1. Ping, P., Zhang, J., Pierce, W. M., Jr., and Bolli, R. (2001) Functional proteomic analysis of protein kinase C epsilon signaling complexes in the normal heart and during cardioprotection. *Circ. Res.* **88,** 59–62.

2. Husi, H., Ward, M. A., Choudhary, J. S., Blackstock, W. P., and Grant, S. G. (2000) Proteomic analysis of NMDA receptor-adhesion signaling complexes. *Nat. Neurosci.* **3,** 661–669.

3. Vondriska, T. M., Klein, J. B., and Ping, P. (2001) Use of functional proteomics to investigate PKCε-mediated cardioprotection: the signaling module hypothesis. *Am. J. Physiol.* **280,** H1434–H1441.

4. Ytrehus, K., Liu, Y., and Downey, J. M. (1994) Preconditioning protects ischemic rabbits heart by protein kinase C activation. *Am. J. Physiol.* **266,** H1145–H1152.

5. Baines, C. P., Pass, J. M., and Ping, P. (2001) Protein kinases and kinase-modulated effectors in the late phase of ischemic preconditioning. *Basic Res. Cardiol.* **96,** 207–218.

6. Subramaniam, A., Jones, W. K., Gulick, J., Wert, S., Neumann, J., and Robbins, J. (1991) Tissue-specific regulation of the α-myosin heavy chain gene promoter in transgenic mice. *J. Biol. Chem.* **266,** 24,613–24,620.

7. Sambrook, J., Fritsch, E. F., and Maniatis, T. (1989) *Molecular Cloning: A Laboratory Manual,* Cold Springs Harbor Laboratory Press, Plainview, New York.

8. Henderson, C. J. and Wolf, C. R. (1992) Immunodetection of proteins by Western blotting, in *Methods in Molecular Biology, Immunochemical Protocols* (Manson, M. M., ed.), Humana Press, Totowa, NJ, pp. 221–233.

9. Jensen, O. N., Wilm, M., Shevchenko, A., and Mann, M. (1999) Sample preparation methods for mass spectrometric peptide mapping directly from 2-DE gels, in *Methods in Molecular Biology, 2-D Proteome Analysis Protocols* (Link, A. J., ed.), Humana Press, Totowa, NJ, pp. 513–530.

10. Ping, P., Zhang, J., Cao, X., Li, R. CX., Kong, D., Tang, X-L., et al. (1999) PKC-dependent activation of p44/p42 MAPKs during myocardial ischemia-reperfusion in conscious rabbits. *Am. J. Physiol.* **276,** H1468–H1481.

11. Pass, J. M., Zheng, Y., Wead, W. B., Zhang, J., Li, R. C., Bolli, R., et al. (2001) PKCε activation induces dichotomous cardiac phenotypes and modulates PKCε-RACK interactions and RACK expression. *Am. J. Physiol.* **280,** H946–H955.

12. Pandey, A. and Mann, M. (2000) Proteomics to study genes and genomes. *Nature* **405,** 837–846.

IX

PHARMACOLOGICAL PROBES FOR PROTEIN KINASE C

32

Pharmacological Probes for Protein Kinase C

An Introduction

Marcelo G. Kazanietz

1. Pharmacological Modulation of PKC

Understanding the regulation and function of signaling molecules requires the use of pharmacological and molecular approaches. In the case of protein kinase C (PKC) isozymes, the use of pharmacological agents proved to be critical to determine their cellular functions. We have been fortunate that Mother Nature has provided us with a family of potent and selective PKC agonists: the phorbol esters. These natural agents are produced by plants of the family *Euphorbiaceae* and probably protect them from herbivores because they cause acute irritation to mucous membranes. Years of research led to the characterization of the biological properties of phorbol esters and their effects on skin tumor promotion. The identification of PKC as the main receptor for the phorbol esters is probably one of the major breakthroughs in molecular carcinogenesis, and is a perfect example of how disciplines such as biochemistry and cancer research met through pharmacology. In this scenario, the phorbol esters took the center stage. Years later, another set of players also derived from natural products joined the cast: the PKC inhibitors. The availability of potent PKC activators and inhibitors proved to be critical for PKC research and for the determination of PKC-mediated cellular functions.

Because of the vast number of cellular responses regulated by PKC isozymes, the use of phorbol esters became the signature of PKC activation. These natural compounds bind to the C1 domain in PKC, a 50–51 amino acid motif duplicated in tandem in conventional PKCs (cPKCs) and novel PKCs (nPKCs). Details of the molecular interactions between PKC C1 domains and the phorbol

From: *Methods in Molecular Biology, vol. 233: Protein Kinase C Protocols*
Edited by: A. C. Newton © Humana Press Inc., Totowa, NJ

esters have been elucidated in the last years through structural, modeling, and mutagenesis analysis. Binding of the phorbol esters to the C1 domain requires membrane phospholipids. A complex set of interactions regulate the phorbol ester-mediated insertion of PKC into the lipid bilayer, a process that at the cellular level is reflected in the subcellular redistribution or "translocation" of phorbol ester-responsive PKC isozymes *(1–5)*.

The biology of phorbol ester analogs at first appears daunting owing in part from their differential effects on cell growth. It is now well established that phorbol esters can stimulate mitogenesis, but in many cell types PKC activators can also inhibit cell growth, induce apoptosis, or induce differentiation. PKC isozymes have been implicated in cell cycle modulation both at G_1/S and G_2/M *(6)*. The diversity of effects of each PKC isozyme in cell growth and apoptosis, as well as the unique pharmacological properties of PKC ligands, make PKC isozymes an attractive target for selective pharmacological exploitation in cancer chemotherapy and other diseases. Indeed, several PKC agonists and antagonists, as well as PKC antisense oligonucleotides, are in clinical trials for different malignancies. The rationale design of PKC activators and inhibitors will provide us with a large number of pharmacological probes and hopefully isozyme-selective modulators. Although the original concept was that inhibition of PKC function could be an efficient approach for cancer chemotherapy, emerging information suggests that PKC agonists may indeed have therapeutic potential *(7–9)*.

2. The Complexity in the Pharmacology of the PKC Activators

Phorbol esters compete with diacylglycerol (DAG) for binding to PKC. Because of their high stability and potency, they are preferred to the lipid second messenger as PKC activators both in cellular and animal studies. DAG is a relatively simple and highly flexible molecule that is transiently generated in membranes, and it binds only weakly to the PKC C1 domain, which is in sharp contrast to the well-defined, high-binding affinity of the phorbol esters. Indeed, the dissociation constant (K_d) of the phorbol esters is several orders of magnitude lower than the corresponding DAG analog. One of the important challenges in the PKC field is the generation of DAGs with high potency. This has been recently achieved through pharmacophore-guided and modeling approaches based on the crystal coordinates of PKC C1domains. For example, a strategy that has generated novel, potent DAG analogs is to impose conformational rigidity of the glycerol backbone by constraining it into a lactone ring *(10)*. The concept is to identify rigid rotamers that would approximate the actual conformation of the physiologically active DAG, which will result in a reduced entropic penalty associated with DAG binding to the receptor. Some of these DAG analogs, such as the DAG lactones, are indeed as potent as phorbol 12,13-dibutyrate (PDBu).

A level of complexity in the pharmacology of PKC is conferred by the varied nature of PKC ligands. A large number of natural products have PKC-activating capabilities, and in many cases these compounds lack the archetypical diterpene structure. PKC ligands include nonphorbol ester diterpenes (e.g., mezereins, octahydromezerein, and thymeleatoxin), triterpenoids (iridals), ingenols, indole alkaloids (e.g., indolactams and teleocidins), and macrocyclic lactones (e.g., bryostatins). Although structurally unrelated, they all possess the requirements for C1 domain recognition and compete with [^{3}H]PDBu for binding to the PKC C1 domains. Numerous studies have provided us with a wealth of information regarding the nature of the interactions of these molecules with PKC C1 domains. Interestingly, the biological responses of these agents do not always resemble those of the typical phorbol esters, and in many cases, functional antagonism can be observed. As an example, 12-deoxyphorbol esters and bryostatin 1 block phorbol ester-induced tumor promotion in the mouse skin and can therefore be defined as anti-tumor promoters despite their abilities to activate PKC isozymes in vitro *(11,12)*. The unique pharmacological properties of 12-deoxyphorbol esters and bryostatins cannot be explained by differential binding recognition properties or isozyme selectivity, but probably involves a differential translocation of PKC isozymes, suggesting a high degree of complexity at the cellular level.

3. PKC Inhibitors: Targeting Either the Catalytic or the Regulatory Domain

The ATP-binding motif is highly conserved in many protein kinases, including PKC isozymes. Deletion of the ATP-binding site or point mutations within this motif abolish kinase activity. A large number of compounds directed to the ATP-binding site in PKC (C3 domain) have been designed, although in many cases they also inhibit the activity of other protein kinases such as cAMP-dependent protein kinase, cGMP-dependent protein kinases, myosin light chain kinase, tyrosine-kinases, and cyclin-dependent kinases. Nevertheless, several specific PKC inhibitors are now available, which belong to different classes (indolocarbazoles, bisindolylmaleimides, balanoids, phenylamino-pyrimidines, and others) *(9,13)*. GF 109203X, a derivative structurally similar to staurosporine, is the most common PKC inhibitor used in cellular studies.

A second approach to inhibit PKC function is by irreversible inactivation of its catalytic core. Peptide-substrate analogs that form intermolecular disulfide bridges with the active-site region of PKC isozymes and sterically hinder catalysis have been designed, resulting in the inactivation of phosphotransferase activity. PKC isozymes can also be irreversibly inhibited by S-glutathiolation *(14,15)*.

PKC inhibitors directed at the regulatory region have also been described. These agents target the phorbol ester/phospholipid binding sites or domains involved in translocation. An example of one such inhibitor targeted to the regulatory region of PKC is calphostin C, a compound that inhibits [^{3}H]PDBu binding to PKCs. The mechanism of action of calphostin C is not fully understood, but because this compound requires light activation and generates free radicals, one may expect that nonspecific effects might occur. Additional effects, such as inhibition of ligand binding to C1 domains of non-PKC phorbol ester receptors (RasGRP or chimaerins), may also contribute to the lack of specificity of calphostin C *(16–18)*. Another agent acting at the PKC regulatory domain is the basic lipid sphingosine. In this latter case, PKC inhibition probably involves charge neutralization of the acidic phospholipid cofactor *(19)*.

4. The Issue of Isozyme Selectivity

In most cases, cells express at least five or more PKC isozymes corresponding to the different PKC subfamilies (cPKCs, nPKCs, and aPKCs). One of the key questions is determining which isozymes mediate DAG/phorbol ester responses in cells. Many of these issues have been approached by expressing dominant positive (constitutively active) or dominant negative (kinase inactive) PKC isozymes in cells. Still, the selectivity of the dominant negative PKCs is controversial, and their precise mechanism of action is unknown. Because of the high homology of the domains targeted by the pharmacological modulators of PKC, lack of isozyme selectivity is expected in most cases. The challenge is to design isozyme-selective PKC activators and inhibitors. To date, this was only partially achieved. Most of the phorbol ester analogs bind with similar affinity to recombinant PKC isozymes in vitro, as determined by competition assays using [^{3}H]PDBu *(20)*. Only modest differences in affinity have been found for mezerein derivatives. Likewise, only a few examples of selective inhibition have been reported, such as Gö6976 (cPKC-selective), rottlerin (PKCδ-selective), CGP53506 (PKCα-selective), and LY333531 (PKCβ-selective).

Novel strategies are emerging for the achievement of isozyme selectivity. Antisense oligonucleotides for PKC isozymes have been developed as tools to inhibit PKC expression. A representative example is Isis 3521, a 20-mer phosphorothioate oligonucleotide targeted to PKCα mRNA. It was found to cause downregulation of PKCα protein and mRNA in various cell types, and has profound effects on proliferative responses *(21)*. Ribozyme inhibition of PKCα has also proven successful to block the expression of this isozyme by selectively targeting its mRNA *(22)*.

One of the novel concepts is that isozyme selectivity can be achieved in cellular models by selective translocation of PKC isozymes to different intracellular compartments, as described in the Blumberg's laboratory *(23,24)*. This can be accomplished by using different classes of analogs or by varying the ligand lipophilicity. A recent study using the DAG lactone HK654 shows selectivity for PKCα vs PKCδ in cellular studies despite its similar potency in binding and in kinase assays for either isozyme *(25)*. Thus, marked discrepancies exist between in vitro effects and cellular effects of ligands directed toward the C1 domain. In vitro pharmacological screenings may certainly underestimate the selectivity observed in cellular assays.

Another interesting approach to achieve isozyme selectivity is the design of pharmacological tools targeted to domains involved in protein–protein interactions that control PKC anchoring *(26)*. Current evidence indicates that specific proteins may interact with individual PKC isozymes through unique targeting motifs. The rationale is that peptides based on the interacting motifs (either in PKCs or in the anchoring proteins) should block PKC translocation and therefore interfere with the function of individual PKC isozymes. Peptides that specifically inhibit the subcellular redistribution of PKCβ, PKCδ and PKCε have been designed using this strategy, which proved to be useful for determining the involvement of individual PKCs in cellular responses such as cardioprotection *(27)*. Similarly, it has been possible to identify peptides that induce PKC translocation independently of DAG/phorbol ester activation, leading to exposure of the catalytic site and to agonism of PKC function. PKC agonism by peptides could also be achieved in an isozyme-specific fashion.

In summary, a large number of pharmacological probes can be used to modulate PKC activity in vitro and in cellular models. A deep understanding of the molecular interactions between PKC isozymes and their ligands will greatly help to rationally design pharmacological modulators of PKC.

References

1. Ono, Y., Fujii, T., Igarashi, K., Kuno, T., Tanaka, C., Kikkawa, U., et al. (1989) Phorbol ester binding to protein kinase C requires a cysteine-rich zinc-finger-like sequence. *Proc. Natl. Acad. Sci. USA* **86,** 4868–4871.
2. Zhang, G., Kazanietz, M. G., Blumberg, P. M., and Hurley, J. H. (1995) Crystal structure of the cys2 activator–binding domain of protein kinase Cδ in complex with phorbol ester. *Cell* **81,** 917–924.
3. Hurley, J. H., Newton, A. C., Parker, P. J., Blumberg, P. M., and Nishizuka, Y. (1997) Taxonomy and function of C1 protein kinase C homology domains. *Protein Sci.* **6,** 477–480.
4. Kazanietz, M. G., Wang, S., Milne, G. W., Lewin, N. E., Liu, H. L., et al. (1995) Residues in the second cysteine-rich region of protein kinase C δ relevant to

phorbol ester binding as revealed by site-directed mutagenesis. *J. Biol. Chem.* **270,** 21,852–21,859.

5. Ron, D. and Kazanietz, M. G. (1999) New insights into the regulation of protein kinase C and novel phorbol ester receptors. *FASEB J.* **13,** 1658–1676.

6. Black, J. D. (2000) Protein kinase C–mediated regulation of the cell cycle. *Front. Biosci.* **5,** D406–D423.

7. Varterasian, M. L., Mohammad, R. M., Shurafa, M. S., Hulburd, K., Pemberton, P. A., Rodriguez, D. H., et al. (2000) Phase II trial of bryostatin 1 in patients with relapsed low-grade non-Hodgkin's lymphoma and chronic lymphocytic leukemia. *Clin. Cancer Res.* **6,** 825–828.

8. Han, Z. T., Zhu, X. X., Yang, R. Y., Sun, J. Z., Tian, G. F., Liu, X. J., et al. (1998) Effect of intravenous infusions of 12-O-tetradecanoylphorbol-13-acetate (TPA) in patients with myelocytic leukemia: preliminary studies on therapeutic efficacy and toxicity. *Proc. Natl. Acad. Sci. USA* **95,** 5357–5361.

9. Barry, O. P. and Kazanietz, M. G. (2001) Protein kinase C isozymes, novel phorbol ester receptors and cancer chemotherapy. *Curr. Pharm. Des.* **7,** 1725–1744.

10. Marquez, V. E., Nacro, K., Benzaria, S., Lee, J., Sharma, R., Teng, K., et al. (1999) The transition from a pharmacophore-guided approach to a receptor-guided approach in the design of potent protein kinase C ligands. *Pharmacol. Ther.* **82,** 251–261.

11. Szallasi, Z., Krsmanovic, L., and Blumberg, P. M. (1993) Nonpromoting 12-deoxy-phorbol 13-esters inhibit phorbol 12-myristate 13-acetate induced tumor promotion in CD-1 mouse skin. *Cancer Res.* **53,** 2507–2512.

12. Hennings, H., Blumberg, P. M., Pettit, G. R., Herald, C. L., Shores, R., and Yuspa, S. H. (1987) Bryostatin 1, an activator of protein kinase C, inhibits tumor promotion by phorbol esters in SENCAR mouse skin. *Carcinogenesis* **8,** 1343–1346.

13. Way, K. J., Chou, E., and King, G. L. (2000) Identification of PKC–isoform–specific biological actions using pharmacological approaches. *Trends Pharmacol. Sci.* **21,** 181–187.

14. Ward, N. E., Pierce, D. S., Stewart, J. R., and O'Brian, C. A. (1999) A peptide substrate-based affinity label blocks protein kinase C-catalyzed ATP hydrolysis and peptide-substrate phosphorylation. *Arch. Biochem. Biophys.* **365,** 248–253.

15. Chu, F., Ward, N. E., and O'Brian, C. A. (2001) Potent inactivation of representative members of each PKC isozyme subfamily and PKD via S-thiolation by the tumor-promotion/progression antagonist glutathione but not by its precursor cysteine. *Carcinogenesis* **22,** 1221–1229.

16. Diwu, Z. and Lown, J. W. (1994) Photosensitization with anticancer agents 19. EPR studies of photodynamic action of calphostin C: formation of semiquinone radical and activated oxygen on illumination with visible light. *Free Radic. Biol. Med.* **16,** 645–652.

17. Areces, L. B., Kazanietz, M. G., and Blumberg, P. M. (1994) Close similarity of baculovirus-expressed *n*-chimaerin and protein kinase Cα as phorbol ester receptors. *J. Biol. Chem.* **269,** 19,553–19,558.

18. Lorenzo, P. S., Behshti, M., Pettit, G. R., Stone, J. C., and Blumberg, P. M. (2000) The guanine nucleotide exchange factor RasGRP is a high affinity target for diacylglycerol and phorbol esters. *Mol. Pharmacol.* **57,** 840–846.

19. Hannun, Y. A. and Bell, R. M. (1989) Regulation of protein kinase C by sphingosine and lysosphingolipids. *Clin. Chim. Acta* **185,** 333–345.

20. Kazanietz, M. G., Areces, L. B., Bahador, A., Mischak, H., Goodnight, J., Mushinski, J. F., et al. (1993) Characterization of ligand and substrate specificity for the calcium-dependent and calcium-independent protein kinase C isozymes. *Mol. Pharmacol.* **44,** 298–307.

21. Geiger, T., Muller, M., Dean, N. M., and Fabbro, D. (1998) Antitumor activity of a PKC-alpha antisense oligonucleotide in combination with standard chemotherapeutic agents against various human tumors transplanted into nude mice. *Anticancer Drug Des.* **13,** 35–45.

22. Leirdal, M. and Sioud, M. (1999) Ribozyme inhibition of the protein kinase C alpha triggers apoptosis in glioma cells. *Br. J. Cancer* **80,** 1558–1564.

23. Wang, Q. J., Bhattacharyya, D., Garfield, S., Nacro, K., Marquez, V. E., and Blumberg, P. M. (1999) Differential localization of protein kinase C δ by phorbol esters and related compounds using a fusion protein with green fluorescent protein. *J. Biol. Chem.* **274,** 37233–37239.

24. Wang, Q. J., Fang, T. W., Fenick, D., Garfield, S., Bienfait, B., Marquez, V. E., et al. (2000) The lipophilicity of phorbol esters as a critical factor in determining the pattern of translocation of protein kinase C δ fused to green fluorescent protein. *J. Biol. Chem.* **275,** 12,136–12,146.

25. Garcia-Bermejo, M. L., Coluccio Leskow, F., Fujii, T., Wang, Q., Blumberg, P. M., Ohba, M., et al. (2002) DAG-lactones, a new class of PKC agonists, induce apoptosis in LNCaP prostate cancer cells by selective activation of PKCα. *J. Biol. Chem.* **277,** 645–655.

26. Schechtman, D. and Mochly-Rosen, D. (2002) Isozyme-specific inhibitors and activators of protein kinase C. *Methods Enzymol.* **345,** 470–489.

27. Mochly-Rosen, D., Wu, G., Hahn, H., Osinska, H., Liron, T., Lorenz, J. N., et al. (2000). Cardiotrophic effects of protein kinase C ε: analysis by in vivo modulation of PKCε translocation. *Circ. Res.* **86,** 1173–1179.

33

Applications of Inhibitors for Protein Kinase C and Their Isoforms

Garry X. Shen, Kerrie J. Way, Judith R. C. Jacobs, and George L. King

1. Introduction

Protein kinase C (PKC) represents a family of phospholipid-dependent serine/threonine kinase. The activation of PKC has been involved in the signal regulation of enormous physiological and pathological cellular processes. PKC is found in almost all types of cells and tissues; however, PKC-mediated cellular functions are tissue and PKC–isoform-specific *(1,2)*. This chapter summarizes up-to-date information on PKC inhibitors, particularly PKC isoform-specific inhibitors, and their applications.

ATP-binding site inhibitors are the most commonly used general PKC inhibitors. An early representative of the type of inhibitors is isoquinoline sulphonamides (H7) *(3)*. H7 also inhibits cAMP- and cGMP-dependent protein kinases. It is often used with HA1004, a structural analogue of H7, which is a relatively stronger inhibitor for cAMP- and cGMP-dependent protein kinases than for PKC *(4)*. Staurosporine, a microbial alkaloid with an indolocarbazole structure, is a potent PKC inhibitor *(5)*, but it also suppresses the activities of many other protein kinases *(6,7)*. Modifications of the structure of staurosporine have resulted in several sets of more potent and specific PKC inhibitors, including CGP 41251 and UCN-01 *(8,9)*. Calphostin C, a perulenequinone, is a PKC-specific inhibitor that acts on the phorbol ester-binding site of PKC *(10)*. Sphingosine competes with phospholipids and phorbol ester on the binding to PKC. Safingol, a homologue of sphingosine, enhanced the antitumor effects of chemotherapeutic agents *(11)*.

From: *Methods in Molecular Biology, vol. 233: Protein Kinase C Protocols*
Edited by: A. C. Newton © Humana Press Inc., Totowa, NJ

1.1. PKC–Isoform-Selective Inhibitors

Isoform-selective PKC inhibitors are expected to have tissue- or cell-specific PKC inhibition and bring less toxic side effects than nonselective PKC inhibitors. Several PKC inhibitors that bind to ATP-binding sites preferably suppressed the activities of cPKCs than other types of PKC isoforms, such as UCN01 and CGP 41251 *(9,12)*. Rottlerin, a natural product derived from *Mallotus philippinensis*, has exhibited some selectivity to inhibit PKCδ as compared with cPKCs, PKCζ and other nPKC isoforms. The inhibitory activities of rottlerin are not limited to PKC; it also inhibits the activity of protein kinase A and calmodulin kinase III *(13,14)*. LY333531 and its analogins are a bisindolylmalemides and belongs to the category of ATP binding site inhibitors. They inhibited the activity of PKC-βI and βII >40 times greater than other cPKCs, nPKCs, aPKC, and other ATP-dependent kinases *(15,16)*. PKC activation requires translocation of PKC from cytosol to the receptor for activated C kinase (RACK). Peptide against RACKs reduced the translocation of PKC in a PKC–isoform-specific manner. These approaches have been limited in cellular experiments to identify the roles of PKC isoforms through transient permeabilization *(17,18)*. PKC–isoform-specific inhibition may be achieved using molecular biological approaches, including antisenses *(19)*, ribozymes *(20)*, or adenovirus-expressing negative dominants of PKC isoforms *(21)*.

1.2. PKC Inhibitors in Humans

LY333531, a PKC β–isoform-specific inhibitor, has been applied in human studies and clinical trials. Oral administration of LY333531 for 7 d inhibited the reduction of endothelium-dependent vasodilation induced by acute hyperglycemia in healthy humans *(22)*. Several clinical trials that are using LY333531 to prevent microvascular complications in diabetic patients are ongoing *(23)*. No significant toxic effect of LY333531 has been found during the early phases of the clinical trials. Other pharmacological effects of LY333531, including antitumor effect, have been investigated in preclinical studies. Antisense oligonucleotide for PKC-α has been applied in cancer patients in phase I clinical trials *(24)*. A few types of nonselective PKC inhibitors have been tested clinically. UCN-01 sensitized the lymphoma to cytotoxic effect of antitumor therapy in patients *(25)*. CGP41251 is currently in a phase I clinical trial for treatment of advanced cancer. Treatment with CGP41251 reduced the release of tumor necrosis factor and interleukin-6 responding to mitogen from lympocytes of advanced cancer patients *(26)*. Safingol was applied in pilot clinical trials in cancer patients for an adjunctive with chemotherapeutic agents *(27)*.

2. Materials
2.1. Compounds

1. Staurosporine, antibiotic AM-2282 (Sigma)
2. Dimethylsulfoxide (DMSO).
3. CGP 41251, 4'-*N*-benzoyl staurosporine (CIBA-GEIGY).
4. Calphostin C, UCN-1028C (Sigma).
5. 8-Watt fluorescent bulb.
6. Safingol, L-threo-dihydrosphingosine, (2S,3S)-2-amino-1,3-octadecanediol (Eli Lilly; Indianapolis, IN/ Kabi Pharmacia; Clayton, NC).
7. 5% Dextrose in water.
8. LY333531, (*S*)-13-[(dimethylamino)methyl]-10,11,14,15-tetrahydro-4,9:16,21-dimetheno-1H,13H-dibenzo[e,k]pyrrolo[3,4-h][1,4,13]oxadiazacyclohexadecene-1,3(2H)-dione (Eli Lilly Research Laboratories, Indianapolis, IN).

2.2. Peptide Translocation Inhibitors of PKC
2.2.1. Microinjection of PKC Peptide

1. Large electrode tip (0.6 to 0.8 M Omega).
2. PKC peptide (0.1 µM).
3. Internal solution for intracellular injection: 139.8 mM CsCl, 0.1 or 10 mM K$_2$ ethylenebis (oxyethylenenitrilo)tetraacetic acid (EGTA) 4 mM MgCl$_2$, 0.062 mM CaCl$_2$, 5 mM Na$_2$ creatine phosphate, 10 mM HEPES, 3.1 mM Na2ATP, and 0.42 mM Na$_2$GTP; at pH 7.1 (adjust with KOH).

2.2.2. Transient Permeabilization Protocol

1. PKC peptide.
2. Phosphate-buffered saline (PBS): 1mM CaCl$_2$, 0.4 mM MgCl$_2$•6H$_2$O, 2.7 mM KCl, 0.15 mM KH$_2$PO$_4$, 8 mM NaH$_2$PO$_4$•7H$_2$O, 136 mM NaCl, at room temperature and on ice.
3. PBS.
4. 2X concentration of sterile permeabilization buffer containing 40 mM HEPES (pH 7.4), 20 mM EGTA, 280 mM KCl, 100 µg/mL saponin (Sigma), 10 mM sodium azide (not for studies assessing *c-fos* mRNA or ^{14}C-phenylalanine accumulation), and oxalic acid dipotassium salt 10 mM.
5. ATP, 200 mM, pH 7.4.

2.3. Antisense Oligonucleotides

1. Antisense or sense oligonucleotides for PKC isoforms (therefore, PKC isoforms α, β, ε, γ, and δ are commercially available from Perkin-Elmer, Wellesley, MA or Biognostic GmbH, Goettingen, Germany).
2. 10 µg/mL lipofectin.

2.4. Adenovirus Expressing Negative Dominants of PKC Isoforms

2.4.1. Growing 293HEK Cells and Amplifying Adenovirus

1. Transformed human embryo kidney (HEK293) cells, containing and expressing the early region 1 (E1) of Ad 5 virus (*see* **Note 1**).
2. Culture media for HEK293 cells: cells are cultured in DMEM high glucose supplemented with penicillin (100 U/mL), streptomycin (100 µg/mL,) and 10% heat-inactivated fetal bovine serum (FBS).
3. Warm (37°C) 1X citric saline (autoclaved) to detach cells during splitting. 10× citric saline: 50 g of potassium chloride, 22 g of sodium citrate in 500 mL of tissue culture-grade water. Dilute citric saline 10× and autoclave.
4. PBS.
5. Culture dishes: 100- and 150-mm Petri dishes, 12-well plates.
6. Adenovirus with PKC constructs.
7. Methanol and dry ice.
8. Blue light or UV light.
9. Microscope with 10X magnification.

2.4.2. Concentrating Adenovirus

1. NP-40.
2. Centrifuge bottles.
3. 20% PEG/2.5 M NaCl (*see* **Note 2**).
4. PBS.
5. Cesium chloride:
   4 M CsCl in 10 mM HEPES.
   2.2 M CsCl in 10 mM HEPES.
6. Dialyzer (Spectra/Por; MWCO: 12000–14000).
7. PBS + 10% glycerol.

3. Methods

3.1. Compounds

3.1.1. Staurosporine

Staurosporine (antibiotic AM-2282, Sigma; $C_{28}H_{26}N_4O_3$; molecular weight: 466) is a nonisoform-specific potent PKC inhibitor and is light sensitive. The inhibition of staurosporine on PKC activity was not competed by phosphatidylserine, calcium, diacylglycerol, or ATP. It did not interfere the binding of 12-*O*-tetradecanoylphorbol-13-acetate or phorbol-12,13-dibutyrate, which suggests it does not interact the regulatory domain of PKC. Staurosporine also inhibited the activity of cAMP-dependent protein kinase in similar extent as that for PKC. The inhibition of staurosporine on cAMP-dependent protein kinase may contribute to the cytotoxicity of staurosporine *(5)*.

Table 1
Compounds Used in Studies to Inhibit PKC

Compound	Cells/species	Effect	Dosage	Reference
Staurosporine	Rat brain	↓ PKC activity	LC_{50}: 2.7 nM	*5*
	HeLa S3 cells	↓ Growth	IC_{50}: 4.08 pM	*5*
CGP 41251	Tumor cells	↓ Tumor growth	1/20 MTD effective	*9*
	Adult patients with solid tumors	Mild side effects: fatigue, nausea, vomiting	25–300 mg/d	*28*
Calphostin C	Cells	↓ PKC activity	IC_{50}: 0.05 μM	*30*
	HeLa S3 cells	↓ Growth	IC_{50}: 0.18 mg/mL	*30*
LY333531	SMC	↓ Phosphorylation PKC βII, not PKC α	10–20 nM, 20 min	*35*
	SMC	↓ Autophosphorylation PKC βII, not PKC α	10 nM	*35*
	Rat, human mesanglial cells	↓ Glucose-induced DAG increase	20 nM	*36*

1. Dissolve staurosporine in DMSO up to 1 mg/mL. (*see* **Note 3**). Dissolve further into final concentration in media (for cell culture) or aqueous working solution.
2. Stimulate cells or tissue with staurosporine. For recommended dosages, *see* **Table 1**.

3.1.2. CGP 41251

The inhibitory activity of CGP 41251 (4′-*N*-benzoyl staurosporine; CIBA-GEIGY; $C_{35}H_{32}N_4O_4$; Molecular weight: 558) on PKC is comparable with the inhibitory effect on phosphorylase kinase (48 nM), but more than 48 times greater than that for PKA, S6 kinase, or tyrosine-specific kinase.

1. Dissolve compound in DMSO (ex. 10 mM) and stored at –20°C. The stock solution may be further diluted in aqueous solution with the concentration of DMSO less than 0.5%.
2. Incubate cells or tissue with Staurosporine. For recommended dosages and duration, *see* **Table 1**.

3.1.3. Calphostin C

The inhibition of calphostin C (UCN-1028C; Sigma; $C_{44}H_{38}O_{14}$; molecular weight: 790) on PKC is 1000 times stronger than that for cAMP-dependent protein kinase or tyrosine-specific protein kinase (IC_{50} >50 μM). The activity of calphostin C is not competed by calcium or phospholipid. Calphostin C inhibited the binding of phorbol-12,13-dibutyrate to PKC, which suggests that the inhibitor may interact with the regulatory domain of PKC *(30)*.

The powders are light sensitive; the compound can be activated by light *(29)*.

1. A stock solution of calphostin C should be freshly made with DMSO up to 1 mg/mL, which may be further diluted into aqueous solution or medium.
2. Incubate cells or tissue with the compound. An 8-watt fluorescent bulb located 15 cm from the culture dishes during the incubation is recommended by the vender, to activate the compound. For recommended dosages and durations, *see* **Table 1**.

3.1.4. Safingol

Safingol(L-threo-dihydrosphingosine, (2S,3S)-2-amino-1,3-octadecanediol; Eli Lilly; Indianapolis, IN/ Kabi Pharmacia; Clayton, NC; $C_{18}H_{39}NO_2$; molecular weight: 301) is a general PKC inhibitor acting on the regulatory domain of the enzyme with a minimal or no activity on other protein kinases *(31,32)*.

In vivo studies in rats and dogs demonstrated that safingol had little antitumor activity when used alone, but the antitumor activity was enhanced by a combination therapy with doxorubicin or other chemotherapy agents *(33)*. A pilot clinical trial with cancer patients demonstrated that safingol, up to the dosage of 120 mg/m², does not cause dose-limiting hematology or other toxicity. Minor antitumor responses were seen in 4 of 17 patients using safingol in combination with 45 mg/m² doxorubicin *(34)*.

1. Lipid emulsion of safingol (5 mg/mL) is provided by the vendors, and can be diluted to desired concentrations with 5% dextrose in water.
2. Apply the emulsion on the surface.

3.1.5 LY333531

LY333531 ((S)-13-[(dimethylamino)methyl]-10,11,14,15-tetrahydro-4,9:16, 21-dimetheno-1H,13H-dibenzo[e,k]pyrrolo[3,4-h][1,4,13]oxadiazacyclo-hexadecene-1,3(2H)-dione) is not commercially available; however, requests for the compound can be made directly to the Eli Lilly Research Laboratories, Indianapolis, IN. LY333531 is a dimethylamine analogue, was one of a series of novel N-N'-bridged bisindolylmaleimide moieties developed (**Fig. 1**), and it acts as a competitive inhibitor for ATP binding to PKC *(15)*. Kinetic analysis with Lineweaver-Burk and Dixon plots showed LY333531 as a reversible competitive inhibitor of ATP binding, with a K_I of 2 nM for PKCβI. Selectively to inhibit the PKCβI and βII isoforms was achieved in the nanomolar range

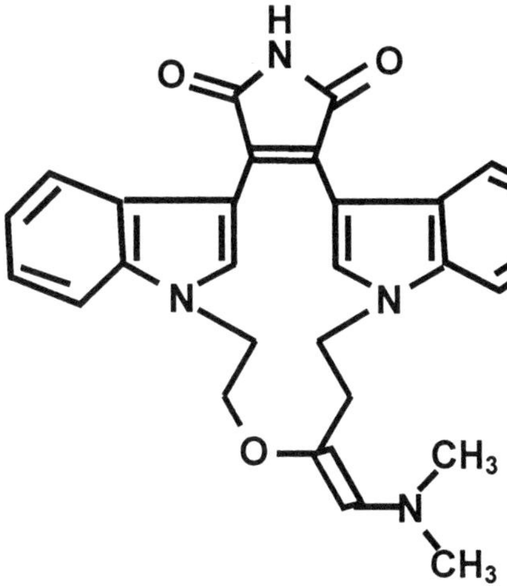

Fig. 1. Macrocyclic bisindolylmaleimide structure of LY333531

Table 2
Isoform Selectivity of LY333531 for PKC

PKC isoform	IC_{50} (nM)
α	360
βI	4.7
βII	5.9
γ	400
δ	250
ε	600
ζ	>100,000
η	52

(IC_{50} = 4.7 and 5.9 nM, respectively), vs other calcium-dependent PKC isoforms, such as PKCα and γ (IC_{50} = 360–400 nM, respectively), or the novel and atypical PKCs (**Table 2**). Although LY333531 was reported to also inhibit calcium calmodulin ATP-dependent kinases (IC_{50} 6.2 μM), its selectivity for PKCβ is three orders of magnitude greater. LY333531 was also found to be inactive against PKA, casein kinase, and src-tyrosine kinase; therefore, compared with the PKC inhibitor staurosporine, LY333531 exhibits greater selectivity for PKC (*15*).

Numerous reports have described the successful use of LY333531 as a selective inhibitor of PKCβ in vitro in cultured cell systems, in vivo by oral administration in chow to experimental animals, and more recently in clinical trials where it has been administered orally to patients with diabetes and to

Table 3
In Vivo Use of LY333531 in Animals and Humans

LY333531 dose	Species	Delivery method	Treatment duration	Plasma concentration	Reference
0.1–10 mg/kg/d	Rat	Oral, chow	2 wk	5.7–19 nM	*35*
10 mg/kg/d	Rat	Oral, chow	12 wk	19 nM	*36*
50 mg/kg/d	Rat	Oral, gavage; suspended in 10% tragacanth gum	4 wk	N/R	*41*
10 mg/kg/d	Rat	Oral, chow	16 wk	N/R	*37*
0.5 mg/kg, BID	Pig	Oral, capsule	12 wk	77–302 ng/mL	*38*
8 mg BID, 16 mg BID, 16 mg/d	Human	Oral	30 d	N/R	*39*
32 mg/d	Human	Oral	7 d	16 nM	*22*

Abbreviations: N/R, not reported in reference provided.

healthy volunteers (*see* **Table 3**). These studies show that similar concentrations of LY333531 can be used in vivo and in vitro to selectively inhibit PKCβ. To date, all reported in vivo studies using LY333531 deliver the compound orally in chow, capsule or solution form. As for in vitro studies using LY333531, doses of the compound that achieve plasma concentrations between 5 and 20 nM, will be most effective to inhibit the PKCβ isoform. The use of doses that elevate plasma LY333531 levels to 100 nM or above should be avoided because inhibition of the PKCα isoform may occur. The serum half-life of LY333531 was determined to be approx 6 h, with its desmethylated metabolite product showing similar potency and inhibitory specificity for PKCβ *(35)*. Koya and coworkers *(36)* were the first to demonstrate that oral treatment of diabetic rats with LY333531 (10 mg/kg, 12 wk) could selectively prevent an increase in phosphorylation of the PKCβI and not PKCα in glomeruli. This study also noted that LY333531 produced a parallel increase in basal PKCα phosphorylation levels in glomeruli from control animals; however, the mechanisms or consequences for this change were unclear. The effectiveness of LY333531 to reduce the diabetes-induced increases in PKC activity in different tissues has been shown to be dependent on dosage. In diabetic rats, increased retina PKC activity was reduced with LY333531 at a dose of 0.1–10 mg/kg, whereas doses of 1.0–10 mg/kg reduced PKC activity in glomeruli *(35)*. These results suggest tissue accumulation of LY333531 may differ within the animals. In these studies, LY333531 had no effect on the diabetes-induced increase in DAG levels. Reports on the use of LY333531 in animals other than rodents are few.

However, the therapeutic benefits of LY333531 have also been assessed in a pig model of branch retinal vein occlusion *(38)*.

Results from a Phase 1 safety and pharmacodynamic clinical trial that administered LY333531 orally to diabetic patients have been reported *(39,40)*. A 1-mo assessment of LY333531 was performed in 29 patients with Type 1 or 2 diabetes. Orally administered LY333531 reached bioactive concentrations, was well tolerated for 30 d at doses up to 16 mg BID, and was not associated with any significant increase in adverse events.

A recent clinical study has used 32 mg LY333531 orally per day for 7 d in 15 healthy human male and female volunteers. This dosage resulted in a plasma concentration of 16 n*M* and did not alter blood glucose levels.

1. The compound is provided in a mesylate form (MW 582.68) as a bright orange powder and is recommended to be dissolved in DMSO, which can be stored at –20°C.
2. Oral administration of LY333531 to animals using available chow, capsule, or solution form. For dosages, *see* **Table 3** (*see* **Note 4**).

3.2. Peptide Translocation Inhibitors of PKC

PKC isoforms are localized to distinct subcellular sites before and after their activation *(42)*. Translocation requires recognition of the kinase by a membrane-bound anchor protein specific for the isoform. Selectivity ensures each PKC isoform will be localized near its defined set of protein substrates. The isoform-selective anchoring proteins for PKC have been termed receptors for activated C kinase (RACKs) *(43)*, and it is likely that separate RACKs exist for each PKC isoform at distinct subcellular regions. Consequently, isoform-selective peptide translocation inhibitors for PKC were developed based on the potential for the interaction site between the kinase and anchoring protein to be modulated *(44,45)*. Mochly-Rosen and coworkers *(46)* were the first to identify and develop short protein sequences, which could inhibit translocation by binding to sites on either the RACK or its PKC isoform *(46)*.

RACK binding to PKC is saturable and there might be more than one type of RACK for each PKC isoform within a cell. The majority of peptide inhibitors developed, are directed to the RACK-binding site on PKC. **Table 4** lists identified peptide translocation inhibitors of various PKC isoforms, their peptide sequence, and PKC-binding site. For cPKCs, the PKC–RACK interaction is mediated in part by the C2 region, whereas the C2-like region is involved for nPKC. Peptide inhibitors for the cPKCs have been directed to the RACK-binding site within the C2 domain *(44,45)*. Recently, the V5 domain has been identified to also contain part of the RACK-binding site for PKCβ2 and PKCβ1, and peptide inhibitors for these isoforms have been developed *(47)*.

Table 4
Peptide Fragments Used in Studies to Inhibit PKC Isoforms

Peptide name	Action	Sequence	Corresponding amino acids	Reference
βC2-1	Inhibits cPKC	KQKTKTIK	PKCβ 209–216	*52,53,66,68,69*
βC2-2		MDPNGLSDPYVKL	PKCβ 186–198	
βC2-4		SLNPEWNET	PKCβ 218–226	
βIV5-3	Inhibits PKCβI	KLFIMN	N/R	*47*
βIIV5-3	Inhibits PKCβII	QEVIRN	N/R	*47*
εV1	Inhibits PKCε	N/R	PKCε 1–144	*53,55,66,69*
εV1-2		EAVSLKPT	PKCε 14–21	
δV1	Inhibits PKCδ	N/R	PKCδ 1–144	*49,54,55*
δV1-1		SFNSYELGSL	PKCδ 8–17	
ηV1-2	Inhibits PKCη	EAVGLQPT	PKCη 18–25	*55*
Control pseudosubstrate	Positive control to inhibit PKC	RFARKGALRQKNV	PKC 19–31	*52,57*
Scrambled pseudosubstrate	Negative control	RALQRAKNEVHKVFK (GNR)	N/R	*52,57*
Scrambled εV1-2	Negative control for PKCε	LSETKPAV	PKCε 14–21	*69*

Abbreviations: PKC, protein kinase C; N/R, not reported in reference provided.

Peptide inhibitors of PKCε were directed toward its variable region, which had homology to the C2 region of cPKCs.

Table 4 provides a list of reported peptide modulators along with their peptide sequence to allow for their synthesis. Structure and purity of amino acids generated must be assessed. Greater than 90% purity is required. These peptide fragments are on average no more then 10 amino acids in length, and functional effectiveness can be achieved at intracellular concentrations of 5 to 50 n*M*. Successful functioning of peptide inhibitors can be determined by inhibition of phorbol ester-induced translocation of a PKC isoform as assessed by immunofluorescence and immunoblot analysis. Important to these studies is the comparison of the peptide inhibitor of interest to the actions of positive and negative control peptides. A scrambled peptide of the same amino acid sequence can be used as a negative control, and should produce no effect, examples are shown in **Table 4**. A control peptide, which acts as a general inhibitor of PKC translocation, should inhibit PKC translocation in a nonspecific manner (**Table 3**).

One distinction regarding peptide use is their inability to cross biological membranes. Therefore, for cultured cell studies, they require introduction into the cell. To date, methods of PKC peptide delivery to living cells include microinjection, transient permeablization, and use of peptide transporters, such as Antennapedia-derived peptide, TAT-derived peptide, or poly-arginine. The majority of the above mentioned studies have used neonatal and adult cardiac myocyte cultures from rat and rabbit. Further discussion of these delivery methods follows.

3.2.1. Microinjection of PKC Peptides

Initial studies, which reported PKC peptide use, used the microinjection technique *(52,56,57)*. This technique is useful for single-cell analysis studies, such as electrophysiology, and when large-sized cells are used, such as *Xenopus* oocytes. Conversely, microinjection can be considered technically difficult with a low percentage of cell survival, and single-cell application may be limiting when biochemical analysis requires a larger sample amount to be obtained.

1. For microinjection experiments, the use of a large electrode tip is preferable (0.6 to 0.8 M Omega).
2. Patch electrodes should contain the PKC peptide (0.1 μM) together with the following internal solution for intracellular injection (mM; **ref. 52**): CsCl 139.8, K_2 EGTA 0.1 or 10, $MgCl_2$ 4, $CaCl_2$ 0.062, Na_2 creatine phosphate 5, HEPES 10, Na_2ATP 3.1, and Na_2GTP 0.42; at pH 7.1 (adjust with KOH).
3. Peptide injection should be over a period of 5 to 10 min to allow adequate diffusion before further experimentation *(52)*.

3.2.2. Transient Permeabilization Protocol for PKC Peptides

Johnson and coworkers *(57)* developed a permeabilization protocol that uses saponin and ATP to effectively introduce peptides into neonatal cardiac myocytes. This technique allows for peptide delivery to a greater number of cells and for a higher percentage of cell survival. It is effective for introduction of fragments up to 25 kDa in size, and the intracellular molar concentration achieved is approx 10% of that applied extracellularly. The following permeablilzation protocol can be applied to plated cardiac myocytes, if other cell types are used conditions may vary.

1. Remove culture media and save at 37°C.
2. Saline (PBS, mM; $CaCl_2$ 1, $MgCl_2 \cdot 6H_2O$ 0.4, KCl 2.7, KH_2PO_4 0.15, $NaH_2PO_4 \cdot 7H_2O$ 8, NaCl 136). Prepare sterile stocks of PBS at room temperature and on ice.
   (i) Apply PBS at room temperature for 2 min then aspirate. Use 2 mL for a 35-mm plate, 5 mL for 60-mm plate, and 7 mL for a 100-mm plate.
   (ii) Apply chilled PBS to plate on ice for 2 min and then aspirate.
3. Apply chilled permeabilization buffer (1X concentration, with or without peptide) to plate on ice, for 10 min. Apply only enough buffer to cover cell layer (1 mL for

35-mm plate, 1.2 mL for 60-mm plate, 3 mL for 100-mm plate). Prepare buffer as described. (1) Freshly prepare a 2× concentration of sterile permeabilization buffer containing (mM): HEPES 40 (pH 7.4), EGTA 20, KCl 280, saponin 100 µg/mL (Sigma), sodium azide 10, and oxalic acid dipotassium salt 10. Omit sodium azide for studies assessing *c-fos* mRNA or 14C-phenylalanine accumulation. (2) Prepare stock of 200 mM ATP, pH 7.4. (3) Permeabilization buffer should be diluted to ×1 concentration with application of peptide (100 nM to 10 µM), 30 µL of ATP stock/mL buffer (final concentration 6 mM, add immediately before buffer application) and deionized water. Buffer should be chilled on ice before use.

4. Aspirate all buffer following incubation and gently wash cells on ice with chilled PBS, four times. Re-apply chilled PBS (fifth application following washes) to cells on ice for 20 min.
5. Aspirate PBS. Re-apply PBS at room temperature, and leave cells at room temperature for 2 min.
6. Aspirate PBS. Re-apply PBS at 37°C and incubate cells at 37°C for 2 min.
7. Aspirate PBS. Re-apply original media at 37°C and incubate cells at 37°C for 15 to 30 min before commencement of experimental study.

Several points are critical to the success of this procedure, these include ensuring culture plates and solutions are chilled on ice to avoid exposure to room temperature, which will encourage proteolysis and cell degradation; the gradual temperature change; inclusion of ATP in the buffer to improve saponin permeablilization; and inclusion of ascorbic acid (80 µM) in the media of cardiac myocytes (*see* **Note 5**).

3.3.3. Peptide Transporters

Peptide transporters are able to translocate across the plasma membrane and deliver large (~100 kDa) biologically active proteins into cells *(58)*. Peptides are cross-linked to these transporters via an N-terminal cysteine-cysteine bond that is then reduced upon entry into the cell. Examples of transporters utilized to carry PKC peptides include: (1) *Drosophila* Antennapedia homeodomain-derived peptide *(59–61)*; (2) TAT-derived peptide *(49,62–64)*; and (3) poly-arginine (R7 or r7; **ref. *65***). Chemical generation of carrier-peptide conjugates and their validation have been described previously at length in the references listed, and will not be included in this section. Rather, a brief overview of carrier-PKC peptide usage is provided.

The Antennapedia-derived peptide sequence C-RQIKIWFQNRRMKWKK (also termed penetratin; **refs. *44–59***), which is identical to the third loop of its homeodomain, can translocate through biological membranes *(49,59–61,64)*. This conjugate has been used to introduce PKC peptides into adult cardiac myocytes *(66,67)*. After adding carrier-peptide (0.1–1 µM) conjugate to media for 20 min at 37°C, intracellular levels of PKC peptide will not exceed 10% of the applied concentration, and a high proportion of cells exposed will contain peptide. Subsequently, the HIV TAT-derived peptide sequence YGRKKRRQRRR *(48–58)* was shown to have similar efficiency to that of Antennapedia in delivery of PKC peptides to cardiac myocytes *(62–64)*. Control experiments should include use of a carrier-carrier dimer or carrier-scrambled peptide conjugates. Entry of carrier conjugates into the cell can be followed by fluorescent labeling of the carrier.

Poly-arginine or R7, a short oligomer of arginine, was recently shown be more efficient at entering cells than other polycationic homopolymers *(65)*. This was confirmed using cardiac myocytes, where R7 was shown to be superior to the above-mentioned carrier peptides, with respect to improved peptide delivery and reduced cell damage, when used at 100 nM for 15 min *(64)*. Interestingly, the R7-peptide conjugate was also effective when administered to isolated rat heart, using retrograde perfusion via coronary arteries for 20 min at 500 nM *(64)*. Thus demonstrating R7 can transport peptides through blood vessels to nearby target organs.

3.3. Antisense Oligonucleotides for PKC Isoforms

Phosphorothioate antisense or sense oligonucleotides for PKC isoforms may be synthesized based on the sequence of initial codons of the PKC isoforms. Certain PKC isoforms (e.g., α, β, ϵ, γ, δ) may be obtained commercially from Perkin Elmer (Wellesley, MA) or Biognostic GmbH (Goettingen, Germany).

Antisense of PKCα, ISIS-3521, is used here as an example. The single-chain DNA is soluble in water or medium. Dosages of antisenses for PKCs may vary between types of PKC isoforms and cells.

1. A preincubation of 2–25 μM phosphorothioate antisense or sense of PKC-α (5′-GTCCCTCGCCTCCTG-3′ and 5′-GTCCTCCGCCGCTCCCTG-3′) with 10 μg/mL of lipofectin for 30 min at room temperature.
2. The mixture will be added to cultured cells for an incubation of 6 h at 37°C. After washing cells with medium to remove lipofectin.
3. The incubation may be continued with serum-containing medium with the presence of the same concentration of antisense or sense oligonucleotides for a required period *(70,71)*.

3.4. Adenovirus Expressing Negative Dominants of PKC Isoforms

Adenovirus vectors have been used often successfully as a delivery system in gene therapy in cell cultures, animal studies, and even in humans. Adenoviruses are large, 80–90 nm in diameter, viruses with linear double-stranded DNA and they posses a protein capsid. In humans, adenoviruses can cause infections of the upper respiratory tract, kerato-conjuctivitis, gastroenteritis, pneumonia, bronchitis, hepatitis and cystitis. They cause mild, self-limiting illnesses.

The advantages of using adenovirus as a delivery vector include the following: (1) they can infect a large number of both dividing and nondividing cells, (2) they can be used for different (human) cell types, (3) it is relatively easy to prepare a huge amount of virus with high titers ($<10^{11}$ pfu/mL), potentially, the size of inserted DNA can be up to 35 kb, and (4) the efficiency of infection is high (often nearly 100%). Furthermore, the genome of common adenoviruses are completely known and the genome can be easily modified. The most commonly used adenovirus in gene therapy is adenovirus type 2 and 5, both belonging to subgroup C. The adenoviral genome consists of four early transcriptional regions (E1–E4) and one major late transcriptional region (L1–L5).

To introduce the gene of interest into the viral gene, one or two regions (E1, E2, E3, and/or E4) have to be deleted to make place. Commonly, the E1- and E3-deleted adenoviruses are used in experiments for gene expression. The E1 deletion is to render the virus replication-defective, and the E3 deletion is mainly to create space for the insertion of the transgene. The E3 region is important to suppress the host immune response during virus infection, but is not required for replication or packaging of the virus in vitro. Furthermore, the promotor for E3 region requires E1 products; therefore the E3 region in E1-deleted cells is not relevant. However, both E1- and E3-region deleted vectors elicited strong inflammatory and immune responses in vivo. The adenovirus enters the host cell through receptor-mediated endocytosis via clathrin-coated vesicles into endosomes. After acidification of the endosome, the virion is released within the cytoplasm and transported, probably actively on microtubules, to the nuclear membrane. The DNA and terminal protein are internalized by an unknown mechanism into the host cell nucleus. The adenovirus starts dominating the protein synthetic machinery of the host cell and promotes translation of its own transcripts, while suppressing the translation of the host cell genes. The viral DNA does not integrate into the host cell DNA, but will stay as an episomal gene and replicates extrachromosomally under control of a replication vector. The viral gene may not segregate to every daughter cell during mitosis, and the viral gene will get lost in long-term experiments with fast-dividing cells.

Several groups have been using dominant negative PKC isoform adenoviruses. Kinase-dead PKC isoform adenoviruses are constructed by changing the ATP-binding site of the genome, a point mutation is commonly used. Resulting in the lack of ATP binding, and therefore the phosphorylated PKC isoforms is not able to stimulate the next step in the signaling cascade. **Table 5** shows an overview of the different mutations used for kinase-dead PKC adenoviruses.

Table 5
Overview of Dominant Negative PKC Adenoviruses Used
in Studies to Inhibit PKC

PKC Isoform	Mutation	Cell type	Reference
α	K368R, ATP-binding site	NIH3T3 fibroblasts, Parotid C5 cells, BAEC, SMC, BREC, BRPC	*72–76*
	K368A, ATP-binding site	CHO	*77*
βI	K371M	COS-7 cells	*78*
	T500V, Activation loop	BAECs, SMCs, BRECs	*76*
βII	T500V, Activation loop	BAECs, SMCs, BRECs	*76*
γ	K273E	CHO cells, 3T3-L1 fibroblasts	
	K275W	NIH 3T3 fibroblasts	
δ	K376R, ATP-binding site	Primary mouse keratinocytes, LNCaP Prostate cancer cells, Parotid C5 cells	*79–81*
	K367A, ATP-binding site	MDCK II epithelial cells, Primairy cultured Skeletal Cells	*82*
	K378R, ATP-binding site	BAEC, SMC, BREC, BRPC	*76*
		Rat Pheochromocytoma PC12 cells, Rat skeletal cells	*83,84*
ε	K436R, ATP-binding site	NIH 3T3 fibroblasts	*85*
	K427R, ATP-binding site	BAEC, SMC, BREC, BRPC	*76*
	K437M, ATP-binding site		
	T566A	CHO cells	*77*
	K436R and A159E, ATP-binding site and pseudosubstrate domain	Rabit cardiomyocytes, Rat ventricular myocytes	*86,87*
η	K384A	Human keratinocytes	*88,89*
θ	K409R	NIH 3T3 fibroblasts	*74,75*
		Murine EL4 thymoma cells	*75*
λ	K273E, Lacks NH2 terminal	HL1C cells; CHO cells; 3T3-L1 adipocytes; L6 myotubes,	*77,90–92*
	K275W	NIH3T3 fibroblasts	*73,74*
		MDCK II epithelial cells	*93*
ζ	K273W, ATP-binding site	BAEC, BREC, BRPC, SMC	*76,94,95*
	K273E, ATP-binding site	L6 myoblasts	*77*
	K275W, ATP-binding site	NIH3T3 fibroblasts	*73*
	K281W, ATP-binding site	MDCK II epithelial cells	*93*
	K282R, ATP-binding site	Hepatoma Fao cells	*96*
		Rat skeletal muscle (myotubes)	*84*

This protocol describes the methods for the use of adenoviruses made with the AdEasy System (Stratagene). Shortly, construction of this adenovirus is as follows. The PKC isoform gene is first cloned into a shuttle vector, the pAdTrack-CMV. The plasmid is then cotransformed into *Escherichia coli* BJ5183 cells with an adenovirus backbone plasmid. Selection of the recombinants takes place with the use of the kanamycin resistance gene. Finally, the linearized recombinant plasmid is transfected into E1-transformed HEK293 cells, an adenovirus packaging cell line. The constructed adenovirus, containing the PKC isoform, is type 5 with deletion of regions E1 and E3. The promoter is Cytomegalovirus and the adenovirus backbone posses a green fluorescence protein. When the adenovirus-infected cells are excited by blue light or UV light, the cells will yield a bright green fluorescence. The green fluorescence is useful for the detection of the adenovirus infection.

Potentially, other adenoviruses constructs can be used, but might need minor changes in the protocol. 293HEK cells are used for the viral replication.

3.4.1. Adenovirus Amplification

1. Culture 293HEK cells to 80% confluency in 100-mm Petri dishes in DMEM-H containing 10% FBS (*see* **Note 6**).
2. Wash cells with serum-free DMEM-H and aspirate medium.
3. Add 1 mL of viral solution and incubate for 1–2 h. Rocking plates by hand every 15 min to prevent cells from drying out.
4. Add 7 mL of DMEM-H containing 10% FBS and incubate for 3–5 d until most cells detached from culture plate.
5. Collect all cells and media (total 8 mL), by pipetting up and down, into vial.
6. Freeze and thaw cells four times. Freezing: use methanol with dry ice and freeze solution for 30 min. Thaw at room temperature, shake/vortex solution during thawing (*see* **Note 7**).
7. Centrifuge solution at 3000*g*, 5 min, collect supernatant, and discard pellet.
8. Culture 293HEK cells to 80% confluency in 150-mm Petri dishes in DMEM-H containing 10% FBS. Wash cells with serum-free medium, aspirate medium and add 8 mL viral solution to the plate. After 1–2 h, add 12 mL of medium. Follow the same methods from **Notes 1**, **2**, **5**, **6**, and **8**. The total volume of viral solution will be 20 mL after this two-step method (*see* **Note 9**).
9. Keep final solution at –80°C, aliquoted into 1 mL vials.

3.4.2. Concentrating Adenovirus for In Vivo Infection

To use adenovirus in vitro, you need to scale up and concentrate the virus to end up with a high titer. To infect mice with adenovirus, commonly a total amount of 3×10^8 pfu per gram mouse body weight is used. Therefore, using a volume of 200 μL, an adenovirus solution concentration around 10^{10}–10^{12} pfu/mL is necessary. To achieve this concentration, you have to concentrate the viral solution by the Cesium Chloride procedure.

1. Amplify the adenovirus as described above. Most experiments require at least 400 mL per adenovirus.
2. Collect the medium, containing adenovirus, into 250-mL centrifuge bottles. Add 0.5% NP-40.
3. Shake gently 10 min at room temperature.
4. Centrifuge 15 min at 2000g and transfer supernatant to new bottle.
5. Add 0.5 volumes of 20% polyethylene glycol/2.5 M NaCl.
6. Shake bottles 30 min (or O/N) on ice.
7. Centrifuge 15 min for 2000g and discard supernatant.
8. Resuspend the pellet in 1.25 mL of PBS (*see* **Note 10**). Transfer to microfuge tube and centrifuge 10 min for 10,000g.
9. Take supernatant and pool into 15-mL conical tube.
10. Add about 0.55 g of cesium chloride (CsCl) per mL of viral suspension to final density (weight) of 1.34 g/mL by adding either PBS or CsCl (*see* **Note 11**).
11. Transfer to centrifuge tube and spin in Ultracentrifuge 20 h at 50,000g at room temperature.
12. Collect the densest white band (*see* **Note 12**).
    Optional, second CsCl step gradient
    a. Put virus solution and saturated CsCl, total 5 mL, into tube for SW41.
    b. Add 2mL of 4 M CsCl in 10 mM HEPES.
    c. Add 4 mL of 2.2 M CsCl in 10 mM HEPES.
    d. Centrifuge at 10,000g for 180 min at 4°C.
    e. Aspirate the virus solution, just below 2.2 M CsCl layer (top layer), using a capillary pipet.
13. Desalt the virus overnight by using a dialyzer in PBS. Change buffer 2 h after starting.
14. Store aliquots in –80°C. Viral solution can be kept for weeks to months.

3.4.3. Titering Virus

1. Culture HEK293 cells into a 12-well plate until 80–90% confluent (*see* **Note 13**).
2. Dilute viral solution using DMEM-H (w/out serum) to 10^{-3}, 10^{-4}, 10^{-5}, and 10^{-6}. Keep the virus solution on ice.
3. Aspirate media from 12-well plate, apply 500 μL of viral solution (different concentrations) to wells. Leave for 1 h in incubator and rock plate every 20 min.
4. Add 1 mL of DMEM-H with 10% FBS after 1 h.
5. Count fluorescing cells 19 h after start infection (18 h after adding medium) in wells infected with 10-4 diluted virus solution. Take the average of four fields at 10× magnification (*see* **Note 14**).
6. Titer (pfu/mL) = Average count × 7.8×10^{-6} (*see* **Note 15**).

3.4.4. Infect Cells with Adenovirus

Methods to infect your cells of interest with PKC isoform adenoviruses may vary. The standard protocol is given below, but adenovirus volume, duration of incubation, titer of the adenovirus and the use of the washing step, can be changed in order to get the optimal infection rate. Use non-PKC isoform-GFP-containing control adenovirus and noninfected cells as control in all experiments.

1. Culture 100-mm Petri dish with cells, confluency, depending on experiment, around 80%.
2. Aspirate media from cells (optional: wash with serum-free medium).
3. Infect cells, using a volume of 200 to 500 μL, with the viral solution diluted to titer 1×10^7 pfu/mL.
4. Incubate cells 1–2 h with the adenovirus, rocking the plate every 15 min.
5. Add medium (including serum) to final volume.
6. After 24–48 h, wash cells and check percentage of infection under microscope using blue light to detect the green fluorescence.
7. Start starvation cells, if necessary for experiment, and stimulate/harvest cells at day 2–3 after infection.

4. Notes

1. E1-deleted adenoviruses are very useful for amplifying because they complement the growth of E1-deleted adenovirus vectors.
2. To make this salt, first make 5 *M* NaCl, add polyethylene glycol, dissolve, and adjust the volume. This can take about 2 h.
3. Staurosporine is light sensitive. The solution is stable for months at 4°C.
4. LY333531 is a reversible competitive-inhibitor of ATP binding to PKC. The reversibility of its actions should be kept in mind, especially when subsequent experimental techniques that employ vigorous washing steps are used. This includes the classic assay of PKC activity in cytosol and membrane fractions, in which LY333531 may be inadvertently removed during the isolation procedure. In addition, it should be kept in mind that although the effectiveness of LY333531 can be determined by its ability to prevent phosphorylation of PKCβ, it does not prevent PKC translocation, thus the effectiveness of this compound cannot be assessed by this assay.
5. When using techniques that involve the disruption of the plasma membranes, it is essential to ensure cell integrity is not compromised. Parameters that should be checked and not altered especially when using cardiac myocytes include contraction rate; gene expression, such as *c-fos* mRNA; protein synthesis,; cell viability; and PKC isoform localization. Sham procedures should also be performed as controls.
6. 293HEK cells are very sensitive. Split cells at 80–95% confluent because cells will die when passing the 100% confluency. Split ratio should be 1 : 3–4. Do not

let the cells dry out. The above rules will avoid the reduction in ability of the cells to package the virus correctly.

7. Do not leave virus solution in the 37°C waterbath to prevent killing the virus. Alternatively, you can sonicate the cells for 10 min instead of the freeze–thawing cycle.
8. While using adenoviruses, work according to Biosafety level 2 practice. E1-deleted adenoviruses cannot replicate, however, mutations in the genome could potentially lead to adenoviruses capable to replicate. Therefore, keep all viral waste separate and autoclave before disposing. Work in a hood approved for Biosafety level 2 when using the virus. And use a separate incubator to prevent virus infection into other cells.
9. This two-step procedure is necessary to keep the titer of the adenovirus.
10. Try to collect the complete pellet because this includes the adenovirus. If necessary, you can increase the volume of the PBS. But, be aware that the titer will be reduced, by increasing the volume.
11. First weight 1 mL to virus solution, add CsCl or PBS depending on the weight to end up with the final density of 1.34 g/mL.
12. Multiple bands can be formed from damaged virus.
13. You need four wells for each virus, including noninfected control wells.
14. Plan your infection in the afternoon, to prevent coming back in the middle of the night.
15. To count titers in the order of 10^8 pfu/mL, count the wells with 10^{-4} diluted adenovirus. For titers in the order of 10^{10}, count the wells with 10^{-6} diluted adenovirus, and adjust the wells (make virus dilutions 10^{-4}, 10^{-5}, 10^{-6}, and 10^{-7}).

Acknowledgments

The authors thank for the support of NIH, grants ROI EY5110 to G.L.K. Iacocca Visiting Scientist Award, operating grants from CIHR and CDA for G.X.S. K.J.W. is supported by a postdoctoral fellowship from the American Diabetes Association and the Juvenile Diabetes Foundation International.

References

1. Inoue, M., Kishimoto, A., Takai, Y., and Nishizuka, Y. (1977) Studies on a cyclic nucleotide-independent protein kinase and its proenzyme in mammalian tissues. *J. Biol. Chem.* **232,** 7610–7616.
2. Mochly-Rosen, D. and Kauvar L. M. (1998) Modulating protein kinase C signal transduction. *Adv. Pharmacol.* **44,** 91–147.
3. Kawamoto, S. and Hidaka, H. (1984) 1-(5-isoquinolinesulfonyl)-s-methy;piperazine (H-1) is a selective inhibitor of protein kinase in rabbit platelets. *Biochem. Biophys. Res. Commun.* **125,** 258–264.
4. Shen, X. Y., Hamilton, T. A., and DiCorleto, P. E. (1989) Lipopolysaccharide-induced expression of the competence gene KC in vascular endothelial cells is mediated through protein kinase C. *J. Cell. Physiol.* **140,** 44–51.

5. Tamaoki, T., Nomoto, H., Takahashi, I., Kato, Y., Morimoto, M., and Tomita, F. (1986) Staurosporine, a potent inhibitor of phospholipid/Ca++ dependent protein kinase. *Biochem. Biophys. Res. Commun.* **135**, 397–402.

6. Casnellie, J. E. (1991) Protein kinase inhibitors: probes for the functions of protein phosphorylation. *Adv. Pharmacol.* **22**, 167–205.

7. Buchholz, R. A., Dundore, R. L., Cumiskey, W. R., Harris, A. L., and Silver, P. J. (1991) Protein kinase inhibitors and blood pressure control in spontaneously hypertensive rats. *Hypertension* **17**, 91–100.

8. Seynaeve, C. M., Stetler-Stevenson, M., Sehers, S., Kaur, G., Sausville, E. A., and Worland, P. J. (1993) Cell cycle arrest and growth inhibition by the protein kinase antagonist UCN-01 in human breast carcinoma cells. *Cancer Res.* **51**, 4888–4892.

9. Meyer, T., Regenass, U., Fabbro, D., Alteri, E., Rosel, J., Muller, M., et al. (1989) A derivative of staurosporine (CGP 41251) shows selectivity for protein kinase C inhibition and in vitro anti-tumor activity. *Int. J. Cancer* **43**, 851–856.

10. Kobayashi, E., Nakano, H., Morimoto, M., and Tamaoki, T. (1989) Calphostin C (UCN-1028C), a novel microbial compound, is a highly potent and specific inhibitor of protein kinase C. *Biochem. Biophys. Res. Commun.* **159**, 548–553.

11. Kedderis, L. B., Borigian, H. P., Kleeman, J. M., Hall, R. L., Palmer, T. E., Harrison, S. D., Jr., et al. (1995) Toxicity of the protein kinase C inhibitor safingol administered alone and in combination with chemotherapeutic agents. *Fund. Appl. Toxicol.* **25**, 201–217.

12. Mizuno, K., Saido, T. C., Ohno, S., Tamaoki, T., and Suzuki, K. (1993) Staurosporine-related compounds, K252a and UCN-01, inhibit both cPKC and nPKC. *FEBS Lett.* **330**, 114–116.

13. Gschwendt, M., Muller, H. J., Kielbassa, K., Zang, R., Kittstein, W., Rincke, G., et al. (1994) Rottlerin, a novel protein kinase inhibitor. *Biochem. Biophys. Res. Commun.* **199**, 93–98.

14. Mandil, R., Ashkenazi, E., Blass, M., Kronfeld, I., Kazimirsky, G., Rosenthal, G., et al. (2001) Protein kinase Calpha and protein kinase Cdelta play opposite roles in the proliferation and apoptosis of glioma cells. *Cancer Res.* **61**, 4612–4619.

15. Jirousek, M. R., Gillig, J. R., Gonzalez, C. M., Heath, W. F., McDonald, J. H. 3rd, Neel, D. A., et al. (1996) (S)-13-[(dimethylamino)methyl]-10,11,14,15-tetrahydro-4,9:16, 21-dimetheno-1H, 13H-dibenzo[e,k]pyrrolo[3,4-h][1,4,13]oxadiazacyclohexadecene-1,3(2H)-d ione (LY333531) and related analogues: isozyme selective inhibitors of protein kinase C beta. *J. Med. Chem.* **39**, 2664–2671.

16. Ren, S., Shatadal, S., and Shen, G. X. (2000) Protein kinase C-β mediates lipoprotein-induced generation of PAI-1 from vascular endothelial cells. *Am. J. Physiol.* **278**, E656–E662.

17. Yedovitzky, M., Mochly-Rosen, D., Johnson, J. A., Gray, M. O., Ron, D., Abramovitch, E., et al. (1997) Translocation inhibitors define specificity of protein kinase C isoenzymes in pancreatic beta-cells. *J. Biol. Chem.* **272**, 1417–1420.

18. Wu, H. L., Albrightson, C., and Nambi, P. (1999) Selective inhibition of rat mesangial cell proliferation by a synthetic peptide derived from the sequence of the C2 region of PKC beta. *Peptides* **20**, 675–678.

19. Farese, R. V., Standaert, M. L., Ishizuka, T., Yu, B., Hernandez, H., Waldron, C., et al. (1991) Antisense DNA downregulates protein kinase C isozymes (beta and alpha) and insulin-stimulated 2-deoxyglucose uptake in rat adipocytes. *Antisense Res. Dev.* **1**, 35–42.
20. Sioud, M. and Sorensen, D. R. (1998) A nuclease-resistant protein kinase C alpha ribozyme blocks glioma cell growth. *Nat. Biotechnol.* **16**, 556–561.
21. Igarashi, M., Wakasaki, H., Takahara, N., Ishii, H., Jiang, Z. Y., Yamauchi, T., et al. (1999) Glucose or diabetes activates p38 mitogen-activated protein kinase via different pathways. *J. Clin. Invest.* **103**, 185–95.
22. Beckman, J. A., Goldfine, A. B., Gordon, M. B., Garrett, L. A., and Creager, M. A. (2002) Inhibition of protein kinase Cbeta prevents impaired endothelium-dependent vasodilation caused by hyperglycemia in humans. *Circ. Res.* **90**, 107–111.
23. Meier, M. and King, G. L. (2000) Protein kinase C activation and its pharmacological inhibition in vascular disease. *Vasc. Med.* **5**, 173–185.
24. Li, K., Zhang, J., Sirois, P., and Hu, Y. (1999) Technology evaluation: ISIS-3521. *Curr. Opin. Mol. Ther.* **1**, 393–398.
25. Wilson, W. H., Sorbara, L., Figg, W. D., Mont, E. K., Sausville, E., Warren, K. E., et al. (2000) Modulation of clinical drug resistance in a B cell lymphoma patient by the protein kinase inhibitor 7-hydroxystaurosporine: presentation of a novel therapeutic paradigm. *Clin. Cancer Res.* **6**, 415–421.
26. Thavasu, P., Propper, D., McDonald, A., Dobbs, N., Ganesan, T., Talbot, D., et al. (1999) The protein kinase C inhibitor CGP41251 suppresses cytokine release and extracellular signal-regulated kinase 2 expression in cancer patients. *Cancer Res.* **59**, 3980–3984.
27. Schwartz, G. K., Ward, D., Saltz, L., Casper, E. S., Spiess, T., Mullen, E., et al. (1997) A pilot clinical/pharmacological study of the protein kinase C-specific inhibitor safingol alone and in combination with doxorubicin. *Clin. Cancer Res.* **3**, 537–543.
28. Thavasu, P., Propper, D., McDonald, A., Dobbs, N., Ganesan, T., Talbot, D., et al. (1999) The protein kinase C inhibitor CGP41251 suppresses cytokine release and extracellular signal-regulated kinase 2 expression in cancer patients. *Cancer Res.* **59**, 3980–3984.
29. Bruns, R. F., Miller, F. D., Merriman, R. L., Howbert, J. J., Heath, W. F., Kobayashi, E., et al. (1991) Inhibition of protein kinase C by calphostin C is light-dependent. *Biochem. Biophys. Res. Commun.* **176**, 288–293.
30. Kobayashi, E., Nakano, H., Morimoto, M., and Tamaoki, T. (1989) Calphostin C (UCN-1028C), a novel microbial compound, is a highly potent and specific inhibitor of protein kinase C. *Biochem. Biophys. Res. Commun.* **159**, 548–553.
31. Hannun, Y. A. and Bell, R. M. (1987) Lysosphingolipids inhibit protein kinase C: implications for the sphingolipidoses. *Science* **235**, 670–674.
32. Kedderis, L. B., Bozigian, H. P., Kleeman, J. M., Hall, R. L., Palmer, T. E., Harrison, S, D. Jr., et al. (1995) Toxicity of the protein kinase C inhibitor safingol administered alone and in combination with chemotherapeutic agents. *Fund. Appl. Toxicol.* **25**, 201–217.

33. Siemann, D. W., Jiang, J. B., Ballas, L., and Janzen, W. (1993) Threodihydro-sphingosine potentiates the in vivo antitumor efficacy of cisplatin and Adriamycin. *Proc. Am. Assoc. Cancer. Res.* **34,** 411.

34. Schwartz, G. K., Ward, D., Saltz, L., Casper, E. S., Spiess, T., Mullen, E., et al. (1997) A pilot clinical/pharmacological study of the protein kinase C-specific inhibitor safingol alone and in combination with doxorubicin. *Clin. Cancer Res.* **3,** 537–43.

35. Ishii, H., Jirousek, M. R., Koya, D., Takagi, C., Xia, P., Clermont, A., et al. (1996) Amelioration of vascular dysfunctions in diabetic rats by an oral PKC beta inhibitor. *Science* **272,** 728–731.

36. Koya, D., Jirousek, M. R., Lin, Y.-W., Ishii, H., Kuboki, K., and King, G. L. (1997) Characterization of protein kinase C β isoform activation on the gene expression of transforming growth factor-β, extracellular matrix components, and prostanoids in the glomeruli of diabetic rats. *J. Clin. Invest.* **100,** 115–126.

37. Koya, D., Haneda, M., Nakagawa, H., Isshiki, K., Sato, H., Maeda, S., et al. (2000) Amelioration of accelerated diabetic mesangial expansion by treatment with a PKC β inhibitor in diabetic db/db mice, a rodent model for type 2 diabetes. *FASEB J.* **14,** 439–447.

38. Danis, R. P., Bingaman, D. P., Jirousek, M., and Yang, Y. (1998) Inhibition of intraocular neovascularization caused by retinal ischemia in pigs by PKCβ inhibition with LY333531. *Invest. Ophthalmol. Vis. Sci.* **39,** 171–179.

39. Aiello, L. P., Bursell, S. E., Devries, T., Alatorre, C., King, G. L., and Ways, K. (1999) Protein kinase C β-selective inhibitor LY333531 ameliorates abnormal retinal hemodynamics in patients with diabetes. *Diabetes* **48,** A19.

40. Aiello, L. P., Bursell, S. E., Devries, T., Alatorre, C., King, G. L., and Ways, D. K. (1999) Amelioration of abnormal retinal hemodynamics by a protein kinase C β selective inhibitor (LY333531) in patients with diabetes: results of a Phase 1 safety and pharmacodynamic clinical trial. *IOVS* **40,** S192.

41. Nakamura, J., Kato, K., Hamada, Y., Nakayama, M., Chaya, S., Nakashima, E., et al. (1999) A protein kinase C-β-selective inhibitor ameliorates neural dysfunction in streptozotocin-induced diabetic rats. *Diabetes* **48,** 2090–2095.

42. Disatnik, M. H., Buraggi, G., and Mochly-Rosen, D. (1994) Localization of protein kinase C isozymes in cardiac myocytes. *Exp. Cell Res.* **210,** 287–297.

43. Mochly-Rosen, D., Khaner, H., and Lopez, J. (1991) Identification of intracellular receptor proteins for activated protein kinase C. *Proc. Natl. Acad. Sci. USA* **88,** 3997–4000.

44. Mochly-Rosen, D. and Gordon, A. S. (1998) Anchoring proteins for protein kinase C: a means for isozyme selectivity. *FASEB J.* **12,** 35–42.

45. Csukai, M., and Mochly-Rosen, D. (1999) Pharmacologic modulation of protein kinase C isozymes: the role of RACKs and subcellular localization. *Pharmacol. Res.* **39,** 253–259.

46. Mochly-Rosen, D. (1995) Localization of protein kinases by anchoring proteins: A theme in signal transduction. *Science* **268,** 247–251.

47. Stebbins, E. G. and Mochly-Rosen, D. (2001) Binding specificity for RACK1 resides in the V5 region of βII protein kinase C. *J. Biol. Chem.* **276**, 29,644–29,650.
48. Souroujon, M. C. and Mochly-Rosen, D. (1998) Peptide modulator of protein–protein interactions in intracellular signaling. *Nat. Biotech.* **16**, 919–924.
49. Chen, L., Hahn, H., Wu, G., Chen, C.-H., Liron, T., Schechtman, D., et al. (2001) Opposing cardioprotective actions and parallel hypertrophic effects of δPKC and εPKC. *Proc. Natl. Acad. Sci. USA* **98**, 11,111–11,119.
50. Schechtman, D. and Mochly-Rosen, D. (2002) Isozyme-specific inhibitors and activators of protein kinase C. *Methods Enzymol.* **345**, 470–489.
51. Mackay, K. and Mochly-Rosen, D. (2001) Localization, anchoring, and functions of protein kinase C isozymes in the heart. *J. Mol. Cell. Cardiol.* **33**, 1301–1307.
52. Zhang, Z.-H., Johnson, J. A., Chen, L., El-Sherif, N., Mochly-Rosen, D., and Boutjdir, M. (1997) C2 Region-derived peptides of beta-protein kinase C regulates cardiac Ca2+ channels. *Circ. Res.* **80**, 720–729.
53. Gray, M. O., Karliner, J. S., and Mochly-Rosen, D. (1997) A selective ε-protein kinase C antagonist inhibits protection of cardiac myocytes from hypoxia-induced cell death. *J. Biol. Chem.* **272**, 30,945–30,961.
54. Liu, G. S., Cohen, M. V., Mochly-Rosen, D., and Downey, J. M. (1999) Protein kinase C-ε is responsible for the protection of preconditioning in rabbit cardio-myocytes. *J. Mol. Cell Cardiol.* **31**, 1937–1948.
55. Johnson, J. A., Gray, M. O., Chen, C.-H., and Mochly-Rosen, D. (1996) A protein kinase C translocation inhibitor as an isozyme-selective antagonist of cardiac function. *J. Biol. Chem.* **271**, 24,962–24,966.
56. Smith, B. L. and Mochly-Rosen, D. (1992) Inhibition of protein kinase C function by injection of intracellular receptors for the enzyme. *Biochem. Biophys. Res. Commun.* **188**, 1235–1240.
57. Johnson, J. A., Gray, M. O., Karliner, J. S., Chen, C.-H., and Mochly-Rosen, D. (1996) An improved permeabilization protocol for the introduction of peptides into cardiac myocytes: application to protein kinase C research. *Circ. Res.* **79**, 1086–1099.
58. Lindgren, M., Hallbrink, M., Prochiantz, A., and Langel, U. (2000) Cell-penetrating peptides. *Trends Pharmacol. Sci.* **21**, 99–103.
59. Derossi, D., Joliot, A. H., Chassaing, G., and Prochiantz, A. (1994) The third helix of the Antennapedia homeodomain translocates through biological membranes. *J. Biol. Chem.* **269**, 10,444–10,450.
60. Theordore, L., Derossi, D., Chassaing, G., Llirbat, B., Kubes, M., Jordon, P., et al. (1995) Intraneuronal delivery of protein kinase C pseudosubstrate leads to growth cone collapse. *J. Neurosci.* **15**, 7158–7167.
61. Derossi, D., Calvet, S., Trembleau, A., Brunissen, A., Chassaing, G., and Pro-chiantz, A. (1996) Cell internalization of the third helix of the Antennapedia homeodomain is receptor-independent. *J. Biol. Chem.* **271**, 18,188–18,193.
62. Vives, E., Brodin, P., and Lebleu, B. (1997) A trancated HIV-1 Tat protein basic domain rapidly translocates throught the plasma membrane and accumulates in the cell nucleus. *J. Biol. Chem.* **272**, 16,010–16,017.

63. Schwarze, S. R., Ho, A., Vocero-Akbani, A., and Downy, S. F. (1999) In vivo protein transduction: delivery of a biologically active protein into the mouse. *Science* **285,** 1569–1572.

64. Chen, L., Wright, L. R., Chen, C.-H., Oliver, S. F., Wender, P. A., and Mochly-Rosen, D. (2001) Molecular transporters for peptides: delivery of a cardioprotective εPKC agonist peptide into cells and intact ischemic heart using a transport system, R_7. *Chem. Biol.* **8,** 1123–1129.

65. Mitchell, D. J., Kim, D. T., Steinman, L., Fathman, C. G., and Rothbard, J. B. (2000) Polyarginine enter cells more efficiently than other polycationic homopolymers. *J. Peptide Res.* **56,** 318–325.

66. Chen, C.-H., Gray, M. O., and Mochly-Rosen, D. (1999) Cardioprotection from ischemia by a brief exposure to physiological levels of ethanol: role of epsilon protein kinase C. *Proc. Natl. Acad. Sci. USA* **96,** 12,784–12,789.

67. Dorn, G. W. II, Souroujon, M. C., Liron, T., Chen, C.-H., Gray, M. O., Zhou, H. Z., et al. (1999) Sustained *in vivo* cardiac protection by a rationally designed peptide that causes ε protein kinase C translocation. *Proc. Natl. Acad. Sci. USA* **96,** 12,798–12,803.

68. Ron, D., Luo, J., and Mochly-Rosen, D. (1995) C2 region-derived peptides inhibit translocation and function of β protein kinase C *in vivo. J. Biol. Chem.* **270,** 24,180–24,187.

69. Mayne, G. C. and Murray, A.W. (1998) Evidence that protein kinase Cε mediates phorbol ester inhibition of calphostin C- and tumor necrosis factor-α-induced apoptosis in U937 histiocytic lymphoma cells. *J. Biol. Chem.* **273,** 2411–24121.

70. Xia, P., Aiello, L. P., Ishii, H., Jiang, Z. Y., Park, D. J., Robinson, G. S., et al. (1996) Characterization of vascular endothelial growth factor's effect on the activation of protein kinase C, its isoforms, and endothelial cell growth. *J. Clin. Invest.* **98,** 2018–2026.

71. Bogatkevich, G. S., Tourkina, E., Silver, R. M., and Ludwicka-Bradley, A. (2001) Thrombin differentiates normal lung fibroblasts to a myofibroblast phenotype via the proteolytically activated receptor-1 and a protein kinase C-dependent pathway. *J. Biol. Chem.* **276,** 45,184–45,192.

72. Matassa, A. A., Carpenter, L., Biden, T. J., Humphries, M. J., and Reyland, M. E. (2001) PKCdelta is required for mitochondrial-dependent apoptosis in salivary epithelial cells. *J. Biol. Chem.* **276,** 29,719–29,728.

73. Uberall, F., Hellbert, K., Kampfer, S., Maly, K., Villunger, A., Spitaler, M., et al. (1999) Evidence that atypical protein kinase C-lambda and atypical protein kinase C-zeta participate in Ras-mediated reorganization of the F-actin cytoskeleton. *J. Cell Biol.* **144,** 413–425.

74. Uberall, F., Giselbrecht, S., Hellbert, K., Fresser, F., Bauer, B., Gschwendt, M., et al. (1997) Conventional PKC-α, novel PKC-ε and PKC-θ, but not atypical PKC-λ are MARCKS kinases in intact NIH 3T3 Fibroblasts. *J. Biol. Chem.* **272,** 4072–4078.

75. Baier-Bitterlich, G., Uberall, F., Bauer, B., Fresser, F., Wachter, H., Grunicke, H., et al. (1996) Protein kinase C-theta isoenzyme selective stimulation of

the transcription factor complex AP-1 in T lymphocytes. *Mol. Cell Biol.* **16,** 1842–1850.

76. Suzuma, K., Naruse, K., Suzuma, I., Takahara, N., Ueki, K., Aiello, L. P., et al. (2000) Vascular endothelial growth factor induces expression of connective tissue growth factor via KDR, Flt1, and phosphatidylinositol 3-kinase-akt-dependent pathways in retinal vascular cells. *J. Biol. Chem.* **275,** 40,725–40,731.

77. Matsumoto, M., Ogawa, W., Hino, Y., Furukawa, K., Ono, Y., Takahashi, M., et al. (2001) Inhibition of insulin-induced activation of Akt by a kinase-deficient mutant of the epsilon isozyme of protein kinase C. *J. Biol. Chem.* **276,** 14,400–14,406.

78. Kuroda, S., Tokunaga, C., Kiyohara, Y., Higuchi, O., Konishi, H., Mizuno, K., et al. (1996) Protein–protein interaction of zinc finger LIM domains with protein kinase C. *J. Biol. Chem.* **271,** 31,029–31,032.

79. Li, L., Lorenzo, P. S., Bogi, K., Blumberg, P. M., and Yuspa, S. H. (1999) Protein kinase C delta targets mitochondria, alters mitochondrial membrane potential, and induces apoptosis in normal and neoplastic keratinocytes when overexpressed by an adenoviral vector. *Mol. Cell Biol.* **19,** 8547–8558.

80. Fujii, T., Garcia-Bermejo, M. L., Bernabo, J. L., Caamano, J., Ohba, M., Kuroki, T., et al. (2000) Involvement of protein kinase C delta (PKCdelta) in phorbol ester-induced apoptosis in LNCaP prostate cancer cells. Lack of proteolytic cleavage of PKCdelta. *J. Biol. Chem.* **275,** 7574–7582.

81. Matassa, A. A., Carpenters, L., Biden, T. J., Humphries, M. J., and Reyland, M. E. (2001) PKCδ is required for mitochondrial-dependent apoptosis in salvary epithelial cells. *J. Biol. Chem.* **276,** 29,719–29,728.

82. Braiman, L., Alt, A., Kuroki, T., Ohba, M., Bak, A., Tennenbaum, T., et al. (2001) Insulin induces specific interaction between insulin receptor and protein kinase C delta in primary cultured skeletal muscle. *Mol. Endocrinol.* **15,** 565–574.

83. Braiman, L., Alt, A., Kuroki, T., Ohba, M., Bak, A., Tennenbaum, T., et al. (2001) Activation of protein kinase C zeta induces serine phosphorylation of VAMP2 in the GLUT4 compartment and increases glucose transport in skeletal muscle. *Mol. Cell Biol.* **21,** 7852–7861.

84. Han, J. M., Kim, J. H., Lee, B. D., Lee, S. D., Kim, Y., Jung, Y. W., et al. (2002) Phosphorylation-dependent regulation of phospholipase D2 by protein kinase C delta in rat Pheochromocytoma PC12 cells. *J. Biol. Chem.* **277,** 8290–8297.

86. Ping, P., Zhang, J., Cao, X., Li, R. C., Kong, D., Tang, X. L., et al. (1999) PKC-dependent activation of p44/p42 MAPKs during myocardial ischemia- reperfusion in conscious rabbits. *Am. J. Physiol.* **276,** H1468–H1481.

87. Strait, J. B., III, Martin, J. L., Bayer, A., Mestril, R., Eble, D. M., and Samarel, A. M. (2001) Role of protein kinase C-epsilon in hypertrophy of cultured neonatal rat ventricular myocytes. *Am. J. Physiol.* **280,** H756–H766.

88. Kuroki, T., Kashiwagi, M., Ishino, K., Huh, N., and Ohba, M. Adenovirus-mediated gene transfer to keratinocytes—a review. (1999) *J. Invest. Dermatol. Symp. Proc.* **4,** 153–157

89. Ohba, M., Ishino, K., Kashiwagi, M., Kawabe, S., Chida, K., Huh, N. H., et al. (1998) Induction of differentiation in normal human keratinocytes by adenovirus-

mediated introduction of the eta and delta isoforms of protein kinase C. *Mol. Cell Biol.* **18,** 5199–5207.

90. Kotani, K., Ogawa, W., Hino, Y., Kitamura, T., Ueno, H., Sano, W., et al. (1999) Dominant negative forms of Akt (protein kinase B) and atypical protein kinase Clambda do not prevent insulin inhibition of phosphoenolpyruvate carboxykinase gene transcription. *J. Biol. Chem.* **274,** 21,305–21,312.

91. Kotani, K., Ogawa, W., Matsumoto, M., Kitamura, T., Sakaue, H., Hino, Y., et al. (1998) Requirement of atypical protein kinase clambda for insulin stimulation of glucose uptake but not for Akt activation in 3T3-L1 adipocytes. *Mol. Cell. Biol.* **18,** 6971–6982.

92. Bandyopadhyay, G., Kanoh, Y., Sajan, M. P., Standaert, M. L., and Farese, R. V. (2000) Effects of adenoviral gene transfer of wild-type, constitutively active, and kinase-defective protein kinase C-lambda on insulin-stimulated glucose transport in L6 myotubes. *Endocrinology* **141,** 4120–4127.

93. Suzuki, A., Yamanaka, T., Hirose, T., Manabe, N., Mizuno, K., Shimizu, M., et al. (2001) Atypical protein kinase C is involved in the evolutionarily conserved par protein complex and plays a critical role in establishing epithelia-specific junctional structures. *J. Cell Biol.* **152,** 1183–1196.

94. Suzuma, I., Suzuma, K., Ueki, K., Hata, Y., Feener, E. P., King, G. L., et al. (2002) Stretch-induced retinal vascular endothelial growth factor expression is mediated by phosphatidylinositol 3-kinase and protein kinase C (PKC)-zeta but not by stretch-induced ERK1/2, Akt, Ras, or classical/novel PKC pathways. *J. Biol. Chem.* **277,** 1047–1057.

95. Igarashi, M., Wakasaki, H., Takahara, N., Ischii, H., Jiang, Z. Y., Yamauchi, T., et al. (1999) Glucose or diabetes activates p38 mitogen-activated protein kinase via different pathways. *J. Clin. Invest.* **103,** 185–195.

96. Liu, Y. F., Paz, K., Herschkovitz, A., Alt, A., Tennenbaum, T., Sampson, S. R., et al. (2001) Insulin stimulates PKCzeta -mediated phosphorylation of insulin receptor substrate-1 (IRS-1). A self-attenuated mechanism to negatively regulate the function of IRS proteins. *J. Biol. Chem.* **276,** 14,459–14,465.

34

Phorbol Esters as Probes for the Study of Protein Kinase C Function

Marcelo G. Kazanietz and Patricia S. Lorenzo

1. Introduction

A wealth of information about protein kinase C (PKC) and its biological functions has been generated with the use of phorbol esters as pharmacological probes. Phorbol esters function as ultrapotent analogs of the second messenger diacylglycerol (DAG), the endogenous ligand of PKC. However, their higher potency and reduced metabolism relative to DAG have made the phorbol esters the most common pharmacological agents for the study of PKC function.

Today, despite the advances in novel molecular methodologies, the use of phorbol esters still represents the traditional approach to investigate PKC function and regulation. In this chapter, we outline the most common experimental approaches using phorbol esters, provide information about the available phorbol ester derivatives, and discuss the potential pitfalls in the use of phorbol esters as unique PKC modulators.

1.1. PMA, Related Diterpenes, and Nonphorbol Analogs of DAG

Phorbol esters are natural products derived from Croton tiglium and other plants from the Euphorbiaceae family (1). They have the ability to activate all PKC isoforms except the atypical ones, which are also insensitive to DAG. Phorbol 12-myristate 13-acetate (PMA, also known as TPA) is the prototype of the phorbol esters and is normally used as the first tool when the involvement of PKC in a biological system needs to be defined. In addition to the phorbol esters, a number of structurally distinct natural products also serve as DAG-mimetics and bind to PKC with high affinity. These compounds include the ingenol esters, the daphnane derivatives mezerein and thymeleatoxin,

From: *Methods in Molecular Biology, vol. 233: Protein Kinase C Protocols*
Edited by: A. C. Newton © Humana Press Inc., Totowa, NJ

teleocidine and their structural simpler congener (–)-indolactam V (ILV), polyacetates like aplysiatoxin, and the bryostatins (**Fig. 1**).

Since the initial characterization of PMA as a tumor promoter in the mouse skin, it has been long clear that the various phorbol ester analogs differ in their ability to cause (or not) tumor promotion. Thus, although PMA and 12-deoxyphorbol 13-tetradecanoate display similar potency in the mouse skin tumor promotion paradigm, 12-deoxyphorbol 13-phenylacetate (dPP), the 12-deoxyphorbol 13-acetate (prostratin), and bryostatin 1 lack tumor-promotion activity *(2–4)*. Other unique biological properties have been also described for these last compounds. At least two mechanisms can explain this heterogeneity: (1) the diversity within the PKC family, and (2) the presence of additional, non-PKC, targets.

With a few exceptions that depend on the cell type and end-point assayed, most of the phorbol ester derivatives exhibit only modest selectivity for PKC isozymes. Thymeleatoxin, for example, has been reported to preferentially activate PKCα, PKCβI, and PKCγ purified from rat brain, with little effect on PKCδ and PKCε *(5)*. The reduced affinity of thymeleatoxin for the nPKCs was also shown in studies using recombinant PKC isozymes *(6)*. Nevertheless, in renal mesangial cells, the same compound did not show selectivity when translocation and downregulation were assessed *(7)*. Another example of variability in isoform selectivity is that of mezerein and ILV, which can preferentially activate PKCα at concentrations between 1 to 100 n*M*, whereas higher doses activate other isotypes as well *(8)*.

Despite the low degree of isozyme selectivity, which limits the use of the individual phorbol esters as isoform-specific activators in vitro, important information has been generated by the analysis of the structure-activity relationships.

1.2. Phorbol Esters as Pharmacological Probes

In addition to the numerous uses of phorbol esters and related compounds to assess PKC function, these agents represent the most important tools for the evaluation of the ligand-binding properties of PKC isozymes in vitro and the assessment of PKC redistribution or translocation. We will discuss these important aspects of the phorbol ester pharmacology.

1.2.1. Use of Phorbol Esters and Related Compounds as PKC Ligands

Binding of phorbol esters (or DAG) is the first step in the activation of PKC. This binding is saturable and occurs through specific interactions within the C1 domain in the regulatory region of the PKC molecule *(9)*.

Because phorbol esters bind to PKC with high affinity, they can be used in in vitro receptor-binding assays to estimate the number and distribution

Phorbol Esters

- Phorbol 12-myristate 13-acetate   (PMA or TPA)
- Phorbol 12, 13-dibutyrate   (PDBu)
- 12-deoxyphorbol 13-tetradecanoate (dPT)
- 12-deoxyphorbol 13-acetate   (Prostratin)
- 12-deoxyphorbol 13-phenylacetate   (dPP)

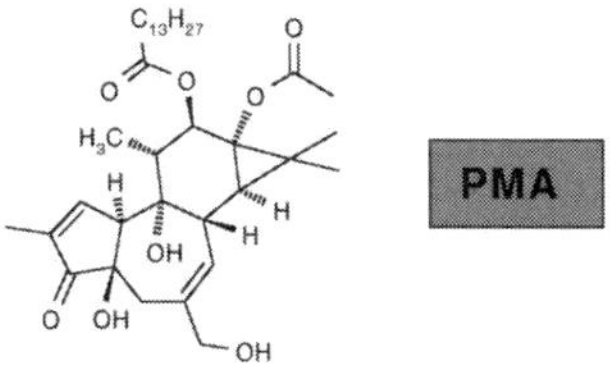

Daphnane derivatives

- Mezerein
- Thymeleatoxin

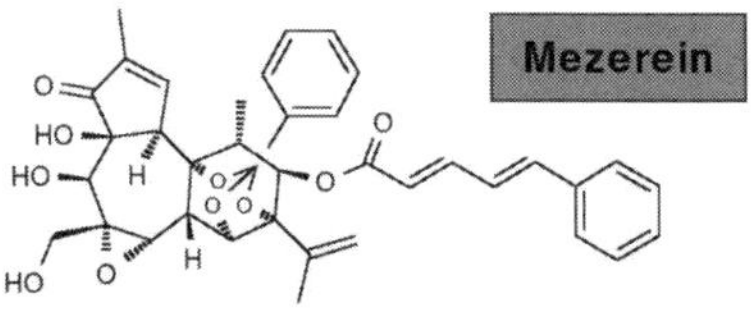

Indole alkaloids

- Teleocidin
- (-)-Indolactam V   (ILV)
- (-)-Octyl-indolactam V   (Octyl-ILV)

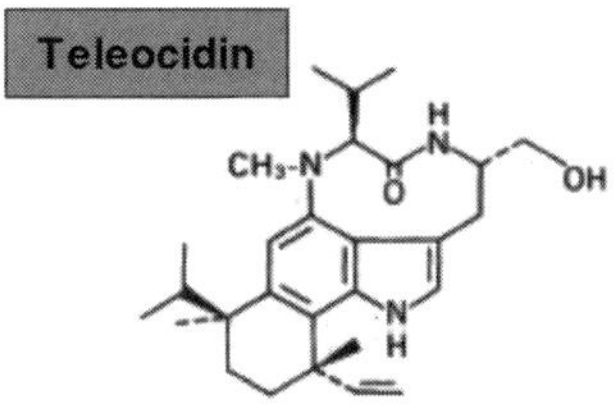

Ingenol Esters

- Ingenol 3-benzoate
- Ingenol 3-tetradecanoate

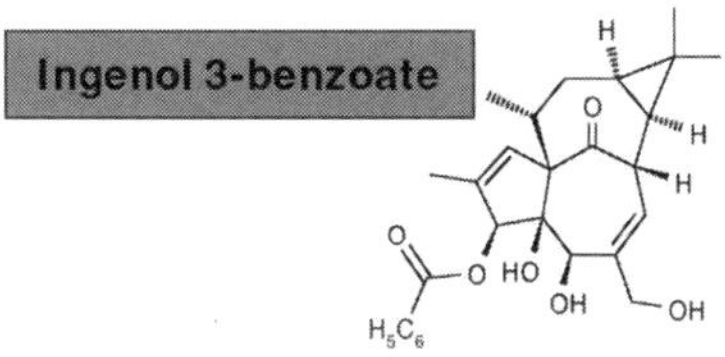

Macrocyclic lactones

- Bryostatin 1
- Bryostatin 2

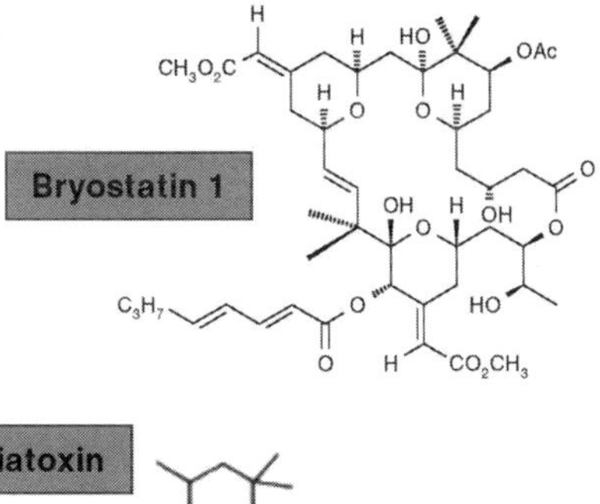

Polyacetates

- Debromoaplysiatoxin
- Aplysiatoxin

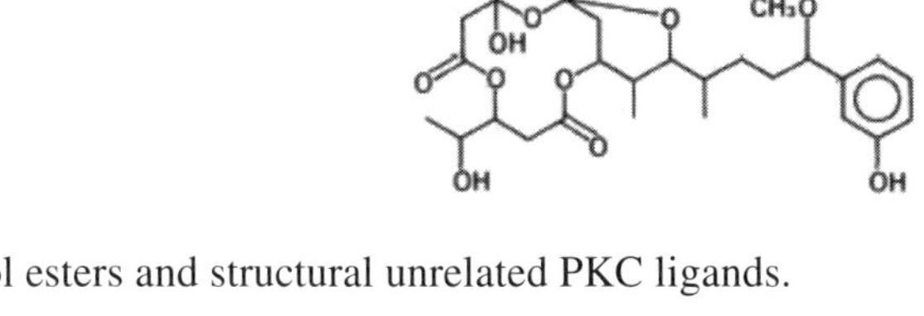

Fig. 1. Phorbol esters and structural unrelated PKC ligands.

of PKC molecules ("receptors") in tissue extracts, cell lysates, or subcellular fractions. In vitro binding assays are usually performed using a radiolabeled ligand, tritiated phorbol 12,13-dibutyrate ([^{3}H]-PDBu). The methodology has been characterized in detail by Dr. Peter M. Blumberg (National Cancer Institute, Bethesda, MD) and is described in Chapter 3.

The affinity for the interaction between ligand and receptor in the binding assays is determined by the equilibrium dissociation constant (K_d), which provides a valuable parameter for structure-activity relation studies. By using a variety of phorbol ester derivatives, two patterns of PKC selectivity have been revealed. Whereas the 12-deoxyphorbol esters and mezerein possess 5 to 10 times higher affinity for the cPKCs α, β, and γ, the indole alkaloids ILV and (–)-octyl-indolactam V (octyl-ILV) show little selectivity between cPKCs and nPKCs *(6)*. Importantly, the lipid cofactor utilized in the in vitro binding assay is critical for the determination of the ligand affinity. Indeed, the differences in K_d observed by different laboratories are probably related to the different lipid compositions or the methodology used to prepare lipid vesicles/micelles. Great care should be taken when comparing affinities obtained by the different methodologies.

1.2.2. Phorbol Esters as Tools to Study Intracellular Localization of PKC Isozymes

One of the most common approaches for evaluating PKC activation is the determination of PKC translocation. Translocation is defined as the subcellular redistribution of PKC in response to physiological stimuli or pharmacological agonists. The subcellular localization of PKC in the resting state depends on the cell or tissue being examined. In numerous cell types, PKC isozymes localize to the cytosol and redistribute to the plasma membrane. In addition, PKC isoforms can also translocate to the perinuclear envelope, organelles, and cytoskeletal structures. It has been found that the pattern of translocation depends on the compound and the isoform being examined. The C1 domains in PKCs are critical for translocation by phorbol esters, although the significance of each C1 domain may not be the same for discrete PKC isozymes. For example, the significance of the two C1 domains in cPKCs and nPKCs has been examined by introducing point mutations and measuring the ability of phorbol esters to induce translocation on the mutated PKC molecules *(10,11)*. One recent study shows that derivatives with different tumor-promoting properties vary in their ability to translocate PKCδ to the plasma membrane in CHO-K1 cells *(12)*. Interestingly, a strong correlation was found between the potency of the derivative to induce tumor promotion and its ability to bring PKCδ to the plasma membrane, supporting the idea that differential targeting of PKC isozymes can influence the biological consequences of their activation.

At least two different approaches can be used in the analysis of PKC translocation: subcellular fractionation and fluorescent microscopy. The first method is based on the separation of different subcellular fractions by solubilization followed by ultracentrifugation. The second method is based on the use of fluorescent probes to determine the distribution of PKC in the intact cells. Indirect immunofluorescence, utilizing antibodies raised against the PKC isoform of interest, and the use of fluorescent phorbol ester derivatives *(13)*, were the first approaches available. However, research on PKC translocation gathered momentum when the green fluorescent fusion protein (GFP) entered into the scene *(14)*. GFP-tagged PKC constructs have contributed enormously to the study of the dynamics of PKC in living cells. Together with the advances in confocal microscopy and imaging, precise visualization of the pattern of intracellular translocation in response to phorbol ester analogs have begun to emerge.

1.3. Some Considerations about Translocation Studies Using Phorbol Esters

The phorbol esters and derivatives are water insoluble and, therefore, they have to be dissolved in organic solvents. In most cases, they can be dissolved in dimethylsulfoxide (DMSO) or ethanol. However, it is important to be sure that neither DMSO nor ethanol will interfere with or affect the translocation studies. As a starting point, it is advisable to prepare 1000× stock solutions of the phorbol ester derivative based on the final concentration required for the studies, so that vehicle would not exceed 0.1% (v/v) final.

Some derivatives, such as bryostatin 1, are particularly susceptible to binding to glass or plastic surfaces when in aqueous solutions. When adding DMSO or ethanol solutions of the compound to a cell culture, the serum present in the medium will act as a protein carrier reducing the nonspecific absorption to the surface of the culture dish. If serum-free medium is used, we suggest that some protein carrier is present in the incubation medium to reduce absorption. Similar considerations should be applied for perfusion during live-cell confocal microscopy studies.

Another aspect that has to be considered during translocation studies is the reversibility of the phorbol ester effect. Subcellular fractionation studies done in primary mouse keratinocytes have shown that phorbol esters like PDBu or dPP are easily removed during the washing steps, causing a rapid re-redistribution of PKC isozymes *(15)*. This rapid, reversible translocation has to be carefully monitored to avoid misinterpretation of the results, particularly if lack of translocation is observed.

Translocation studies using phorbol esters provide not only important information regarding the intracellular location of PKC isozymes, but they can also serve to estimate the potency of the compound in cellular models. Indeed, the potency of phorbol ester analogs for PKC translocation is normally lower than the affinity observed in in vitro binding assays, which are usually performed under nonphysiological conditions and high phospholipid (phosphatidylserine) concentrations. Concentration-response curves can be generated by monitoring the disappearance of cytosolic PKC by densitometry upon treatment with different concentration of phorbol esters (*see*, e.g., **refs. 10,16,** and **17**). In most cases, a concentration-dependent curve will follow a sigmoidal shape. Sigmoid curves can be fitted using non-linear regression programs, and the EC_{50} (effective concentration 50%) of the phorbol ester can be estimated from the curve. In addition to concentration-dependent curves, time-course analysis (translocation as a function of time) also provides useful information. It is important to determine a time in which the corresponding PKC isozyme is not downregulated by the phorbol ester. Because concentration-response curves are normally done after a short period of incubation with the phorbol ester, PKC downregulation should not be a problem.

1.4. PKC Ligands That Function as Inhibitors

There are at least two classes of compounds that exhibit high affinity for binding to PKC in in vitro assays, which behave in the intact cells as partial antagonists of PKC function. These two classes are represented by bryostatins (such as bryostatin 1) and the deoxyphorbol esters (such as dPP and prostratin). Although these ligands activate PKC in vitro and induce translocation of PKC in cellular models, they inhibit some of the biological effects of phorbol ester derivatives such as PMA. For example, they all suppress tumor promotion induced by PMA in the mouse skin model *(2–4)*.

Although these ligands may represent a valuable alternative to PKC inhibition, which has been conventionally approached by targeting the kinase domain of the enzyme, their use is still limited because the molecular basis for this antagonism is poorly understood. Among these derivatives, bryostatin 1 is the one that has been most extensively studied, largely because of its potential chemotherapeutic activity, as being assessed in clinical trials *(18)*. In in vitro assays, bryostatin 1 displays greater affinity for binding to PKC when compared to the typical phorbol esters *(19)*. In cellular models, however, it only induces a subset of the biological effects of the phorbol esters, and antagonizes those phorbol ester actions that it cannot produce *(20)*. By analyzing the pattern of translocation and downregulation induced by bryostatin 1 in cells, it is evident that this compound causes a unique pattern of down-regulation for PKCδ *(21,22)*. Although low nanomolar concentrations of bryostatin 1 lead to degradation of PKCδ, higher concentrations do not induce downregulation and in fact can protect PKCδ from being downregulated by PMA. This unique effect on PKCδ may explain the antagonistic effect of bryostatin 1 in cellular models. Interestingly, recent work done on chimeric PKC molecules has demonstrated that although the regulatory domain of PKCδ drives bryostatin 1 to cause the biphasic downregulation, the catalytic domain is essential for the protection from downregulation *(23)*, suggesting a level of complexity yet to be deciphered.

Selective downregulation, as that produced by bryostatin 1, represents *per se* an attractive approach in the study of PKC function. Similarly, complete PKC depletion, as that achieved by some phorbol esters (PMA), is another strategy for inhibition of PKC. However, it should be recognized that a universal treatment for inducing selective or complete depletion of PKCs does not exist. Indeed, this phenomenon not only depends on the PKC ligand but also on the isoform and the cell type being examined. As an example, we can highlight the behavior of PKCδ and PKCε: whereas PMA is able to downregulate both isoforms in rat C6 glioma cells, only PKCδ is sensitive to PMA-induced degradation in mouse 3T3-F442A adipocytes *(24,25)*. Understanding the molecular basis for this type of heterogeneity should contribute to the development of derivatives with specific patterns of PKC activation and downregulation.

1.5. Phorbol Esters as Activators of Non-PKC Targets: A Re-Evaluation of PKC Selectivity

The discovery of non-PKC receptors for DAG has challenged the use of phorbol esters as exclusive PKC ligands. These novel targets include the mammalian Munc13 proteins and its *Caenorhabditis elegans* homologue Unc-13, the chimaerins, and the recently characterized RasGRP family (**Fig. 2**). All of them possess a single C1 domain with the structural features required for recognition of DAG and phorbol esters. The Munc13/Unc-13 proteins participate in neurotransmission, acting as vesicle-priming proteins *(26–28)*. Chimaerins possess a GTPase activator protein domain that accelerates GTP hydrolysis from the small GTP-binding protein Rac *(29,30)*. Finally, RasGRP represents a novel family of guanine nucleotide exchange factors for Ras-like GTPases *(31–33)*. Several reviews focusing on the pharmacology and biology of these novel phorbol esters/DAG receptors have been published recently *(34–37)*.

Like the DAG-sensitive PKCs, Munc13/Unc-13, chimaerins, and RasGRPs bind phorbol esters in vitro and in cells. They exhibit nanomolar affinities for binding to phorbol esters in the same range as PKCs, and ligand binding is phospholipid dependent *(38–41)*. In cellular models, they all translocate after phorbol ester treatment, as judged by subcellular fractionation and fluorescent microscopy studies *(27,31,41–43)*. With a few exceptions *(39)*, structure-activity analysis reveals striking similarities in binding recognition relative to PKCs.

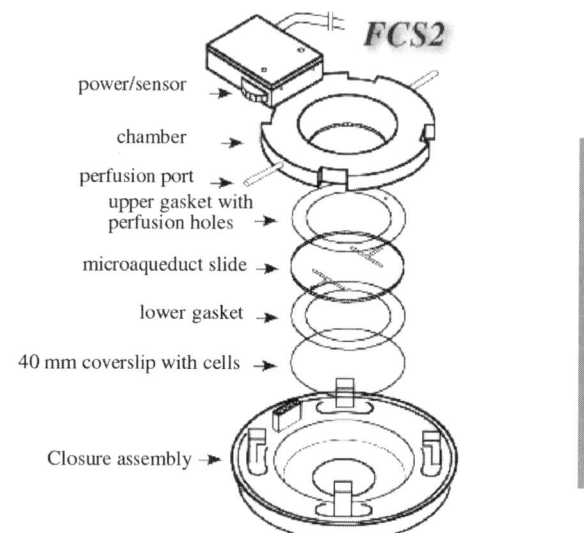

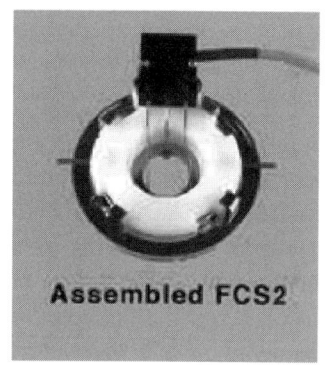

Fig. 2. Live-cell micro-observation chamber (reprinted with permission from Bioptechs, Inc.).

PKD/PKCμ represents a distinct family of phorbol ester receptors *(44)*. Because PKD/PKCμ is a serine/threonine kinase, there was at the beginning some controversy as whether it should be included as part of the PKC family. However, PKD/PKCμ shows enzymological properties distinct from those of PKC. Moreover, the recent identification of additional cDNA clones similar in overall structure and substrate specificity to PKD/PKCμ *(45,46)* supported the concept that PKD/PKCμ is a distant relative of PKC. Although PKD/PKCμ possesses two C1 domains, binding studies indicate that the second (C1b) domain is responsible for the high-affinity interactions with phorbol esters, both in vitro and in cells *(47)*. In addition to binding recognition, PKD/PKCμ also redistributes in response to phorbol esters in vivo *(48)*.

Although a comprehensive review of the phorbol ester properties of these non-PKC targets is beyond the scope of this chapter, the information presented here should serve as an indication that phorbol esters are not unique PKC activators. Therefore, when using phorbol esters as probes to examine the potential involvement of PKC in a biological effect, proper controls should be used. For example, a PKC inhibitor such as GF 109203X (bisindolylmaleimide I) may be useful *(49)*. Unfortunately, selective inhibitors for the "non-PKC" phorbol ester receptors are not available yet. Regarding the use of specific PKC inhibitors, one should consider those acting on the kinase domain. Inhibitors like calphostin C, which targets the C1 domain of PKC, have also been reported to inhibit Unc-13, chimaerins, and RasGRP *(40,50)*.

The following sections will describe the general procedures for analyzing PKC-GFP translocation in live cells. In addition, we will review a protocol for subcellular fractionation, as this method is still widely used in the study of PKC translocation induced by phorbol esters.

2. Materials

2.1. Materials for Assessing GFP-PKC Translocation by Phorbol Esters

1. Mammalian expression vector with GFP tag (pEGFP, Clontech; pQBI25, Qbiogene).
2. cDNA of a PKC isoform.
3. NIH 3T3 cell culture (or other cell line).
4. Stock solution of the phorbol ester analog.
5. Transfection reagent.
6. Coverslips.
7. Phenol red free Dulbecco's modified Eagle's medium with 4 mM L-glutamine and 4.5 g/L glucose, containing 10% fetal bovine serum (formulation may need to be changed depending on the cell type being used).
8. 70% ethanol.
9. Scanning confocal microscope.
10. Chamber for live cell imaging with temperature probe and controller.
11. Microperfusion pump with connection tubings (refer to the manufacturer's instructions).
12. Video capturing system.

2.2. Materials for Assessing Phorbol Ester-Induced Translocation by Subcellular Fractionation

1. Lysis Buffer: 20 mM Tris-HCl (pH 7.4) containing 5 mM ethylenebis (oxy-ethylenenitrilo) tetraacetic acid (EGTA), 1 mM 4-(2-aminoethyl) benzenesulfonyl fluoride, and 20 µM leupeptin.
2. 1X phosphate-buffered saline (PBS): 20 mM NaCl, 2.7 mM KCl, 10 mM Na$_2$HPO$_4$, and 1.8 mM KH$_2$PO$_4$, pH 7.4.
3. 10% Triton-X 100 solution (in distilled water).
4. Sonicator with probe.
5. Ultracentrifuge type Beckman TL-100 with TLA-100 rotor.
6. 2X sample buffer: 950 mL of Laemmli sample buffer (62.5 mM Tris-HCl, pH 6.8, 2% sodium dodecyl sulfate, 25% glycerol, 0.01% bromophenol blue) plus 50 mL of β-mercaptoethanol.
7. Phorbol ester stock solution.
8. 8 or 10% polyacrylamide gel (materials required: 40% acrylamide/bisacrylamide, 10% SDS; 1.5 M Tris-HCl, pH 8.8; 0.5 mM Tris-HCl, pH 6.8; 10% ammonium persulfate; TEMED; distilled water).

3. Methods
3.1. GFP-PKC Translocation by Phorbol Esters

The following protocol outlines the analysis of translocation of a GFP-PKC by phorbol ester analogs in live NIH 3T3 cells using confocal microscopy. This protocol can be adapted to other cell lines, after optimization of the transfection conditions. Translocation experiments using GFP and phorbol esters can also be done on fixed GFP-transfected cells instead of live cells (*see* **Note 1**).

3.1.1. Construction of the GFP-PKC Mammalian Expression Vector

At least three companies, Clontech (Palo Alto, CA), Qbiogene (Carlsbad, CA), and Invitrogen (Carlsbad, CA), offer variants of epitope-tagged GFP plasmids, for cloning at either the N- or the C-termini of GFP. Two of the variants that have been successfully used in our laboratories are the pEGFP-N1 from Clontech, and the pQBI25 from Qbiogene, which produce tagged proteins by fusion to the N-terminus of GFP.

Important aspects to be considered in the selection of the vector and in the cloning of the PKC include the nature of the promoter, the selectable marker for bacterial growth, the presence of a stop codon in the PKC cDNA if a GFP-N-terminal construct is to be generated, and the preservation of the frame between GFP and PKC during the subcloning. Restriction maps and sequences for the GFP vectors can be obtained at http://www.clontech.com and http://www.qbiogene.com. Please note that the bacterial colonies containing the pEGFP-N1 vector are kanamycin resistant, while pQBI25 confers ampicillin resistance.

3.1.2. Transient Transfection of Mammalian Cells and Confocal Analysis

We have most often used the liposome-mediated transfection method (*51*), but other methods such as calcium phosphate may also be suitable.

1. Seed 5×10^4 NIH 3T3 cells into a 60-mm dish containing a 40-mm glass coverslip (*see* **Note 2**). Incubate overnight at 37°C in a CO_2 incubator.
2. Prepare the GFP-PKC vector for transfection. The DNA should be of high purity (A280/260 ≥ 1.8) and sterile. We strongly recommend the use of CsCl purification or Qiagen kits for plasmid preparation. Follow the instruction that accompanies the transfection reagent to be used.
3. Forty-eight hours after transfection, examine the dishes under an inverted fluorescent microscope using a FITC filter to confirm GFP expression before proceeding with confocal microscopy.
4. Take the coverslip from the 60-mm dish and position it into the live-cell chamber (*see* **Note 2**). Attach the thermoregulator to the chamber and adjust the temperature to 37°C (*see* **Note 3**). The chamber should have two perfusion ports: connect

the entrance port to a micro-perfusion pump using a piece of Tygon tubing. The pump will allow for the medium to be exchanged continuously or by steps with adjustable flow rates. Collect the perfusate waste at the second perfusion port.

5. Initiate the perfusion with complete, phenol red free medium (*see* **Note 4**). After an equilibration period of 10–15 min, switch the perfusate to complete medium containing the phorbol ester at the final concentration desired.

6. Excite the GFP chromophore and visualize the fluorescence signal using a FITC filter set (or equivalent). Record the GFP distribution. Take into account that there is a lag (time) for the medium containing the phorbol ester to reach the chamber, which depends on the flow rate and the size of the Tygon tubing. This time should be calculated in advance and needs to be considered when analyzing the time-course for the phorbol ester-induced translocation.

7. At the end of the experiment, wash tubing and chamber several times with 70% ethanol and distilled water.

3.2. Phorbol Ester-Induced Translocation Assessed by Subcellular Fractionation

1. Treat the cells with the phorbol ester analog, or vehicle, for the desired amount of time. Stop the incubation by placing the dishes on ice. After washing the cells twice with ice-cold 1X PBS, harvest them by scrapping in ice-cold lysis buffer. Keep the dishes on ice all the time. For 60-mm dishes, use at least 60 μL of lysis buffer.

2. Using a sonicator, subject the cells to one pulse of 6- to 8-s duration. Take an aliquot of the lysate and save it as total fraction.

3. Centrifuge the remaining lysate at $100,000g$ for 1 h at 4°C. This supernatant represents the "Soluble Fraction" (Cytosolic Fraction).

4. Resuspend the pellet in lysis buffer, using the same volume as originally used during the harvesting. Sonicate for 6 s, add 10% Triton X-100 to a final 1% concentration, and incubate at 4°C during 2 h for solubilization of the membrane-bound proteins.

5. Centrifuge the tubes for 1 h at $100,000g$ at 4°C. Collect the supernatant as the detergent-soluble fraction (Triton X-100-soluble). Resuspend the pellet in lysis buffer and sonicate as described before. This suspension represents the detergent-insoluble fraction (Triton X-100-insoluble).

6. Prepare the four fractions (total, soluble, detergent-soluble, and detergent-insoluble) for electrophoresis (sodium dodecyl sulfate polyacrylamide gel electrophoresis) by diluting the samples 1:2 with 2X sample buffer and boiling for 5 min. Load equal amounts of protein for each sample (*see* **Note 5**) on a 8 or 10% polyacrylamide gel. Blot gels and proceed with immunostaining according to standard techniques.

4. Notes

1. Fixation should be done with paraformaldehyde solutions (freshly made in phosphate buffered saline). Most of the solvents commonly used as fixatives,

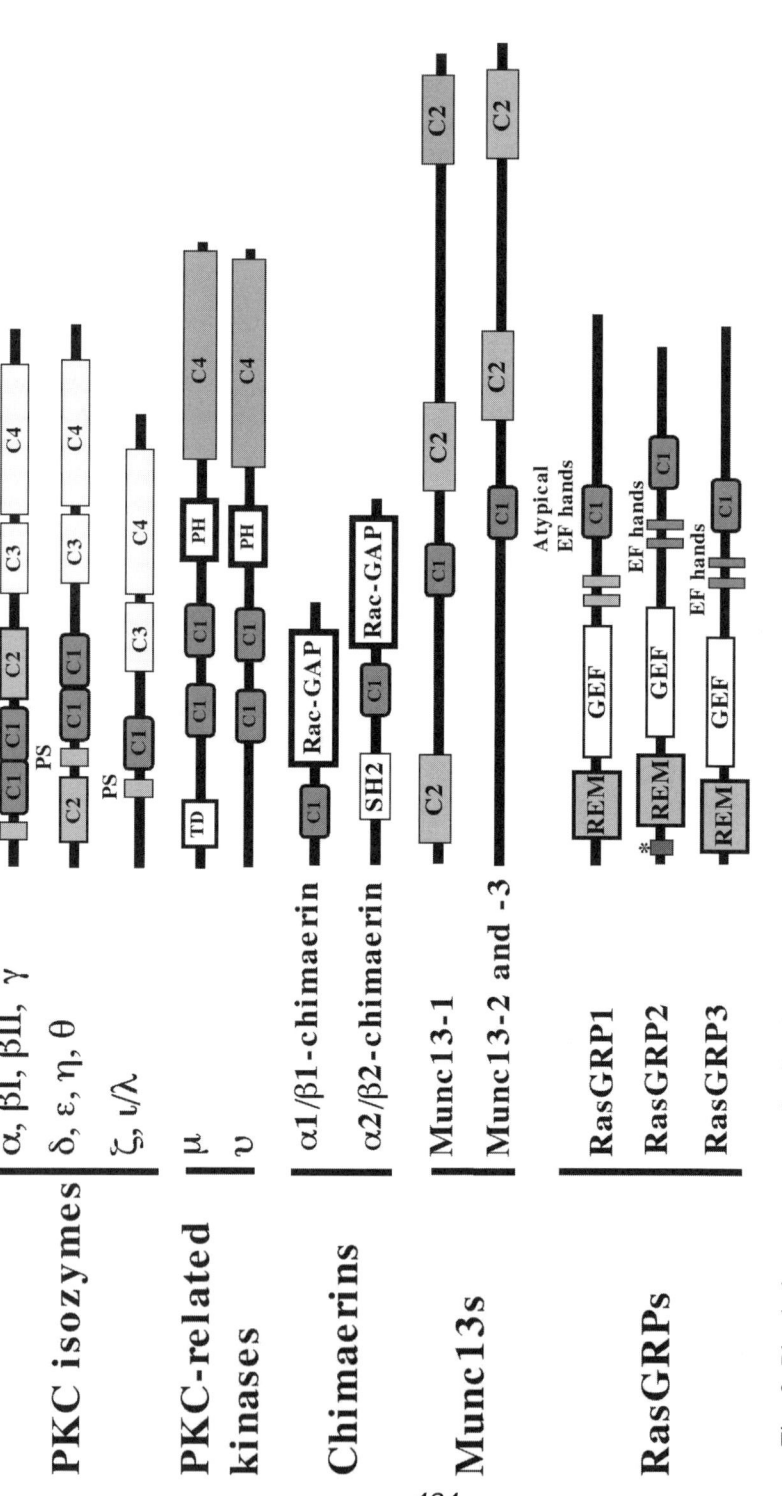

Fig. 3. Phorbol ester receptor family. C1, diacylglycerol/phorbol ester-binding domain; C2, phospholipid (Ca²⁺)-binding motif; C3, ATP-binding site; C4, serine/threonine protein kinase domain; EF, hands, Ca²⁺-binding motif; Rac-GAP, GTPase activator protein domain for Rac; GEF, guanine nucleotide exchange factor domain; PH, pleckstrin homology domain; PS, pseudosubstrate region; REM, Ras exchange motif; SH2, Src homology 2 domain; TD, transmembrane domain; *, acylation motif.

such as methanol or ethanol, severely affect the GFP chromophore. In addition, it is advisable to seal the coverslip with rubber cement or molten agarose because nail polish can interfere with the GFP fluorescence.

2. The type and size of the coverslip employed depend on the chamber used in the experiment. We have used a live-cell micro-observation chamber from Bioptechs (**Fig. 3**, http://www.bioptechs.com), which requires 40-mm coverslips.

3. Depending on the numeric aperture of the lenses, an objective heater may be needed in addition to the chamber thermoregulator.

4. The autofluorescence generated by the phenol red usually present in all media formulations will severely affect the background, so phenol-red free media should be used instead.

5. Depending on the method used for protein determination, Triton X-100 present in the Detergent-soluble and Detergent-insoluble fractions could interfere with the quantitation. Select a method accordingly (for example, Bio-Rad DC Protein Assay, Bio-Rad, Hercules, CA). After Western blotting, stain the blots with 0.1% Ponceau S Solution in 5% acetic acid, for the estimation of protein content in the individual lanes by densitometry. This type of staining has been found to be linear up to 30 µg of protein. The Ponceau staining can be easily removed by washing with 1X PBS.

Acknowledgments

The laboratory of M.G.K is supported by grants from NCI (NIH), American Cancer Society (ACS) and Department of Defense (DOD). The laboratory of P.S.L. is supported in part by a grant from NCI (NIH).

References

1. Hecker, E. (1978) Structure-activity relationships in diterpene esters irritant and cocarcinogenic to mouse skin, in *Carcinogenesis, Mechanism of Tumor Promotion and Cocarcinogenesis*, Vol. 2 (Slaga, T. J., Sivak, A., and Boutwell, R. K., eds.), Raven Press, New York, pp. 11–48.

2. Szallasi, Z., and Blumberg, P. M. (1991) Prostratin, a nonpromoting phorbol ester, inhibits induction by phorbol 12-myristate 13-acetate of ornithine decarboxylase, edema, and hyperplasia in CD-1 mouse skin. *Cancer Res.* **51,** 5355–5360.

3. Szallasi, Z., Krsmanovic, L, and Blumberg, P. M. (1993) Nonpromoting 12-deoxy-phorbol 13-esters inhibit phorbol 12-myristate 13-acetate induced tumor promotion in CD-1 mouse skin. *Cancer Res.* **5,** 2507–2512.

4. Hennings, H., Blumberg, P. M., Pettit, G. R., Herald, C. L., Shores, R., and Yuspa, S. H. (1987) Bryostatin 1, an activator of protein kinase C, inhibits tumor promotion by phorbol esters in SENCAR mouse skin. *Carcinogenesis* **8,** 1343–1346.

5. Ryves, W. J., Evans, A. T., Olivier, A. R., Parker, P. J., and Evans, F. J. (1991) Activation of the PKC-isotypes α, $\beta1$, γ, δ, and ε by phorbol esters of different biological activities. *FEBS Lett.* **288,** 5–9.

6. Kazanietz, M. G., Areces, L. B., Bahador, A., Mischak, H., Goodnight, J., Mushinski, J. F., et al. (1993) Characterization of ligand and substrate specificity for the calcium-dependent and calcium-independent PKC isozymes. *Mol. Pharmacol.* **44,** 298–307.

7. Huwiler, A., Fabbro, D., and Pfeilschifter, J. (1994) Comparison of different tumour promoters and bryostatin 1 on protein kinase C activation and down-regulation in rat renal mesangial cells. *Biochem. Pharmacol.* **48,** 689–700.

8. Geiges, D., Meyer, T., Marte, B., Vanek, M., Weissgerber, G., Stabel, S., et al. (1997) Activation of protein kinase C subtypes α, γ, δ, ϵ, ζ, and η by tumor-promoting and nontumor-promoting agents. *Biochem. Pharmacol.* **53,** 865–875.

9. Ono, Y., Fujii, T. Igarashi, K., Kuno, T., Tanaka, C., Kikkawa, U., et al. (1989) Phorbol ester binding to protein kinase C requires a cysteine-rich zinc-finger-like sequence. *Proc. Natl. Acad. Sci. USA* **86,** 4868–4871.

10. Bogi, K., Lorenzo, P. S., Szallasi, Z., Acs, P., Wagner, G. S., and Blumberg, P. M. (1998) Differential selectivity of ligands for the C1a and C1b phorbol ester binding domains of protein kinase Cδ: possible correlation with tumor-promoting activity. *Cancer Res.* **58,** 1423–1428.

11. Bogi, K., Lorenzo, P. S., Acs, P., Szallasi, Z., Wagner, G. S., and Blumberg, P. M. (1999) Comparison of the roles of the C1a and C1b domains of protein kinase Cα in ligand induced translocation in NIH 3T3 cells. *FEBS Lett.* **456,** 27–30.

12. Wang, Q. J., Bhattacharyya, D., Garfield, S., Nacro, K., Marquez, V. E., and Blumberg, P. M. (1999) Differential localization of protein kinase Cδ by phorbol esters and related compounds using a fusion protein with green fluorescent protein. *J. Biol. Chem.* **274,** 37,233–37,239.

13. Khalil, R. A. and Morgan, K. G. (1991) Imaging of protein kinase C distribution and translocation in living vascular smooth muscle cells. *Circ. Res.* **69,** 1626–1631.

14. Prasher, D. C. (1995) Using GFP to see the light. *Trends Genet.* **11,** 320–323.

15. Szallasi, Z., Smith, C. B., and Blumberg, P. M. (1994) Dissociation of phorbol esters leads to immediate redistribution to the cytosol of protein kinases Cα and Cδ in mouse keratinocytes. *J. Biol. Chem.* **269,** 27,159–27,162.

16. Acs, P., Bogi, K., Lorenzo, P. S., Marquez, A. M., Biro, T., Szallasi, Z., et al. (1997) The catalytic domain of protein kinase C chimeras modulates the affinity and targeting of phorbol ester-induced translocation. *J. Biol. Chem.* **272,** 22,148–22,153.

17. Caloca, M. J., Wang, H., Delemos, A., Wang, S., and Kazanietz, M. G. (2001) Phorbol esters and related analogs regulate the subcellular localization of β2–chimaerin, a nonprotein kinase C phorbol ester receptor. *J. Biol. Chem.* **276,** 18,303–18,312.

18. Mutter, R. and Wills, M. (2000) Chemistry and clinical biology of the bryostatins. *Bioorg. Med. Chem.* **8,** 1841–1860.

19. Kazanietz, M. G., Lewin, N. E., Gao, F., Pettit, G. R., and Blumberg, P. M. (1994) Binding of [26-^{3}H]bryostatin 1 and analogs to calcium-dependent and calcium-independent protein kinase C isozymes. *Mol. Pharmacol.* **46,** 374–379.

20. Blumberg, P. M. and Pettit, G. R. (1992) The bryostatins, a family of protein kinase C activators with therapeutic potential, in *New Leads and Targets in Drug*

Research, Alfred Benzon Symposium 33 (Krogsgaard-Larsen P., Christensen, S. B., and Kofod, H., eds.), Munksgaard International, Copenhagen, Denmark, pp. 273–285.

21. Szallasi, Z., Smith, C. B., Pettit, G. R., and Blumberg, P. M. (1994) Differential regulation of protein kinase C isozymes by bryostatin 1 and phorbol 12-myristate 13-acetate in NIH 3T3 fibroblasts. *J. Biol. Chem.* **269,** 2118–2124.

22. Szallasi, Z., Denning, M. F., Smith, C. B., Dlugosz, A. A., Yuspa, S. H., Pettit, G. R., et al. (1994) Bryostatin 1 protects protein kinase C-δ from down-regulation in mouse keratinocytes in parallel with its inhibition of phorbol ester-induced differentiation. *Mol. Pharmacol.* **46,** 840–850.

23. Lorenzo, P. S., Bogi, K., Acs, P., Pettit, G. R., and Blumberg, P. M. (1997) The catalytic domain of protein kinase Cδ confers protection from down-regulation induced by bryostatin 1. *J. Biol. Chem.* **272,** 33,338–33,343.

24. Chen, C. C. (1993) Protein kinase C α, δ, ϵ, and ζ in C6 glioma cells. TPA induces translocation and down-regulation of conventional and new PKC isoforms but not atypical PKC ζ. *FEBS Lett.* **332,** 169–173.

25. MacKenzie, S., Fleming, I., Houslay, M. D., Anderson, N. G., and Kilgour, E. (1997) Growth hormone and phorbol esters require specific protein kinase C isoforms to activate mitogen-activated protein kinases in 3T3-F442A cells. *Biochem. J.* **324,** 159–165.

26. Maruyama, I. N. and Brenner, S. (1991) Phorbol ester/diacylglycerol-binding protein encoded by the *unc-13* gene of *Caenorhabditis elegans*. *Proc. Natl. Acad. Sci. USA* **88,** 5729–5733.

27. Betz, A., Ashery, U., Rickmann, M., Augustin, I., Neher, E., Sudhof, T. C., et al. (1998) Munc13-1 is a presynaptic phorbol ester receptor that enhances neurotransmitter release. *Neuron* **21,** 123–136.

28. Ashery, U., Varoqueaux, F., Voets, T., Betz, A., Thakur, P., Koch, H., et al. (2000) Munc13-1 acts as a priming factor for large dense-core vesicles in bovine chromaffin cells. *EMBO J.* **19,** 3586–3596.

29. Hall, C., Monfries, C., Smith, P., Lim, H. H., Kozma, R., Ahmed, S., et al. (1990) Novel human brain cDNA encoding a 34,000 Mr protein n-chimaerin, related to both the regulatory domain of protein kinase C and BCR, the product of the breakpoint cluster region gene. *J. Mol. Biol.* **5,** 11–16.

30. Ahmed, S., Lee, J., Kozma, R., Best, A., Monfries, C., and Lim, L. (1993) Novel functional target for tumor-promoting phorbol ester and lysophosphatidic acid: The p21rac-GTPase activating protein n-chimaerin. *J. Biol. Chem.* **268,** 10,709–10,712.

31. Ebinu, J. O., Bortoff, D. A., Chan, E. Y. M., Stang, S. L., Dunn, R. J., and Stone, J. C. (1998) RasGRP, a Ras guanylnucleotide-releasing protein with calcium- and diacylglycerol-binding motifs. *Science* **260,** 1082–1086.

32. Kawasaki, H., Springett, G. M., Toki, S., Canales, J. J., Harlan, P., Blumenstiel, J. P., et al. (1998) A Rap guanine nucleotide exchange factor enriched highly in the basal ganglia. *Proc. Natl. Acad. Sci. USA* **95,** 13,278–13,283.

33. Yamashita, S., Mochizuki, N., Ohba, Y., Tobiume, M., Okada, Y., Sawa, H., et al. (2000) CalDAG-GEFIII activation of Ras, R-ras, and Rap1. *J. Biol. Chem.* **275,** 25,488–25,493.

34. Ron, D. and Kazanietz, M. G. (1999) New insights into the regulation of protein kinase C and novel phorbol ester receptors. *FASEB J.* **13,** 1658–1676.

35. Kazanietz, M. G. (2000) Eyes wide shut: protein kinase C isozymes are not the only receptors for the phorbol ester tumor promoters. *Mol. Carcinogenesis* **2,** 5–11.

36. Kazanietz, M. G., Caloca, M. J., Eroles, P., Fujii, T., Garcia-Bermejo, M. L., Reilly, M., et al. (2000) Pharmacology of the receptors for the phorbol ester tumor promoters. Multiple receptors with different biochemical properties. *Biochem. Pharmacol.* **60,** 1417–1424.

37. Kazanietz, M. G. (2002) Novel "non-kinase" phorbol ester receptors: the C1 domain connection. *Mol. Pharmacol.* **61,** 759–767.

38. Kazanietz, M. G., Lewin, N. E., Bruns, J. D., and Blumberg, P. M. (1995) Characterization of the cysteine-rich region of the *Caenorhabditis elegans* protein Unc-13 as a high affinity phorbol ester receptor. Analysis of ligand-binding interactions, lipid cofactor requirements, and inhibitor sensitivity. *J. Biol. Chem.* **270,** 10,777–10,783.

39. Caloca, M. J., Fernandez, N., Lewin, N. E., Ching, D., Modali, R., Blumberg, P. M., et al. (1997) β2-chimaerin is a high affinity receptor for the phorbol ester tumor promoters. *J. Biol. Chem.* **272,** 26,488–26,496.

40. Lorenzo, P. S., Beheshti, M., Pettit, G. R., Stone, J. C., and Blumberg, P. M. (2000) The guanine nucleotide exchange factor RasGRP is a high-affinity target for diacylglycerol and phorbol esters. *Mol. Pharmacol.* **57,** 840–846.

41. Lorenzo, P. S., Kung, J. W., Bottorff, D. A., Garfield, S. H., Stone, J. C., and Blumberg, P. M. (2001) Phorbol esters modulate the Ras exchange factor RasGRP3. *Cancer Res.* **61,** 943–949.

42. Tognon, C. E., Kirk, H. E., Passmore, L. A., Whitehead, F. P., Der, C. J., and Kay, R. J. (1998) Regulation of RasGRP via a phorbol ester-responsive C1 domain. *Mol. Cell Biol.* **18,** 6995–7008.

43. Caloca, M. J., Garcia-Bermejo, M. L., Blumberg, P. M., Lewin, N. E., Kremmer, E., Mischak, H., et al. (1999) β2-chimaerin is a novel target for diacylglycerol: binding properties and changes in subcellular localization mediated by ligand binding to its C1 domain. *Proc. Natl. Acad. Sci. USA* **96,** 11,854–11,859.

44. Valverde, A. M., Sinnett-Smith, J., Van Lint, J., and Rozengurt, E. (1994) Molecular cloning and characterization of protein kinase D: a target for diacylglycerol and phorbol esters with a distinctive catalytic domain. *Proc. Natl. Acad. Sci. USA* **91,** 8572–8576.

45. Hayashi, A., Seki, N., Hattori, A., Kozuma, S., and Saito, T. (1999) PKC ν, a new member of the protein kinase C family, composes a fourth subfamily with PKC μ. *Biochim. Biophys. Acta* **1450,** 99–106.

46. Sturan, S., Van Lint, J., Muller, F., Wilda, M., Hameister, H., Hocker, M., et al. (2001) Molecular cloning and characterization of the human protein kinase

D2. A novel member of the protein kinase D family of serine threonine kinases. *J. Biol. Chem.* **276,** 3310–3318.

47. Iglesias, T., Matthews, S., and Rozengurt, E. (1998) Dissimilar phorbol ester binding properties of the individual cysteine-rich motifs of protein kinase D. *FEBS Lett.* **437,** 19–23.

48. Matthews, S., Iglesias, T., Cantrell, D., and Rozengurt, E. (1999) Dynamic re-distribution of protein kinase D (PKD) as revealed by GFP-PKD fusion protein: dissociation from PKD activation. *FEBS Lett.* **457,** 515–521.

49. Blumberg, P. M., Acs, P., Bhattacharyya, D., and Lorenzo, P. S. (2000) Inhibitors of protein kinase C and related receptors for the lipophilic second-messenger sn-1,2-diacylglycerol, in *Signaling Networks and Cell Cycle Control: the Molecular Basis of Cancer and Other Diseases* (Gutkind, S. S., ed.), Humana Press, Totowa, NJ, pp. 347–364.

50. Areces, L. B., Kazanietz, M. G., and Blumberg, P. M. (1994) Close similarity of baculovirus expressed n-chimaerin and protein kinase Cα as phorbol ester receptors. *J. Biol. Chem.* **269,** 19,553–19,558.

51. Felgner, P. L., Gadek, T. R., Holm, M., Roman, R, Chan, H. W., Wenz, M., et al. (1987) Lipofection: a highly efficient, lipid-mediated DNA-transfection procedure. *Proc. Natl. Acad. Sci. USA* **84,** 7413–7417.

35

Irreversible Inactivation of Protein Kinase C Isozymes by Thiol-Reactive Peptide Substrate Analogs

Catherine A. O'Brian, Nancy E. Ward, Feng Chu, and Jubilee R. Stewart

1. Introduction

Protein kinases play diverse and essential roles in the regulation of cell growth, survival, and differentiation, and in other critical biological functions, for example, neurotransmission *(1,2)*. Substrate selectivity is a major factor in the division of labor among the members of the protein kinase superfamily *(1)*. Typically, local sequences surrounding phospho-acceptor residues are highly important in the recognition of protein substrates by protein kinase C (PKC), cAMP-dependent protein kinase (PKA), and a number of other Ser/Thr protein kinases *(1)*. This has served as the basis for the use of synthetic peptide substrates of defined sequence to identify structural determinants that govern the substrate selectivities of these kinases *(1,3,4)*. This approach was first developed in the laboratory of Edwin G. Krebs, with the use of extensive series of synthetic peptides that corresponded to known phospho-acceptor sites in protein substrates of PKA *(1,5)*. The synthetic peptide Leu-Arg-Arg-Ala-Ser-Leu-Gly (kemptide) was identified in those studies as an excellent artificial PKA substrate (K_m= 16 μM) *(5)*. By applying the same approach, the synthetic peptide substrate Arg-Arg-Lys-Ala-Ser-Gly-Pro-Pro-Val, which corresponds to the sequence of a major PKC phosphorylation site in lysine-rich histone, was later identified as an artificial substrate suitable for analysis of PKC catalysis, based on its K_m (130 μM) and on the lipid cofactor dependence of its phosphorylation by purified rat brain PKC (>10-fold stimulation by phorbol

From: *Methods in Molecular Biology, vol. 233: Protein Kinase C Protocols*
Edited by: A. C. Newton © Humana Press Inc., Totowa, NJ

12-myristate 13-acetate/phosphatidyl serine or Ca^{2+}/phosphatidyl serine) *(6)*. More recently, synthetic peptide substrates have been used to define subtle differences in substrate selectivities among the isozymes in the PKC family *(4)*.

Reactive peptide-substrate analogs that covalently modify protein kinases offer an approach for mapping the phospho-acceptor substrate-binding domain in protein kinase active sites *(7)*. This approach, which was first developed by the laboratory of E. Thomas Kaiser in studies focused on PKA, can be used to identify active-site residues that are proximal to bound phospho-acceptor substrate and also to titrate the active site, that is, to directly measure the moles of active site available for modification. In studies of the catalytic subunit of PKA, which contains two Cys residues, modification by an S→C kemptide analog delivered as the mixed disulfide Leu-Arg-Arg-Ala-Cys (3-nitro-2-pyridinesulfenyl)-Leu-Gly was chemically restricted to disulfide linkage with Cys residues *(7)*. This narrow reactivity resulted in peptide labeling of a single residue, Cys 199, correctly placing that residue in the substrate-binding region of the active site *(7)*, as revealed almost a decade later by the resolution of the crystal structure of the PKA catalytic subunit in complex with a peptide-substrate analog *(8)*. These studies validated the use of reactive peptide-substrate analogs as active-site affinity labels of protein kinases.

Studies with synthetic peptides have revealed that substrate recognition by PKC isozymes generally requires the presence of multiple basic residues proximal to the phospho-acceptor residue of the substrate and, for several of the isozymes, at positions both *N*- and *C*-terminal with respect to that residue *(4)*. An example of an excellent PKC substrate is Arg-Lys-Arg-Thr-Leu-Arg-Arg-Leu (RKRTLRRL) (K_m= 20 μM), which corresponds to a prominent PKC phosphorylation site in the epidermal growth factor receptor (Thr-654) *(9)*. Because Cys residues that occur in highly basic sequences have been observed to have vastly accelerated thiol-disulfide exchange reaction rates compared with Cys residues that occur in uncharged local sequences *(10,11)*, we considered that T→C analogs of RKRTLRRL might be Cys-reactive with PKC and potentially could label the substrate-binding region of the active site. We determined that RKRCLRRL and various *N*-biotinylated analogs of the peptide spontaneously modify PKC isozymes through disulfide linkage with cysteine residue(s) in the active-site region *(12–14)*. The purpose of this chapter is to describe in detail the methodology used to demonstrate that *N*-biotinyl-RKRCLRRL represents a novel class of active-site affinity labels for PKC isozymes. Methods employed to characterize the covalent modification and associated irreversible inactivation of PKC isozymes by *N*-biotinyl-RKRCLRRL and related peptides are described.

2. Materials

1. PKC-inactivating Cys-reactive synthetic peptides were prepared as described in **Subheading 3.**; they include RKRCLRRL, RKRCLRR, and RKR-D-CLRRL. Analogs of the peptides without PKC-inactivating capability include KRCLRRL, RKRCLR, KRCLRR, and KRCLR *(13)*.
2. Tris-HCl and Equilibration Buffers: Tris-HCl Buffer is 20 mM Tris-HCl, pH 7.5. Equilibration Buffer is 20 mM Tris-HCl, pH 7.5, 1 mM ethylenediamine tetra-acetic acid, 1 mM ethylenebis(oxyethylenenitrilo)tetraacetic acid, 10 μg/mL leupeptin, 0.4 mM phenylmethylsulfonyl fluoride.
3. The PKC-inactivating N-biotinylated peptides N-biotinyl-RRRCLRRL, N-biotinyl-RKRCLRRL, and N-biotinyl-KKKCLKKL and the corresponding C→T peptide-substrates are prepared by solid-phase synthesis and high-performance liquid chromatography-purified to >98% purity.
4. Nonreducing sodium dodecyl sulfate polyacrylamide gel electrophoresis (SDS-PAGE) sample buffer: 60 mM Tris-HCl, pH 6.8, 2% SDS,10% glycerol.
5. Tris-buffered saline (TBS): 20 mM Tris-HCl, 0.5 M NaCl, pH 7.5.
6. TBST: TBS containing 0.05% Tween-20.

3. Methods

3.1. Design of Cys-Reactive Peptide-Substrate Analogs for PKC Isozymes

Candidate Cys-reactive peptide-substrate analogs were designed for PKC isozymes based on the following rationale. We first considered, based on observations by Snyder et al. *(10)*, that basic residues that are proximal to cysteine in a peptide or protein sequence may increase the nucleophilicity of the cysteine residue by enhancing its thiolate anion character (RS⁻), that is, by lowering the pKa of the thiol. Second, we took into consideration the successful design of active site-directed, Cys-reactive peptide-substrate analogs for PKA, which is closely related to PKC in the protein kinase superfamily *(2)*, through the substitution of the phospho-acceptor residue of a synthetic peptide substrate with a cysteine residue bearing a disulfide-linked leaving group *(7)*. Next, we reasoned that because phospho-acceptor residues are often surrounded by several basic residues in PKC substrates *(4)*, it might be possible to design T/S→C synthetic peptide-substrate analogs of PKC that spontaneously react with PKC isozymes, with consequent covalent attachment of the peptide at the active-site region via a disulfide linkage. To test this hypothesis, T→C analogs of the PKC-substrate RKRTLRRL were synthesized by standard solid-phase methodology using the Vega coupler 250 peptide synthesizer and purified to greater than 98% purity by reverse-phase high-performance liquid chromatography using a Vydac C4 column with acetonitrile gradient elution at the M.D. Anderson Cancer Center Synthetic Antigen Facility.

3.2. Measurement of the Irreversible Inactivation of PKC Isozymes by S-Thiolating Peptide-Substrate Analogs

The methodology that we used to analyze the irreversible inactivation of PKC isozymes by Cys-reactive peptide-substrate analogs *(12–14)* is described under **Subheading 3.2., steps 1–4**. In the first step, thiol-protective (reducing) agents are removed from the PKC stock, to allow PKC thiols to undergo oxidative reactions. Next, PKC is preincubated with Cys-reactive peptides and analyzed for peptide-induced inactivation. Finally, the role of disulfide formation in the inactivation mechanism is demonstrated by the dithiothreitol (DTT) reversibility of inactivation.

3.2.1. Removal of Reducing Agent from Purified PKC

1. PKC isozyme stock solutions are typically stored at $-20°C$ in the presence of the reducing agent β-mercaptoethanol or DTT (5–15 mM) to protect the isozymes from air oxidation. Over long-term storage (weeks or months), air oxidation of the reducing agent may lead to partial oxidation of PKC thiols. We have found that renaturing PKC thiols to their reduced state through incubation with DTT is a necessary preconditioning step to obtain reproducible results in analyses of oxidative mechanisms of regulation of purified PKC. Preconditioning with DTT restores native PKC from heterogeneous, partially oxidized forms of the enzyme. We typically incubate 5 µg of a commercial purified PKC isozyme (from Pan Vera Corp, Calbiochem Corp, or other commercial suppliers), which is approx 50 µL of the commercial stock, in equilibration buffer containing 2 mM DTT (final volume = 500 µL) for 20 min at 4°C.

2. This is followed by gel filtration of the PKC sample on a small desalting column at 4°C (G-25 resin equilibrated with equilibration buffer; bed volume = 2 mL and height = 4 cm) to remove excess DTT *(13)*. PKC can be recovered with minimal residual DTT and a greater than 90% yield after identifying elution positions of PKC and DTT, based on PKC activity and 5,5′-dithio-bis(2-nitrobenzoic acid) reactivity respectively; the calibrated column may be re-used many times. The PKC sample should routinely be eluted into a single vial that can be capped with limited air volume, kept at 4°C, and used for analysis of oxidative PKC regulation within 2–3 h.

3.2.2. Preincubation of PKC with Cys-Reactive Peptide-Substrate Analogs

1. The reaction of Cys-reactive peptides with PKC is conducted by preincubating gel-filtered PKC (~1 µg of a commercial purified PKC isozyme, e.g., human PKCα, or a purified rat brain PKC isozyme mixture; 95 µL) with the Cys-reactive peptide (final concentration, 0.1–50 µM; 5 µL) in a total volume of 105 µL (20 mM Tris-HCl, pH 7.5). In some cases, DTT (final concentration, 2.0 mM; 5 µL) is included to quench the oxidative reaction (*see* **Note 1**).

2. Each preincubation mixture is incubated in a capped 0.2 mL polymerase chain reaction tube for 5 min at 30°C and then placed on ice. We have found that the reactions of PKC isozymes with the Cys-reactive peptides that we have investigated are complete under these conditions within seconds at 30°C *(13)*, so that a 5-min reaction period at 30°C assures detection of the reactivity of this peptidic class of PKC inactivators with PKC.

3.2.3. Measurement of the Inactivating Effects of Cys-Reactive Peptide-Substrate Analogs on PKC Activity

1. To measure the inactivation of PKC by the Cys-reactive peptide-substrate analogs, a 10-µL aliquot of the PKC-peptide preincubated mixture is added to a PKC assay mixture (total volume, 120 µL), which contains histone as the substrate *(13)*. The final peptide concentration in the assay mixture is less than 5 µ*M*, which is far below concentrations required to reversibly inhibit PKC by competition with the substrate histone. Thus, the observed peptide effects on PKC activity are restricted to oxidative inactivation; this can be demonstrated by the quenching of peptide-mediated PKC inactivation through the inclusion of 2.0 m*M* DTT in the preincubation mixtures (*see* **Note 2**).
2. Although various standard PKC assay methods can be used for this analysis, it is obvious that any reducing agent in the assay mixture recipe must be excluded. For assays of Ca^{2+}-dependent PKC isozymes (cPKCs) in peptide inactivation analyses, we typically employ assay mixtures composed of 20 m*M* Tris-HCl, pH 7.5; 10 m*M* $MgCl_2$; 0.2 m*M* $CaCl_2$;, 30 µg/mL phosphatidylserine; 6 µ*M* [γ-^{32}P]ATP (5000–8000 cpm/pmol); 0.67 mg/mL histone IIIS; and 100 ng of PKC; a 5- to 10-min reaction at 30°C produces linear kinetics *(12–14)*.

3.2.4. Demonstration that PKC Inactivation by the Cys-Reactive Peptides Is DTT-Reversible

Although quenching of peptide-mediated PKC inactivation by DTT indicates an oxidative mechanism, DTT reversal of inactivation must be shown to demonstrate involvement of disulfide bridge formation in the inactivation mechanism *(14)*. We accomplished this by modifying the preincubation conditions for the PKC-peptide mixture *(14)*.

1. PKC is first preincubated with the inactivator-peptide as described under **Subheading 3.2.2.** for 2 min at 25°C, which achieves full peptide-induced PKC inactivation.
2. The PKC:peptide mix is then incubated for an additional 2 min at 25°C with/without 2.0 m*M* DTT. These conditions achieve full DTT reversal of peptide-mediated PKC inactivation *(14)*. We note that because the PKC thiols are renatured to their native reduced state in the initial step of the procedure (*see* **Subheading 3.2.1.**), DTT treatment of PKC in the second preincubation step has negligible effects on PKC activity in the absence of inactivator-peptide.

3.3. Detection of Disulfide-Linked Complexes of PKC Isozymes and S-Thiolating Peptides: Correlation of Covalent Modification with Inactivation of PKC by N-Biotinylated, Cys-Reactive Peptide-Substrate Analogs

1. To demonstrate PKC inactivation by disulfide-linkage of Cys-reactive peptide-substrate analogs to PKC isozymes, that is, *S*-thiolation of PKC by the peptides, we use *N*-biotinylated peptides. As a first step, it is necessary to design active site-directed, *N*-biotinylated peptide-substrate analogs that are Cys-reactive (*see* **Subheading 3.3.1.**).

2. Next, DTT-reversible conjugation of the peptides to PKC is detected using avidin-conjugated to horseradish peroxidase. This is accomplished by probing peptide-inactivated PKC samples that have been subjected to nonreducing SDS PAGE and transferred to nitrocellulose membranes with avidin-horseradish peroxidase (*see* **Subheading 3.3.2.**). The detection of peptide-modified PKC allows the correlation of PKC inactivation with labeling by the peptide, to demonstrate PKC inactivation by an *S*-thiolation mechanism (*see* **Subheading 3.3.2.**). The protection against PKC inactivation and labeling that is afforded by active-site ligands supports the use of the *N*-biotinylated Cys-reactive substrate analogs as active-site affinity labels of PKC (*see* **Subheading 3.3.3.**).

3.3.1. Detection of N-Biotinylated Peptide Labeling of PKC: Correlation of Labeling with PKC Inactivation

We previously developed *N*-myristoylated peptide-substrate analogs with potent inhibitory activity against PKC that involved reversible binding at the active-site region *(9)*. The accommodation of peptide-substrate analogs with bulky *N*-terminal lipid modifications by the active site of PKC suggested the potential usefulness of the *N*-terminus of Cys-reactive peptide-substrate analogs as an attachment position for a tag to track oxidative peptide conjugation to PKC. We therefore designed *N*-biotinylated synthetic peptide substrates corresponding to the Cys-reactive substrate analog RKRCLRRL, and found that *N*-biotinylation actually increased the ability of the peptides to productively bind to the active site of PKC, for example, the K_m of *N*-biotinyl-RKRTLRRL and RKRTLRRL are 5 μ*M* and 20 μ*M*, respectively *(13)*. This indicated the suitability of *N*-biotinyl-RKRCLRRL and variants (*N*-biotinyl-RRRCLRRL and *N*-biotinyl-KKKCLKKL) as candidate active-site affinity labels for PKC.

1. To detect the covalent modification of PKC isozymes by *N*-biotinylated Cys-reactive inactivator-peptides and to allow correlation of the modification with PKC inactivation, PKC was incubated with *N*-biotinylated Cys-reactive peptides (5 min, 30°C) (*see* **Subheading 3.2.2.**) for inactivation analysis (*see* **Subheading 3.2.3.**), in parallel with analysis of the covalent conjugation of the peptide to PKC (*see* **Note 3**) *(13)*.

2. To detect PKC labeling by the peptide, a fraction of each preincubation mixture is boiled for 90 s in nonreducing (β-mercaptoethanol-free) SDS-PAGE sample buffer. The denaturing conditions of sample preparation, followed by electrophoretic separation of the samples by 10% SDS-PAGE, ensure disruption of any reversible PKC-peptide binding interactions, while the disulfide linkage of the peptide to PKC is stable throughout the entire analysis.

3. Next, the samples are electrophoretically transferred to nitrocellulose membranes, and nonspecific binding sites in the membranes are blocked by incubation in 3% BSA in TBS for 15–20 h, that is, by standard Western-analysis blocking methodology.

4. The membranes are then probed for biotinylated polypeptide species, that is, *N*-biotinyl peptide-conjugated PKC isozymes, using avidin conjugated to horseradish peroxidase (Bio-Rad product at a dilution of 1:3000) in 1% BSA in TBST for 2 h at 25°C, followed by washing with TBST.

5. Avidin-reactive bands, which appear at migration positions that closely correspond to the PKC isozymes in the analysis, are visualized by enhanced chemiluminescence (Amersham Corp.) and quantitated by densitometric analysis of the xerograms *(13)*.

6. To confirm equivalent loading of PKC among the samples under comparison in the analysis, membranes are stripped using 62.5 mM Tris-HCl, pH 6.7, 100 mM β-mercaptoethanol, 2% SDS (30 min, 25°C), and then reprobed by standard Western analysis with the appropriate PKC isozyme Abs.

7. We have found that *N*-biotinyl-RRRCLRRL-modified forms of PKC-α, β, γ, ε, and ζ co-migrate with the unmodified forms of the isozymes in nonreducing 10% SDS-PAGE *(13)*. The full reversal of biotinylated-peptide labeling of PKC produced by the addition of 1–5 mM DTT to PKC-peptide incubations mixtures (*see* **Subheading 3.2.2.**) is used to demonstrate that the peptide is conjugated to PKC via a disulfide linkage *(13)*.

3.3.2. Use of Cys-Reactive Peptide-Substrate Analogs as PKC Active-Site Affinity Labels

The design of Cys-reactive PKC-inactivator peptides, based on a minimal structural modification (T/S→C) of synthetic peptide-substrates that are efficiently recognized by the enzyme (K_m<50 μM), can be anticipated to involve an active site-directed inactivation mechanism. Two approaches can be employed to test this.

1. First, the design of a series of structurally related synthetic peptide substrates and a corresponding set of T→C PKC-inactivator peptides allows a structure-activity analysis to compare the rank order of potencies of the T peptides as PKC substrates vs the T→C peptides as PKC inactivators. For example, we have shown that *N*-biotinyl-RRR(T,C)LRRL is much more potent as a PKC substrate (T)/ PKC inactivator (C) than *N*-biotinyl-KKK(T,C)LKKL *(13)*.

2. Second, the question of whether the peptides inactivate and label PKC by an active site-directed mechanism can also be tested by investigating the abilities of active-site ligands, such as nucleotide (MgATP), protein, and peptide substrate/ substrate analogs, to afford protection against modification/inactivation. This is accomplished by including active-site ligands in the PKC-peptide preincubation mixtures (*see* **Subheading 3.2.2.**) before analysis of peptide inactivation (*see* **Subheading 3.2.3.**) and covalent modification of PKC (*see* **Subheading 3.3.2.**). We have thus shown, for example, that MgATP and the pseudosubstrate-peptide FARKGALRQ are each similarly effective against *N*-biotinyl-RRRCLRRL-mediated inactivation and labeling of PKC isozymes *(13)*.

4. Notes

1. In our experience, the lyophilized, redox-sensitive Cys peptides, for example, RKRCLRRL, are stable for 2–3 yr when stored at –70°C, based on inactivation potencies against PKC. Use of the lyophilized peptides within 6 mo is preferred. For studies with PKC, the peptides are prepared as 1.0 mM stock solutions in 20 mM Tris-HCl, pH 7.5. The peptide stock solutions are stable for up to several months at 4°C, when stored in tightly capped vials with limited air volume; use of the stock solutions within 2–3 wk is recommended.
2. To assay the irreversible inactivation of PKC produced by preincubation of the enzyme with Cys-reactive inactivator peptides, we routinely employ histone as the substrate. This is because in some cases, when synthetic peptide substrates are used instead of histone, the Cys-reactive peptides show activity as reversible competitive inhibitors. Use of histone in the PKC assays simplifies the analysis of irreversible PKC inactivation by avoiding measurable reversible inhibition of PKC by the peptidic inactivators.
3. Arg-rich Cys-reactive peptides are generally much more effective than Lys-rich peptides in labeling PKC isozymes. For example, 1 μM of the Arg-rich octapeptide *N*-biotinyl-RKRCLRRL is sufficient to produce a prominent labeled band with nanogram quantities of commercially available purified PKC isozymes.

References

1. Edelman, A. M., Blumenthal, D. K., and Krebs, E. G. (1987) Protein serine/ threonine kinases. *Annu. Rev. Biochem.* **56,** 567–613.
2. Hanks, S. K. and Hunter, T. (1995) The eukaryotic protein kinase superfamily: kinase (catalytic) domain structure and classification. *FASEB J.* **9,** 576–596.
3. Colbran, J. L., Francis, S. H., Leach, A. B., Thomas, M. K., Jiang, H., McAllister, L. M., et al. (1992) A phenylalanine in peptide substrates provides for selectivity between cGMP-and cAMP-dependent protein kinases. *J. Biol. Chem.* **267,** 9589–9594.
4. Nishikawa, K., Toker, A., Johannes, F.-J., Songyang, Z., and Cantley, L. C. (1997) Determination of the specific substrate sequence motifs of protein kinase C isozymes. *J. Biol. Chem.* **272,** 952–960.

5. Kemp, B. E., Graves, D. J., Benjamini, E., and Krebs, E. G. (1977) Role of multiple basic residues in determining the substrate specificity of cyclic AMP-dependent protein kinase. *J. Biol. Chem.* **252,** 4888–4894.
6. O'Brian, C. A., Lawrence, D. S., Kaiser, E. T., and Weinstein, I. B. (1984) Protein kinase C phosphorylates the synthetic peptide Arg-Arg-Lys-Ala-Ser-Gly-Pro-Pro-Val in the presence of phospholipid plus either Ca^{2+} or a phorbol ester tumor promoter. *Biochem. Biophys. Res. Commun.* **124,** 296–302.
7. Bramson, H. N., Thomas, N., Matsueda, R., Nelson, N. C., Taylor, S. S., and Kaiser, E. T. (1982) Modification of the catalytic subunit of bovine heart cAMP-dependent protein kinase with affinity labels related to peptide substrates. *J. Biol. Chem.* **257,** 10,575–10,581.
8. Knighton, D. R., Zheng, J., Eyck, L. F. T., Ashford, V. A., Xuong, N.-H., Taylor, S. S., et al. (1991) Crystal structure of the catalytic subunit of cyclic adenosine monophosphate-dependent protein kinase. *Science* **253,** 407–414.
9. Ward, N. E. and O'Brian, C. A. (1993) Inhibition of protein kinase C by N-myristoylated peptide substrate analogs. *Biochemistry* **32,** 11,903–11,909.
10. Snyder, G. H., Cennerazzo, M. J., Karalis, A. J., and Field, D. (1981) Electrostatic influence of local cysteine environments on disulfide exchange kinetics. *Biochemistry* **20,** 6509–6519.
11. Abate, C., Patel, L., Rauscher, F. J., III, and Curran, T. (1990) Redox regulation of Fos and Jun DNA-binding activity *in vitro. Science* **249,** 1157–1161.
12. Ward, N. E., Gravitt, K. R., and O'Brian, C. A. (1995) Irreversible inactivation of protein kinase C by a peptide-substrate analog. *J. Biol. Chem.* **270,** 8056–8060.
13. Ward, N. E., Gravitt, K. R., and O'Brian, C.A. (1996) Covalent modification of protein kinase C isozymes by the inactivating peptide substrate analog *N*-biotinyl-Arg-Arg-Arg-Cys-Leu-Arg-Arg-Leu. *J. Biol. Chem.* **271,** 24,193–24,200.
14. Ward, N. E., Pierce, D. S., Stewart, J. R., and O'Brian, C. A. (1999) A peptide substrate-based affinity label blocks protein kinase C-catalyzed ATP hydrolysis and peptide-substrate phosphorylation. *Arch. Biochem. Biophys.* **365,** 248–253.

X

GENETIC APPROACHES TO STUDYING PROTEIN KINASE C

36

Genetic Approaches to Studying Protein Kinase C

An Introduction

Robert O. Messing

The mammalian protein kinase C (PKC) super family is comprised of 10 gene products, and this large number of isozymes has presented a challenge to understanding the function of individual family members. Nonselective pharmacological approaches have implicated PKCs in a number of physiological processes and have identified numerous PKC substrates. However, it has been difficult to assign individual functions and substrates to specific isozymes. This is partly because few cell-permeable molecules are available for use as isozyme-selective PKC inhibitors. In addition, PKC isozymes exhibit broadly overlapping substrate specificities in vitro. This has led many investigators to suggest that there is redundancy of function within the PKC family. On the other hand, several studies have identified second messengers that activate subgroups of PKC isozymes, whereas others have identified isozyme-specific patterns of subcellular localization before and after PKC activation. These findings suggest that there is functional specificity among different members of the PKC family.

Because most pharmacological agents lack isozyme specificity, genetic approaches have provided a useful and powerful alternative to study the function of individual PKC isozymes. Genetic strategies used to inhibit isozyme function include elimination or down-regulation of isozyme abundance through gene targeting, RNA interference, antisense RNA, generation of kinase-inactive mutants that act as dominant negative inhibitors, or expression of kinase fragments or peptides that act as inhibitors of isozyme translocation to sites of activation. Genetic strategies that have been used to promote activation of

From: *Methods in Molecular Biology, vol. 233: Protein Kinase C Protocols*
Edited by: A. C. Newton © Humana Press Inc., Totowa, NJ

individual PKC isozymes include isozyme over-expression and expression of pseudosubstrate domain mutants that are constitutively active. These manipulations have involved transient or stable transgenic expression of cDNA constructs in cells, generation of transgenic animals, or elimination of genes in mice by homologous recombination in embryonic stem cells. Invertebrate animal models have proven valuable in the study of loss of function mutants and in identification of proteins that interact with specific PKC isozymes. Expression of mammalian isozymes in yeast increases cell-doubling time in proportion to the level of active PKC, thereby providing a rapid cell-based assay of PKC activation. This approach has recently been used to map domains on PKCα that interact with different PKC activators and to screen PKC isozyme mutations for their effects on kinase activity. This part will describe studies of PKC isozymes using genetic techniques. Chapter 37 will review invertebrate and mouse models used to identify isozyme-specific functions in whole animals. Chapter 38 will focus on the use of transgenic approaches to the study of PKC isozymes in mammalian cell culture. Chapter 39 will discuss using yeast as an assay system for the identification of PKC activators.

37

Animal Models in the Study of Protein Kinase C Isozymes

Doo-Sup Choi and Robert O. Messing

1. Introduction

The decoding of genomes for several species has provided investigators with an unprecedented wealth of DNA sequence information to identify, mutate, and overexpress genes in animals, and study phenotypes that result from these manipulations. This is particularly valuable for the study of the protein kinase C (PKC) family because the pharmacological tools available to examine individual PKC isozymes in whole animals are few. This chapter reviews some of the animal models and approaches that have been used and discusses some of the limitations inherent in these studies, particularly those involving knockout and transgenic mice. A protocol for identifying founder lines of transgenic mice that express the tetO-minimal CMV (Ptet) promoter driving a mouse PKCε cDNA transgene is also described.

1.1. Invertebrate Models

1.1.1. C. elegans

The nematode *C. elegans* offers several advantages as a model system for study of PKC function. There are methods for generating mutant, transgenic, and null strains that allow for analysis of gene function at both the organism and single-cell levels. *C. elegans* contains only 959 somatic cells and its developmental and cell biology are extremely well characterized. Several signal transduction pathways have been genetically mapped and found to use proteins and mechanisms similar to those found in higher eukaryotes. Four PKC genes have been identified: *tpa-1*, *pkc-1*, *pkc-2*, and *pkc-3*. Both *tpa-1* and *pkc-1* encode novel PKC isozymes, *pkc-2* encodes several splice variants with

From: *Methods in Molecular Biology, vol. 233: Protein Kinase C Protocols*
Edited by: A. C. Newton © Humana Press Inc., Totowa, NJ

homology to mammalian conventional PKCs, and *pkc-3* encodes an atypical PKC *(1–4)*.

Genetic approaches have been used to identify functions for *tpa-1* and *pkc-3* gene products. The *tpa-1* gene was discovered in a screen of transposon Tc1-induced mutants resistant to the growth inhibiting and incoordinating effects of phorbol esters *(4)*. Tc1 was found to be inserted in a region encoding the kinase domain of the *tpa-1* gene, resulting in the loss of the two alternatively spliced products TPA1A and TPA1B, which are novel PKCs most homologous to mammalian PKCδ and PKCθ. Introduction of genomic DNA containing the wild-type *tpa-1* locus into a Tc1-insertion mutant restored TPA-1 production and sensitivity to phorbol ester *(5)*, indicating that wild-type TPA-1 is necessary and sufficient for developmental and locomotor sensitivity to phorbol ester in *C. elegans*.

The *pkc-3* gene product PKC3 was isolated from a *C. elegans* cDNA library using probes derived from ESTs homologous with human PKCζ *(3)*. PKC3 shares substantial identity with mammalian PKC λ/ι (57%) and PKCζ (55%). In adult worms, PKC3 mRNA and protein are present in approximately 85 muscle, epithelial, and hypodermal cells involved in feeding, digestion, excretion, and reproduction. PKC3 is also present early in embryonic development. Ablation of PKC3 expression using capped antisense RNA injected into the gonadal syncytium of young adult worms disrupts embryonic development *(4)*. *C. elegans* embryos depleted of PKC3 by RNA interference die and show defects resembling those observed in *par-3* (partitioning-defective) and *par-6* mutants, which show defects in early asymmetric cell divisions *(6)*. A link between atypical PKCs and *par* gene products was suggested when screening of a mammalian cDNA expression library with PKCζ identified an atypical PKC-interacting protein homologous to PAR-3 *(7)*. PKC3 has been found to form a ternary complex with PAR-3 *(6)* and PAR-6 *(8)*, and the absence of any one of these proteins disrupts localization of the other two, producing very similar defects in early development in *C. elegans*.

1.1.2. D. melanogaster

Five PKC genes have been identified in *D. melanogaster*. The first one described, *Pkc53E*, maps to position 53E2 on the second chromosome *(9)*, is expressed in neural tissues of the head, and is most homologous to conventional mammalian PKC isozymes *(10)*. A second, highly related, conventional PKC gene, *inaC*, is located at position 53E1 on the second chromosome and is expressed only in photoreceptor cells *(10)*. *Pkc98F* was cloned from a genomic library and is most related to novel mammalian PKC isozymes. It is expressed during embryonic development and is present in the head of adult flies *(10)*. *Pkcdelta* was identified from the *Drosophila* genome database and maps to position 11A10-11; the predicted protein sequence (SWP:P83099) shows 53% identity with mouse PKCδ (SWP:P28867). A single atypical PKC gene, *aPKC*, was identified from the *Drosophila* genome database and maps to 51D7-8 on the right arm of chromosome 2 *(11)*. The predicted amino acid sequence shows 68% identity to mouse PKCλ, 63% identity to mouse PKCζ, and 58% identity to *C. elegans* PKC-3.

Mutants have been described for only two of these genes. Studies with *inaC* mutants have shown that this PKC is involved in deactivation of rhodopsin-mediated signaling *(12)*. Examination of a P-element insertion mutant in *aPKC* revealed that disruption of this gene is lethal early in development *(11)*. This PKC isozyme binds to the multi-PDZ domain protein Bazooka, which is a homolog of Par-3 and atypical PKC-interacting protein. Examination of loss-of-function mutants shows loss of apico-basal polarity and abnormal layering of epithelia, and abnormal spindle orientation in neuroblasts. Therefore, *aPKC*, like PKC-3 and mammalian atypical PKC, participate in a conserved mechanism the controls cell polarity and early development.

1.2. Gene Targeting in Mice

1.2.1. Phenotypes in PKC Knockout Mice

In attempting to understand isozyme-specific functions in humans, use of invertebrate models presents limitations because there are a small number of corresponding behaviors and fewer PKC family members. However, gene targeting in mice has proven to be a very useful tool for understanding the function of mammalian genes, and all human PKC isozymes have closely related counterparts in mice. To date, phenotypes have been identified for six mouse PKC isozymes (**Table 1**). In addition, mice lacking PKCα appear to have no obvious developmental or neurobehavioral abnormalities, but show a deficit in hippocampal LTP (J. David Sweatt, personal communication). Deletion of the PKCλ gene appears to be lethal in mice (M. Leitges, personal communication), as is the deletion of the related atypical PKC isozymes PKC3 in *C. elegans* (*see* **Subheading 1.1.1.**) and aPKC in *D. melanogaster* (*see* **Subheading 1.1.2.**). This suggests an important role for PKCι/λ in mammalian development.

So far, all published work on PKC-null mice has involved conventional gene targeting in which one or more critical exons are replaced by an antibiotic resistance cassette. In addition to loss of isozyme-specific signaling in adult tissues, other factors can contribute to the phenotypes observed in these conventional knockout mice. Chronic loss of the isozyme can also evoke homeostatic mechanisms that compensate for the lost gene and contribute to the phenotype. This can complicate attempts to fully understand isozyme function in a normal animal. However, the chronic nature of the knockout model

Table 1
Phenotypes in PKC-Null Mice

Isozyme	Phenotype	Reference
β	• B cell immunodeficiency.	*(38)*
	• Enhanced glucose transport in adipocytes, reversed by transgenic βI.	*(39)*
	• Inhibited mast cell degranulation and IL-6 production.	*(40)*
	• Reduced PMA-stimulated NADPH oxidase activity in neutrophils.	*(16)*
	• Impaired hypoxia-induced expression of Egr-1 and tissue factor in lung (procoagulant cascade).	*(41)*
	• Deficits in cued and contextual fear conditioning.	*(42)*
	• Reduced retinal neovascularization induced by hypoxia.	*(24)*
γ	• Diminished LTP mild defecits in spatial and contextual learning, depressed glutamate-stimulated R3 phosphorylation in hippocampus.	*(43–45)*
	• Impaired motor coordination and impaired synapse elimination in cerebellum.	*(46,47)*
	• Reduced neuropathic pain, reduced desensitization of mu opioid receptors, enhanced mu-opioid antinociception.	*(48–50)*
	• Decreased ethanol-induced hypothermia, hypnotic effect of ethanol, ethanol enhancement of $GABA_A$ receptor function in vitro, and tolerance to hypnotic effect of ethanol; increased ethanol consumption.	*(51–53)*
	• Decreased anxiety.	*(54)*
	• Increased infarct size after focal cerebral ischemia.	*(55)*
δ	• Enhanced pervanadate-induced tyrosine phosphorylation in mast cells.	*(56)*
	• Increased vein graft arteriosclerosis.	*(57)*
ε	• Decreased inflammatory nociceptor sensitization.	*(18)*
	• Decreased alcohol consumption and enhanced positive allosteric modulation of GABA-A receptors; reduced operant self-administration of alcohol, ethanol-induced mesolimbic dopamine release, and alcohol withdrawal severity.	*(19,58,59)*
	• Decreased anxiety.	*(60)*
	• Impaired macrophage activation by LPS and IFN gamma and decreased survival after *Escherichia coli* or *Staphalococcus aureus* injection.	*(61)*
ζ	• Delayed B cell maturation and impaired NFκB activation in by tumor necrosis factorα and Il-1 in embryonic fibroblasts and lung, by lymphotoxin β receptor activation in embryonic fibroblasts, and by lipopolysaccharide in lung.	*(62)*
θ	• Impaired T cell receptor activation of NFκB in mature T cells	*(63)*

can provide useful predictions about responses to chronic treatment with an isozyme-selective PKC inhibitor. Loss of a PKC isozyme in one organ may exert remote effects on other organ systems, particularly if it regulates secretion of an endocrine hormone. If the targeted PKC plays a role in development, then the null allele may be lethal, or the adult phenotype will show a developmental change. This can complicate attempts to understand the function of the isozyme in adult cells. Finally, polymorphisms in background genes can contribute to the phenotype.

Many of the phenotypes described in **Table 1** are subtle and have required careful study of individual organ systems for detection. As more investigators examine different organ systems in these animals, additional phenotypes are likely to be discovered. Therefore, the availability of these conventional knockout models provides a valuable resource to the scientific community. Here, we suggest some guidelines that we have adopted for dealing with issues related to genetic background and for discerning developmental from adult signaling events in conventional PKC knockout mice.

1.2.2. Genetic Background in Conventional Knockout Mice

Heterogeneity in genetic background is a particularly important problem that has received attention in several reviews *(13,14)*. Polymorphisms in other genes can mask or enhance the null phenotype. Most investigators use hybrid C57BL/6 × 129 mice for studies because they breed well and perform well in behavioral assays. However, differential fixation of 129 and C57BL/6 alleles in populations of wild-type and knockout mice can lead to false conclusions about the knockout phenotype.

One can study congenic mice to eliminate this problem. It is easy to generate inbred 129 mice by crossing chimeric animals with 129 mice of the substrain from which the ES cells were derived. Because there are genetic differences between 129 substrains *(14)*, crossing chimeras into a different substrain will not yield a congenic line. Congenic C57BL/6 mice can be generated through repeated backcrossing for at least eight generations, and may be used for studies in which 129 mice perform poorly. Congenic C57BL/6 and inbred 129 mice heterozygous for the null allele can be bred to produce F1 hybrid wild-type and knockout littermates for studies. Because F1 mice inherit one C57BL/6 chromosome and one 129 chromosome of each chromosome pair, the genetic background is nearly identical in all mice. However, it takes several months to generate congenic C57BL/6 mice. In the meantime, one can use wild-type and null littermate F2 progeny of F1 heterozygotes generated from chimeric and C57BL/6 mice. This still produces an equal representation of C57BL/6 and 129 alleles in the general F2 population, but each individual F2 mouse may have a different proportion of C57BL/6 and 129 alleles. This can make the experimental results more variable than results from studies that use inbred, congenic, or F1 hybrid mice.

In hybrid mice, genes near the targeted locus will be 129 alleles in the knockouts and C57BL/6 alleles in the wild-type animals. This also occurs in congenic C57BL/6 mice since, even after 10 generations of backcrossing, there are still nearly 50 genes with 129 alleles linked to the targeted locus in the null mice *(15)*. However, if studies using inbred 129 mice give results similar to those observed with hybrid or backcrossed C57BL/6 mice, it is not likely that neighboring 129 alleles contribute to the phenotype.

In some studies, 129 mice cannot be used or may not be available. As an alternative, some investigators have compared phenotypes in wild-type C57BL/6 mice with those in wild-type 129 mice, and if the results are similar, assume that there is no influence of genetic background. However, this does not account for potential epistatic interactions between different background alleles that could influence the phenotype. It also does not provide a solution if there are differences observed between C57BL/6 and 129 wild-type animals. One can instead use genetic markers to identify C57BL/6 and 129 polymorphisms closely linked to the targeted locus. By crossing wild-type C57BL/6 and 129 mice, one can identify wild-type F1 or F2 progeny with C57BL/6 alleles surrounding the locus and compare these mice with wild-type mice that carry 129 alleles in that region. If their phenotypes are similar, it is unlikely that polymorphisms near the targeted PKC locus contribute to differences between wild-type and null hybrid mice.

1.2.3. Pharmacological Studies to Confirm Phenotypes

A simpler approach to confirming that phenotypes observed in adult PKC-null mice are not due to a confounding effect of genetic background or to developmental changes is to use pharmacological agents that inhibit the targeted PKC isozyme. Although the number of isozyme-selective inhibitors is limited, some do exist. Examples include LY333531 and 379196, which are selective, small molecule inhibitors of PKCβ *(16)*. In addition, peptides derived from the C2-like domains of PKCε and PKCδ *(17)* also appear to be isozyme selective, and modifications that permit cellular uptake have made it possible to use them in cell culture and isolated organ preparations. Thus, LY333531 and 379196 have been used to confirm that the loss of PKCβ leads to inhibited mast cell degranulation and interleukin-6 production *(16)*, whereas we have used the peptide εV1-2 to confirm that loss of PKCε reduces pain sensitization by inflammatory mediators *(18)* and enhances the sensitivity of $GABA_A$ receptors to positive allosteric modulators *(19)*.

1.3. Transgenic Mice

1.3.1. Transgenic Overexpression of PKC Isozymes

Another approach to study the function of individual PKC isozymes is to generate transgenic mice that overexpress an isozyme. Generally, it takes less time to generate transgenic mice than knockout mice, and there are a wide variety of promoters available for tissue-selective expression. Thus, PKC isozymes have been selectively overexpressed in skin *(20)* and the heart *(21,22)* using tissue-specific promoters. Transgenic expression can also be useful for determining functions of an isozyme splice variant when gene targeting eliminates all splice variants of an isozyme. For example, overexpression of PKCβII is sufficient to cause colonic hyperproliferation and increased risk of colon carcinogenesis *(23)*. Phenotypes observed in transgenic mice can be compared to those observed in gene-targeted mice, and, if opposite, may provide some confidence that the phenotypic changes are caused by isozyme expression and not spurious effects of genetic background. Thus, overexpression of PKCβII increases the angiogenic response to oxygen deprivation in the retina, which is opposite to the phenotype observed in PKCβ-null mice *(24)*.

1.3.2. Transgenic Expression of PKC Peptides

Transgenic methods can also be used to express peptides that activate or inhibit PKC isozymes. This approach has been cleverly used to demonstrate trophic effects of PKCε in cardiomyocytes using transgenes encoding the inhibitory εV1 fragment of PKCε or the activating, eight-amino acid, pseudo-receptor for activated C kinase sequence of PKCε driven by a mouse α myosin-heavy chain promoter *(25)*. Transgenic overexpression or activation of a PKC isozyme does not always generate phenotypes opposite to null mutants. For example, although PKCβ overexpression causes cardiomyopathy *(21,22)*, loss of PKCβ does not protect against cardiomyopathy induced by phenylephrine or aortic binding *(26)*. Likewise, unlike εV1-expressing mice, which develop a dilated cardiomyopathy *(25)*, PKCε-null mice have normal cardiac size and function *(27)*. The reasons for this discrepancy may involve different compensatory changes provoked by the εV1 transgene and the null mutation, or differences in genetic background (FVB/N in the transgenic mice and C57BL/6 × 129SvJae in the null mice).

1.3.3. Transgenic Restoration of Targeted PKC Isozymes

Another approach that can be used to confirm that the knockout phenotype is caused by loss of the targeted PKC involves restoring the missing PKC isozyme in PKC null mice using an inducible system. This approach also allows for within subjects comparisons, minimizing variability resulting from individual differences. We have successfully used a tetracycline suppressible system *(28)* that allows for regulated expression of mouse PKCε in neurons, such that when tetracycline is absent, PKCε is expressed (tet-off system) *(29,30)*. This approach requires generating two lines of transgenic mice, one expressing a tetracycline transactivator that binds tetracyclines, and another that expresses a PKC transgene driven by a promoter composed of a tet operon heptamer fused to a minimal CMV promoter. These mice are then intercrossed to yield bigenic mice, which are bred with PKC-null mice to yield knockout mice expressing both transgenes. Expression of the transgenic PKC is restricted to cells that contain both transgenes. The promoter used to drive expression of the transactivator determines the cell type in which transgenic PKCε is expressed. **Figure 1** shows the breeding scheme that was used to generate PKCε-null mice that express transgenic mouse PKCε in neurons when fed a diet lacking doxycycline. A similar strategy can be used to generate mice in which transgenic expression is induced by doxycycline (tet-on system) *(31,32)*.

A major goal in generating mice for inducible expression is to identify mice with proper integration of each transgene such that there is little leak but robust expression under appropriate treatment conditions. This requires a two-phase process. The first phase involves generating several lines of transgenic mice expressing the ligand-binding transgene. This can be monitored by Western analysis and immunohistochemistry to choose mice with the highest levels of expression. Once an appropriate tissue-specific ligand-binding mouse line has been identified, one can use it in the second phase to screen mice expressing the transgenic PKC construct to identify bigenic mice with low leak and high levels of inducibility. If performed in vivo, this is a tedious process because one must compare gene expression before and after induction. The process of screening mice carrying the PKC construct can be greatly accelerated by assessing PKC transgene expression in vitro rather than in vivo using mouse ear fibroblasts by the method described below originally developed by Tremblay and colleagues *(33)*.

1.3.4. Limitations in Using Transgenic Mice

Most transgenic mice are generated using promoters that do not produce the same pattern of expression as that observed for the endogenous promoter. If expression is restricted to a subset of cells that normally express the isozyme, then it may be possible to gain information that links function with an anatomical structure. However, if transgenic expression occurs outside the normal pattern of expression, then phenotypes may be observed that have little to do with normal isozyme function. Greater fidelity in matching endogenous

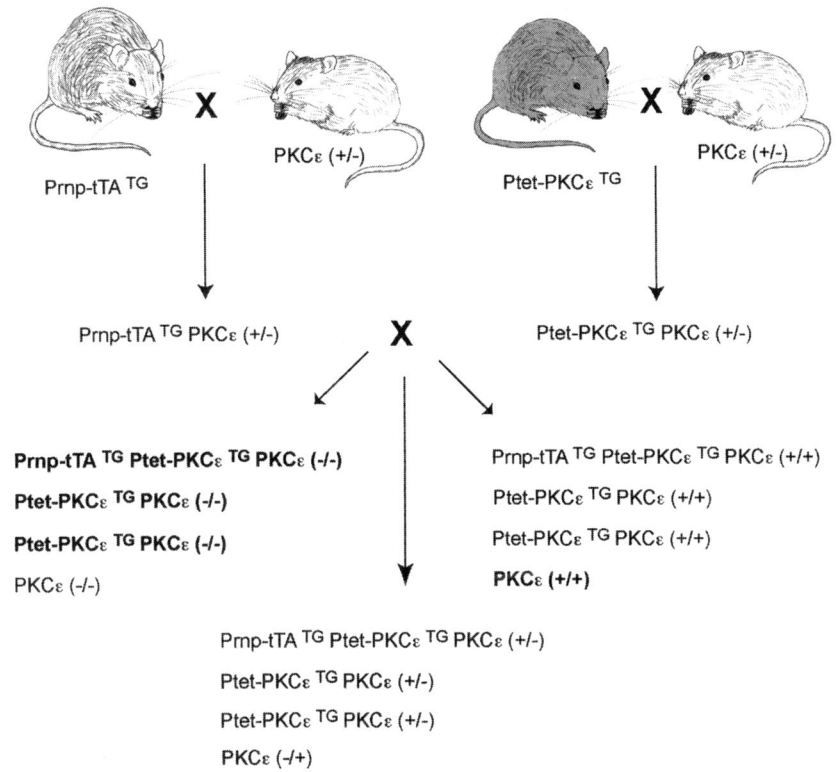

Fig. 1. Breeding scheme for tetracycline regulated expression of transgenic PKCε on the PKCε-null background. Mice heterozygous for the PKCε null mutation (PKCε (+/–) are crossed with transgenic mice expressing the tTA driven by the prion promoter for neuronal expression of tTA (Prnp-tTA) *(33)* or mice carrying the mouse PKCε cDNA sequence under control of the tetO hepatamer-minimal CMV promoter (Ptet-PKCε). Mice are selected that are heterozygous for the null mutation and carry either the Prnp-tTA or the Ptet-PKCε transgene. These mice are crossed to generate mice that lack PKCε and express both transgenes [Prnp-tTA[TG] Ptet[TG] PKCε (–/–)]. Mice that lack PKCε and express only one of the transgenes [Prnp-tTA[TG] PKCε (–/–) or Ptet[TG] PKCε (–/–)] are used as null mutant controls. Wild-type littermates or wild-type mice that express both transgenes [Prnp-tTA[TG] Ptet[TG] PKCε (+/+)] can be used as controls for the wild-type locus. Progeny we used for experiments are in boldface.

expression patterns can be achieved by expressing transgenes under control of endogenous promoters using bacterial artificial chromosomes, as has been described, for example, for expression of neuropeptide Y *(34)*.

Integration of a transgene may produce a loss of function mutation in a nearby locus that could contribute to the phenotype. This can be explored by generating and examining a second transgenic line because integration usually is unlikely to occur at the same locus in both lines.

Overexpression can be a useful model for studying increased function of an isozyme because increased levels of the protein provide increased functional capacity. However, endogenous levels of the isozyme may not be normally rate-limiting, and high levels of transgenic overexpression may lead to the presence of the isozyme in cellular compartments in which it is not normally found. This could produce phenotypes that have little to do with normal function of the isozyme. Therefore, careful analysis of gross and cellular expression patterns is important in interpreting phenotypes in transgenic mice that overexpress PKC isozymes.

1.4. Future Directions

In addition to generating null alleles, homologous recombination can be used to generate mutant alleles with altered function. Generation of such knock-in mutants could be very useful in studying PKC processing by phosphorylation in vivo and in substrate mapping with mutant isozymes that utilize novel ATP analogs and ATP-binding site inhibitors (*see* chapter 21).

RNA interference (RNAi) has proven useful for decreasing the expression of specific gene products in invertebrates *(35)*. Recently, it has become possible to use this approach to knock down genes in mammalian cells *(36)*. If appropriate RNAi constructs can be identified for PKC isozymes in mammalian cells, it may be possible to express these in transgenic mice. Inducible expression could allow for the level of the RNAi, and in turn the degree of isozyme knock down, to be titrated. This could allow correlation of protein level with phenotype expression, and provide additional evidence for a causal relationship.

Several investigators are currently using recombinases to generate null mutant mice and to avoid some of the limitations associated with conventional knockout technology. Cre recombinase is a bacteriophage enzyme that can excise regions of DNA lying between pairs of its site-specific, 34-nucleotide, recognition sites known as loxP sequences. A targeting vector is designed with loxP sequences flanking one or more exons. ES cells carrying this floxed gene are transfected with a Cre recombinase transgene to induce the null mutation. Alternatively, the ES cells can be used to generate floxed mice carrying the loxP-flanked gene. These are crossed with mice expressing a Cre recombinase transgene to yield null mutant mice in vivo. Wild-type mice are generated by crossing floxed mice with mice from the same genetic background as that of the Cre transgenic mice. This generates populations of wild type and null progeny with nearly identical genetic background. The only background difference is that wild-type animals carry two loxP sites and null mice have one. If the loxP sites do not alter RNA processing or stability, this difference will be silent.

A very useful application of in vivo recombination is in production of tissue-specific knockout mice using tissue-specific promoters to drive Cre expression. A potentially powerful application would be to use inducible systems to regulate Cre activity and generate inducible knockout mice. As discussed above for inducible transgenic mice, this would permit a within-subjects experimental design whereby each animal can serve as its own control. It also would allow one to determine if the phenotype in an adult knockout results from developmental changes, and could permit the study of gene function at different developmental stages or when a conventional knockout is lethal during development.

Several technical issues have made it difficult to generate inducible knockout mice. Cre-mediated excision does not happen in all cells at the same time and there can be a delay of several days between induction of Cre transcription and gene excision. Therefore, induced Cre-mediated excision is not very useful for detailed studies of rapidly changing events. There may be leaky expression of Cre activity in the absence of the inducing agent and mosaic expression of activity following induction *(37)*. If the mosaic expression pattern is consistent among individuals in the mouse line, however, it could be used to identify specific cells that contribute to the null phenotype. This could be very informative in studying a heterogeneous tissue, such as the brain. In the future, improvements in inducible systems will hopefully yield transgenic mice expressing robust inducible recombinase activity in specific cell types. This should allow investigators to use their floxed mouse lines for studies of tissue-specific, inducible knockout mice.

1.5. Mouse Ear Fibroblast Assay for Screening Ptet Founder Lines of Mice

We describe our experience using this method to identify founder lines of transgenic mice that express the tetO-minimal CMV (Ptet) promoter driving a mouse PKCε cDNA transgene. The approach involves transfecting fibroblasts derived from the founders with a CMV-tTA cDNA vector to allow for expression of transgenic PKCε that can be suppressed by doxycycline. Founders with cells showing robust expression without doxycycline and complete suppression with doxycycline can be rapidly identified.

2. Materials

1. Dulbecco's Modified Eagle Media (DMEM): Gibco-BRL (cat. no. 11965-092).
2. Fetal Bovine Serum (FBS): JRH Bioscience (cat. no. 12103-78P).
3. L-Glutamine, 200 mM (100X: Gibco-BRL (cat. no. 11965-092).
4. Penicillin-streptomycin (100X, 10,000 units of penicillin and 10,000 µg/mL of streptomycin: Gibco-BRL (cat. no. 15140-122).

5. Collagenase/dispase, 0.1 units/mL of collagenase and 0.8 units/mL dispase: Roche Applied Science (cat. no. 269638; *see* **Note 1**).
6. pTet-off plasmid, CMV-tTA: Clontech (cat. no. K1620-A).
7. SuperFect Transfection Reagent: Qiagen (cat. no. 301307).
8. G418, Geneticin: Gibco-BRL (cat. no. 10131-027).
9. Doxycycline: Clontech (cat. no. 8634-1) (*see* **Note 2**).
10. Cell lysis buffer: 20 mM Tris-HCl (pH 7.4), 2 mM ethylenediamine tetraacetic acid , 10 mM ethylenebis(oxyethylenenitrilo)tetraacetic acid, 40 μg/mL leupeptin, 40 μg/mL aprotinin, 25 μg/mL soybean-trypsin inhibitor, and 1 mM phenlymethylsulfonyl fluoride (*see* **Note 3**).
11. PKCε Antibody: BD Transduction Laboratories (cat. no. P14820-050) (*see* **Note 4**).

3. Methods
3.1. Mouse Ear Fibroblast Cultures

1. Cut half of the ear from transgenic founder mice expressing Ptet-PKCε and put the tissue in DMEM medium with 10% FBS and 2X L-glutamine and 1X penicillin-streptomycin.
2. Wash the tissue twice with 70% ethanol. Place a drop of DMEM media on the tissue and cut with a single-edge razor blade in a sterile plate (*see* **Note 5**).
3. Add 1 mL of media to the macerated tissue and pipette it into one well of a six-well culture plate. Add another 0.5 mL to the remaining cells that were not pipetted initially and transfer these to the well. Repeat with another 0.5 mL. Try to pick up as much of the ear tissue as possible.
4. Add 2 μL of collagenase/dispase and incubate the fibroblasts and the enzyme at 37°C overnight.
5. Collect the media and cells and place in a 15 (or 50 mL) conical Falcon tube. Wash the well with another 1–2 mL of media to ensure all cells are removed.
6. Disaggregate the cells at least 5 times through a 20-gauge needle and about 10 times through a 22-gauge needle (*see* **Note 6**).
7. Spin the tubes 15 min at 900g to pellet the cells. Aspirate the media gently using a pipetter (not a vacuum aspirator) and wash the cells with phosphate-buffered saline (PBS).
8. Repeat the spin and gently remove the PBS with a pipetter once again.
9. Resuspend the cells in 2 mL of DMEM/10% FBS/2X L-glutamine/1X penicillin-streptomycin using a syringe and a 22-gauge needle.
10. Place in one well of a six-well plate and let grow at 37°C for 24–36 h. Wash the cells with media twice to remove cell debris. Add 2 mL of fresh media.
11. To split the cells, aspirate the media, wash once with PBS, add 1 mL of trypsin, and incubate at 37°C for 1 min. Then, add more media and transfer the cells to a larger flask (*see* **Note 7**).

3.2. Transient Transfection of Cultured Mouse Fibroblast Cells with Ptet-Off Plasmid

1. The day before transfection, transfer $2–8 \times 10^5$ cells to each 60-mm dish in 5 mL of DMEM growth media.
2. Incubate the cells for 2–3 d at 37°C, 5% CO_2 incubator until approx 80% confluent.
3. Dilute 5 µg of CMV-tTA vector (10 µL of 0.5 µg/µL) with 140 µL of DMEM growth medium containing no serum, proteins, or antibiotics, to a total volume of 150 µL. Mix and spin down the solution for a few seconds to remove drops from the top of the tube.
4. Add 30 µL of SuperFect Transfection Reagent (Qiagen) to the DNA solution. Mix by pipetting up and down five times.
5. Incubate the samples for 10 min at room temperature to allow complex formation.
6. While complex formation takes place, gently aspirate the growth medium from the dish, and wash cells once with 4 mL of PBS.
7. Add 1 mL of cell growth medium containing serum and antibiotics (1× penicillin-streptomycin) to the reaction tube containing the transfection complexes. Mix by pipetting, and immediately transfer the total volume to the cells in the 60-mm dishes.
8. Incubate with complexes for 3 h at 37°C, in a 5% CO_2 incubator.
9. Remove medium containing the remaining complexes from the cells by gentle aspiration, and wash cells once with 4 mL of PBS.

3.3. Suppression of Gene Expression Using Doxycycline

1. Add cell growth medium containing serum and antibiotics (1X penicillin-streptomycin) and 200 µg/mL of G418. To examine regulation of PKCε gene expression, treat some cultures with 0.5 µg/mL doxycycline (*see* **Note 8**).
2. Incubate the cells for 3–4 d at 37°C, 5% CO_2 incubator.
3. Remove culture media and wash twice with ice-cold PBS.
4. Add 250 µL of cell lysis buffer to each plate, and collect the cells using a cell scraper.
5. Prepare proteins for immunoblot, and determine PKCε expression level (*see* **Note 9**).

4. Notes

1. Prepare 100X stock solution in PBS, frozen in 100-µL aliquots at –20°C until use.
2. Doxycycline (Dox) is a derivative of tetracycline. Dox can be used at 100-fold lower concentrations than tetracycline, yielding inducing concentrations ranging from 10 pg/mL to 1 µg/ml. Dissolve at 1 mg/mL in H_2O and filter sterilize. Dox

is light sensitive and can be stored at 4°C in the dark up to 4 wk for short-term storage. Aliquots can be frozen at –20°C for long-term storage.

3. Prepare 1000× protease inhibitor mixture in H_2O containing 40 mg/mL leupeptin, 40 mg/mL aprotinin, and 25 mg/mL soybean-trypsin inhibitor. Prepare 1 M stock of phenylmethylsufonyl flouride in DMSO. Freeze the stock solution at –20°C for long-term storage.

4. Other PKC antibodies also are available from many vendors (e.g., BD Transduction Lab, Santa Cruz Biotech., Zymed Lab or Upstate Biotech). In case of PKCε, the monoclonal antibody from BD Transduction works well.

5. Chop the ear tissue until it becomes a paste. The tissue can be dragged across the plate with short strokes to ensure that it is as disaggregated as possible.

6. This step may be critical to get good cultures. Thus, it is important to disaggregate tissues totally.

7. Usually transfer to a T25 culture flask, and then a T75 flask for maintenance of the cultures.

8. To examine inducibility of the PKCε gene in Ptet-PKCε transgenic mice, compare the expression level between Dox treated and non-treated pTet-off (CMV-tTA) plasmid transfected cells in triplicate plates. It is also important to test for leakage of PKCε gene expression by comparison of Dox treated and non-treated pTet-off plasmid nontransfected with CMV-tTA.

9. Examine at least two mice per founder line. Expression level may be little variable between mice. Choose the lines with the highest expression and the lowest leakage when suppressed by Dox. Then, breed with tissue-specific promoter driven tTA or rtTA transgenic mice to achieve Tet-Off or Tet-On bigenic PKC inducible mice.

Acknowledgments

The authors wish to thank Jennifer Cabbage for artwork in Figure 1. This work was supported by funds provided by the State of California for medical research on alcohol and substance abuse through the University of California at San Francisco, and by NIH grant AA08117 (to R.O.M.).

References

1. Islas-Trejo, A., Land, M., Tcherepanova, I., Freedman, J. H., and Rubin, C. S. (1997) Structure and expression of the Caenorhabditis elegans protein kinase C2 gene. Origins and regulated expression of a family of Ca2+-activated protein kinase C isoforms. *J. Biol. Chem.* **272,** 6629–66240.

2. Land, M., Islas, T. A., Freedman, J. H., and Rubin, C. S. (1994) Structure and expression of a novel, neuronal protein kinase C (PKC1B) from *Caenorhabditis elegans.* PKC1B is expressed selectively in neurons that receive, transmit, and process environmental signals. *J. Biol. Chem.* **269,** 9234–9244.

3. Wu, S. L., Staudinger, J., Olson, E. N., and Rubin, C. S. (1998) Structure, expression, and properties of an atypical protein kinase C (PKC3) from Caenorhabditis

elegans. PKC3 is required for the normal progression of embryogenesis and viability of the organism. *J. Biol. Chem.* **273,** 1130–1143.

4. Tabuse, Y., Nishiwaki, K., and Miwa, J. (1989) Mutations in a protein kinase C homolog confer phorbol ester resistance on Caenorhabditis elegans. *Science* **243,** 1713–1716.

5. Tabuse, Y., Sano, T., Nishiwaki, K., and Miwa, J. (1995) Molecular evidence for the direct involvement of a protein kinase C in developmental and behavioural susceptibility to tumour–promoting phorbol esters in Caenorhabditis elegans. *Biochem. J.* **312,** 69–74.

6. Tabuse, Y., Izumi, Y., Piano, F., Kemphues, K. J., Miwa, J., and Ohno, S. (1998) Atypical protein kinase C cooperates with PAR-3 to establish embryonic polarity in Caenorhabditis elegans. *Development* **125,** 3607–3614.

7. Izumi, Y., Hirose, T., Tamai, Y., Hirai, S., Nagashima, Y., Fujimoto, T., et al. (1998) An atypical PKC directly associates and colocalizes at the epithelial tight junction with ASIP, a mammalian homologue of Caenorhabditis elegans polarity protein PAR-3. *J. Cell Biol.* **143,** 95–106.

8. Hung, T. J. and Kemphues, K. J. (1999) PAR-6 is a conserved PDZ domain-containing protein that colocalizes with PAR-3 in Caenorhabditis elegans embryos. *Development* **126,** 127–135.

9. Rosenthal, A., Rhee, L., Yadegari, R., Paro, R., Ullrich, A., and Goeddel, D. V. (1987) Structure and nucleotide sequence of a Drosophila melanogaster protein kinase C gene. *EMBO J.* **6,** 433–441.

10. Schaeffer, E., Smith, D., Mardon, G., Quinn, W., and Zuker, C. (1989) Isolation and characterization of two new Drosophila protein kinase C genes, including one specifically expressed in photoreceptor cells. *Cell* **57,** 403–412.

11. Wodarz, A., Ramrath, A., Grimm, A., and Knust, E. (2000) Drosophila atypical protein kinase C associates with Bazooka and controls polarity of epithelia and neuroblasts. *J. Cell Biol.* **150,** 1361–1374.

12. Hardle, R. C., Peretz, A., Suss-Toby, E., Rom-Glas, A., Bishop, S. A., Selinger, Z., et al. (1993) Protein kinase C is required for light adaptation in *Drosophila* photoreceptors. *Nature* **363,** 634–637.

13. Banbury Conference on Genetic Background in Mice. (1997) Mutant Mice and Neuroscience: Recommendations Concerning Genetic Background. *Neuron* **19,** 755–759.

14. Simpson, E. M., Linder, C. C., Sargent, E. E., Davisson, M. T., Mobraaten, L. E., and Sharp, J. J. (1997) Genetic variation among 129 substrains and its importance for targeted mutagenesis in mice. *Nat. Genet.* **16,** 19–27.

15. Lathe, R. (1996) Mice, gene targeting and behaviour: more than just genetic background. *Trends Neurosci.* **19,** 183–186; discussion 188–189.

16. Dekker, L. V., Leitges, M., Altschuler, G., Mistry, N., McDermott, A., Roes, J., and Segal, A. W. (2000) Protein kinase C–beta contributes to NADPH oxidase activation in neutrophils. *Biochem. J.* **347,** 285–289.

17. Chen, L., Hahn, H., Wu, G., Chen, C. H., Liron, T., Schechtman, D., et al. (2001) Opposing cardioprotective actions and parallel hypertrophic effects of delta PKC and epsilon PKC. *Proc. Natl. Acad. Sci. USA* **98,** 11,114–11,119.

18. Khasar, S. G., Lin, Y.-H., Martin, A., Dadgar, J., McMahon, T., Wang, D., et al. (1999) A novel nociceptor signaling pathway revealed in protein kinase C ε mutant mice. *Neuron* **24,** 253–260.

19. Hodge, C. W., Mehmert, K. K., Kelley, S. P., McMahon, T., Haywood, A., Olive, M. F., et al. (1999) Supersensitivity to allosteric GABA$_A$ receptor modulators and alcohol in mice lacking PKCe. *Nat. Neurosci.* **2,** 997–1002.

20. Jansen, A. P., Dreckschmidt, N. E., Verwiebe, E. G., Wheeler, D. L., Oberley, T. D., and Verma, A. K. (2001) Relation of the induction of epidermal ornithine decarboxylase and hyperplasia to the different skin tumor-promotion susceptibilities of protein kinase C alpha, -delta and -epsilon transgenic mice. *Int. J. Cancer* **93,** 635–643.

21. Bowman, J. C., Steinberg, S. F., Jiang, T., Geenen, D. L., Fishman, G. I., and Buttrick, P. M. (1997) Expression of protein kinase C beta in the heart causes hypertrophy in adult mice and sudden death in neonates. *J. Clin. Invest.* **100,** 2189–2195.

22. Wakasaki, H., Koya, D., Schoen, F. J., Jirousek, M. R., Ways, D. K., Hoit, B. D., et al. (1997) Targeted overexpression of protein kinase C b2 isoform in myocardium causes cardiomyopathy. *Proc. Natl. Acad. Sci. USA* **94,** 9320–9325.

23. Murray, N. R., Davidson, L. A., Chapkin, R. S., Clay Gustafson, W., Schattenberg, D. G., and Fields, A. P. (1999) Overexpression of protein kinase C betaII induces colonic hyperproliferation and increased sensitivity to colon carcinogenesis. *J. Cell Biol.* **145,** 699–711.

24. Suzuma, K., Takahara, N., Suzuma, I., Isshiki, K., Ueki, K., Leitges, M., et al. (2002) Characterization of protein kinase C beta isoform's action on retinoblastoma protein phosphorylation, vascular endothelial growth factor-induced endothelial cell proliferation, and retinal neovascularization. *Proc. Natl. Acad. Sci. USA* **99,** 721–726.

25. Mochly-Rosen, D., Wu, G., Hahn, H., Osinska, H., Liron, T., Lorenz, J. N., et al. (2000) Cardiotrophic effects of protein kinase C epsilon: analysis by in vivo modulation of PKCepsilon translocation [see comments]. *Circ. Res.* **86,** 1173–1179.

26. Roman, B. B., Geenen, D. L., Leitges, M., and Buttrick, P. M. (2001) PKC–beta is not necessary for cardiac hypertrophy. *Am. J. Physiol.* **280,** H2264–H2270.

27. Jin, Z.-Q., Zhou, H.-Z., Zhu, P., Honbo, N., Mochly-Rosen, D., Messing, R. O., et al. (2002) Cardioprotection mediated by sphingosine-1-phosphate and the ganglioside GM-1 in wild type and εPKC knockout mouse hearts. *J. Appl. Physiol.* **282,** H1970–H1977.

28. Choi, D.-S., Wang, D., Dadgar, J., Chang, W. S., and Messing, R. O. (2002) Ethanol consumption and sensitivity regulated by conditional expression of protein kinase C ε. *J. Neurosci.* **22,** 9905–9911.

29. Gossen, M. and Bujard, H. (1992) Tight control of gene expression in mammalian cells by tetracycline-responsive promoters. *Proc. Natl. Acad. Sci. USA* **89,** 5547–5551.

30. Furth, P. A., St. Onge, L., Böger, H., Gruss, P., Gossen, M., Kistner, A., et al. (1994) Temporal control of gene expression in transgenic mice by a tetracycline-responsive promoter. *Proc. Natl. Acad. Sci. USA* **91,** 9302–9306.
31. Mansuy, I. M., Winder, D. G., Moallem, T. M., Osman, M., Mayford, M., Hawkins, R. D., et al. (1998) Inducible and reversible gene expression with the rtTA system for the study of memory. *Neuron* **21,** 257–65.
32. Malleret, G., Haditsch, U., Genoux, D., Jones, M. W., Bliss, T. V., Vanhoose, A. M., et al. (2001) Inducible and reversible enhancement of learning, memory, and long-term potentiation by genetic inhibition of calcineurin. *Cell* **104,** 675–686.
33. Tremblay, P., Meiner, Z., Galou, M., Heinrich, C., Petromilli, C., Lisse, T., et al. (1998) Doxycycline control of prion protein transgene expression modulates prion disease in mice. *Proc. Natl. Acad. Sci. USA* **95,** 12,580–12,585.
34. Thiele, T. E., Marsh, D. J., Ste. Marie, L., Bernstein, I. L., and Palmiter, R. D. (1998) Ethanol consumption and resistance are inversely related to neuropeptide Y levels. *Nature 396,* 366–369.
35. Caplen, N. J. (2002) A new approach to the inhibition of gene expression. *Trends Biotechnol.* **20,** 49–51.
36. Paddison, P. J., Caudy, A. A., and Hannon, G. J. (2002) Stable suppression of gene expression by RNAi in mammalian cells. *Proc. Natl. Acad. Sci. USA* **99,** 1443–1448.
37. Kellendonk, C., Tronche, F., Casanova, E., Anlag, K., Opherk, C., and Scütz, G. (1999) Inducible site-specific recombination in the brain. *J. Mol. Biol.* **285,** 175–182.
38. Leitges, M., Schmedt, C., Guinamard, R., Davoust, J., Schaal, S., Stabel, S., et al. (1996) Immunodeficiency in protein kinase Cβ-deficient mice. *Science* **273,** 788–791.
39. Standaert, M. L., Bandyopadhyay, G., Galloway, L., Soto, J., Ono, Y., Kikkawa, U., et al. (1999) Effects of knockout of the protein kinase C beta gene on glucose transport and glucose homeostasis. *Endocrinology* **140,** 4470–4477.
40. Nechushtan, H., Leitges, M., Cohen, C., Kay, G., and Razin, E. (2000) Inhibition of degranulation and interleukin–6 production in mast cells derived from mice deficient in protein kinase Cbeta. *Blood* **95,** 1752–1757.
41. Yan, S. F., Lu, J., Zou, Y. S., Kisiel, W., Mackman, N., Leitges, M., et al. (2000) Protein kinase C-beta and oxygen deprivation. A novel Egr-1-dependent pathway for fibrin deposition in hypoxemic vasculature. *J. Biol. Chem.* **275,** 11,921–11,928.
42. Weeber, E. J., Atkins, C. M., Selcher, J. C., Varga, A. W., Mirnikjoo, B., Paylor, R., et al. (2000) A role for the beta isoform of protein kinase C in fear conditioning. *J. Neurosci.* **20,** 5906–5914.
43. Abeliovich, A., Chen, C., Goda, Y., Silva, A. J., Stevens, C. F., and Tonegawa, S. (1993) Modified hippocampal long-term potentiation in PKCγ-mutant mice. *Cell* **75,** 1253–1262.

44. Abeliovich, A., Paylor, R., Chen, C., Kim, J. J., Wehner, J. M., and Tonegawa, S. (1993) PKCγ mutant mice exhibit mild deficits in spatial and contextual learning. *Cell* **75,** 1263–1271.

45. Ramakers, G. M., Gerendasy, D. D., and de Graan, P. N. (1999) Substrate phosphorylation in the protein kinase Cgamma knockout mouse. *J. Biol. Chem.* **274,** 1873–1874.

46. Chen, C., Kano, M., Abeliovich, A., Chen, L., Bao, S., Kim, J. J., et al. (1995) Impaired motor coordination correlates with persistent multiple climbing fiber innervation in PKCγ mutant mice. *Cell* **83,** 1233–1242.

47. Kano, M., Hashimoto, K., Chen, C., Abeliovich, A., Aiba, A., Kurihara, H., et al. (1995) Impaired synapse elimination during cerebellar development in PKCγ mutant rice. *Cell* **83,** 1223–1231.

48. Malmberg, A. B., Chen, C., Tonegawa, S., and Basbaum, A. I. (1997) Preserved acute pain and reduced neuropathic pain in mice lacking PKCγ. *Science* **278,** 279–283.

49. Narita, M., Mizoguchi, H., Nagase, H., Suzuki, T., and Tseng, L. F. (2001) Involvement of spinal protein kinase Cgamma in the attenuation of opioid mu-receptor-mediated G-protein activation after chronic intrathecal administration of [D-Ala2,N-MePhe4,Gly-Ol(5)]enkephalin. *J. Neurosci.* **21,** 3715–3720.

50. Narita, M., Mizoguchi, H., Suzuki, T., Dun, N. J., Imai, S., Yajima, Y., et al. (2001) Enhanced mu-opioid responses in the spinal cord of mice lacking protein kinase Cgamma isoform. *J. Biol. Chem.* **276,** 15,409–15,414.

51. Bowers, B. J., Owen, E. H., Collins, A. C., Abeliovich, A., Tonegawa, S., and Wehner, J. M. (1999) Decreased ethanol sensitivity and tolerance development in gamma-protein kinase C null mutant mice is dependent on genetic background. *Alcoholism Clin. Exp. Res.* **23,** 387–397.

52. Bowers, B. J. and Wehner, J. M. (2001) Ethanol consumption and behavioral impulsivity are increased in protein kinase cgamma null mutant mice. *J. Neurosci.* **21,** RC180.

53. Harris, R. A., McQuilkin, S. J., Paylor, R., Tonegawa, S., and Wehner, J. M. (1995) Mutant mice lacking the γ isoform of protein kinase C show decreased behavioral actions of ethanol and altered function of γ-aminobutyrate type A receptors. *Proc. Natl. Acad. Sci. USA* **92,** 3658–3662.

54. Bowers, B. J., Collins, A. C., Tritto, T., and Wehner, J. M. (2000) Mice lacking PKC gamma exhibit decreased anxiety. *Behav. Genet.* **30,** 111–121.

55. Aronowski, J., Grotta, J. C., Strong, R., and Waxham, M. N. (2000) Interplay between the gamma isoform of PKC and calcineurin in regulation of vulnerability to focal cerebral ischemia. *J. Cereb. Blood Flow Metab.* **20,** 343–349.

56. Leitges, M., Elis, W., Gimborn, K., and Huber, M. (2001) Rottlerin-independent attenuation of pervanadate-induced tyrosine phosphorylation events by protein kinase C-delta in hemopoietic cells. *Lab. Invest.* **81,** 1087–1095.

57. Leitges, M., Mayr, M., Braun, U., Mayr, U., Li, C., Pfister, G., et al. (2001) Exacerbated vein graft arteriosclerosis in protein kinase Cdelta-null mice. *J. Clin. Invest.* **108,** 1505–1512.

58. Olive, M. F., Mehmert, K. K., Messing, R. O., and Hodge, C. W. (2000) Reduced operant ethanol self-administration and in vivo mesolimbic dopamine responses to ethanol in PKCε-deficient mice. *Eur. J. Neurosci.* **12,** 4131–4140.

59. Olive, M. F., Mehmert, K. K., Nannini, M. A., Camarini, R., Messing, R. O., and Hodge, C. W. (2001) Reduced ethanol withdrawal severity and altered withdrawal-induced c-fos expression in various brain regions of mice lacking protein kinase C-epsilon. *Neuroscience* **103,** 171–179.

60. Hodge, C. W., Raber, J., Walter, H. T. M., Sanchez-Perez, A. M., Olive, M. F., Mehmert, K., et al. (2000) Decreased anxiety, reduced stress hormones and neurosteroid supersensitivity in mice lacking protein kinase Cε. *J. Clin. Invest.* **110,** 1003–1010.

61. Castrillo, A., Pennington, D. J., Otto, F., Parker, P. J., Owen, M. J., and Bosca, L. (2001) Protein kinase Cepsilon is required for macrophage activation and defense against bacterial infection. *J. Exp. Med.* **194,** 1231–1242.

62. Leitges, M., Sanz, L., Martin, P., Duran, A., Braun, U., Garcia, J. F., et al. (2001) Targeted disruption of the zetaPKC gene results in the impairment of the NF-kappaB pathway. *Mol. Cell* **8,** 771–780.

63. Sun, Z., Arendt, C. W., Ellmeier, W., Schaeffer, E. M., Sunshine, M. J., Gandhi, L., et al. (2000) PKC-theta is required for TCR-induced NF-kappaB activation in mature but not immature T lymphocytes. *Nature* **404,** 402–7.

38

Genetic Manipulation of Protein Kinase C In Vivo

Alex Toker

1. Introduction

Studies since 1980 have established that protein kinase C (PKC) regulates a plethora of downstream signaling pathways leading to numerous cellular responses. Although much is known concerning PKC regulation by lipid cofactors and phosphorylation, the direct protein substrates that relay the PKC signal remain largely undescribed, although proteins such as myristoylated alanine-rich C kinase (MARCKS) and pleckstrin are well-known examples. Much of the work aimed at defining substrates of PKC has relied on the use of phorbol esters (*see* Chapter 34) as well as small molecule inhibitors (*see* Chapter 33). Because of concerns with lack of specificity and cytotoxicity associated with these approaches, new methodologies have been developed to more accurately and specifically manipulate PKC activity in cells and thus tackle the mechanisms by which PKC regulates cell function.

There are four distinct approaches that are commonly used to genetically manipulate PKC in cells. Two of these are described in this chapter. The first makes use dominant negative and constitutively active cDNA alleles of PKC isoforms that are transiently transfected into cells to provide a combination of a loss-of-function and gain-of-function approach. These mutant PKC alleles are widely used, although there are some concerns with specificity such that one dominant negative PKC allele may affect the activity of other related PKCs when expressed in cells. Therefore, it is often advantageous to combine a dominant negative approach such as this one, with a complementary technique, such as drug inhibitors. The second methodology makes use of modified phosphorothioate antisense oligonucleotides to reduce or eliminate expression of individual PKC isoforms, again by transient transfection of cells. This

From: *Methods in Molecular Biology, vol. 233: Protein Kinase C Protocols*
Edited by: A. C. Newton © Humana Press Inc., Totowa, NJ

technique is less prone to specificity issues and again is widely used to implicate distinct PKC isoforms in the regulation of downstream signaling events. With both techniques, the goal is to specifically and potently eliminate a distinct PKC isoform without affecting other isoforms. There are two additional experimental approaches that have recently been developed and that will likely provide even greater specificity in these loss-of-function experiments. The first is the generation of PKC knockout animals, particularly mice. All PKCs have recently been knocked out using conventional strategies, and both embryonic stem cells as well as mouse embryonic fibroblasts from these mice have been used to study various signaling pathways and cellular responses when one distinct PKC isoform is absent. Knockout mice for PKCα *(1)*, PKCβ *(2–5)*, PKCγ *(6)*, PKCδ *(7,8)*, PKCε *(9,10)*, PKCθ *(11)*, and PKCζ *(12)* have been reported but will not be further described here. The power of this technique is exemplified by the use of PKCζ –/– cells to study the involvement of this PKC in the activation of the NF-κB pathway and B cell development *(12)*, and the use of PKCδ –/– mice to implicate this isoform in B cell tolerance *(7,8)*. Therefore, the use of cells derived from PKC knockout mice will undoubtedly prove to be a valuable weapon in the arsenal of tools to be used to study PKC biology. An equally exciting prospect will be the development of RNAi (RNA interference, also known as siRNA, small interfering RNA) methods for PKC isoforms. This novel technique relies on the transfection of a small 21 nucleotide RNA duplex against the amino terminus of a given mRNA to specifically target them for degradation and thus reduce or eliminate protein expression *(13)*. This technique has proven to be successful for a number of proteins, such as lamin *(13)*, p53 *(14)*, and ShcA *(15)*. However, to date no information exists for RNAi duplexes designed against PKCs. The design of the RNAi duplex is critical to the success of silencing and often several duplexes must be attempted for a given mRNA before one is found to work efficiently at reducing protein expression. Because RNAi duplexes for PKCs have not yet been reported, this technique will not be described here. The reader is directed to the other references for the use and design of RNAi *(13,14)*.

2. Materials

2.1. Dominant Negative PKC and Constitutively Active PKC

1. PKC cDNA in a mammalian expression vector (e.g., pcDNA3).
2. Oligonucleotide primers (**Tables 1** and **2**; Integrated DNA Technologies, IDT, Coralville, IA; http://www.idtdna.com).
3. QuickChange™ Mutagenesis kit (Stratagene, La Jolla, CA; http://www.stratagene.com).
4. Plasmid isolation kit (e.g., Concert™ kit, Invitrogen Life Technologies, Carlsbad, CA; http://www.lifetech.com).

Table 1
Dominant Negative PKC Mutants

PKC isoform	SwissProt acc. no.	Subdomain II Lys motif	Mutation	Oligonucleotide primers
HPKCα	P17252	ELYAIKILKKD	Lys368→Trp	GAGCTGTACGCCATCTGGATCCTGAAGAAGGAC GTCCTTCTTCAGGATCCAGATGGCGTACAGCTC
HPKCβ	P05771 (βI) P05127 (βII)	ELYAVKILKKD	Lys371→Trp	GAGCTCTATGCTGTGGATCCTGAAGAAGGAC GTCCTTCTTCAGGATCCACACAGCATAGAGCTC
HPKCγ	P05129	ELYAIKILKKD	Lys380→Trp	GAGCTCTACGCCATCTGGATCTTGAAAAAGGAC GTCCTTTTTCAAGATCCAGATGGCGTAGAGCTC
HPKCδ	Q05655	EYSAIKALKKD	Lys378→Trp	GAGTACTCTGCCATCTGGGCCCTCAAGAAGGAT ATCCTTCTTGAGGGCCCAGATGGCAGAGTACTC
HPKCε	Q02156	EVYAVKVLKKD	Lys437→Trp	GAAGTATATGCTGTGTGGTCTTAAAGAAGGAC GTCCTTCTTTAAGACCCACACAGCATATACTTC
HPKCη	P24723	DLYAVKVLKKD	Lys383→Trp	GACCTCTATGCTGTGTGGGTGCTGAAGAAGGAC GTCCTTCTTCAGCACCCACACAGCATAGAGGTC
HPKCθ	Q04759	QFFAIKALKKD	Lys409→Trp	CAATTTTCGCAATATGGCCTTAAAGAAAGAT ATCTTTCTTTAAGGCCCATATTGCGAAAAATTG
HPKCζ	Q05513	QIYAMKVVKKE	Lys281→Trp	CAGATTTACGCCATGTGGGTGGTCAAGAAGGAG CTCCTTCTTGACCACCCACATGGCGTAAATCTG
HPKCλ	Q62074	RIYAMKVVKKE	Lys273→Trp	CGTATTTATGCAATGTGGGTTGTGAAAAAGAG CTCTTTTTTCACAACCCACATTGCATAAATACG

Table 1 shows the nine human PKC isotypes with corresponding SwissPROT database accession numbers. The motif surrounding the critical Lys residue in the kinase subdomain II is also shown along with the amino acid position of the respective Lys in each PKC. The oligonucleotide primers to be made for mutation of each Lys is also shown, for both top and bottom strands, where the Lys is mutated to a Trp. The top primer sequence corresponds to the top strand (5′ → 3′), the bottom primer is reversed and complementary to the top primer (also shown 5′ → 3′).

Table 2
Constitutively Active Pseudosubstrate PKC Mutants

PKC isoform	SwissProt acc. no.	Pseudosubstrate sequence	Mutation	Oligonucleotide primers
HPKCα	P17252	RFARKGALRQKNV	Ala25→Glu	TTCGCCCGCAAAGGGGACCTGAGGCAGAAGAAC
				GTTCTTCTGCCTCAGGTCCCCTTTGCGGGCGAA
HPKCβ	P05771 (βI) P05127 (βII)	RFARKGALRQKNV	Ala25→Glu	TTCGCCCGCAAAGGCGACCTCCGGCAGAAGAAC
				GTTCTTCTGCCGGAGGTCGCCTTTGCGGGCGAA
HPKCγ	P05129	LFCRKGALRQKVV	Ala24→Glu	TTCTGCAGAAAGGGGACCTGAGGCAGAAGGTG
				CACCTTCTGCCTCAGGTCCCCTTTCTGCAGAA
HPKCδ	Q05655	TMNRRGAIKQAKI	Ala147→Glu	ATGAACCGCCGCGGAGACATCAAACAGGCCAAA
				TTTGGCCTGTTTGATGTCTCCGCGGCGGTTCAT
HPKCε	Q02156	PRKRQGAVRRRVH	Ala159→Glu	AGGAAGCGGCAGGGGACGTCAGGCGCCAGGGTC
				GACCCTGGCGCCTGACGTCCCCCTGCCGCTTCCT
HPKCη	P24723	TRKRQRAMRRRVH	Ala160→Glu	AGGAAGCGCCAAAGGACCATGCGAAGGCGAGTC
				GACTCGCCTTGCGCATGTCCCCTTTGGCGCTTCCT
HPKCθ	Q04759	LHQRRGAIKQAKV	Ala148→Glu	CATCAGCGCCGGGGTGACATCAAGCAGGCAAAG
				CTTTGCCTGCTTGATGTCACCCCGGCGCTGATG
HPKCζ	Q05513	SIYRRGARRWRKL	Ala119→Glu	ATCTACCGCCGGGGAGACAGAAGATGGAGGAAG
				CTTCCTCCATCTTCTGTCTCCCCGGCGGTAGAT
HPKCλ	Q62074	SIYRRGARRWRKL	Ala120→Glu	ATCTACCGCTAGAGGTGACCGCCGCTGGAGAAAG
				CTTTCTCCAGCGGCGGTCACCTCTACGGTAGAT

This table shows the nine human PKC isotypes with corresponding SwissPROT database accession numbers. The motif surrounding the conserved Ala residue in the pseudosubstrate sequence is shown along with the amino acid position. The oligonucleotide primers to be made for mutation of each Ala to Glu is also shown, for both top and bottom strands. The top primer sequence corresponds to the top strand ($5' \rightarrow 3'$), the bottom primer is reversed and complementary to the top primer (also shown $5' \rightarrow 3'$).

Table 3
Antisense PKC Oligonucleotides

PKC Isoform	Ref.	Oligonucleotide Sequence	
HPKCα	*(28)*	GGG AAA ACG TCA GCC ATG GTC	**(AS)**
		GAC CAT GGC TGA CGT TTT CCC	**(S)**
HPKCβ	*(29)*	AAG ATG GCT GAC CCG GCT CGC	**(AS)**
		GCG AGC CGG GTC AGC CAT CTT	**(S)**
HPKCγ	*(29)*	CAA GAT GGC TGA CCC GGC CGC	**(AS)**
		GCG GCC GGG TCA GCC ATC TTG	**(S)**
HPKCδ	*(30)*	ACG GCG CCA TGG TGG G	**(AS)**
		CCC ACC ATG GCG CCG T	**(S)**
HPKCε	*(31)*	GAA CAC TAC CAT GGT CGG	**(AS)**
		CCG ACC ATG GTA GTG TTC	**(S)**
HPKCη	*(32)*	CGA GAT CCA CCC AAC CCT CG	**(AS)**
		CGA GGG TTG GGT GGA TCT CG	**(S)**
HPKCθ	*(33)*	AGA AAT GGC GAC ATG GTT	**(AS)**
		AAC CAT GTC GCC ATT TCT	**(S)**
HPKCζ	*(30)*	GCT CCC TTC CAT CTT GGG	**(AS)**
		CCC AAG ATG GAA GGG AGC	**(S)**
HPKCι/λ	N.D.	GGG TCG GCA TCT CCT C	**(AS)**
		GAG GAG ATG CCG ACC C	**(S)**

The sequence of the sense (**S**)and antisense (**AS**) oligonucleotides to be made with the phosphorothioate modification are shown. The published reference for each oligonucleotide is also shown (Ref.). For PKCι/λ, there are no documented cases describing the use of a synthetic antisense oligonucleotide (N.D., not determined). The sequence shown for the sense and antisense PKCι/λ oligonucleotide was derived from the translation-initiation sequence of PKCι.

2.2. Antisense Oligonucleotides

1. Antisense and sense phosphorothioate oligonucleotides (**Table 3**; Integrated DNA Technologies, IDT, Coralville, IA; http://www.idtdna.com).
2. Lipofectamine (Invitrogen Life Technologies) and other transfection reagents, *see* Chapter 15, **Subheading 2.1.**
3. PKC isoform specific antibodies (*see* Chapter 15, **Table 1**).

3. Methods
3.1. Dominant Negative PKC

The strategy to specifically block PKC isoform function and signaling in cells using dominant negative PKC mutants is widely used and has provided much insight into the mechanisms by which PKCs contribute to the activation of secondary signaling pathways. As with many other kinases of the AGC family, PKC kinase inactive alleles in which the critical Lys residue in the ATP-binding site is mutated have been used as dominant negative mutants. This section will describe the construction of kinase-inactive mutants of all PKC isoforms by mutation of the ATP-binding site. The PKC cDNA is mutated at the critical Lys residue in each PKC isoform using standard PCR reaction conditions, and the resulting mutant cDNA is introduced into cells by transient transfection.

It is important to note a number of potential caveats in the use of kinase-inactive mutants of PKC as dominant negative alleles. By definition, a dominant negative kinase acts by suppressing the activation of the endogenous enzyme such that it is not able to phosphorylate substrates. The use of kinase-inactive alleles ensures that the transfected protein is also not able to function as a protein kinase. The commonly held assumption as to why kinase-inactive alleles of protein kinases function as dominant negatives is that they will be subjected to the same upstream regulatory signals as the endogenous protein, thus preventing its efficient activation. In the case of PKC, its upstream regulators, diacylglycerol (DAG), phosphatidylserine, and phosphoinositide-dependent kinase (PDK-1) will therefore be limiting to the endogenous kinase upon overexpression of the dominant negative allele. However, because all PKCs have near identical regulatory requirements, such as DAG and PDK-1, it is reasonable to assume that any one PKC kinase-inactive allele will also function to block other PKCs in the same cells. The lack of specificity using this dominant negative approach must be taken into consideration when evaluating the consequence of overexpression of such alleles. This point is exemplified by a study in which the dominant negative properties of several PKC mutant alleles were tested against active transfected PKC isoforms representing the three subclasses, conventional, novel, and atypical PKCs (16). For example, a dominant negative, kinase-inactive PKCζ allele not only inhibited co-transfected wild-type PKCζ activity but also PKCα and PKCε (16). This is despite the fact that there are documented examples in which distinct kinase-inactive PKC mutants have been shown to mediate specific responses in transfected cells (17). Because cellular localization is an equally important determinant in the regulation and activation of PKCs (see Chapter 26), it is likely that specificity in this approach is achieved through the appropriate intracellular location of the expressed dominant negative PKC without affecting other PKCs in distinct locations. Because of these important issues of specificity, it is essential to use a complementary loss of function approach when analyzing the contribution of a particular PKC isoform to a cellular response (e.g., antisense oligonucleotides or drug inhibitors). The most widely used dominant negative PKC strategy is to mutate the conserved Lys residue in the ATP-binding site, typically to a conserved Arg residue. We favor the muta-

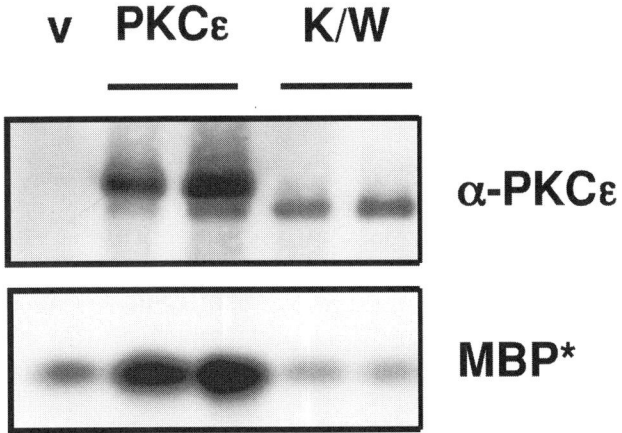

Fig. 1. Protein kinase activity of wild-type and kinase-inactive PKCε. HEK293 cells were transiently transfected with vector alone (v), FLAG-PKCε, or FLAG-PKCε.Lys437Trp (K/W) in duplicate, as indicated. Cell lysates were immunoprecipitated with FLAG antibody and an immune complex kinase assay using myelin basic protein (MBP) was carried out using [γ-^{32}P]ATP. Phosphorylated MBP (MBP*) was detected by autoradiography. PKCε expression was detected by western immunoblotting of total cell lysate with an anti-PKCε-specific antibody (α-PKCε). Adapted from **ref.** *18*.

tion of this Lys to a Trp residue such that the resulting mutant is quantitatively kinase inactive toward exogenous substrates in vitro and in vivo (*see* **Note 1**). **Figure 1** shows the activity of wild-type PKCε and a kinase-inactive PKCε Lys437Trp mutant when expressed in HEK293 cells and assayed in an immune-complex kinase assay *(18)*. The mutation is introduced by standard PCR reaction techniques using synthetic oligonucleotides. We favor the use of the QuickChange site-directed mutagenesis strategy (Stratagene, La Jolla, CA; http://www.stratagene.com). The specific residue to be mutated, as well as its position in all PKC isoforms is shown in **Table 1**.

1. Obtain PKC cDNA in an appropriate mammalian expression vector (e.g., pcDNA3).
2. Obtain oligonucleotide primer pairs for the corresponding PKC isoform (**Table 1**).
3. Carry out site-directed mutagenesis by PCR (the primer pairs in **Table 1** are specifically designed for use with the QuickChange PCR mutagenesis kit, Stratagene; *see* **Note 2**).
4. Screen resulting transformants by DNA sequencing of at least four individual colonies.

5. Carry out a large-scale plasmid isolation of the mutant PKC cDNA.

6. The mutated cDNA can now be introduced by transient transfection into mammalian cells. For details on cell types and transfection protocol, *see* Chapter 15, **Subheadings 3.1.–3.3.**

3.2. Constitutively Active PKC

Transient transfection of constitutively active PKC is often used as a complementary approach to dominant negative strategies to implicate distinct PKC isoforms in a particular cellular response. In addition, this gain of function approach does not suffer from the specificity issues that are a concern with loss of function mutations in dominant negative alleles. One of the most commonly used methods to generate a constitutively active PKC mutant is to introduce a point mutation in the pseudosubstrate domain of PKC. The pseudosubstrate domain of PKC binds to the active site in the inactive kinase thus preventing substrate interaction *(19,20)*, and binding of DAG supplies the free energy necessary to release the pseudosubstrate from the substrate-binding cavity *(21,22)*. Importantly, it has been shown that either deletion of the pseudosubstrate domain, or point mutation of a conserved Ala residue to a negatively charged Glu renders the kinase constitutively active toward substrates in vitro and in vivo *(23,24)*. This effector-independent mutant can therefore be used as a gain of function allele when compared to the wild-type enzyme, which still requires DAG binding for efficient activation. We have used a distinct but equally efficient method to render PKC isoforms constitutively active, by the addition of a membrane-targeting module to the amino terminus of the kinase to artificially target it to the plasma membrane when expressed in cells *(18,25)*. This approach has been successfully used for several lipid signaling enzymes such as phosphoinositide 3-kinase and the Akt/PKB and PDK-1 serine/threonine kinases. We have shown that membrane targeting of PKCε and PKCζ by the addition of the first six amino acids of the pp60Src sequence that targets the enzyme for myristoylation also renders them constitutively active in vivo *(18,25)*. The rationale for this strategy is that artificial membrane targeting will position the kinase in close proximity to its upstream regulators, namely DAG and PDK-1, and allow efficient activation in the absence of additional activating stimuli.

This section describes the steps necessary to generate constitutively active PKC isoforms by mutation of the pseudosubstrate domain, or by the addition of a myristoylation sequence at the amino terminus. Both techniques require site-directed mutagenesis using oligonucleotide primers encompassing the desired mutations.

3.2.1. Pseudosubstrate Mutants

PKC cDNA is mutated by polymerase chain reaction (PCR) at the conserved Ala residue in the pseudosubstrate region. The specific residue to be mutated, as well as its position in all PKC isoforms is shown in **Table 2**.

1. Obtain PKC cDNA in an appropriate mammalian expression vector (e.g., pcDNA3).
2. Obtain oligonucleotide primer pairs for the corresponding PKC isoform (**Table 2**).
3. Carry out site-directed mutagenesis by PCR (the primer pairs in **Table 2** are specifically designed for use with the QuickChange PCR mutagenesis kit, Stratagene).
4. Screen resulting transformants by DNA sequencing of at least four individual colonies.
5. Carry out a large-scale plasmid isolation of the mutant PKC cDNA.
6. The mutated cDNA can now be introduced by transient transfection into mammalian cells. For details on cell types and transfection protocol, *see* Chapter 15, **Subheadings 3.1.–3.3.**

3.2.2. Membrane-Targeted Mutants

In this protocol, the myristoylation sequence of the pp60Src tyrosine kinase is added in frame to the amino terminus of PKC cDNA, targeting the kinase for lipid modification and subsequent membrane localization. The sequence is added using standard PCR reaction conditions using oligonucleotide primers encompassing the myristoylation sequence. As an example, the sequences used to add a myristoylation sequence to the amino terminus of human PKCζ are shown below:

Human PKCζ amino terminus:	ATG	CCC	AGC	AGG	ACC	
	Met	Pro	Ser	Arg	Thr	
Src myristoylation sequence:	ATG	GGG	AGC	AGC	AAG	AGC
	Met	Gly	Ser	Ser	Lys	Ser

5′ oligonucleotide primer:

*Hind*III	Src Myr Signal	PKCζ amino term.

5′ GCG AAG CTT ATG GGG AGC AGC AAG AGC AAG CCC ATG CCC AGC AGG ACC 3′

We also recommend that the oligonucleotide primer to the carboxyl terminus includes an epitope tag such as HA, Myc, or FLAG. *see* Chapter 15, **Subheading 3.2.** for more details.

3′ oligonucleotide primer:

*Xba*I	Stop	HA tag	PKCζ carboxyl term.

5′ CGC TCT AGA TCA AGC ATA GTC CGG GAC GTC GTA AGG GTA CAC CGA CTC CTC 3′

1. Obtain PKC cDNA in an appropriate mammalian expression vector (e.g., pcDNA3).
2. Obtain forward (5′) and reverse (3′) oligonucleotide primer pairs for the corresponding PKC isoform (shown above).
3. Carry out PCR under standard conditions (the primer pairs above are specifically designed for use with human PKCζ cDNA).
4. Ligate resulting PCR product to mammalian expression vector (e.g., pcDNA3). Screen resulting transformants by DNA sequencing of at least four individual colonies.
5. Carry out a large-scale plasmid isolation of the mutant PKC cDNA.
6. The mutated cDNA can now be introduced by transient transfection into mammalian cells. For details on cell types and transfection protocol, *see* Chapter 15, **Subheadings 3.1.–3.3.**
7. Verify expression of the mutant PKC by Western immunoblotting with either anti-HA or anti-PKC isoform-specific antibodies (*see* Chapter 15, **Subheading 3.4.**).

Figure 2 shows the expression and kinase activity of a Myr.PKCζ mutant when expressed in HEK293 cells compared with wild-type PKCζ *(25)*. Similar results were obtained with a Myr.PKCε mutant *(18)*.

3.3. Antisense Oligonucleotides

Synthetic antisense oligonucleotides have been widely used to implicate distinct PKC isozymes in biological processes. The advantage of this loss-of-function approach is that it is less prone to lack of specificity encountered with dominant negative, kinase-inactive mutants. The reason for this is that an antisense oligonucleotide directed against a particular PKC isoform will typically not interfere with the expression or function of other PKCs in the same cell. However, this must be determined empirically for each antisense and cell line to be used. The most widely used antisense method is the chemical synthesis of a phosphorothioate-modified oligonucleotide. The phosphorothioate modification reduces degradation by nucleases of the transfected oligonucleotide and permits more quantitative inhibition of protein expression (*see* **Note 3**). Antisense oligonucleotides are typically designed to the translation initiation of the PKC amino terminus, and both sense and antisense oligonucleotides should be used to control for specificity. Antisense oligonucleotides to many PKC isoforms have been widely used, and are summarized in **Table 3**. It is also worth noting that many studies have also made use of the antisense approach to reduce or eliminate PKC isoform expression using antisense RNA

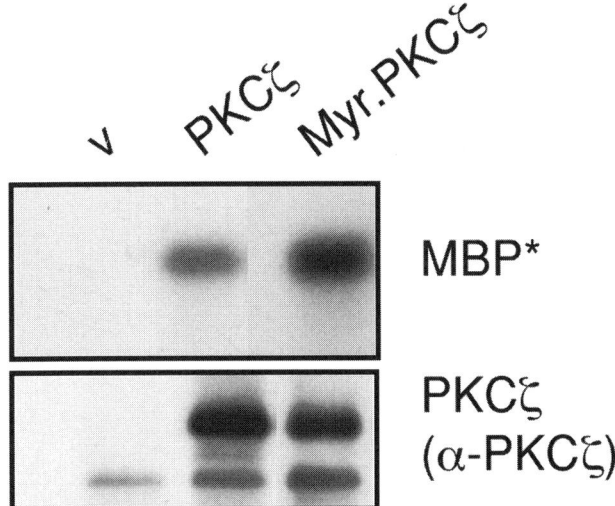

Fig. 2. Protein kinase activity of wild-type and membrane-targeted, Myr.PKCζ. HEK293 cells were transiently transfected with vector alone (v), FLAGPKCz, or Myr.PKCζ.FLAG as indicated. Cell lysates were immunoprecipitated with FLAG antibody and an immune complex kinase assay using myelin basic protein (MBP) was carried out using [γ-^{32}P]ATP. Phosphorylated MBP (MBP*) was detected by autoradiography. PKCζ expression was detected by Western immunoblotting of total cell lysate with an anti-PKCζ-specific antibody (α-PKCζ). Adapted from **ref. 25**.

cloned in a plasmid expression vector, either to a fragment of the PKC or the entire open reading frame. For example, antisense RNA to PKCα, PKCε, PKCδ, and PKCλ in a plasmid expression vector have been used to implicate these isoforms in the transcriptional activation of *c-fos (26)*. Similarly, antisense RNA to PKCι has been used to implicate this isoform in cell survival mediated by the NF-κB transcription factor *(27)*. Because synthetic phosphorothioate antisense oligonucleotides can be readily obtained from commercial sources, this methodology will be described here.

1. Obtain antisense PKC phosphorothioate oligonucleotide against the desired PKC isoform (**Table 3**).
2. Dissolve oligonucleotides in sterile H_2O at a final concentration of 1 m*M*, aliquot, and store at –20°C.
3. Seed cells at a density of 2×10^5 cells per well in a 6-well plate and allow to reach 40–50% confluency at the time of transfection.
4. Transfect oligonucleotide using the lipofectamine procedure described in Chapter 15, **Subheading 3.2.** Use 10 μg of each oligonucleotide and 40 μL of lipofectamine.

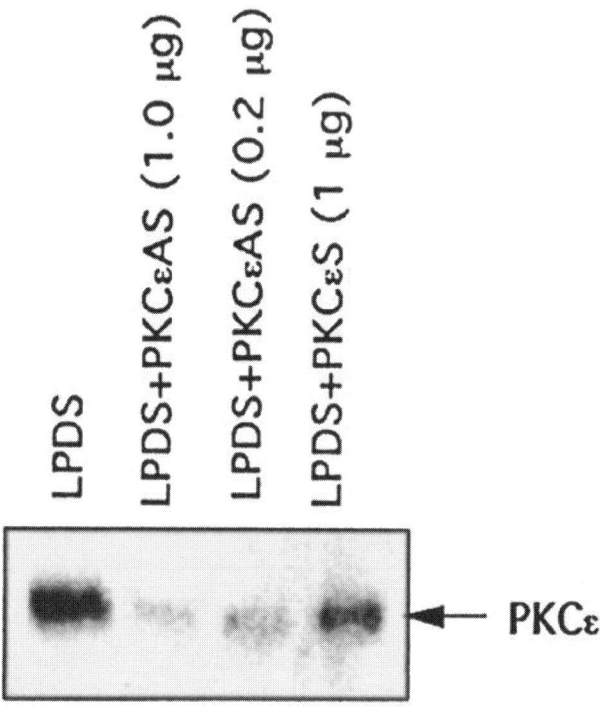

Fig. 3. Effect of PKCε-specific antisense oligonucleotide on PKCε expression. HepG2 cells were transfected with the indicated concentrations of PKCε phosphorothioate sense (S) and antisense (AS) oligonucleotides from **Table 3** for 36 h. Cells were then lysed and 50 μg of total cell lysate was resolved by sodium dodecyl sulfate polyacrylamide gel electrophoresis and immunoblotted with a PKCε-specific antibody. LPDS, lipoprotein-deficient serum. Reproduced with permission from **ref. *31***.

5. Harvest cells transfected with AS and S oligonucleotides at different times points, that is, 12, 24, 36, and 48 h.
6. Harvest cells, lyse, and separate total cell lysates by sodium dodecyl sulfate polyacrylamide gel electrophoresis (*see* Chapter 15, **Subheading 3.3.–3.4.**).
7. Verify that expression of the relevant PKC has been reduced by immunoblotting with isoform-specific PKC antibodies (Chapter 15, **Table 1**). To evaluate specificity, verify that the procedure has not resulted in a reduction of expression of other PKCs.

An example of antisense-induced reduction in expression of PKCε is shown in **Fig. 3** using sense and antisense phosphorothioate oligonucleotides in transfected HepG2 cells *(31)*.

4. Notes

1. We have noticed that mutation of the critical Lys residue to the commonly used Arg in the PKC ATP-binding site does not always result in a quantitatively inactive kinase, depending on the isoform. For this reason, we favor mutation to the less conserved residue, Trp. This point is exemplified by a study of PKCι, which showed that mutation of Lys274 to Arg did not result in any appreciable loss of protein kinase activity *(34)*. In contrast, a Lys274Trp mutant was completely kinase inactive toward exogenous substrates, and was found to act as a dominant negative allele in transfected cells *(34)*.
2. The PCR site-directed methodology makes use of the Quickchange technology (Stratagene), and the manufacturer's protocols and guidelines should be followed. Critical parameters are the use of a high-fidelity DNA Polymerase (e.g., Pfu), GC content and Tm of the oligonucleotide. Alternatively, the required PKC Ala to Glu mutation can be introduced using conventional PCR mutagenesis.
3. The most common modification used for antisense oligonucleotides is the phosphorothioate substitution in the oligonucleotide backbone. Although this renders the oligo nuclease resistant, the modified oligos have a lower Tm than their phosphodiester counterparts and can exhibit significant nonsequence specific activity. Therefore, an option is to use chimeric oligonucleotides. These oligos have modified 3′ and 5′ ends with a normal phosphodiester core allowing nuclease resistance at the termini, while limiting the phosphorothioate content. This reduces nonsequence specific artifacts. As with all oligonucleotides, chemical byproducts generated in the synthesis can be toxic to cells and the oligonucleotides should be purified, typically by reverse-phase high-performance liquid chromatography or by acrylamide gel electrophoresis.

References

1. Letiges, M., Plomann, M., Standaert, M. L., Bandyopadhyay, G., Sajan, M. P., Kanoh, Y., et al. (2002) Knockout of PKCalpha Enhances Insulin Signaling Through PI3K. *Mol. Endocrinol.* **16,** 847–858.
2. Leitges, M., Schmedt, C., Guinamard, R., Davoust, J., Schaal, S., Stabel, S., et al. (1996) Immunodeficiency in protein kinase cbeta-deficient mice. *Science* **273,** 788–791.
3. Standaert, M. L., Bandyopadhyay, G., Galloway, L., Soto, J., Ono, Y., Kikkawa, U., et al. (1999) Effects of knockout of the protein kinase C beta gene on glucose transport and glucose homeostasis. *Endocrinology* **140,** 4470–4477.
4. Weeber, E. J., Atkins, C. M., Selcher, J. C., Varga, A. W., Mirnikjoo, B., Paylor, R., et al. (2000) A role for the beta isoform of protein kinase C in fear conditioning. *J. Neurosci.* **20,** 5906–5914.
5. Suzuma, K., Takahara, N., Suzuma, I., Isshiki, K., Ueki, K., Leitges, M., et al. (2002) Characterization of protein kinase C beta isoform's action on retinoblastoma protein phosphorylation, vascular endothelial growth factorinduced endothelial cell proliferation, and retinal neovascularization. *Proc. Natl. Acad. Sci. USA* **99,** 721–726.

6. Narita, M., Aoki, T., Ozaki, S., Yajima, Y., and Suzuki, T. (2001) Involvement of protein kinase Cgamma isoform in morphine-induced reinforcing effects. *Neuroscience* **103,** 309–314.

7. Miyamoto, A., Nakayama, K., Imaki, H., Hirose, S., Jiang, Y., Abe, M., et al. (2002) Increased proliferation of B cells and auto-immunity in mice lacking protein kinase Cdelta. *Nature* **416,** 865–869.

8. Mecklenbrauker, I., Saijo, K., Zheng, N. Y., Leitges, M., and Tarakhovsky, A. (2002) Protein kinase Cdelta controls self-antigen-induced B-cell tolerance. *Nature* **416,** 860–865.

9. Olive, M. F., Mehmert, K. K., Messing, R. O., and Hodge, C. W. (2000) Reduced operant ethanol self-administration and in vivo mesolimbic dopamine responses to ethanol in PKCepsilon–deficient mice. *Eur. J. Neurosci.* **12,** 4131–4140.

10. Castrillo, A., Pennington, D. J., Otto, F., Parker, P. J., Owen, M. J., and Bosca, L. (2001) Protein kinase Cepsilon is required for macrophage activation and defense against bacterial infection. *J. Exp. Med.* **194,** 1231–1242.

11. Sun, Z., Arendt, C. W., Ellmeier, W., Schaeffer, E. M., Sunshine, M. J., Gandhi, L., et al. (2000) PKC-theta is required for TCR-induced NF-kappaB activation in mature but not immature T lymphocytes. *Nature* **404,** 402–407.

12. Leitges, M., Sanz, L., Martin, P., Duran, A., Braun, U., Garcia, J. F., et al. (2001) Targeted disruption of the zetaPKC gene results in the impairment of the NF-kappaB pathway. *Mol. Cell.* **8,** 771–780.

13. Elbashir, S. M., Harborth, J., Lendeckel, W., Yalcin, A., Weber, K., and Tuschl, T. (2001) Duplexes of 21-nucleotide RNAs mediate RNA interference in cultured mammalian cells. *Nature* **411,** 494–498.

14. Brummelkamp, T. R., Bernards, R., and Agami, R. (2002) A system for stable expression of short interfering RNAs in mammalian cells. *Science* **296,** 550–553.

15. Kisielow, M., Kleiner, S., Nagasawa, M., Faisal, A., and Nagamine, Y. (2002) Isoform–specific knockdown and expression of adaptor protein ShcA using small interfering RNA. *Biochem. J.* **363,** 1–5.

16. Garcia-Paramio, P., Cabrerizo, Y., Bornancin, F., and Parker, P. J. (1998) The broad specificity of dominant inhibitory protein kinase C mutants infers a common step in phosphorylation. *Biochem. J.* **333,** 631–636.

17. Soh, J. W., Lee, E. H., Prywes, R., and Weinstein, I. B. (1999) Novel roles of specific isoforms of protein kinase C in activation of the c-fos serum response element. *Mol. Cell. Biol.* **19,** 1313–1324.

18. Cenni, V., Doppler, H., Sonnenburg, E. D., Maraldi, N., Newton, A. C., and Toker, A. (2002) Regulation of novel protein kinase C epsilon by phosphorylation. *Biochem. J.* **363,** 537–545.

19. House, C. and Kemp, B. E. (1987) Protein kinase C contains a pseudosubstrate prototope in its regulatory domain. *Science* **238,** 1726–1728.

20. House, C. and Kemp, B. E. (1990) Protein kinase C pseudosubstrate prototope: structure–function relationships. *Cell Signal.* **2,** 187–190.

21. Orr, J. W., Keranen, L. M., and Newton, A. C. (1992) Reversible exposure of the pseudosubstrate domain of protein kinase-C by phosphatidylserine and diacylglycerol. *J. Biol. Chem.* **267,** 15,263–15,266.

22. Dutil, E. M. and Newton, A. C. (2000) Dual role of pseudosubstrate in the coordinated regulation of protein kinase C by phosphorylation and diacylglycerol. *J. Biol. Chem.* **275,** 10,697–10,701.
23. Pears, C. J., Kour, G., House, C., Kemp, B. E., and Parker, P. J. (1990) Mutagenesis of the pseudosubstrate site of protein kinase C leads to activation. *Eur. J. Biochem.* **194,** 89–94.
24. Genot, E. M., Parker, P. J., and Cantrell, D. A. (1995) Analysis of the role of protein kinase C–alpha, –epsilon, and –zeta in T cell activation. *J. Biol. Chem.* **270,** 9833–9839.
25. Chou, M. M., Hou, W., Johnson, J., Graham, L. K., Lee, M. H., Chen, C. S., et al. (1998) Regulation of protein kinase C zeta by PI 3-kinase and PDK-1. *Curr. Biol.* **8,** 1069–1077.
26. Kampfer, S., Hellbert, K., Villunger, A., Doppler, W., Baier, G., Grunicke, H. H., et al. (1998) Transcriptional activation of c-fos by oncogenic Ha-Ras in mouse mammary epithelial cells requires the combined activities of PKC-lambda, epsilon and zeta. *EMBO J.* **17,** 4046–4055.
27. Lu, Y., Jamieson, L., Brasier, A. R., and Fields, A. P. (2001) NF–kappaB/RelA transactivation is required for atypical protein kinase C iota-mediated cell survival. *Oncogene* **20,** 4777–4792.
28. Wang, A., Nomura, M., Patan, S., and Ware, J. A. (2002) Inhibition of protein kinase Calpha prevents endothelial cell migration and vascular tube formation in vitro and myocardial neovascularization in vivo. *Circ. Res.* **90,** 609–616.
29. Lu, D., Yang, H., Lenox, R. H., and Raizada, M. K. (1998) Regulation of angiotensin II–induced neuromodulation by MARCKS in brain neurons. *J. Cell. Biol.* **142,** 217–227.
30. Liedtke, C. M. and Cole, T. (1997) Antisense oligodeoxynucleotide to PKC-delta blocks alpha 1-adrenergic activation of Na-K-2Cl cotransport. *Am. J. Physiol.* **273,** C1632–C1640.
31. Mehta, K. D., Radominska-Pandya, A., Kapoor, G. S., Dave, B., and Atkins, B. A. (2002) Critical role of diacylglycerol- and phospholipid-regulated protein kinase C epsilon in induction of low-density lipoprotein receptor transcription in response to depletion of cholesterol. *Mol. Cell. Biol.* **22,** 3783–3793.
32. Hussaini, I. M., Karns, L. R., Vinton, G., Carpenter, J. E., Redpath, G. T., Sando, J. J., et al. (2000) Phorbol 12-myristate 13-acetate induces protein kinase ceta-specific proliferative response in astrocytic tumor cells. *J. Biol. Chem.* **275,** 22,348–22,354.
33. Sparatore, B., Patrone, M., Passalacqua, M., Pedrazzi, M., Pontremoli, S., and Melloni, E. (2000) Human neuroblastoma cell differentiation requires protein kinase C-theta. *Biochem. Biophys. Res. Commun.* **279,** 589–594.
34. Spitaler, M., Villunger, A., Grunicke, H., and Uberall, F. (2000) Unique structural and functional properties of the ATP-binding domain of atypical protein kinase C-iota. *J. Biol. Chem.* **275,** 33,289–33,296.

39

Yeast as a Host to Screen for Modulators and Regulatory Regions of Mammalian Protein Kinase C Isoforms

Amadeo M. Parissenti and Heimo Riedel

1. Introduction

Expression of various mammalian protein kinase C (PKC) isoforms in yeast including *Saccharomyces cerevisiae* and *Schizosaccharomyces pombe* results in phenotypic changes such as substantial increases in the cell doubling time, alterations in cell morphology, enhanced production of cytoplasmic vesicles, and elevated uptake of external Ca^{2+}. These phenotypes are the consequence of mammalian PKC catalytic activity. They require the expression of a mammalian PKC isoform, its proper domain structure including a functional catalytic domain, and typically activators such as phorbol esters. The observed phenotype is proportional to mammalian PKC catalytic activity and can be easily evaluated by observing the yeast cell doubling time. As a result, mammalian PKC activation by cell membrane-permeable activators such as tumor promoting phorbol esters, co-expressed protein ligands, or changes in PKC structure such as by random cDNA mutagenesis can be estimated through the resulting yeast colony size on agar plates or yeast culture density on microtiter plates. In this chapter, yeast is introduced as a host to screen libraries of PKC mutants, pharmacologic compounds, or PKC-binding proteins, for their impact on PKC catalytic activity. These libraries may originate from natural or synthetic pools of reagents or random cDNA mutagenesis strategies.

1.1. Expression of Mammalian Protein Kinase C Isoforms in Yeast

The molecular dissection of PKC function in mammalian cells is complicated by the genetic complexity of mammals, multiple PKC isoforms, and cross-

From: *Methods in Molecular Biology, vol. 233: Protein Kinase C Protocols*
Edited by: A. C. Newton © Humana Press Inc., Totowa, NJ

talk among numerous mammalian signaling pathways *(1–4)*. The analysis of PKC structure and function is facilitated in simple and genetically accessible eukaryotic experimental models such as budding and fission yeast *(5–13)*. Numerous mammalian proteins have been successfully expressed in yeast and disruptions of specific biochemical pathways in yeast can often be complemented by expression of the corresponding mammalian protein homologue. Significant compatibility has been demonstrated between yeast and mammalian regulatory mechanisms and signaling pathways *(14)*. Treatment with phorbol ester specifically activated bovine PKCα and rat PKCβI upon expression in *Saccharomyces cerevisiae (5,6)*. Several mammalian PKC isoforms have been expressed in *S. cerevisiae* and *Schizosaccharomyces pombe* with functional characteristics similar to those observed in mammalian cells *(5–13)*. In contrast, endogenous yeast PKC activity appears to be insensitive to mammalian PKC activators, including phorbol ester *(15–20)*. (*see* **Notes 1** and **2** for the properties of endogenous PKC activities.) Consequently, yeast represents a simple experimental host system for the specific evaluation of mammalian PKC function because heterologous, recombinant mammalian PKC is activated by phorbol treatment without affecting yeast PKC activity. In the approach we are presenting in this chapter (**Fig. 1**), an expression plasmid encoding a specific mammalian PKC isoform is introduced into *S. cerevisiae* cells. Transformants are isolated on selective medium and treated with galactose to induce PKC expression through cDNA transcription that is regulated by a galactose-inducible promoter. The expressed mammalian PKC is activated by phorbol ester and the resulting phenotype is scored (**Fig. 2A**).

Fig. 1. (*see facing page*) Strategies to screen for modulators and regulatory regions of mammalian PKC isoforms. The introduction of PKC expression plasmids into yeast cells, selection of transformants, induction of expression, and screening for PKC activators or inhibitors by observing the resulting phenotype has been outlined in sequential steps. Plasmids (black circles) carrying a *LEU2* gene (boxed LEU) are shown with wild-type (open box) or mutant (solid box) PKC cDNA inserts. Transformed *S. cerevisiae* cells are represented by large grey circles with a bulge (a characteristic of budding yeast) at the top right side. Expressed PKC proteins are shown with the two major regulatory and catalytic domains (ovals), joined by a hinge region (thin line). A closed conformation (adjacent ovals) depicts an inactive enzyme whereas increasing catalytic activation is represented by proportional separation of the domains (ovals) for wild-type (open ovals) and mutant (solid ovals) PKC forms. PKC activators (solid waved bar) and inhibitors (solid pentagon) are shown free or bound to the PKC regulatory or catalytic domains, respectively. The number of small (black) yeast cells shown at the various stages represents the relative resulting cell densities to reflect one example of a possible experimental outcome as it affects the cell doubling time. Experimental strategies to analyze PKC expression and its consequences are listed at bottom.

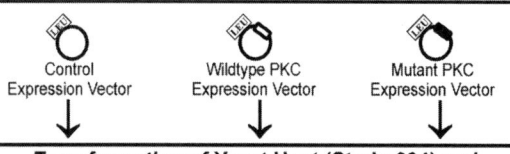

Introduction of PKC Protein-coding cDNA into Leu+ Expression Vector

Control Expression Vector   Wildtype PKC Expression Vector   Mutant PKC Expression Vector

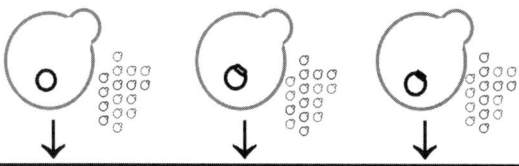

Transformation of Yeast Host (Strain 334) and Selection of Plasmid-Containing Cells in Leu- Medium

[Potential Co-Transformation with Expression Vectors (Ura+) Encoding PKC-Interacting Proteins]

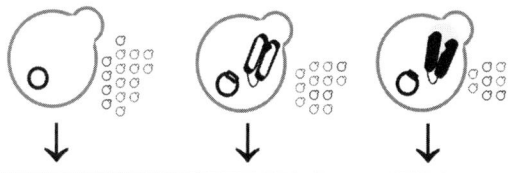

Induction of PKC Expression by Galactose

Phenotypic Screen for PKC Activity

Addition of PKC Activator

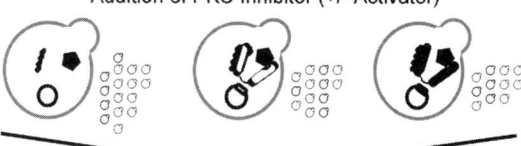

Addition of PKC Inhibitor (+/- Activator)

Extract Preparation for Protein Expression/Size and Catalytic Analysis

Growth Assays

Calcium Uptake Tests

Microscopic Observation of Cell Morphology

A — **G a l**

— **PMA** — **PMA**

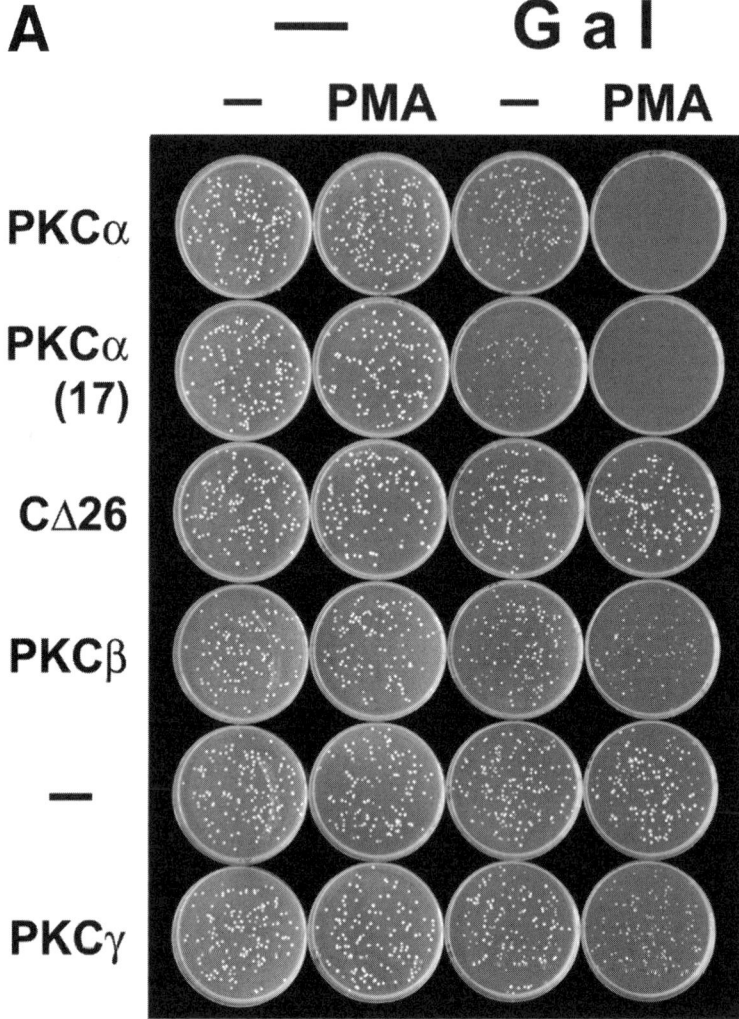

PKCα

PKCα
(17)

CΔ26

PKCβ

—

PKCγ

Fig. 2A. Correlation of *S. cerevisiae* phenotype with expression of various mammalian PKC isoforms or mutants. Bovine PKCα, rat PKCβI, rat PKCγ, and various bovine PKCα mutants were expressed from YEp51/52 in *S. cerevisiae* strain 334. Equal numbers of yeast cells have been plated on synthetic minimal selective medium (Leu⁻) and the resulting colonies are displayed after 2 d of growth. For each PKC isoform or mutant, agar plates are shown with any combination of galactose (Gal or –) to induce PKC expression and phorbol ester (PMA or –) to activate PKC. Plasmids encoding PKCα, βI, or γ are compared with control plasmids (–), with a plasmid resulting in higher PKCα expression levels (carrying 10 nucleotides of the 3′ untranslated sequence, PKCα17) (described in **ref. 5**), and with a plasmid encoding a truncated, catalytically inactive PKCα mutant (CΔ26) lacking 26 amino acids of the carboxyl terminus.

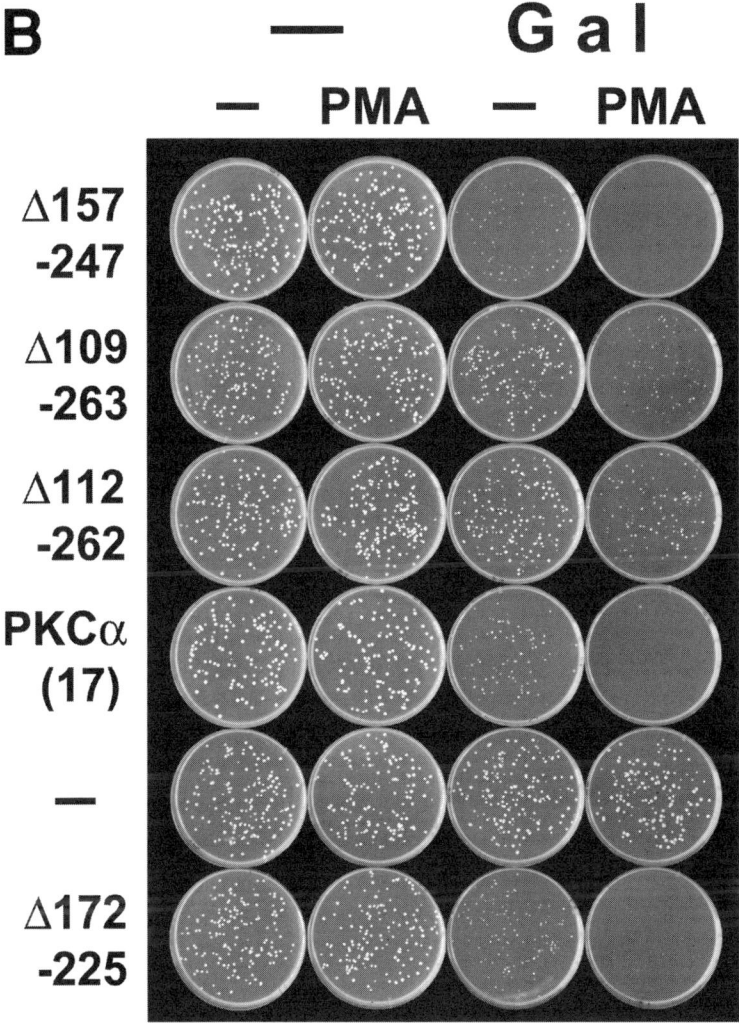

Fig. 2B. Plasmids encoding high-level expression of PKCα17 are compared with control plasmids (–), and with derivatives of plasmid PKCα17 encoding various internal deletion mutants of PKCα lacking the aa indicated: Δ157–247, Δ109–263, Δ112–262, or Δ172–225 mutants. Δ157–247 and Δ172–225 still carry both Cys-rich regions (Cys1, Cys2) in the PKCα regulatory domain whereas Δ109–263 and Δ112–262 lack Cys2.

Alternatively, PKC catalytic activity may be induced by other ligands (including heterologous proteins that are expressed from cDNA) or by structural changes of PKC, such as those resulting from CDNA mutagenesis. In a similar fashion, a number of mammalian PKC isoforms (ε, γ, δ, ζ, and η) have been

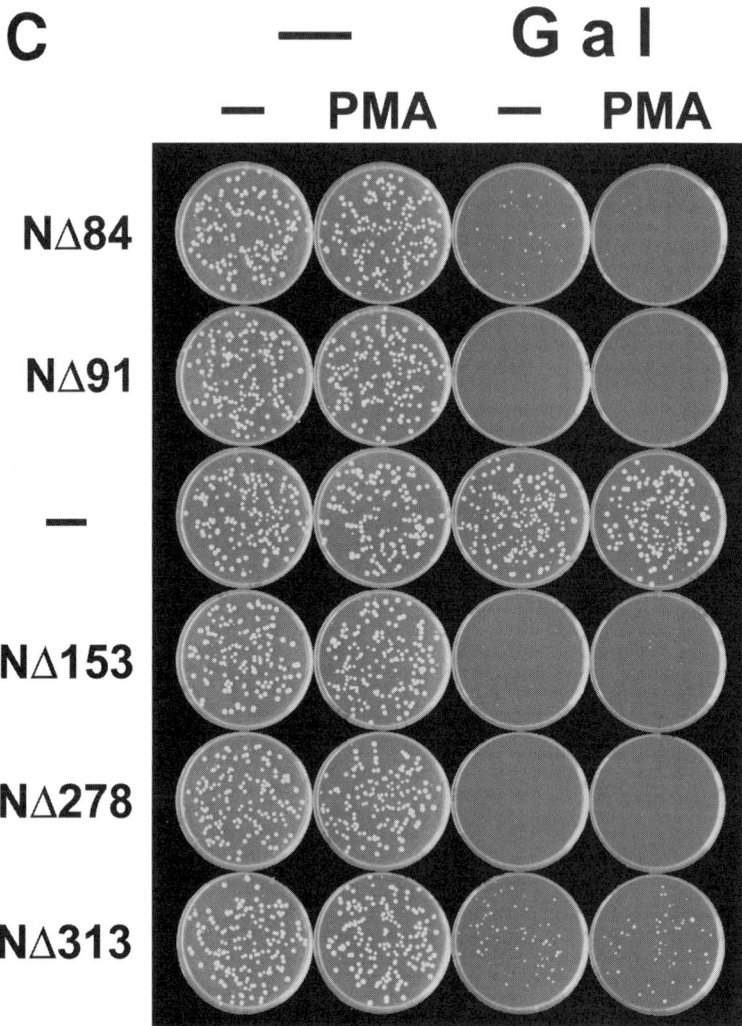

Fig. 2C. Control plasmids are compared with PKCα amino terminal truncation mutants lacking the aa indicated: NΔ84, NΔ91, NΔ153, NΔ278, or NΔ313. NΔ84 carries both Cys-rich regions, NΔD91 lacks part of Cys2, and NΔ153, NΔ278, and NΔ313 lack both Cys-rich regions.

successfully expressed in the yeast *S. pombe* *(8,9)*. As expected, PKCγ and δ were responsive to phorbol ester; however, in contrast to observations in *S. cerevisiae*, PKCα- or PKCβI-expressing *S. pombe* transformants have not been found *(8)* (*see* **Note 3** for PKC isoforms that have been expressed in either yeast host.)

1.2. Mammalian PKC-Dependent Phenotypic Changes in Yeast

Specific phenotypes are induced in *S. cerevisiae* in response to expression and activation of specific mammalian PKC isoforms. Responses include a pronounced increase in the cell doubling time as shown in **Fig. 2A** *(6,7,13,21)*, changes in cell morphology, including multicellular strings *(6,11,13)*, Ca^{2+} dependence of cell viability *(5)*, and enhanced external Ca^{2+} uptake *(5,6,13)*. These phenotypes are triggered by galactose-induction of PKC expression and are fully dependent on mammalian PKC catalytic activity *(5,6,13,22)*. PKCα- and PKCβI-dependent phenotypes are most pronounced when the enzymes are activated by phorbol ester *(5–7,22)*. The induced phenotypes are abolished by changes in PKC structure that interfere with catalytic function, indicating that the phenotypes are a result of the catalytic activity of the enzyme *(5–7,21)*. Responses, including the increase in cell doubling time, are reversible after mammalian PKC activation by PMA has ceased, indicating that physiologic changes are transient *(5–7,21)*. Mammalian PKC isoforms expressed in fission yeast (*S. pombe*) have resulted in overall comparable inhibitory effects on the cell doubling time and in enhanced production of cytoplasmic vesicles *(8,9)*.

1.3. Screening for PKC Activators and Inhibitors

The correlation of the described yeast phenotypes with mammalian PKC catalytic activity has been exploited as a tool to screen for activators and to some degree for inhibitors of PKC catalytic activity as outlined in **Fig. 1**. Libraries of natural or synthetic compounds can be screened in multiwell microtiter plates to rapidly identify agents that alter the cell doubling time in yeast strains expressing mammalian PKC isoforms. The phenotypic readout correlates sensitively with catalytic activation. As shown in **Fig. 2B,C**, subtle functional differences between PKC mutants can be detected based on average colony size. Differences have been demonstrated between the diterpene esters mezerein, phorbol esters PMA and dPPA, and the indol alkaloid indolactam V in terms of their ability to activate PKCα via specific PKC domains *(10,22)*. Under both in vitro and in vivo conditions, phorbol esters activated PKCα via either Cys-rich region (Cys1 or Cys2) in the PKC regulatory domain, whereas mezerein activation was specifically dependent on the Cys1 region *(10,22)*. Libraries of compounds can be screened for their PKC-inhibitory potential in the presence of activators (or PKC-activating mutations) as long as they do not display toxicity that affects the yeast doubling time (**Fig. 1**). Tested PKC inhibitors include calphostin C, dequalinium, chelerythrine chloride, and bryostatin *(9,10,22–24)*. Similar approaches have been used in *S. pombe* to show functional differences among various PKC activators and inhibitors (*see* **Note 4**) *(9)*.

1.4. Screening for PKC Regions That Regulate Catalytic Activity

The PKC-dependent phenotype can be exploited to screen for regulatory regions involved in PKC catalytic activity through activation or autoinhibition. Random mutagenesis by exonuclease truncation, degenerate oligonucleotides, or linker scanning can produce libraries of structural PKC mutants that are rapidly scored for their functional characteristics based on the resulting phenotype when expressed in yeast. A number of PKCα and βI deletion mutants have been created by Bal31 exonuclease treatment and were scored for their functional consequences (**Fig. 2B,C**) based on their impact on the cell doubling time of *S. cerevisiae (6,7,21)*. This strategy resulted in the functional definition of two independent Cys-rich regions of which at least one is required for phorbol ester activation and in the definition of regions that are dispensable for phorbol ester activation *(25)*. It helped define auto-inhibitory sequences in the regulatory domain that result in PKC activation when truncated *(23)* and helped define the boundaries of the PKC catalytic domain at the carboxyl terminus *(21)*. It defined differences between distinct classes of PKC activators in their PKC sequence requirements to stimulate catalytic activation *(10,22)*. It helped identify PKC regulatory regions that interact with specific inhibitors *(23,24)*. In most of these approaches, functionally relevant PKC mutants were identified based on their modulation of the yeast cell doubling time on agar plates. Even subtle changes in PKC structure and function can be distinguished based on the resulting size differences of expressing yeast colonies (**Fig. 2B,C**).

1.5. Screening for PKC-Modulating Peptide Ligands

The PKC-dependent phenotype can be exploited to screen for PKC-interacting proteins that regulate PKC activity. For this purpose, a library encoding PKC-interacting proteins is co-expressed from a second plasmid in yeast cells expressing a mammalian PKC isoform. Sequences that modulate the PKC-mediated increase in cell doubling time are expected to encode peptides with a potential PKC regulatory role in vivo. In support of this concept, the regulatory domain of human PKCα, when co-expressed with rat PKCβI, interfered with the PKCβI-mediated phenotype, which included an increase in the cell doubling time and alterations in yeast cell morphology *(11)*.

2. Materials

2.1. Equipment

1. 30°C Incubator (may require cooling at high ambient temperature).
2. Culture roller.
3. Absorbance spectrophotometer.

4. Table-top (clinical) low-speed centrifuge.
5. Protein gel electrophoresis equipment.
6. Immunoblotting system.
7. Phase contrast microscope.
8. Multiscreen Assay System including MADV 0.6 µm 96-well filtration plates (Millipore).
9. Scintillation counter.
10. Replica plating system (base and velvet cloth).

2.2. Reagents

Reagents were obtained from Sigma Laboratories (St. Louis, MO) unless otherwise stated.

1. *S. cerevisiae* strain 334 (*see* **Notes 1** and **5**).
2. *S. cerevisiae* dropout mixture: 2 g each of alanine, arginine, aspartic acid, cysteine, glutamine, glutamic acid, glycine, inositol, isoleucine, leucine, lysine, phenylalanine, proline, serine, threonine, tyrosine, valine, histidine, methionine, tryptophan, and uracil; 0.2 g of *p*-aminobenzoic acid; and 1 g of adenine hemisulfate. The mixture was ground with a mortar and pestle to similar-sized crystals and stored desiccated at 4°C.
3. Synthetic complete medium (for *S. cerevisiae* growth): 2 g *S. cerevisiae* dropout mixture, and 1.7 g of yeast nitrogen base (per liter) without amino acids and ammonium sulphate (Difco Laboratories, Detroit MI). Carbon sources glucose, glycerol, or galactose were added to final concentrations of 2, 3, and 2%, respectively, from filter-sterilized 10× stocks (after autoclaving the media).
4. Synthetic complete agar: synthetic complete medium with 20 g of agar (Difco Laboratories, Detroit MI) per liter of medium, autoclaved, and divided into approx 30 (8 cm) Petri dishes.
5. Leu⁻ medium: synthetic complete medium prepared with yeast dropout mixture lacking leucine.
6. Leu⁻ agar: Leu⁻ medium with 20 g of agar (per liter), autoclaved, and divided into approx 30 (8 cm) Petri dishes.
7. *S. cerevisiae* expression vectors: YEp51, YEp52 *(26)*, and derivatives *(5–7)* or pYES2 (Clontech Laboratories) (*see* **Note 6**) were purified from *E. coli* DH5α with a plasmid isolation kit (Midi, Qiagen Laboratories).
8. PKC cDNAs (*see* **Note 3**) *(1,3,4)*.
9. YPAD medium: 10 g of yeast extract, 20 g of peptone, 20 g of glucose, and 0.035 g of adenine hemisulfate in 300 mL of deionized water, autoclaved, and stored at room temperature.
10. Lithium acetate solution: 100 mM lithium acetate dihydrate, 10 mM Tris-HCl, pH 7.5, 1 mM ethylenediamine tetraacetic acid (EDTA), autoclaved and stored at room temperature.

11. PEG solution: 40% polyethylene glycol 4000, 10 mM Tris-HCl, pH 7.5, 1 mM EDTA, and 100 mM lithium acetate, autoclaved, and stored at room temperature.

12. TE solution: 10 mM Tris-HCl, pH 7.5, and 1 mM EDTA, autoclaved and stored at room temperature.

13. PMA stock solution: 1 mM phorbol-12-myristate-13-acetate (PMA) dissolved in dimethylsulfoxide and stored in 50-µL aliquots at –20°C.

14. Protein extraction buffer: 0.5 mM EDTA, 0.5 mM ethylenebis(oxyethylenenitrilo) tetraacetic acid (EGTA), 25 µg/mL leupeptin, 10 mM 2-mercaptoethanol, 0.5% Triton X-100, 20 mM Tris-HCl, pH 7.5, stored in 10-mL aliquots at –20°C.

15. Glass bead slurry: equal volumes of acid-washed glass beads (425 to 600 µm in diameter) and protein extraction buffer.

16. Column wash buffer: 20 mM Tris-HCl, pH 7.5, 0.5 mM EDTA, 0.5 mM EGTA, and 10 mM 2-mercaptoethanol.

17. Column elution buffer: 200 mM NaCl in column wash buffer.

18. TBS: 20 mM Tris-HCl, pH 7.6, 137 mM NaCl, stored at room temperature.

19. Blocking buffer: 5% instant skim milk powder (Carnation instant milk) in TBS.

20. TBST: 0.1% Tween 20 in TBS.

21. Antibody probing buffer: 5% instant skim milk powder (Carnation instant milk) in TBST.

22. PKC substrates: [Ser-25]PKCα_{19-31} or [Ser-159]PKC$\varepsilon_{153-164}$ from Life Technologies (Burlington, ON) and Upstate Biotechnology Inc. (Lake Placid, NY), respectively; prepared as 1 mM stock solutions and stored in 20-mL aliquots at –20°C.

23. PKC reaction buffer (5X): 820 µM ATP, 2.5 µCi [γ-^{32}P]ATP (Mandel Scientific, Guelph, ON), 90 mM MgCl$_2$, 100 mM Tris-HCl, pH 7.5, 250 µM PKC substrate (prepared just prior to the assay).

24. Phosphatidylserine micelle solution (4X): phosphatidylserine (from Avanti Polar Lipids) in chloroform (10 mg/mL) supplemented with either dimethyl sulfoxide (control) or PMA at final concentrations of 660 µg/mL phosphatidylserine and 4 µM PMA (prepared just prior to the assay).

25. Phosphocellulose filter squares: 1-inch squares cut from Whatman P81 phosphocellulose paper and prelabeled with a soft pencil.

26. PKC assay washing solution: 1% phosphoric acid.

27. Phosphate-buffered saline (without CaCl$_2$ or MgSO$_4$) from Life Technologies (Burlington, ON).

28. Calcium uptake buffer: 50 mM MES, pH 6.5, 5 mM MgSO$_4$.

29. ^{45}Ca^{2+} uptake solution: 0.23 mCi/mL ^{45}Ca^{2+} (5–30 Ci/g Ca^{2+}) in 1 mM CaCl$_2$.

30. Calcium uptake wash buffer: 10 mM MES, 10 mM Tris-HCl, pH 6.0, 35 mM CaCl$_2$, 100 mM NaCl, 25 mM KCl.

31. Bal 31 cleavage buffer (5X): 3 M NaCl, 60 mM CaCl$_2$, 60 mM MgCl$_2$, 100 mM Tris-HCl, pH 8.0, 1 mM EDTA.

3. Methods

3.1. Preparation of Mammalian PKC-Expressing Yeast Strains

3.1.1. Construction of Yeast Expression Plasmids
for Mammalian PKC Isoforms

A prerequisite for the phenotypic screen for PKC activity as outlined in **Fig. 1** are shuttle expression vectors that can be propagated and selected for in *E. coli* (via resistance to ampicillin) and in *S. cerevisiae* (via Leu- or Ura-independent growth). A critical advantage of YEp51/52 or pYES2 lies in the regulation of heterologous cDNA expression by a galactose-inducible promoter region. Regulation of expression allows the rapid propagation of transformed cells in the absence of PKC expression and the direct comparison of phenotypic changes in response to the induction of PKC expression.

The first step in the phenotypic screen requires the preparation of mammalian PKC expression constructs. Mammalian PKC cDNAs encoding bovine PKCα, human PKCα, rat PKC βI, human PKCγ, and murine PKCε (*see* **Note 3**) can be obtained from a number of sources (*1–4*), including John Knopf at The Genetics Institute (Cambridge, MA) or Research Genetics (Carlsbad, CA). The cDNAs are released from their host plasmids by restriction enzyme digestion and introduced into the galactose-inducible yeast expression vectors YEp51, YEp52, or pYES2 (*see* **Note 6**) at appropriate restriction sites using standard molecular biological approaches (*27*). Some cDNAs can be inserted into yeast expression plasmids via specifically designed oligonucleotide adapters that facilitate ligation of incompatible DNA termini and can provide five adenine residues upstream of the ATG start codon to mimic sequences found in highly expressed *S. cerevisiae* genes (*5,7,21,28*). The ligation product is introduced into *E. coli* DH5α and transformants are selected on Luria broth (LB) agar plates in the presence of 200 µg/mL ampicillin using standard procedures (*27*). Plasmids are isolated from selected colonies using an isolation kit ("Mini" Prep, Qiagen Laboratories). Proper expression constructs carrying a PKC cDNA insert are identified by restriction endonuclease analysis and confirmed by DNA sequencing.

PKC cDNAs can be expressed in *S. pombe* (*see* **Note 7**) with shuttle vector pREP3X (*8*). This plasmid is a derivative of pREP1 (*29*) in which the *Nde*I restriction site was replaced with an *Xho*I site to allow translation initiation from the ATG start codon of the inserted cDNA. A strong promoter (nmt 1, reportedly eightfold stronger than the alcohol dehydrogenase constitutive promoter) regulates cDNA transcription which is repressed 80-fold by thiamine during normal culture (*29,30*).

3.1.2. Lithium Acetate Transformation of Yeast Cells
with PKC Expression Plasmids

Expression plasmids carry a selectable marker, *Leu2/* for YEp51/52 and *Ura3* for pYES2, that permits the continued selection of transformed yeast cells. PKC expression vectors are introduced into *S. cerevisiae* cells by lithium acetate transformation *(31)*. To prepare competent cells 1 mL of a saturated overnight culture of *S. cerevisiae* (strain 334 grown in synthetic complete medium) is used to inoculate 300 mL of YPAD medium. (*see* **Note 5** for a description of *S. cerevisiae* strain 334.) Cells are propagated overnight under agitation at 30°C until the cell density reaches 0.4 OD_{600}. Cells are sedimented at 5000*g* for 10 min at room temperature, rinsed with 10 mL of sterile, distilled water, and resuspended in 1.5 mL of lithium acetate solution. For transformation 200-µL aliquots of cells are placed in 2.2 mL microfuge tubes and 25 µL of solution containing 5 µg of yeast expression vector and 200 µg of herring sperm DNA is layered on top (*see* **Note 8**). Expression vectors lacking a PKC cDNA insert serve as controls. After addition of 1.2 mL of PEG solution the mixture is agitated for 30 min at 30°C followed by a 15-min (heat-shock) incubation at 42°C. Cells are sedimented by centrifugation for 5 min at room temperature, resuspended in 200 µL of TE (100 µL of cells per plate), and spread evenly on two Leu⁻ or Ura⁻ agar plates (depending on the plasmid, *see* **Note 6**) containing 2% glucose for selection of transformants. Plates are incubated at 30°C for about 2 d or until colonies of yeast transformants are well visible. Five random colonies are selected and transferred to new Leu⁻ agar plates to separate pure clones for further analysis.

For transformation of *S. pombe* strains (*see* **Note 7**) PKC expression vectors are introduced by electroporation with a pulse controller (Bio-Rad, based on the manufacturer's instructions). Transformants are selected on Leu⁻ agar plates containing 1 *M* sorbitol and 1 µ*M* thiamine to suppress PKC expression *(8)*. Stable clones are selected by diluting colonies 1000-fold sequentially three times in complete medium followed by re-selection on Leu⁻ plates in the presence of 1 µ*M* thiamine.

3.2. Characterization of Mammalian PKC Expression in Vitro

3.2.1. Preparation of Yeast Detergent Cell Extracts

After yeast transformation, it is critical to verify proper expression of mammalian PKC isoforms or mutants with regard to expected molecular size and catalytic activity. The protocol outlined below specifically addresses the disruption of the yeast cell wall in the preparation of detergent cell extracts. Whole cell detergent extracts from *S. cerevisiae* transformed with PKC-expression plasmids or from control-transformed cells are prepared to evaluate PKC expression products by immunoblotting with mammalian PKC-specific antibodies (*see* **Subheading 3.2.2.**) and by PKC catalytic activity analysis (*see* **Subheading 3.2.3.**), based on a modification of a previously described protocol *(5)*. Two milliliters of a freshly saturated overnight culture serve to inoculate a 300-mL culture containing a yeast carbon source in selective medium with any combination of 2% galactose and 1 μM phorbol ester. After incubation to a cell density of 0.8–1.0 OD_{600} cells are sedimented by centrifugation at $5000g$ for 10 min, washed in phosphate buffered saline without Ca^{2+} and Mg^{2+}, and resuspended in 100 µL of protein extraction buffer. (*see* **Note 9** on the importance of EGTA in this buffer.) After addition of 100 µL glass bead slurry, cells are lysed by vortexing 6 times for 30 s with a 2-min cooling period on ice between all vortexing steps. Cell lysates are clarified by centrifugation at 10,000g for 30 min at 2°C. For some PKC constructs crude extracts from expressing cells are partially purified by diethylaminoethyl cellulose chromatography (DE52, Whatmann Laboratories) to enhance the signal for subsequent PKC catalytic analysis. DE52 powder is hydrated with 50 mM Tris-HCl buffer, pH 7.5, charged with 0.1 N NaOH, and the pH of the resin restored with 1 M Tris-HCl, pH 7.5. A 6-mL column of 50% DE52 slurry is rinsed thoroughly with column wash buffer prior to and again after 3 mL of crude extract is added. Bound proteins are eluted with 5 mL of column elution buffer and the solution is mixed well before storage in aliquots at –80°C.

S. pombe crude detergent cell extracts are prepared as described above except that the extraction buffer contains higher concentrations (2 mM) of EDTA and EGTA and additional protease inhibitors (0.1 mM phenylmethylsulfonyl fluoride and 10 mM benzamidine). After preparation, extracts are precipitated with 45% (wt/vol) ammonium sulphate and re-suspended in extraction buffer before use.

3.2.2. Immunoblot Analysis of PKC with Specific Antibodies

Expression of mammalian PKC constructs in yeast is verified after poly-acrylamide gel electrophoresis by immunoblotting with isoform-specific PKC antibodies. Proteins from yeast detergent cell extracts (prepared as described in **Subheading 3.2.1.**) are separated on sodium dodecyl sulfate polyacrylamide gels and transferred electrophoretically to nitrocellulose membranes using standard techniques *(27)*. All subsequent steps are performed at room temperature. After transfer, membranes are rinsed three times for 15 min in TBS and treated with blocking buffer for 1 h. Membranes are subsequently incubated for 1.5 h at room temperature in antibody probing buffer with a 1/1000 dilution of a mouse monoclonal antibody against human PKCα (Upstate Biotechnology;

Lake Placid, NY) β, γ, or ε (Transduction Laboratories, Lexington, KY). After three wash steps for 15 min with TBST, membranes are incubated for 1 h with horseradish peroxidase-conjugated rabbit antimouse IgG, diluted 10,000-fold in antibody probing buffer. After five 15-min wash steps with TBST, membrane-bound, immunoreactive proteins are visualized by chemiluminescence in 10 mL of reaction volumes (ECL substrates, Amersham). Immunoblots are quickly placed in K-pak sealable plastic pouches after excess liquid had been removed with paper towels and exposed for 5–30 min by autoradiography (X-OMAT, Kodak).

S. *pombe*-denatured protein extracts *(32)* are similarly analyzed by immunoblotting with rabbit polyclonal isoform-specific PKC antibodies directed against carboxyl terminal sequences *(33–35)*, followed by visualization with horseradish peroxidase-conjugated antibodies.

3.2.3. Catalytic Analysis of PKC with Specific Substrates

It is critical to evaluate the catalytic function of mammalian PKC isoforms that have not been previously expressed in yeast and any resulting phenotype before the start of a screen. The yeast experimental host system enables the evaluation of a single mammalian PKC isoform in the absence of functional interference by other PKC isoforms. Catalytic activity of the expressed PKC isoforms is measured at 30°C for 30 min as described previously *(5)* by monitoring the transfer of radiolabeled phosphate from [γ-^{32}P]ATP to positively charged mammalian PKC substrates such as [Ser-25]PKCα$_{19–31}$ (Life Technologies Burlington, ON) or [Ser-159]PKCε$_{153–164}$ (Upstate Biotechnology Inc., Lake Placid, NY). For each 100-µL reaction, 20 µL of PKC reaction buffer (5X is added to 25 µL of phosphatidylserine micelle solution (4X) with or without PMA. Five microliters of a 250 µ*M* PKC substrate stock solution (or the equivalent volume of H$_2$O as a control) is added and all samples are diluted to a final volume of 70 µL. Reactions are initiated by adding 30 µL of crude or partially purified yeast cell extract (*see* **Subheading 3.2.1.**). Assays are staggered and terminated sequentially by spotting 90 µL of the reaction mixture onto prelabeled Whatman P81 phosphocellulose filter squares that are immediately immersed in 1% phosphoric acid. Filters are rinsed thoroughly, six times with 500 mL of 1% phosphoric acid to remove unincorporated [γ-^{32}P]ATP. The efficiency of the wash steps is monitored with a control filter. The amount of radiolabeled substrate bound to the phosphocellulose filter squares is quantified by liquid scintillation spectrometry. Radioactivity associated with the "no substrate" control reaction is subtracted from the experimental results to adjust for the signal obtained by phosphorylation of endogenous yeast substrates in the extract. The concentration of PKC substrates in the reaction can be varied and pharmacological activators or inhibitors of PKC can be included.

S. pombe cell extracts are similarly analyzed for PKC catalytic activity except that reactions contain 0.75 mM CaCl$_2$, 250 µg/mL substrates, 1.25 mg/mL phosphatidylserine, and 250 ng/mL PMA.

3.3. Analysis of the Yeast Phenotype

3.3.1. Cell Doubling Time

In response to the expression and activation of several mammalian PKC isoforms in yeast, the cell doubling time is dramatically increased. The magnitude of this growth-inhibitory response varies with the carbon source (*see* **Note 10**) and shows some isoform specificity (*see* **Note 11**). Subtle differences in PKC catalytic activity can be detected in a direct comparison (**Fig. 2**). Screening of large numbers of samples is possible by yeast colony size on agar plates. This should be followed by quantification of phenotypic differences by comparing cell densities (OD$_{600}$) in liquid culture. The impact of mammalian PKC expression on the yeast cell doubling time is quantified by measuring culture densities during the growth of control and PKC-expressing cells in the absence and presence of PKC-activating phorbol esters as previously described *(22)*. Freshly saturated cultures of five clonal isolates of control and PKC-expressing yeast strains are serially diluted 2×10^6-fold in 3 mL of Leu⁻ medium, aiming at starting conditions of about 100 viable cells per culture. Typically, 10 µL of a freshly saturated overnight culture are diluted in 1 mL of sterile distilled water followed by a second dilution step of a 10-µL aliquot in 1 mL of water. Fifteen microliters of the second dilution is then introduced into Leu⁻ medium, containing any combination of the PKC activator PMA (1 µM) and 2% galactose including controls lacking either or both components (*see* **Fig. 1**). Glucose (2%) or glycerol (3%) serves as a carbon source. Cell proliferation is evaluated by measuring culture densities at 600 nm (OD$_{600}$) at regular time intervals. Because cell doubling times vary considerably depending on the expression and activation of PKC isoforms, accurate culture density readings may be delayed for several days until readable cell densities are reached, initially as judged by visual inspection in bright light.

The impact of mammalian PKC expression on the doubling time of *S. pombe* is determined with comparable strategies *(8)*. (*see* **Note 12** about the strong isoform specificity observed for PKC-induced growth inhibition in *S. pombe*.) Freshly saturated cultures propagated in the presence of thiamine are thoroughly rinsed with Leu⁻ medium, diluted 1:100 into fresh medium and incubated for 24 h. Cultures are diluted in Leu⁻ medium in the presence or absence of 100 ng/mL PMA and/or 1 µM thiamine. At various times aliquots are transferred to 96-well microtiter plates and measured for their optical density (OD_{595}) in an automated reader (Molecular Devices, Menlo Park, CA).

3.3.2. Cell Uptake of External Ca^{2+} or Lucifer Yellow

The impact of mammalian PKC expression on external Ca^{2+} uptake into yeast cells is examined as previously described (5). Control and PKC-expressing yeast cells are propagated to an optical density of approx 0.5 OD_{600} in 100 mL of Leu⁻ medium containing 2% glucose only (to prevent mammalian PKC expression) or both glucose and 2% galactose (to induce mammalian PKC expression). Cells are sedimented by centrifugation at 5000g for 20 min at room temperature, washed twice in phosphate-buffered saline (Ca^{2+} and Mg^{2+}-free), and resuspended in calcium uptake buffer at a concentration of 2.0×10^8 cells/mL. Four 100 microliter samples of each cell suspension are transferred to a microtiter plate and 11 µL of 10 µM PMA or 11 µL of dimethyl sulfoxide as a control are added to two of the samples. After 45-min incubation under agitation at room temperature, 10 µL of $^{45}Ca^{2+}$ uptake solution is added and the incubation continued for 2.5 h at room temperature. Samples are transferred to 96-well filtration plates (Multiscreen MADV, 0.6 µm pore size, Millipore) using a 12-channel pipet. The solution is immediately removed through the filter by negative pressure and cells are quickly rinsed on the filters four times with 200 µL of calcium uptake wash buffer. After blotting to a paper towel to remove excess liquid (see **Note 13**), membranes at the bottom of each well are dried overnight, released from their wells using a custom-made steel cylinder, and transferred to scintillation vials. Radiolabeled Ca^{2+} associated with the membranes is then quantified by liquid scintillation spectroscopy in 1 mL of scintillation fluid. In controls, $^{45}Ca^{2+}$ solution is added to a duplicate sample of cells and immediately rinsed on filters as described above. The radioactivity associated with these samples is considered to represent $^{45}Ca^{2+}$ nonspecifically bound to the samples or membrane and its value is subtracted from all uptake readings.

For *S. pombe* the impact of PKC expression on cellular uptake of lucifer yellow was addressed rather than the uptake of external Ca^{2+} (8). Cultures of 20 mL are propagated in Leu⁻ medium into the logarithmic growth phase in the presence or absence of 100 ng/mL PMA and/or 1 µM thiamine, washed twice in ice-cold Leu⁻ medium, and resuspended in medium containing 5 mg/mL lucifer yellow carbonyl hydrazine and 1 mM NaN₃. Cultures of 0.5 mL are subsequently incubated at 32°C for 90 min, whereas control cultures are kept on ice during the incubation period. Subsequently, all cultured cells are rinsed five times with 2 mL of ice-cold medium, sedimented, and vortexed for 1 min with 1 mL of acid-washed glass beads. An additional 2 mL of medium is added and the liquid (excluding the glass beads) removed and clarified by centrifugation for 15 min at 5000g. Fluorescence of the released lucifer yellow is measured in a fluorimeter (Perkin-Elmer, Norwalk, CT). The ability of PKC expression and activation to modulate lucifer yellow uptake is highly isoform specific (*see* **Note 14**) *(8)*. Conceivably, lucifer yellow uptake and radiolabeled Ca^{2+} uptake may measure a similar biological phenomenon in the two different yeasts (*see* **Note 15**) *(5,6,8,13)*.

3.3.3. Microscopic Evaluation of Cell Morphology

The impact of mammalian PKC expression on yeast cell morphology is evaluated by phase contrast microscopy. Freshly saturated (overnight) cultures of control and PKC-expressing yeast cells in Leu⁻ medium containing 2% glucose are diluted to 0.02 OD_{600} in synthetic Leu⁻ medium containing 3% glycerol in the presence or absence of 2% galactose and/or 1 μM PMA. Cultures are incubated for 2 d at 30°C before 10 mM EDTA is added to prevent Ca^{2+}-mediated intercellular adhesion. Cells are vigorously vortexed and visualized by phase contrast microscopy (*see* **Notes 11** and **14**). Ten randomly selected fields containing a total of approx 500 cells per field are scored for alterations in cell morphology.

3.4. Screening for Modulators of Mammalian PKC Activity in Yeast

3.4.1. Screening for PKC Activators or Inhibitors

Figure 1 outlines a strategy to screen for PKC activators or inhibitors. In the first step a mammalian PKC isoform is introduced into an expression vector such as YEp51 or YEp52, which should preferably allow the regulation of protein expression (*see* **Subheading 3.1.1.**). Yeast host cells are either transformed with a control (lacking PKC DNA) or with a PKC expression plasmid and cells are continuously propagated in Leu⁻ medium to maintain selection for plasmid-bearing cells (*see* **Subheading 3.1.2.**). Aliquots of a defined number of cells are transferred to galactose-containing medium (control: without galactose) to induce mammalian PKC expression. Cell growth is monitored for control and PKC-expressing cells (*see* **Subheading 3.3.1.**). If classic mammalian PKC isoforms (such asPKCα, β, or γ) are expressed upon galactose induction, cell growth will be little affected when compared to control plasmids or cultures without galactose, because basal PKC catalytic activity is minimal (*see* **Subheading 1.2.**, **Fig. 2A**). In contrast, mammalian PKCε expression in glycerol-containing medium affects yeast growth, cell morphology, and Ca^{2+} uptake in the absence of phorbol ester activators *(13)*. For the expression of other PKC isoforms the putative phenotypes in *S. cerevisiae* or their potential phorbol ester dependence remain to be determined.

The subsequent steps of the screening protocol depend on the phenotype observed upon galactose induction. For expressed classic PKC isoforms, cultures grown in selective medium with galactose can be subjected to activators to assess their potency to inhibit yeast growth. Pools of putative inhibitors can also be screened based on their modulation of catalytically activated PKC forms as long as they do not display toxicity to yeast cells that may increase the doubling time. For PKC isoforms that increase the cell doubling without external stimulation it is important to confirm whether catalytic activity of the expressed PKC is constitutive in the yeast host (protocol outlined in **Subheading 3.2.3.**). If the phenotype is correlated with catalytic activity, PKC inhibitors can in principle be identified in a screen. It may also be possible to identify activators if constitutive reduction of cell doubling time can still be significantly enhanced. On the other hand, if the observed phenotype were caused by an effect on the host cell that is unrelated to catalytic activity, this would preclude a meaningful screen for modulators of the specific underlying PKC isoform.

The PKC-dependent increase in yeast cell doubling time can be exploited as a phenotype to screen random libraries of PKC ligands or cDNA mutants for their ability to modulate PKC catalytic activity. The resulting PKC catalytic activity can be scored by colony size on agar plates (**Fig. 2**) or cell density in liquid cultures (*see* **Subheading 3.1.1.**) *(7,22)*. To screen for activators or inhibitors of PKC, a variety of cultures of control and PKC-expressing yeast cells (grown in Leu$^-$ medium with 2% glucose or 3% glycerol) are set up in 96-well plates in the presence or absence of 2% galactose. To screen for PKC-activating compounds, test agents are added to control and to PKC-expressing yeast cells both of which are compared with control samples lacking activator. Agents that induce a significant increase in the cell doubling time in a galactose-dependent manner, specifically in cells transformed with a yeast expression vector containing the PKC cDNA, represent strong candidates for PKC activators. To screen agents for PKC inhibitory potential, samples are prepared as described above except that 1 μ*M* PMA is added to all cultures. Agents that are able to significantly lower the observed galactose-dependent increase in the cell doubling time, specifically in cells transformed with PKC expression plasmids, represent likely PKC inhibitors.

3.4.2. Screening for PKC Regulatory Regions

The PKC-dependent increase in the yeast cell doubling time can be exploited as a phenotype to screen for PKC cDNA mutants that result in altered PKC catalytic activity. Such mutants have the potential to identify regulatory regions within the PKC primary structure *(21,23–25)* and can be created with various standard protocols *(27)*, including degenerate oligonucleotide, linker scanning, or deletion mutagenesis. A protocol for staggered deletion mutagenesis is described below as an example.

To generate amino terminal or carboxyl terminal PKC deletion mutants, the expression plasmid is linearized by cleavage with a restriction endonuclease at a site upstream or downstream of the translation start or stop codon, respectively. The linearized expression plasmid is digested with exonuclease Bal 31 for 30 min at 37°C. Several parallel samples are incubated with a series of twofold dilutions of Bal 31 in cleavage buffer starting with 2 µL of undiluted enzyme. All reactions are terminated by the addition of EGTA (pH 8.0) to a final concentration of 20 mM and heated for 5 min at 65°C to remove residual nuclease activity. The extent of the truncations is then evaluated by agarose gel electrophoresis of appropriate restriction fragments. Pools of truncated fragments of a desirable size range are isolated from preparative agarose gels and joined with complementary PKC sequences in expression plasmids to recreate large pools of individual PKC truncation mutants.

Ligated constructs encoding pools of amino terminal, carboxyl terminal, and internal PKC deletion mutants are introduced into *S. cerevisiae* strain 334 and selected on Leu⁻ agar plates containing 2% glucose as described in **Subheading 3.1.2.** Conditions are adjusted to result in approx 200 colonies per 8-cm plate that are replica-plated onto alternative carbon sources in the presence or absence of 2% galactose. The procedure for yeast colony replica plating is described in **Note 16**.

Generally, (1) minor carboxyl terminal truncations match the normal PKC phenotype, whereas progressive truncations result in the sudden loss of the PKC phenotype and match the control yeast phenotype *(21)* because of inactivation of the PKC catalytic domain (**Fig. 2A**). (2) Amino terminal truncations (**Fig. 2C**), however, result in PKC catalytic activation (decreased colony size) in the absence of phorbol ester as a result of the removal of the pseudosubstrate auto-inhibitory region. Catalytic activity remains partially phorbol ester responsive as long as one Cys-rich region stays intact. Partial truncation of the second Cys-rich region, however, results in PKC catalytic activity (decreased colony size) that is completely phorbol ester-unresponsive (**Fig. 2C**) *(22)*. This phenotype largely persists with increasing amino terminal truncations until the amino terminal region of the PKC catalytic domain is ultimately affected. At this point, complete inactivation of the expressed mutant PKC results in a wild-type yeast phenotype. (3) Increasing internal PKC deletions near the second Cys-rich region (**Fig. 2B**) result in progressive reduction of phorbol ester-stimulated PKC catalytic activity (increasing colony size when compared to normal PKC) but remain in principle phorbol ester-responsive as long as the first Cys-rich region stays intact *(25)*.

Consequently, PKC catalytic activity and its phorbol ester dependence are proportionally represented by the resulting yeast colony size and minor functional differences are distinguishable on agar plates (**Fig. 2**). Because individual yeast colony size varies somewhat based on parameters other than expressed PKC activity, subsequent cell density analysis in liquid culture provides a more quantitative approach to discern subtle differences in catalytic activity (*see* **Subheading 3.3.1.**). Similarly, the effect of other PKC isoforms and their correlation to specific PKC sequence requirements can be screened by this approach.

3.4.3. Screening for PKC-Modulating Binding Proteins

Collections of agents, including proteins, can be screened in yeast for their potential to interact with mammalian PKC and to modulate its catalytic activity (**Fig. 1**) *(11)*. Potential PKC-interacting protein-encoding cDNAs from an expression library are inserted into a yeast expression vector pYES2 (Invitrogen, Carlsbad, CA) containing the *URA3* gene, while mammalian PKC cDNA is co-expressed from YEp51/YEp52 (as described above in **Subheading 3.1.1.**). YEp51/YEp52 and pYES2 are suitable expression plasmids for this approach because they carry complementary *LEU2* and *URA3* genes, respectively, that allow co-selection for the presence of both plasmids in Leu⁻/Ura⁻ medium (*see* **Note 6**). *S. cerevisiae* strain 334, which carries a leu⁻ and ura⁻ genotype (*see* **Note 5**), is transformed sequentially, selecting first for mammalian PKC expression from YEp51/YEp52 in Leu⁻ medium. cDNAs encoding potential PKC-interacting proteins are liberated from the library by digestion with specific restriction endonucleases and inserted into pYES2. Pools of plasmids carrying various cDNAs are amplified and purified from transformed *E. coli* and introduced into a mammalian PKC-expressing *S. cerevisiae* strain by lithium acetate transformation (as described in **Subheading 3.1.2.**) followed by coselection in Leu⁻/Ura⁻ medium. Pools of yeast transformants are isolated and serial dilutions are plated on Leu⁻/Ura⁻ agar plates containing 2% glucose. Plates containing approx 200 well-resolved colonies are replica plated on Leu⁻/Ura⁻ agar containing various carbon sources in the absence or presence of 2% galactose and/or 1 µ*M* PMA as described in **Note 16**. Sequential transformation of the two plasmids is preferable to cotransformation. It ensures that any observed phenotypic differences are caused by the expression of an interacting protein rather than by variation in PKC expression between clones. Isoform-specific antibodies to carboxyl terminal epitopes of PKC isoforms serve to verify similar levels of PKC in the isolated clones (*see* **Note 17**). Colonies are candidates for the expression of PKC-regulating proteins if their doubling time (colony size) is significantly altered when compared with PKC-expressing cells carrying control pYES2. Further control experiments to verify a candidate sequence include liquid growth assays (*see* **Subheading 3.3.1.**), catalytic analysis (*see* **Subheading 3.2.3.**), and the characterization of expression products (*see* **Subheading 3.2.2.**).

4. Notes

1. Endogenous PKC activity has been reported in *S. cerevisiae*, including a gene *PKC1* that led to cell cycle arrest upon disruption *(16,17)*. Mutagenesis of the gene *PKC1* resulted in a cell cycle-specific osmotic stability defect *(19)* and enhanced mitotic recombination *(36)*, suggesting separate roles of *PKC1* in the regulation of osmotic stability as well as in DNA metabolism. PKC1 appears to function upstream of mitogen-activated protein kinase to regulate high osmolarity glycerol response pathways *(14)*. Existence of an additional PKC gene (*PKC2*) implicated in the cellular response to amino acid starvation has been reported *(37)* but could not be verified *(38)*.

2. Two PKC-related genes, essential for cell viability and implicated in cell shape control have been identified in *S. pombe (39)*. It is intriguing to speculate that the growth-interfering phenotype in this organism may be the result of PKC-mediated changes in cytoskeletal structure *(40)*. Consistent with this view, a mitogen-activated protein kinase homolog (*Pmk1*) was identified that regulates cell shape and cytokinesis, and functions coordinately with the PKC pathway *(41)*.

3. In *S. cerevisiae* expression of bovine PKCα *(5,7)*, rat PKCβI *(6)*, rat PKCγ *(10)*, and murine PKCε *(13)* has been described. In *S. pombe* expression of mammalian PKC isoforms γ, δ, ε, ζ, and η has been shown *(8,9)*.

4. Mammalian PKC-dependent phenotypes in *S. pombe* uncovered functional differences between PKC activators and inhibitors. Calphostin C reduced PKCδ-induced growth inhibition to a greater extent than chelerythrine chloride, whereas the opposite was observed for PKCγ *(9)*. Bryostatin 1 stimulated the growth of PKCγ-expressing *S. pombe* cells but inhibited growth of PKC δ-expressing cells *(9)*. These observations implicate significant PKC isoform preferences for several pharmacological agents that modulate PKC activity.

5. cDNA expression in yeast under the control of a galactose-inducible promoter is greatly facilitated by *S. cerevisiae* strain 334 *(42)*. Its genotype includes mutations in yeast proteases (*pep4-3*; *prb1-1122*) to prevent degradation of expressed heterologous proteins. Mutations in pathways involved in uracil and leucine biosynthesis (*ura3-52* and *leu2-3112*) enable coselection of distinct expression plasmids. A mutation in the gene involved in glucose repression of the galactose-inducible promoter (*reg1-501*) allows induction of gene expression by galactose in the presence of glucose. This mutation permits short yeast cell doubling times in glucose-containing medium during galactose-induced gene expression. A mutation in a gene involved in galactose catabolism (*gal1*) prevents metabolism of the inducer galactose as a carbon source and helps maintain high

induction levels, particularly during growth on a poor carbon source such as glycerol *(42)*.

6. A variety of *S. cerevisiae* expression plasmids can be used for the expression of mammalian PKC isoforms in yeast such as YEp51, YEp52 *(26)* and derivatives, or pYES2 (Invitrogen, Carlsbad, CA) that were used in the experiments reported here. These plasmids carry origins of replication for *S. cerevisiae* and *E. coli*, and selectable markers for stable propagation in both hosts, including an ampicillin resistance marker (*E. coli*) and the *LEU2* or *URA3* genes to select for *S. cerevisiae* transformants in Leu⁻ and Ura⁻ synthetic media, respectively.

7. *S. pombe* (*h⁻,ade6-704, leu1-32, ura4-Δ18*) served as a host to express mammalian PKC cDNAs in fission yeast *(8)*.

8. For the lithium acetate transformation protocol, it is critical that herring sperm DNA is sheered prior to use such as by sonication. Sonicated preparations of herring sperm DNA are available commercially from Life Technologies (Burlington, ON).

9. The presence of EGTA in the PKC protein extraction buffer is essential to avoid loss of PKC catalytic activity, likely caused by the presence of Ca^{2+}-dependent proteases in the *S. cerevisiae* extract.

10. The doubling time of mammalian PKC-expressing *S. cerevisiae* cells greatly varies depending on the expressed PKC isoform, the carbon source, and the presence of PKC activators. Expression of PKCα or PKCβI resulted in a dramatic increase in the cell doubling time on glucose (**Fig. 2A**) or glycerol as carbon source upon phorbol ester activation. Expression of PKCε resulted in an increase in the cell doubling time on glycerol but not on glucose as carbon source which was phorbol ester-independent *(13)*. Typically, the impact of the PKC response was most pronounced on a poor carbon source such as glycerol *(13)*.

11. Some of the responses to mammalian PKC expression and activation in *S. cerevisiae* appear to be isoform specific. In addition to an increase in the cell doubling time, PKCβI activation results in the formation of multicellular strings *(6,11)* that were not observed in response to PKCα activation *(5,7)*. The increase in the *S. cerevisiae* cell doubling time that was noted in glucose-containing medium in response to activation of several mammalian PKC isoforms, including α *(7)*, βI *(6)*, and marginally for γ (**Fig. 2A**) was not observed after activation of PKCε *(13)*.

12. Isoform-specific growth-inhibitory phenotypes have been described in response to mammalian PKC expression and activation in *S. pombe* *(8,9)*. Expression and activation of PKCγ, δ, and η induced dramatic growth inhibition, whereas PKCε and PKCζ exhibited little to no effect, respectively.

13. Blotting of the Millipore 96-well filtration plates to a paper towel immediately after the wash steps in the calcium uptake protocol is essential to prevent loss of cell-associated radiolabeled Ca^{2+}.

14. Marked isoform-specific vesicle formation in *S. pombe* was revealed by electron microscopy upon expression and activation of PKCγ, δ, and η but not for PKCε and ζ *(8)*. Because expression of PKCε exhibited some growth suppression

but no vesicle accumulation, it appears that both responses can be functionally separated. Vesicle accumulation was attributed to enhanced endocytosis, induced by specific PKC isoforms. Expression of PKCδ and η by thiamine removal led to enhanced uptake of lucifer yellow and the same isoforms were localized to endocytotic, clathrin-associated vesicles. Enhanced uptake of lucifer yellow was also observed in PKCγ-expressing cells, but only after catalytic activation of the enzyme with phorbol ester *(8,9)*.

15. Conceivably, stimulation of external Ca^{2+} uptake in response to activation of PKCα, βI, and ε in *S. cerevisiae (5,6,13)* may reflect enhanced endocytosis of external Ca^{2+} ions, comparable with lucifer yellow uptake observed in *S. pombe (8)*. Alternatively, Ca^{2+} uptake may reflect enhanced endocytosis facilitated by increased vacuolization of the cytoplasm.

16. Replica plating of yeast colonies *(27)* involved attaching a plexiglas disk (8 cm diameter, 1 cm high) using glue to a hollow plexiglas tube (8 cm in diameter, 8 cm long). A sterile velveteen sheet was draped over the Plexiglas disk and held in place using an adjustable tube clamp. A culture plate containing yeast colonies on its agar surface was then lightly touched to the velveteen surface and the velveteen was marked for orientation using a sterile eight-gauge needle. Replica plates containing yeast nutrient agar with various carbon sources and/or PKC modulators were touched to the velvet surface, marking the position of the needle marks on the bottom of the replica plate for orientation. Plates were incubated at 30°C until colonies were visible.

17. Specific antibodies directed against epitopes within the catalytic domain are advantageous in immunoblotting experiments to compare the relative expression levels of mammalian PKC isozymes during coexpression of various PKC regulatory regions or other interacting proteins. Isoform-specific antibodies to carboxyl terminal epitopes of various PKC isozymes are available from Panvera Corp. (Madison, WI).

Acknowledgments

We are grateful to Dr. Nora Riedel for many ideas, the critical discussion of the manuscript, and her invaluable editorial support. We thank numerous colleagues who participated in experiments described in this chapter or kindly made reagents available as cited in the text and references. Part of the effort that resulted in this chapter was supported by the National Science Foundation (Grant MCB-9808795 to H. R.), the National Institutes of Health (Grant CA77837 to H. R.), the Canadian Institutes of Health (Grant MOP-15037 to A. P.), and Cancer Care Ontario (support funds to A. P.).

References

1. Parker, P. J., Coussens, L., Totty, N., Rhee, L., Young, S., Chen, E., et al. (1986) The complete primary structure of protein kinase C—the major phorbol ester receptor. *Science* **233**, 853–859.

2. Coussens, L., Parker, P. J., Rhee, L., Yang-Feng, T. L., Chen, E., Waterfield, M. D., et al. (1986) Multiple, distinct forms of bovine and human protein kinase C suggest diversity in cellular signaling pathways. *Science* **233,** 859–866.
3. Knopf, J. L., Lee, M. H., Sultzman, L. A., Kriz, R. W., Loomis, C. R., Hewick, R. M., et al. (1986) Cloning and expression of multiple protein kinase C cDNAs. *Cell* **46,** 491–502.
4. Housey, G. M., O'Brian, C. A., Johnson, M. D., Kirschmeier, P., and Weinstein, I. B. (1987) Isolation of cDNA clones encoding protein kinase C: evidence for a protein kinase C-related gene family. *Proc. Natl. Acad. Sci. USA* **84,** 1065–1069.
5. Riedel, H., Parissenti, A. M., Hansen, H., Su, L., and Shieh, H.-L. (1993) Stimulation of calcium uptake in Saccharomyces cerevisiae by bovine protein kinase C alpha. *J. Biol. Chem.* **268,** 3456–3462.
6. Riedel, H., Hansen, H., Parissenti, A. M., Su, L., Shieh, H.-L., and Zhu, J. (1993) Phorbol ester activation of functional rat protein kinase C beta-1 causes phenotype in yeast. *J. Cell Biochem.* **52,** 320–329.
7. Riedel, H., Su, L., and Hansen, H. (1993) Yeast phenotype classifies mammalian protein kinase C cDNA mutants. *Mol. Cell Biol.* **13,** 4728–4735.
8. Goode, N. T., Hajibagheri, M. A. N., Warren, G., and Parker, P. J. (1994) Expression of mammalian protein kinase C in Schizosaccharomyces pombe: isotype-specific induction of growth arrest, vesicle formation, and endocytosis. *Mol. Biol. Cell* **5,** 907–920.
9. Keenan, C., Goode, N., and Pears, C. (1997) Isoform specificity of activators and inhibitors of protein kinase C gamma and delta. *FEBS Lett.* **415,** 101–108.
10. Shieh, H.-L., Hansen, H., Zhu, J., and Riedel, H. (1996) Activation of conventional mammalian protein kinase C isoforms expressed in budding yeast modulates the cell doubling time—a potential in vivo screen for protein kinase C activators. *Cancer Detect. Prev.* **20,** 576–589.
11. Parissenti, A. M., Kim, S. A., Colantonio, C. M., Snihura, A. L., and Schimmer, B. P. (1996) Regulatory domain of human protein kinase C alpha dominantly inhibits protein kinase C beta-I-regulated growth and morphology in Saccharomyces cerevisiae. *J. Cell Physiol.* **166,** 609–617.
12. Parissenti, A. M., Kirwan, A. F., Kim, S. A., Colantonio, C. M., and Schimmer, B. P. (1996) Molecular strategies for the dominant inhibition of protein kinase C. *Endocr. Res.* **22,** 621–630.
13. Parissenti, A. M., Villeneuve, D., Kirwan-Rhude, A., and Busch, D. (1999) Carbon source-dependent regulation of cell growth by murine protein kinase C epsilon expression in Saccharomyces cerevisiae. *J. Cell Physiol.* **178,** 216–226.
14. Errede, B., and Levin, D. E. (1993) A conserved kinase cascade for MAP kinase activation in yeast. *Curr. Opin. Cell Biol.* **5,** 254–260.
15. Kaibuchi, K., Fukumoto, Y., Oku, N., Takai, Y., Arai, K., and Muramatsu, M. (1989) Molecular genetic analysis of the regulatory and catalytic domains of protein kinase C. *J. Biol. Chem.* **264,** 13,489–13,496.

16. Levin, D. E., Fields, F. O., Kunisawa, R., Bishop, J. M., and Thorner, J. (1990) A candidate protein kinase C gene, PKC1, is required for the S. cerevisiae cell cycle. *Cell* **62,** 213–224.
17. Simon, A. J., Milner, Y., Saville, S. P., Dvir, A., Mochly-Rosen, D., and Orr, E. (1991) The identification and purification of a mammalian-like protein kinase C in the yeast Saccharomyces cerevisiae. *Proc. R. Soc. Lond. B* **243,** 165–171.
18. Iwai, T., Fujisawa, N., Ogita, K., and Kikkawa, U. (1992) Catalytic properties of yeast protein kinase C: difference between the yeast and mammalian enzymes. *J. Biochem.* **112,** 7–10.
19. Levin, D. E. and Bartlett-Heubusch, E. (1992) Mutants in the S. cerevisiae PKC1 gene display a cell cycle-specific osmotic stability defect. *J. Cell Biol.* **116,** 1221–1229.
20. Antonsson, B., Montessuit, S., Friedli, L., Payton, M. A., and Paravicini, G. (1994) Protein kinase C in yeast. Characteristics of the Saccharomyces cerevisiae PKC1 gene product. *J. Biol. Chem.* **269,** 16,821–16,828.
21. Su, L., Parissenti, A. M., and Riedel, H. (1993) Functional carboxyl terminal deletion map of protein kinase C alpha. *Receptors Channels* **1,** 1–9.
22. Shieh, H.-L., Hansen, H., Zhu, J., and Riedel, H. (1995) Differential protein kinase C ligand regulation detected in vivo by a phenotypic yeast assay. *Mol. Carcinogen.* **12,** 166–176.
23. Rotenberg, S. A., Huang, M. H., Zhu, J., Su, L., and Riedel, H. (1995) Deletion analysis of protein kinase C inactivation by calphostin C. *Mol. Carcinogen.* **12,** 42–49.
24. Rotenberg, S. A., Zhu, J., Hansen, H., Li, X.-D., Sun, X.-G., Michels, C. A., et al. (1998) Deletion analysis of protein kinase C alpha reveals a novel regulatory segment. *J. Biochem.* **124,** 756–763.
25. Zhu, J., Hansen, H., Su, L., Shieh, H.-L., and Riedel, H. (1994) Ligand regulation of bovine protein kinase C alpha response via either cysteine-rich repeat of conserved region C1. *J. Biochem.* **115,** 1000–1009.
26. Broach, J. R. (1983) Construction of high copy yeast vectors using 2-μm circle sequences. *Methods Enzymol.* **101,** 307–325.
27. Ausubel, F. M., Brent, R., Kingston, R. E., Moore, D. D., Seidman, J. G., Smith, J. A., et. al, eds. (1994–2002) *Current Protocols in Molecular Biology, Vol. I–IV,* Wiley Interscience, New York.
28. Cigan, A. M. and Donahue, T. F. (1987) Sequence and structural features associated with translational initiator regions in yeast—a review. *Gene* **59,** 1–18.
29. Maundrell, K. (1990) nmt1 of fission yeast. A highly transcribed gene completely repressed by thiamine. *J. Biol. Chem.* **265,** 10,857–10,864.
30. Basi, G., Tommasino, M. S., and Maundrell, K. (1992) Thiamine repressible expression vectors for fission yeast. *Yeast* **8,** S597.
31. Ito, H., Fukuda, Y., Murata, K., and Kimura, A. (1983) Transformation of intact yeast cells treated with alkali cations. *J. Bacteriol.* **153,** 163–168.
32. Moreno, S., Klar, A., and Nurse, P. (1991) Molecular genetic analysis of fission yeast Schizosaccharomyces pombe. *Methods Enzymol.* **194,** 795–823.

33. Marais, R. M. and Parker, P. J. (1989) Purification and characterisation of bovine brain protein kinase C isotypes alpha, beta and gamma. *Eur. J. Biochem.* **182,** 129–137.
34. Dekker, L. V., Parker, P. J., and McIntyre, P. (1992) Biochemical properties of rat protein kinase C-eta expressed in COS cells. *FEBS Lett.* **312,** 195–199.
35. Olivier, A. R. and Parker, P. J. (1992) Identification of multiple PKC isoforms in Swiss 3T3 cells: differential down-regulation by phorbol ester. *J. Cell Physiol.* **152,** 240–244.
36. Huang, K. N. and Symington, L. S. (1994) Mutation of the gene encoding protein kinase C 1 stimulates mitotic recombination in Saccharomyces cerevisiae. *Mol. Cell Biol.* **14,** 6039–6045.
37. Simon, A. J., Saville, S. P., Jamieson, L., Pocklington, M. J., Donnelly, S. F. H., Ron, D., et al. (1993) Characterization of PKC2, a gene encoding a second protein kinase C isotype of Saccharomyces cerevisiae. *Curr. Biol.* **3,** 813–821.
38. Levin, D. E., Stevenson, W. D., and Watanabe, M. (1994) Evidence against the existence of the purported Saccharomyces cerevisiae PKC2 gene. *Curr. Biol.* **4,** 990–995.
39. Toda, T., Shimanuki, M., and Yanagida, M. (1993) Two novel protein kinase C-related genes of fission yeast are essential for cell viability and implicated in cell shape control. *EMBO J.* **12,** 1987–1995.
40. Kobori, H., Toda, T., Yaguchi, H., Toya, M., Yanagida, M., and Osumi, M. (1994) Fission yeast protein kinase C gene homologues are required for protoplast regeneration: a functional link between cell wall formation and cell shape control. *J. Cell Sci.* **107,** 1131–1136.
41. Toda, T., Dhut, S., Superti-Furga, G., Gotoh, Y., Nishida, E., Sugiura, R., et al. (1996) The fission yeast pmk1+ gene encodes a novel mitogen-activated protein kinase homolog which regulates cell integrity and functions coordinately with the protein kinase C pathway. *Mol. Cell Biol.* **16,** 6752–6764.
42. Hovland, P., Flick, J., Johnston, M., and Sclafani, R. A. (1989) Galactose as a gratuitous inducer of GAL gene expression in yeasts growing on glucose. *Gene* **83,** 57–64.

XI

PROTEIN KINASE C IN DISEASE

40

Protein Kinase C in Disease

Cancer

Alan P. Fields and W. Clay Gustafson

1. Introduction

Protein kinase C (PKC) is a family of serine/threonine kinases with a broad range of cellular targets. Members of the PKC family participate at many levels in diverse signal transduction pathways involved in cellular proliferation, differentiation, and survival. Because of the diversity of function and broad specificity of the various PKC family members, disruption or activation of different PKC isozymes can have pleiotropic effects in cellular systems. Aberrations in PKC signaling have been implicated in the development of multiple human diseases including the metabolic and cardiovascular complications of diabetes, central nervous system dysfunctions (such as bipolar disorder), Alzheimer's disease, neuronal degeneration, and cardiovascular disorders such as cardiac hypertrophy, cardiac ischemic preconditioning, and atherosclerosis. However, by far the most prominent association of PKC with disease has been in the promotion and progression of cancer.

Table 1 (refs. *1–13*) lists the major disease processes in which PKC is a known player and relevant references to reviews and important articles. The list is by no means comprehensive but the primary areas of interest in PKC and disease are covered. In the interest of space, we limit our discussion to the involvement of PKC in cancer. Furthermore, we emphasize in vivo studies of PKC function in cancer in animal models and human tumors wherever possible. However, we touch briefly on seminal studies using purified PKC isozymes in vitro or cell model systems.

From: *Methods in Molecular Biology, vol. 233: Protein Kinase C Protocols*
Edited by: A. C. Newton © Humana Press Inc., Totowa, NJ

Table 1
Areas of Major Focus in PKC and Disease States

Diabetes	
Metabolic and Cardiovascular complications	Idris et al., *Diabetologia*, 2001 *(1)*; Ishii et al., *Science*, 1996 *(2)*.
Glucose transport	Standaert et al., *Diabetes*, 2002 *(3)*.
Brain/neuronal disease	Reviewed in Battaini, *Pharm Res*, 2001 *(4)*.
Bipolar disorder	Hahn and Friedman, *Bipolar Disord*, 1999 *(5)*.
Cerebral hypertension	Hughes-Darden et al., *Cell Mol Biol*, 2001 *(6)*.
Alzheimers	Battaini, *Pharm Res*, 2001 *(4)*.
Ethanol intoxication	Bowers and Wehner, *J Neurosci*, 2001 *(7)*.
Cardiovascular disease	
Cardiac ischemic preconditioning	Mackay and Mochly-Rosen, *J Mol Cell Cardiol*, 2001 *(8)*; Kawata et al., *Circ Res*, 2001 *(9)*.
Arteriosclerosis	Leitges et al., *J Clin Invest*, 2001 *(10)*.
Cardiac hypertrophy	Roman et al., *Am J Physiol Heart Circ Physiol*, 2001 *(11)*.
PKC and other disease	
Human Immunodeficiency Virus Infection	Badou et al., *J Virol*, 2000 *(12)*; Zidovetzki et al., *AIDS Res Hum Retroviruses*, 1998 *(13)*.

2. PKC and Cancer

Early studies on the molecular basis of cancer focused on specific genetic changes in proto-oncogenes and tumor suppressors that cause the transformed phenotype. However, the effects of many tumor promoters appeared to be epigenetic, effecting cellular processes such as proliferation, differentiation, and apoptosis through nongenetic mechanisms, rather than through direct modification of DNA. The prototypical tumor promoter, the phorbol ester 12-*O*-tetradecanoylphorbol-13-acetate (TPA), has such pleiotropic effects on these cellular processes. The discovery that PKC is the primary receptor for TPA provided a direct link between PKC and oncogenesis *(14)* and provided a plausible molecular mechanism for the tumor-promoting activity of TPA.

A next major step in the evolving theory that PKC is important in tumorigenesis emerged when PKC was overexpressed in rat R6 and mouse NIH 3T3 fibroblasts *(15,16)*. The result was disordered growth in response to TPA in PKC-overexpressing cells, as well as altered tumorigenicity when PKC-overexpressing 3T3 cells were injected into nude mice. These experiments provided direct evidence that aberrant PKC activity could cause dysregulation of such cellular processes as proliferation, differentiation, and programmed cell death. Further evidence for the connection between PKC and tumor promotion came when it was observed that co-expression of PKC with several different known oncogenes (ras, myc, and fos) exacerbates the transformed phenotype. These studies solidified the link between PKC and tumorigenesis *(17)*.

Our understanding of the role of PKC in carcinogenesis has progressed over the last decade with the characterization of the individual members of the PKC family. The PKC family is composed of 12 distinct isozymes, each containing a highly homologous kinase domain and more variable regulatory domains. Differential regulation of the expression, intracellular localization and activity of specific PKC isozymes have emerged as primary mechanisms by which individual PKC isozymes can regulate key aspects of the transformed phenotype including hyperproliferation, altered or blocked differentiation, migration, invasion and metastasis, and resistance to apoptotic stimuli, such as anticancer chemotherapeutic drugs.

A demonstration of PKC isozyme-specific function can be seen from the characterization of the function of the major PKC isozymes expressed in the K562 chronic myelogenous leukemia cell line *(18,19)*. K562 cells express three major PKC isozymes, the classic PKCs—PKCα and PKCβI, and the atypical PKCι. We have demonstrated that PKC βII is required for K562 cell proliferation, whereas PKCι is necessary for the extreme resistance to drug-induced apoptosis characteristic of chronic myelogenous leukemia *(18–20)*. In contrast, PKCα is important for cell cycle arrest, cytostasis, and cellular differentiation. Therefore, even in this in vitro cell model system, it is evident that different PKC isozymes are involved in distinct, nonoverlapping characteristics of the transformed phenotype (**Fig. 1**). Although the K562 model is useful, it is only one example of a large body of literature in which PKC isozyme function has been studied in isolated cell systems. Interestingly, whereas these studies provide a consensus view of the function of at least some PKC isozymes, for others there are apparent inconsistencies in their function, indicating that PKC isozyme function is likely very complex and may exhibit tissue- and cell type-specificity. Below, we will examine and summarize what is known about PKC isozyme function, primarily in vivo cancer models. Where possible, a consensus view is discussed for each PKC isozyme that emerges from both in vitro and in vivo studies. We will consider separately the classical, novel, and atypical PKC subfamilies.

3.1. Classic PKC Isozymes

The classic PKC subfamily, PKC α, βI, βII, and γ, are distinct from the other PKC subfamilies in that they are regulated by both calcium and the regulatory lipids diacylglycerol and phosphatidylserine. PKCγ is primarily expressed in the nervous system and extensive studies of PKCγ in vivo carcinogenesis models have not been reported. The involvement of classic PKC isozymes in cancer was first suspected when it was found that these enzymes are the major

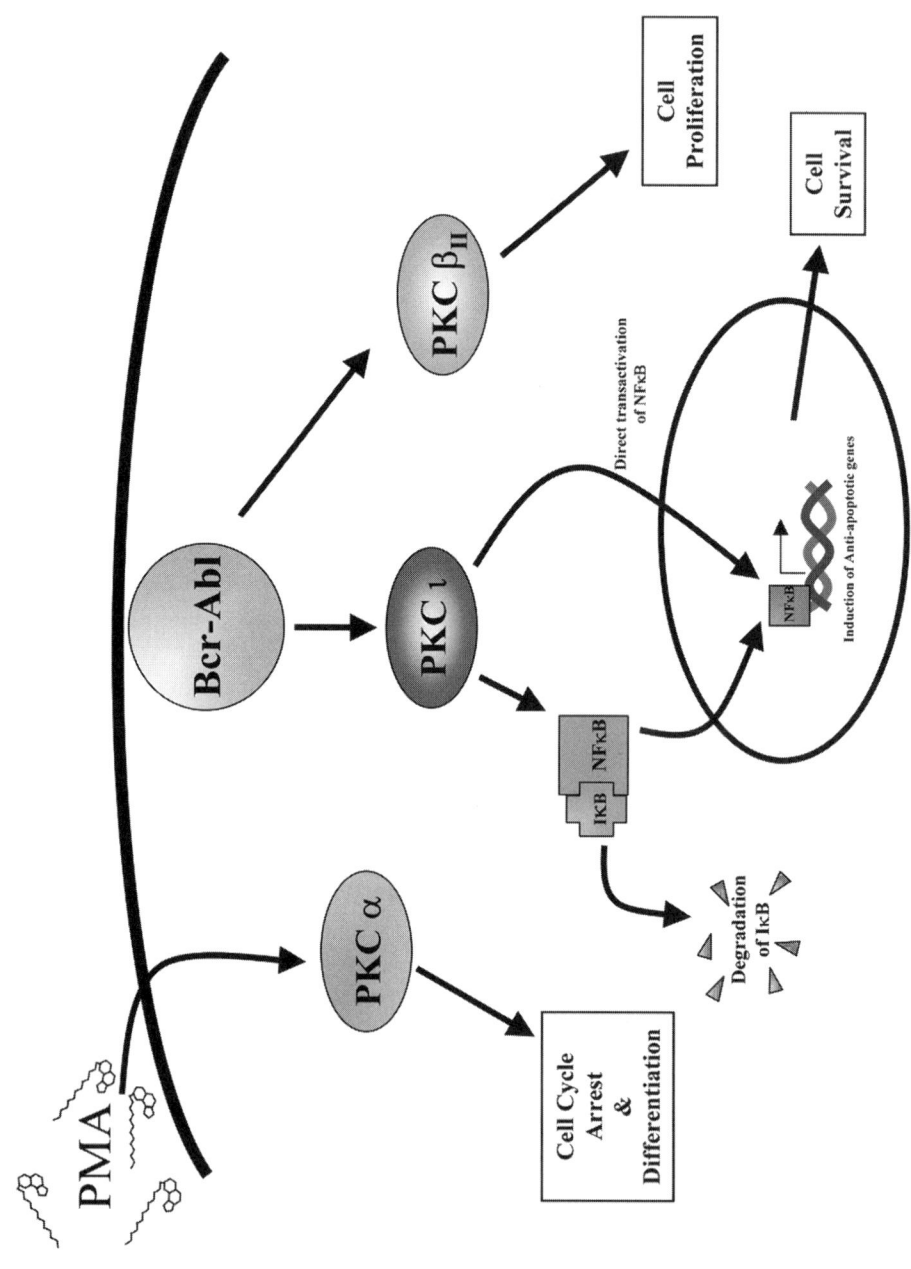

522

intracellular targets for TPA. TPA is a potent activator of classic PKC isozymes in vitro and has been used extensively in skin cancer models of tumorigenesis *(21)*. In cells, phorbol esters first activate classical and novel PKCs (as indicated by membrane translocation). However, when chronically applied, phorbol esters down-regulate these enzymes. This biphasic response to phorbol esters has complicated the interpretation of the role of PKC isozymes in mediating the effects of phorbol esters in cellular and animal studies. This has led to the question of whether phorbol esters are tumor promoters as a consequence of PKC activation, chronic down-regulation with concomitant loss of PKC activity, or both? This question is particularly interesting when one considers the historical significance of phorbol esters in whole animal models, in which TPA is a very strong tumor promoter. The use of genetically engineered animal models that mimic phorbol ester effects may provide more definitive answers to this question.

3.1.1. PKCα: A Regulator of Cell Cycle Progression and Differentiation

The literature on the role of PKCα in cellular physiology and cancer is conflicting, perhaps indicating a degree of cell type and tissue specificity of PKCα function. Many studies on PKCα in cultured cancer cells indicate that PKCα has a cytostatic effect, causing cell cycle arrest and even apoptosis, when overexpressed. Similar cytostatic or pro-differentiation effects of PKCα have been observed in colon, pancreatic, prostate, breast, and melanoma cancer cells *(22–25)*. Consistent with a role in postmitotic cell functions, the expression of PKCα appears to track with postmitotic epithelial cells within the colon and small intestine in vivo *(26)*. The literature indicates that PKCα plays a critical role in the inhibition of cell cycle progression through the G1/S phase transition. Black et al. *(27,28)* have demonstrated that PKCα can cause G1/S phase cell cycle arrest in intestinal epithelial cells through a mechanism that involves induction of the cyclin-dependent kinase inhibitor p21 cip1/waf1. Similar results have been obtained in pancreatic cancer cells in culture *(29)*. These studies provide a plausible mechanism for the cytostatic effects of PKCα observed in cultured cells and the pattern of expression of PKCα in the intestinal epithelium. In both human prostate cancer and gastric cancer cells,

---

Fig. 1. *(see facing page)* Diagrammatic representation of PKC isozyme action in chronic myelogenous leukemia (CML) cells. In K562 CML cells the three PKC isozymes have dramatically different actions. Cytostasis and differentiation functions are mediated by PKCα, resistance to apoptosis is mediated by PKCι via an NFκB-dependent mechanism, and hyperproliferation is mediated in a PKCβII-dependent manner.

PKCα activation induces apoptosis (30,31), a common response of tumor cells to cell cycle arrest. In addition, PKCα enhances the differentiating effects of retinoids on melanoma cells (32). Consistent with a role for PKCα as a growth suppressing gene, the activity (33,34) and protein levels (35,36) of PKCα are lower in both preneoplastic lesions of the colon and in colon carcinomas. Similar results have been obtained in preneoplastic lesions of APC^{min} mice, a genetic animal model of human familial adenomatous polyposis, a rare genetic disorder that predisposes afflicted individuals to colon cancer (37). Similarly, the level of membrane-bound PKCα is lower in human hepatocellular carcinoma specimens when compared to normal adjacent tissue (38), and PKCα is also lower in carcinogen-induced rat hepatocellular carcinoma (39), suggesting that, at least in liver carcinomas, PKCα activity is lower than in normal hepatocytes.

In contrast with the above-mentioned studies indicating a role for PKCα as a tumor suppressor, there are also a number of reports indicating that PKCα can play a proproliferative role in certain cancer cells. Antisense inhibition of PKCα expression leads to reversal of the transformed phenotype in human lung carcinoma cells (40), and induction of apoptosis in glioblastoma cells in response to insulin-like growth factor binding protein 3 (40). Furthermore, overexpression of PKCα in glioma cells leads to increased proliferation and reduced sensitivity to differentiation (40).

Several in vivo studies indicate an association of PKCα with tumorigenesis. Both breast cancers (41–43) and malignant gliomas (44) appear to have increased levels of PKCα when compared with surrounding normal tissue. Furthermore, PKCα expression shows an inverse relationship to estrogen receptor expression in endometrial cancers, suggesting a promotive role for PKCα in estrogen-independent endometrial cancers (45). In contrast, overexpression of PKCα in the skin of transgenic mice has little effect on carcinogen-induced papilloma burden or other measures of tumorigenesis (46), suggesting that PKCα plays only a minimal role in skin carcinogenesis. Taken together, these studies do not provide a clear picture of PKCα function. On the one hand, PKCα is an antiproliferative protein in many cancer cell types, whereas in several systems, it appears to stimulate proliferation. These data suggest that PKCα function is highly cell type and tissue dependent, serving as a molecule that can influence cellular proliferation, differentiation, and survival, depending on the cell type and stimulus.

PKCα has been specifically targeted for inhibition using a 20-mer phosphorothioate antisense oligonucleotide, ISIS3521, designed to hybridize to the 3′-untranslated region of PKCα. ISIS3521 (known clinically as LY90003) is currently in phase II and phase III clinical trials for the treatment of non-small-cell lung cancer in combination with several other chemotherapeutic agents

(47). It is also being evaluated for treatment of non-Hodgkin's lymphoma *(48)*. Initial results have shown moderate success, indicating some efficacy of LY90003. Further phase III trials using LY90003 in combination chemotherapy regimens are currently underway to determine whether it represents a definitive improvement over current treatment regimens.

In conclusion, whereas the majority of the data from studies in both tumor cell lines and animal models suggest a primary role for PKCα in cytostasis, cell cycle arrest and sensitization to differentiating agents, clinical trials with a PKCα-targeted antisense strategy indicates it may have efficacy, at least in combination with other agents, against several types of cancer. Given the conflicting evidence on the role of PKCα in different tumor types, it will be interesting to follow the progress of the clinical trials of LY90003 to assess how the clinical efficacy of PKCα inhibition correlates with the experimental data on PKCα function.

3.2.2. PKCβ: Growth Control (PKCβI) vs Hyperproliferation (PKCβII)

The role of PKCβ in carcinogenesis is clearer than PKCα, although in vivo studies of PKCβ and cancer have focused primarily on colon carcinoma. PKCβ is composed of two distinct mRNA splicing variants, PKCβI and βII, which differ only in their extreme c-terminal region. Interestingly, these two variants appear to differ dramatically in cellular function. PKCβII has been implicated in cellular proliferation whereas PKCβI is implicated in nonreplicative, differentiation-associated phenotypes *(18,49,50)*. In studies of normal colonic epithelium and carcinogen-induced pre-neoplastic lesions and colon carcinomas, PKCβII was shown to be dramatically up-regulated early in the carcinogenesis process. PKCβI, however, showed the inverse pattern, being downregulated in colonic neoplasms *(36,51)*. Furthermore, PKCβI is markedly reduced in the APC[min] mouse colon carcinogenesis model *(37)*. Increased PKCβII mRNA in the fecal matter of carcinogen-induced rats was positively correlated with tumor status *(52)*, suggesting the possible utility of screening for elevated PKCβII expression as the basis for noninvasive diagnostic detection of neoplastic events in the colon. Perhaps most convincingly, transgenic mice overexpressing PKCβII specifically in the colonic epithelium showed an increase in colonic epithelial cell proliferation in vivo and increased susceptibility to carcinogen-induced colon carcinogenesis *(49)* (**Fig. 2**). Seemingly contradictory results have been found in malignant melanomas, where the loss of PKCβ was found in malignant melanomas when compared with normal melanocytes *(53,54)*. However, these studies did not differentiate between the PKCβI and PKCβII splice variants, making it difficult to assess the relative role of these splicing variants in melanoma. Given the differential functions of PKCβI and PKCβII in multiple cells and tissues, careful discrimination of

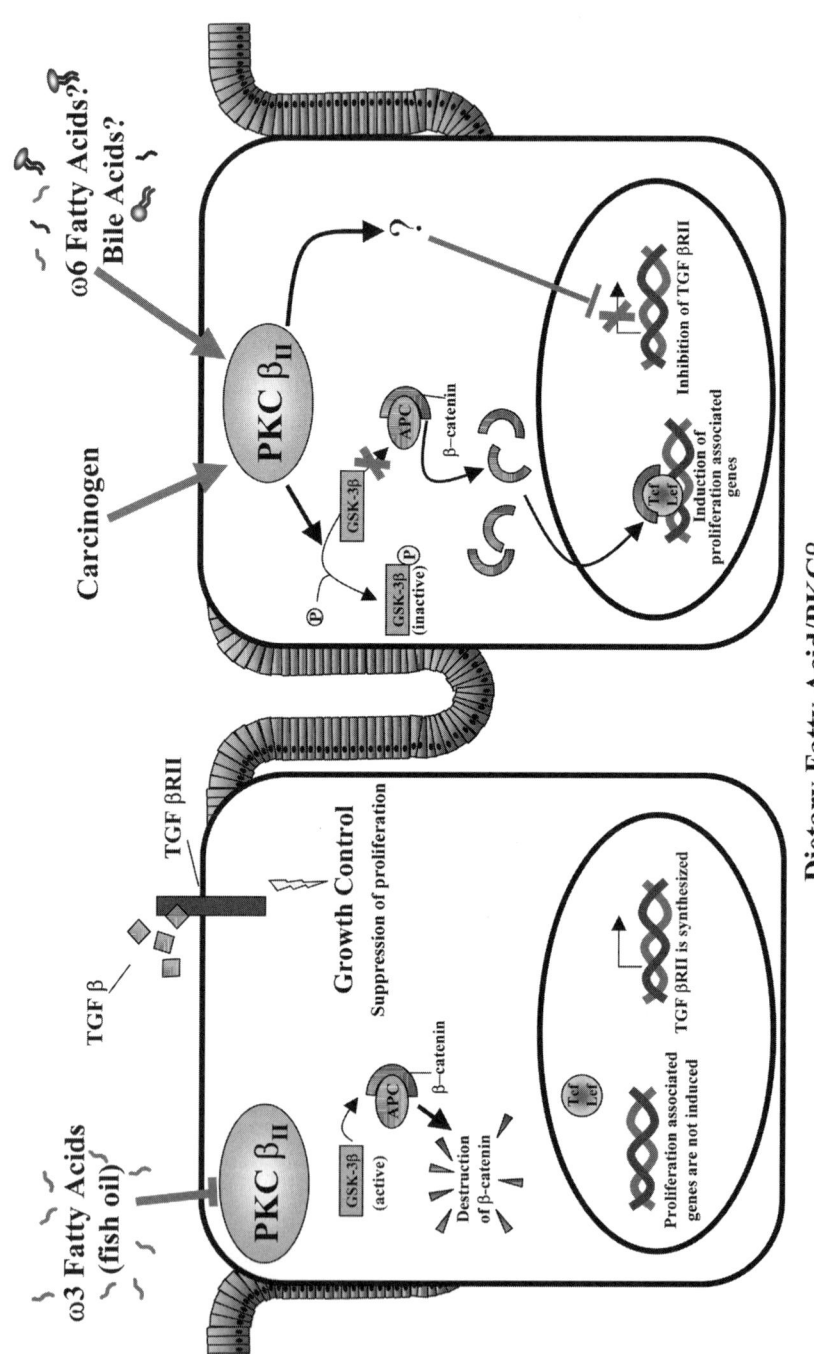

Dietary Fatty Acid/PKCβ$_{II}$ Model for Colon Carcinogenesis

these isozymes is necessary in order to reach meaningful conclusions regarding their roles in tumorigenesis.

Another promising chemotherapeutic agent, LY317615-2HCl, is a highly specific small molecule inhibitor of PKCβ. This agent has been shown to have anti-angiogenic and antitumor effects in xenograft studies of glioblastoma *(55)*, hepatocellular cancer and gastric cancer *(56)*, and non-small-cell lung carcinoma *(57)*. LY317615 is planned for phase I clinical trials for the treatment of patients with solid tumors in combination with other chemotherapeutic agents, such as taxol. A very similar compound, also targeted against PKCβ, is currently in clinical trials for treatment of diabetic complications, and has shown much promise for this indication. In this case, the clinical use of inhibitors of PKCβ, particularly PKCβII, is supported by the data generated in the vast majority of preclinical models.

3.3. Novel PKC Isozymes

The novel PKC subfamily (nPKCs) includes PKCδ, ε, θ, and η. They are distinct from classic PKCs in that they are not dependent on calcium for their activity. However, they are regulated by both diacylglycerol and phosphatidylserine, because of the presence of C1- and C2-like regions within their regulatory domains. The nPKC isozymes that have been most closely examined in relation to cancer are PKCδ and PKCε.

3.3.1. PKCδ: A Proapoptotic PKC

PKCδ has been widely implicated as a pro-apoptotic and/or growth inhibitory isozyme in tumor cell culture systems *(58–62)*. PKCδ protein levels were shown to be reduced in colon carcinomas *(63)* and PKCδ mRNA levels were shown to be decreased in colon adenomas when compared to neighboring normal epithelium *(33)*. Consistent with a role for PKCδ in tumor suppression,

---

Fig. 2. *(see facing page)* PKCβII action in the colonic epithelium in-vivo. Colonocytes in normal colonic epithelium express low levels of PKCβII, and in mice on a diet high in ω3-fatty acids (fish oil), this colonic PKCβII is inhibited, resulting in an overall lower PKCβII kinase activity. Inhibition of PKCβII results in inactivation of the GSK/β-catenin/Tcf-Lef pathway, as well as high level expression of TGFβRII. Both of these actions of ω3-fatty acids results in inhibition of proliferation. When exposed to a carcinogen (or when ω6-fatty acids are consumed), the level and activity of PKCβII increases). Increased PKCβII activity results in an increase in Tcf-Lef-stimulated transcription of proliferation-associated genes, such as c-myc and cyclin D, and the inhibition of TGFβRII expression, resulting in hyperproliferation and loss of growth control.

transgenic mice overexpressing PKCδ specifically in the epidermis exhibited a 76% decrease in papilloma burden in a skin carcinogenesis model when compared to wild-type animals *(46,64,65)*.

In apparent contradiction to these results, elevated levels of PKCδ have been associated with metastasis of mammary tumor cells *(66,67)*, and of murine melanoma cells *(68,69)*. Likewise, the levels of PKCδ were shown to be increased in human hepatocellular carcinomas compared to adjacent normal hepatic tissue *(38)*. This apparent discrepancy might be explained by differences in the regulation of PKCδ activity in these different tissues, though this has not been directly investigated.

3.3.2. PKCε: Role in Proliferation Signaling in Cancer

The literature concerning the role of PKCε in cancer provides convincing evidence that PKCε is a procarcinogenic, growth-modulating protein. PKCε has been shown to stimulate oncogenesis in fibroblast cells, colonic epithelial cells, and prostate cells. Several reports indicate that PKCε functions in the Raf/Ras proliferative signaling pathway to mediate its effects. In addition to its effects on proliferation, PKCε has been implicated in cell survival from apoptosis in glioma and thyroid cancer cells *(70,71)*. Consistent with these in vitro studies, PKCε has been found to be elevated in carcinogen-treated rat liver *(39)*, human colon cancers *(72)*, astroglial cancers *(73)*, and thyroid cancers *(74)*. Furthermore, transgenic mice overexpressing PKCε in the epidermis show a significant decrease in papilloma burden, but a concomitant increase in the development of squamous cell carcinoma of the skin *(46,65)*. Interestingly, these mice develop metastatic squamous cell carcinomas, indicating a role for PKCε in tumor progression to metastatic disease. In addition, an aberrant, truncated form of PKCε has been described in a thyroid cancer cell line that protects these cells from apoptosis *(71)*. However, an analysis of multiple primary thyroid cancers demonstrated that whereas PKCε expression is elevated, no mutations were identified. It remains to be determined whether PKCε mutations are a common event in tumorigenesis. Taken together, current studies indicate that PKCε is an important player in the development and progression of multiple types of cancer, and could be a potentially attractive therapeutic target.

3.4. Atypical PKC Isozymes

The atypical PKC ζ and ι/λ (PKCλ being the mouse homolog of human PKCι) are characterized by their lack of dependence on either calcium or phospholipids for activity. Both of these closely related PKC isozymes have been implicated in cell survival signaling and cellular transformation. However, the literature regarding the in vivo function of these PKC isozymes in cancer is not nearly as developed as that for other PKC isozymes.

3.4.1. PKCζ: Resistance to Apoptosis

The role of PKCζ in carcinogenesis appears to be complex and probably cell type specific. PKCζ has been implicated in resistance to apoptosis *(75)*, suppression of migration in mouse melanoma cells *(76)*, and in proliferation in glioblastoma cells *(77)*. Overexpression of PKCζ has been shown to inhibit invasion and metastasis in prostate cancer cells *(78)*. Increased fecal PKCζ mRNA levels have been implicated as a potential marker for early detection of colonic neoplasms *(52)*. PKCζ levels have been characterized in vivo in colonic carcinogen models by several different groups with conflicting results *(79,80)*. Most recently PKCζ levels have been shown to decrease in the APC^min mouse model for intestinal adenomas *(37)*. Therefore, the exact contribution of PKCζ to oncogenesis appears to be highly cell type dependent. The presence of the highly related PKCι in most, if not all, tissues raises the possibility that many of the effects of these PKC isozymes are either shared or have been misassigned. Indeed, in some systems PKCζ and PKCι appear to be interchangeable, whereas in others they are clearly distinct *(81)*.

3.4.2. PKCι/λ: Resistance to Apoptosis

The literature on PKCι, which is limited largely to in vitro studies in cell lines, reveals a consensus role in cellular survival. In chronic myelogenous leukemia (CML) cells, PKCι has been shown to be necessary for BcrAbl-mediated chemotherapeutic drug resistance *(19,20)*. In CML cells, PKCι mediates cell survival through activation of the NFκB transcriptional pathway *(82)*. Interestingly, whereas PKCι leads to only modest activation of the canonical IκB/NFκB activation pathway, it potently transactivates nuclear NFκB *(82)*. PKCι/λ has been observed to be elevated in carcinogen-induced colonic adenoma (Fields et al., unpublished observations), giving a preliminary indication that PKCι is important in colon carcinogenesis in vivo. Analysis of PKCι/λ function in the colonic epithelium is currently ongoing to further characterize the role of PKCι/λ in colon carcinogenesis.

4. PKC Isozymes: Cellular Mediators of Environmental Cancer Risk

As described above, multiple PKC isozymes have been implicated in various aspects of the transformed phenotype (**Fig. 3**). An important conclusion that can be derived from these studies is that PKC isozyme regulation is very complex. Gaining an in-depth understanding of the role of a particular PKC

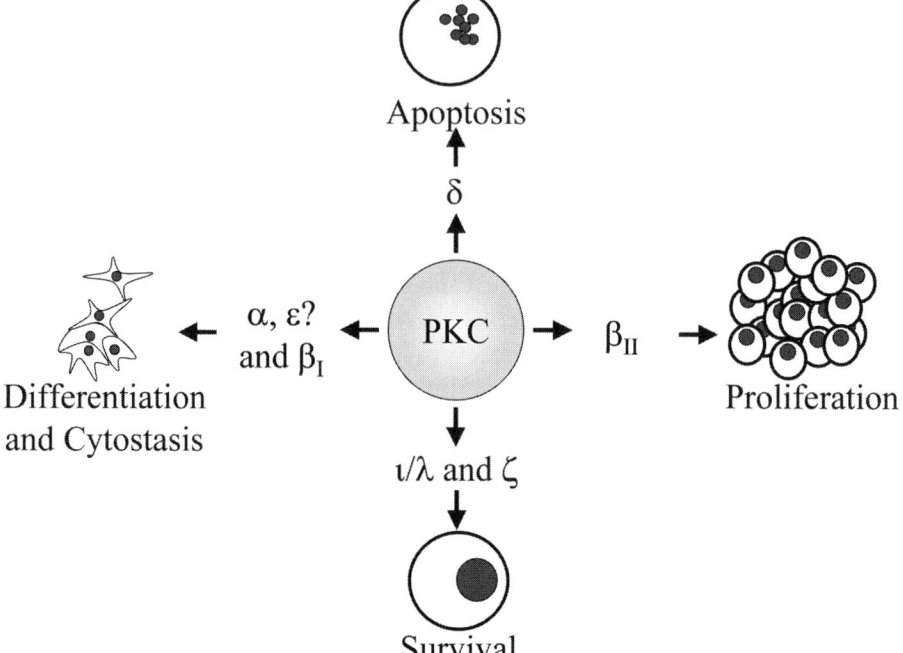

Fig. 3. Known functions of PKC isozymes in in vivo systems.

isozyme in a specific tissue or disease process requires multiple complimentary approaches. It is not sufficient to merely determine the pattern of PKC isozyme expression within a tissue during the process of carcinogenesis, although this clearly provides important clues about which PKC isozymes may be involved in the process under study. Both targeted overexpression of PKC isozymes in specific tissues and genetic knockout or expression of dominant negative mutants of specific PKC isozymes are important complimentary approaches that can often allow one to gain new insight into the specific role of a particular PKC isozyme in the carcinogenic process. Perhaps a more important aspect of these in vivo genetic models in the study of PKC isozyme function is the ability to gain new insight into how PKC activity is regulated in the intact organism.

For example, it has been know for many years that certain dietary components can regulate the activity of PKC isozymes in vitro. Interestingly, many of these compounds, including various fatty acids, bile acids, and calcium are known to be important factors that determine the risk of many epithelial cancers, including colon, breast, and prostate cancer. However, a direct demonstration of the importance of the interaction between a specific dietary factor and

PKC signaling in the carcinogenesis process has been lacking. We have taken advantage of our genetic model of colon cancer, in which PKCβII is specifically overexpressed in the colon of transgenic mice *(49)*, to determine whether PKCβII is an important cellular target of the colon cancer-preventive effects of dietary ω3-fatty acid in vivo. Transgenic PKCβII mice exhibit hyperproliferation and enhanced susceptibility to colon cancer *(49)*. Interestingly, feeding these animals an ω3-fatty acid diet reverses the hyperproliferation and increased susceptibility to colon cancer of transgenic PKCβII mice and reduces membrane-associated PKCβII in the colonic epithelium *(83)*. To identify gene targets of PKCβII, we established intestinal epithelial cells that overexpress PKCβII. Using gene expression profiling, we identified a cohort of genes that are either up or down-regulated by PKCβII. One of these genes, the transforming growth factor β receptor type II (TGFβRII) is transcriptionally repressed by PKCβII in vitro *(70)*. Furthermore, TGFβRII expression is similarly repressed in the colonic epithelium of transgenic PKCβII mice, demonstrating that TGFβRII is a target of PKCβII regulation in vivo, and validating both our in vitro and in vivo models. Finally, transgenic PKCβII mice fed an ω3-fatty acid diet exhibit re-expression of TGFβRII to levels comparable with control mice. These data provide direct evidence that ω3-fatty acids inhibit colon carcinogenesis through inhibition of PKCβII signaling *(83)*. These findings are particularly interesting in light of the epidemiologic data regarding these dietary lipids in colon carcinogenesis. Our data provide direct evidence that dietary lipids can potently inhibit PKCβII activity in vivo to reduce colon cancer risk. Therefore, dietary lipids appear to act as epigenetic tumor promoters and chemopreventive agents through regulation of PKC isozyme signaling in-vivo.

References

1. Idris, I., Gray, S., and Donnelly, R. (2001) Protein kinase C activation: isozyme-specific effects on metabolism and cardiovascular complications in diabetes. *Diabetologia* **44,** 659–673.
2. Ishii, H., Jirousek, M. R., Koya, D., Takagi, C., Xia, P., Clermont, A., et al. (1996) Amelioration of vascular dysfunctions in diabetic rats by an oral PKC beta inhibitor. *Science* **272,** 699–700.
3. Standaert, M. L., Ortmeyer, H. K., Sajan, M. P., Kanoh, Y., Bandyopadhyay, G., Hansen, B. C., et al. (2002) Skeletal muscle insulin resistance in obesity-associated type 2 diabetes in monkeys is linked to a defect in insulin activation of protein kinase C-zeta/lambda/iota. *Diabetes* **51,** 2936–43.
4. Battaini, F. (2001) Protein kinase C isoforms as therapeutic targets in nervous system disease states. *Pharmacol. Res.* **44,** 353–61.
5. Hahn, C. G., Friedman, E. (1999) Abnormalities in protein kinase C signaling and the pathophysiology of bipolar disorder. *Bipolar Disord.* **1,** 81–6.

6. Hughes-Darden, C. A., Wachira, S. J., Denaro, F. J., Taylor, C. V., Brunson, K. J., Ochillo, R. et al. (2001) Expression and distribution of protein kinase C isozymes in brain tissue of spontaneous hypertensive rats. *Cell Mol. Biol.* **47,** 1077–88.

7. Bowers, B. J., Wehner, J. M. (2001) Ethanol consumption and behavioral impulsivity are increased in protein kinase Cgamma null mutant mice. *J. Neurosci.* **21,** RC180.

8. Mackay, K., Mochly-Rosen, D. (2001) Localization, anchoring, and functions of protein kinase C isozymes in the heart. *J. Mol. Cell Cardiol.* **33,** 1301–7.

9. Kawata, H., Yoshida, K., Kawamoto, A., Kurioka, H., Takase, E., Sasaki, Y.,et al. (2001) Ischemic preconditioning upregulates vascular endothelial growth factor mRNA expression and neovascularization via nuclear translocation of protein kinase C epsilon in the rat ischemic myocardium. *Circ. Res.* **88,** 696–704.

10. Leitges, M., Mayr, M., Braun, U., Mayr, U., Li, C., Pfister, G. et al. (2001) Exacerbated vein graft arteriosclerosis in protein kinase Cdelta-null mice. *J. Clin. Invest.* **108,** 1505–12.

11. Roman, B. B., Geenen, D. L., Leitges, M., Buttrick, P. M. (2001) PKC-beta is not necessary for cardiac hypertrophy. *Am. J. Physiol. Heart Circ. Physiol.* **280,** H2264–70.

12. Badou, A., Bennasser, Y., Moreau, M., Leclerc, C., Benkirane, M., Bahraoui, E. (2000) Tat protein of human immunodeficiency virus type 1 induces interleukin-10 in human peripheral blood monocytes: implication of protein kinase C-dependent pathway. *J. Virol.* **74,** 10551–62.

13. Zidovetzki, R., Wang, J. L., Chen, P., Jeyaseelan, R., Hofman, F. (1998) Human immunodeficiency virus Tat protein induces interleukin 6 mRNA expression in human brain endothelial cells via protein kinase C- and cAMP-dependent protein kinase pathways. *AIDS Res. Hum. Retroviruses* **14,** 825–3

14. Castagna, M., Takai, Y., Kaibuchi, K., Sano, K., Kikkawa, U., and Nishizuka, Y. (1982) Direct activation of calcium-activated, phospholipid-dependent protein kinase by tumor-promoting phorbol esters. *J. Biol. Chem.* **257,** 7847–7851.

15. Housey, G. M., Johnson, M. D., Hsiao, W. L., O'Brian, C. A., Murphy, J. P., Kirschmeier, P., et al. (1988) Overproduction of protein kinase C causes disordered growth control in rat fibroblasts. *Cell* **52,** 343–354.

16. Persons, D. A., Wilkison, W. O., Bell, R. M., and Finn, O. J. (1988) Altered growth regulation and enhanced tumorigenicity of NIH 3T3 fibroblasts transfected with protein kinase C-I cDNA. *Cell* **52,** 447–458.

17. Weinstein, I. B. (1991) Nonmutagenic mechanisms in carcinogenesis: role of protein kinase C in signal transduction and growth control. *Environ. Health Perspect.* **93,** 175–179.

18. Murray, N. R., Baumgardner, G. P., Burns, D. J., and Fields, A. P. (1993) Protein kinase C isotypes in human erythroleukemia (K562) cell proliferation and differentiation. Evidence that beta II protein kinase C is required for proliferation. *J. Biol. Chem.* **268,** 15,847–15,853.

19. Murray, N. R. and Fields, A. P. (1997) Atypical protein kinase C iota protects human leukemia cells against drug-induced apoptosis. *J. Biol. Chem.* **272,** 27,521–27,524.

20. Jamieson, L., Carpenter, L., Biden, T. J., and Fields, A. P. (1999) Protein kinase Ciota activity is necessary for Bcr-Abl-mediated resistance to drug-induced apoptosis. *J. Biol. Chem.* **274,** 3927–3930.
21. Yuspa, S. H., Dlugosz, A. A., Denning, M. F., and Glick, A. B. (1996) Multistage *Carcinogenesis* in the skin. *J. Invest. Dermatol. Symp. Proc.* **1,** 147–150.
22. Verstovsek, G., Byrd, A., Frey, M. R., Petrelli, N. J., and Black, J. D. (1998) Colonocyte differentiation is associated with increased expression and altered distribution of protein kinase C isozymes. *Gastroenterology* **115,** 75–85.
23. Day, M. L., Zhao, X., Vallorosi, C. J., Putzi, M., Powell, C. T., Lin, C., et al. (1999) E-cadherin mediates aggregation-dependent survival of prostate and mammary epithelial cells through the retinoblastoma cell cycle control pathway. *J. Biol. Chem.* **274,** 9656–9664.
24. Gruber, J. R., Ohno, S., and Niles, R. M. (1992) Increased expression of protein kinase C alpha plays a key role in retinoic acid-induced melanoma differentiation. *J. Biol. Chem.* **267,** 13,356–13,360.
25. Scaglione-Sewell, B., Abraham, C., Bissonnette, M., Skarosi, S. F., Hart, J., Davidson, N. O., et al. (1998) Decreased PKC-alpha expression increases cellular proliferation, decreases differentiation, and enhances the transformed phenotype of CaCo-2 cells. *Cancer Res.* **58,** 1074–1081.
26. Davidson, L. A., Jiang, Y. H., Derr, J. N., Aukema, H. M., Lupton, J. R., and Chapkin, R. S. (1994) Protein kinase C isoforms in human and rat colonic mucosa. *Arch. Biochem. Biophys.* **312,** 547–553.
27. Black, J. D. (2000) Protein kinase C-mediated regulation of the cell cycle. *Front. Biosci.* **5,** D406–D423.
28. Frey, M. R., Saxon, M. L., Zhao, X., Rollins, A., Evans, S. S., and Black, J. D. (1997) Protein kinase C isozyme-mediated cell cycle arrest involves induction of p21(waf1/cip1) and p27(kip1) and hypophosphorylation of the retinoblastoma protein in intestinal epithelial cells. *J. Biol. Chem.* **272,** 9424–9435.
29. Detjen, K. M., Brembeck, F. H., Welzel, M., Kaiser, A., Haller, H., Wiedenmann, B., et al. (2000) Activation of protein kinase Calpha inhibits growth of pancreatic cancer cells via p21(cip)-mediated G(1) arrest. *J. Cell Sci.* **113,** 3025–3035.
30. Gschwend, J. E., Fair, W. R., and Powell, C. T. (2000) Bryostatin 1 induces prolonged activation of extracellular regulated protein kinases in and apoptosis of LNCaP human prostate cancer cells overexpressing protein kinase C alpha. *Mol. Pharmacol.* **57,** 1224–1234.
31. Okuda, H., Adachi, M., Miyazawa, M., Hinoda, Y., and Imai, K. (1999) Protein kinase C alpha promotes apoptotic cell death in gastric cancer cells depending upon loss of anchorage. *Oncogene* **18,** 5604–5609.
32. Niles, R. M. and Combs, R. (1996) The relationship between susceptibility to retinoic acid treatment and protein kinase C alpha expression in murine melanoma cell lines. *Exp. Cell Res.* **223,** 20–28.
33. Assert, R., Kotter, R., Bisping, G., Scheppach, W., Stahlnecker, E., Muller, K. M., et al. (1999) Anti-proliferative activity of protein kinase C in apical compartments

of human colonic crypts: evidence for a less activated protein kinase C in small adenomas. *Int. J. Cancer* **80,** 47–53.

34. Batlle, E., Verdu, J., Dominguez, D., del Mont, L. M., Diaz, V., Loukili, N., et al. (1998) Protein kinase C-alpha activity inversely modulates invasion and growth of intestinal cells. *J. Biol. Chem.* **273,** 15,091–15,098.

35. Gokmen-Polar, Y., Murray, N. R., Velasco, M. A., Gatalica, Z., and Fields, A. P. (2001) Elevated protein kinase C betaII is an early promotive event in colon *Carcinogenesis. Cancer Res.* **61,** 1375–1381.

36. Suga, K., Sugimoto, I., Ito, H., and Hashimoto, E. (1998) Down-regulation of protein kinase C-alpha detected in human colorectal cancer. *Biochem. Mol. Biol. Int.* **44,** 523–528.

37. Klein, I. K., Ritland, S. R., Burgart, L. J., Ziesmer, S. C., Roche, P. C., Gendler, S. J., et al. (2000) Adenoma-specific alterations of protein kinase C isozyme expression in Apc(MIN) mice. *Cancer Res.* **60,** 2077–2080.

38. Tsai, J. H., Hsieh, Y. S., Kuo, S. J., Chen, S. T., Yu, S. Y., Huang, C. Y., et al. (2000) Alteration in the expression of protein kinase C isoforms in human hepatocellular carcinoma. *Cancer Lett.* **161,** 171–175.

39. Lee, Y. S., Hong, S. I., Lee, M. J., Kim, M. R., and Jang, J. J. (1998) Differential expression of protein kinase C isoforms in diethylnitrosamine-initiated rat liver. *Cancer Lett.* **126,** 17–22.

40. Shen, L., Dean, N. M., and Glazer, R. I. (1999) Induction of p53-dependent, insulin-like growth factor-binding protein- 3-mediated apoptosis in glioblastoma multiforme cells by a protein kinase C alpha antisense oligonucleotide. *Mol. Pharmacol.* **55,** 396–402.

41. Carey, I., Williams, C. L., Ways, D. K., and Noti, J. D. (1999) Overexpression of protein kinase C-alpha in MCF-7 breast cancer cells results in differential regulation and expression of alphavbeta3 and alphavbeta5. *Int. J. Oncol.* **15,** 127–136.

42. Carey, I. and Noti, J. D. (1999) Isolation of protein kinase C-alpha-regulated cDNAs associated with breast tumor aggressiveness by differential mRNA display. *Int. J. Oncol.* **14,** 951–956.

43. Ng, T., Squire, A., Hansra, G., Bornancin, F., Prevostel, C., Hanby, A., et al. (1999) Imaging protein kinase C alpha activation in cells. *Science* **283,** 2085–2089.

44. Leirdal, M. and Sioud, M. (1999) Ribozyme inhibition of the protein kinase C alpha triggers apoptosis in glioma cells. *Br. J. Cancer* **80,** 1558–1564.

45. Fournier, D. B., Chisamore, M., Lurain, J. R., Rademaker, A. W., Jordan, V. C., and Tonetti, D. A. (2001) Protein kinase C alpha expression is inversely related to ER status in endometrial carcinoma: possible role in AP-1-mediated proliferation of ER-negative endometrial cancer. *Gynecol. Oncol.* **81,** 366–372.

46. Jansen, A. P., Dreckschmidt, N. E., Verwiebe, E. G., Wheeler, D. L., Oberley, T. D., and Verma, A. K. (2001) Relation of the induction of epidermal ornithine decarboxylase and hyperplasia to the different skin tumor-promotion susceptibilities of protein kinase C alpha, -delta and -epsilon transgenic mice. *Int. J. Cancer* **93,** 635–643.

47. Mani, S., Rudin, C. M., Kunkel, K., Holmlund, J. T., Geary, R. S., Kindler, H. L., et al. (2002) Phase I clinical and pharmacokinetic study of protein kinase C-alpha antisense oligonucleotide ISIS 3521 administered in combination with 5-fluorouracil and leucovorin in patients with advanced cancer. Clin. *Cancer Res.* **8,** 1042–1048.

48. Goekjian, P. G. and Jirousek, M. R. (2001) Protein kinase C inhibitors as novel anticancer drugs. *Expert. Opin. Invest. Drugs* **10,** 2117–2140.

49. Murray, N. R., Davidson, L. A., Chapkin, R. S., Clay, G. W., Schattenberg, D. G., and Fields, A. P. (1999) Overexpression of protein kinase C betaII induces colonic hyperproliferation and increased sensitivity to colon carcinogenesis. *J. Cell Biol.* **145,** 699–711.

50. Black, J. D. (2001) Protein kinase C isozymes in colon carcinogenesis: guilt by omission. *Gastroenterology* **120,** 1868–1872.

51. Davidson, L. A., Brown, R. E., Chang, W. C., Morris, J. S., Wang, N., Carroll, R. J., et al. (2000) Morphodensitometric analysis of protein kinase C beta(II) expression in rat colon: modulation by diet and relation to in situ cell proliferation and apoptosis. *Carcinogenesis* **21,** 1513–1519.

52. Davidson, L. A., Aymond, C. M., Jiang, Y. H., Turner, N. D., Lupton, J. R., and Chapkin, R. S. (1998) Non-invasive detection of fecal protein kinase C betaII and zeta messenger RNA: putative biomarkers for colon cancer. *Carcinogenesis* **19,** 253–257.

53. Yamanishi, D. T., Graham, M., Buckmeier, J. A., and Meyskens, F. L., Jr. (1991) The differential expression of protein kinase C genes in normal human neonatal melanocytes and metastatic melanomas. *Carcinogenesis* **12,** 105–109.

54. Gilhooly, E. M., Morse-Gaudio, M., Bianchi, L., Reinhart, L., Rose, D. P., Connolly, J. M., et al. (2001) Loss of expression of protein kinase C beta is a common phenomenon in human malignant melanoma: a result of transformation or differentiation? *Melanoma Res.* **11,** 355–369.

55. Teicher, B. A., Menon, K., Alvarez, E., Galbreath, E., Shih, C., and Faul, M. (2001) Antiangiogenic and antitumor effects of a protein kinase C beta inhibitor in human T98G glioblastoma multiforme xenografts. *Clin. Cancer Res.* **7,** 634–640.

56. Teicher, B. A., Menon, K., Alvarez, E., Liu, P., Shih, C., and Faul, M. M. (2001) Antiangiogenic and antitumor effects of a protein kinase C beta inhibitor in human hepatocellular and gastric cancer xenografts. *In Vivo* **15,** 185–193.

57. Teicher, B. A., Menon, K., Alvarez, E., Galbreath, E., Shih, C., and Faul, M. M. (2001) Antiangiogenic and antitumor effects of a protein kinase C beta inhibitor in murine lewis lung carcinoma and human Calu-6 non-small-cell lung carcinoma xenografts. *Cancer Chemother. Pharmacol.* **48,** 473–480.

58. Lu, Z., Hornia, A., Jiang, Y. W., Zang, Q., Ohno, S., and Foster, D. A. (1997) Tumor promotion by depleting cells of protein kinase C delta. *Mol. Cell Biol.* **17,** 3418–3428.

59. Cerda, S. R., Bissonnette, M., Scaglione-Sewell, B., Lyons, M. R., Khare, S., Mustafi, R., et al. (2001) PKC-delta inhibits anchorage-dependent and -independent

growth, enhances differentiation, and increases apoptosis in CaCo-2 cells. *Gastroenterology* **120,** 1700–1712.

60. Buchner, K. (2000) The role of protein kinase C in the regulation of cell growth and in signalling to the cell nucleus. *J. Cancer Res. Clin. Oncol.* **126,** 1–11.
61. da Rocha, A. B., Mans, D. R., Regner, A., and Schwartsmann, G. (2002) Targeting protein kinase C: new therapeutic opportunities against high- grade malignant gliomas? *Oncologist* **7,** 17–33.
62. Schechtman, D. and Mochly-Rosen, D. (2001) Adaptor proteins in protein kinase C-mediated signal transduction. *Oncogene* **20,** 6339–6347.
63. Craven, P. A. and DeRubertis, F. R. (1994) Loss of protein kinase C delta isozyme immunoreactivity in human adenocarcinomas. *Dig. Dis. Sci.* **39,** 481–489.
64. Reddig, P. J., Dreckschmidt, N. E., Ahrens, H., Simsiman, R., Tseng, C. P., Zou, J., et al. (1999) Transgenic mice overexpressing protein kinase C delta in the epidermis are resistant to skin tumor promotion by 12-*O*-tetradecanoylphorbol-13-acetate. *Cancer Res.* **59,** 5710–5718.
65. Reddig, P. J., Dreckschmidt, N. E., Zou, J., Bourguignon, S. E., Oberley, T. D., and Verma, A. K. (2000) Transgenic mice overexpressing protein kinase C epsilon in their epidermis exhibit reduced papilloma burden but enhanced carcinoma formation after tumor promotion. *Cancer Res.* **60,** 595–602.
66. Kiley, S. C., Clark, K. J., Goodnough, M., Welch, D. R., and Jaken, S. (1999) Protein kinase C delta involvement in mammary tumor cell metastasis. *Cancer Res.* **59,** 3230–3238.
67. Kiley, S. C., Clark, K. J., Duddy, S. K., Welch, D. R., and Jaken, S. (1999) Increased protein kinase C delta in mammary tumor cells: relationship to transformtion and metastatic progression. *Oncogene* **18,** 6748–6757.
68. La Porta, C. A., Di Dio, A., Porro, D., and Comolli, R. (2000) Overexpression of novel protein kinase C delta in BL6 murine melanoma cells inhibits the proliferative capacity in vitro but enhances the metastatic potential in vivo. *Melanoma Res.* **10,** 93–102.
69. La Porta, C. A. and Comolli, R. (2000) Overexpression of nPKC delta in BL6 murine melanoma cells enhances TGFbeta1 release into the plasma of metastasized animals. *Melanoma Res.* **10,** 527–534.
70. Shinohara, H., Kayagaki, N., Yagita, H., Oyaizu, N., Ohba, M., Kuroki, T., et al. (2001) A protective role of PKC epsilon against TNF-related apoptosis-inducing ligand (TRAIL)-induced apoptosis in glioma cells. *Biochem. Biophys. Res. Commun.* **284,** 1162–1167.
71. Knauf, J. A., Elisei, R., Mochly-Rosen, D., Liron, T., Chen, X. N., Gonsky, R., et al. (1999) Involvement of protein kinase Cepsilon (PKCepsilon) in thyroid cell death. A truncated chimeric PKCepsilon cloned from a thyroid cancer cell line protects thyroid cells from apoptosis. *J. Biol. Chem.* **274,** 23,414–23,425.
72. Perletti, G. P., Concari, P., Brusaferri, S., Marras, E., Piccinini, F., and Tashjian, A. H., Jr. (1998) Protein kinase C epsilon is oncogenic in colon epithelial cells by interaction with the ras signal transduction pathway. *Oncogene* **16,** 3345–3348.

73. Sharif, T. R. and Sharif, M. (1999) Overexpression of protein kinase C epsilon in astroglial brain tumor derived cell lines and primary tumor samples. *Int. J. Oncol.* **15,** 237–243.

74. Knauf, J. A., Ward, L. S., Nikiforov, Y. E., Nikiforova, M., Puxeddu, E., Medvedovic, M., et al. (2002) Isozyme-specific abnormalities of PKC in thyroid cancer: evidence for post-transcriptional changes in PKC epsilon. *J. Clin. Endocrinol. Metab.* **87,** 2150–2159.

75. Moscat, J., Sanz, L., Sanchez, P., and Diaz-Meco, M. T. (2001) Regulation and role of the atypical PKC isoforms in cell survival during tumor transformation. *Adv. Enzyme Regul.* **41,** 99–120.

76. Sanz-Navares, E., Fernandez, N., Kazanietz, M. G., and Rotenberg, S. A. (2001) Atypical protein kinase C zeta suppresses migration of mouse melanoma cells. *Cell Growth Differ.* **12,** 517–524.

77. Donson, A. M., Banerjee, A., Gamboni-Robertson, F., Fleitz, J. M., and Foreman, N. K. (2000) Protein kinase C zeta isoform is critical for proliferation in human glioblastoma cell lines. *J. Neurooncol.* **47,** 109–115.

78. Powell, C. T., Gschwend, J. E., Fair, W. R., Brittis, N. J., Stec, D., and Huryk, R. (1996) Overexpression of protein kinase C-zeta (PKC-zeta) inhibits invasive and metastatic abilities of Dunning R-3327 MAT-LyLu rat prostate cancer cells. *Cancer Res.* **56,** 4137–4141.

79. Kahl-Rainer, P., Karner-Hanusch, J., Weiss, W., and Marian, B. (1994) Five of six protein kinase C isoenzymes present in normal mucosa show reduced protein levels during tumor development in the human colon. *Carcinogenesis* **15,** 779–782.

80. Wali, R. K., Frawley, B. P., Jr., Hartmann, S., Roy, H. K., Khare, S., Scaglione-Sewell, B. A., et al. (1995) Mechanism of action of chemoprotective ursodeoxycholate in the azoxymethane model of rat colonic carcinogenesis: potential roles of protein kinase C-alpha, -beta II, and -zeta. *Cancer Res.* **55,** 5257–5264.

81. Parker, P. J., and Dekker, L. V. (eds). (1997) *Protein Kinase C.* Springer-Verlag, Heidelberg, Germany.

82. Lu, Y., Jamieson, L., Brasier, A. R., and Fields, A. P. (2001) NF-kappaB/RelA transactivation is required for atypical protein kinase C iota-mediated cell survival. *Oncogene* **20,** 4777–4792.

83. Murray, N. R., Weems, C., Chen, L., Leon, J., Yu, W., Davidson, L. A., et al. (2002) Protein kinase C betaII and TGFbetaRII in omega 3 fatty acid-mediated inhibition of colon carcinogenesis. *J. Cell Biol.* **157,** 915–920.

41

Characterization of the Role of Protein Kinase C Isozymes in Colon Carcinogenesis Using Transgenic Mouse Models

Alan P. Fields, Nicole R. Murray, and W. Clay Gustafson

1. Introduction

Protein kinase C (PKC) is a family of related serine/threonine lipid-dependent kinases that have been implicated in intestinal epithelial cell proliferation, apoptosis, cellular transformation, and colon carcinogenesis in rodent models and in humans *(1–12)*. Early studies revealed that PKC activity is higher in actively proliferating colonic epithelial cells than in their quiescent counterparts *(2)*, suggesting a role for PKC activation in proliferation of these cells. More recent studies have demonstrated that colonic epithelial cells express multiple PKC isozymes and that expression of these isozymes is differentially modulated during colon carcinogenesis *(1,13–15)*. All of these studies indicate a direct connection between PKC and colon carcinogenesis.

1.1. Murine Models for Colon Carcinogenesis

Many studies investigating genetic, dietary, and other environmental factors involved in colon carcinogenesis utilize a well-characterized rodent carcinogenesis model. Mice are excellent models for studying colon cancer for several reasons. First, purebred strains of mice are available, eliminating genetic variability in cancer susceptibility between individuals. Second, environmental factors, such as diet, can be strictly controlled and individual components systematically varied. Third, organ-specific carcinogens, such as 1,2-dimethylhydrazine and its metabolite, azoxymethane (AOM), have been extensively characterized for their ability to selectively induce colon cancer in rodents *(16,17)*. Fourth, AOM reproducibly induces colon tumors that exhibit

From: *Methods in Molecular Biology, vol. 233: Protein Kinase C Protocols*
Edited by: A. C. Newton © Humana Press Inc., Totowa, NJ

many of the same genetic and signal transduction defects identified in human colon carcinomas *(18–21)*. Fifth, AOM also induces aberrant crypt foci (ACF), which represent well-established preneoplastic colonic lesions in both rodents and humans *(22–25)*. Both the number and multiplicity (i.e., number of crypts/focus) of ACF are highly predictive of subsequent tumor development and therefore can be used as reliable biomarkers of colon carcinogenesis *(26–28)*. This chapter describes the methods used to characterize and analyze transgenic mice that express altered PKC expression and/or activity in the colonic epithelium.

2. Materials

1. Surgical equipment, forceps, scissors, gavage needles (Roboz Inc.).
2. RNAse Zap (Ambion).
3. Tissue-embedding cassettes and sponges (Fischer).
4. Pathology core facility.
5. Dissecting microscope.
6. Polytron orbital tissue homogenizer.
7. Bio-Rad Protran II system.
8. Trizol Reagent (Gibco).
9. RNAse free H_2O.
10. Primers: Oligo-(dT) (Promega), transgene-specific oligonucleotide primers.
11. Superscript II reverse transcriptase (Gibco).
12. Easy Start PCR mix-in-a-tube (Molecular Bioproducts).
13. Taq polymerase (Promega).
14. Thermal cycler.
15. Inverted dissecting microscope.
16. Fluorescence microscope.
17. Modified Lysis Buffer (MLB): 20 mM Tris-HCl (pH 7.5), 10% glycerol, 1% NP-40, 10 mM ethylenediamine tetraacetic acid, 150 mM NaCl, Protease Inhibitor cocktail pellet (Roche).
18. Kinase Assay Buffer (KAB) (1×): 50 mM Tris-HCl (pH 7.5), 10 mM MgCl$_2$ 0.5 mM ethylenebis(oxyethylenenitrilo)tetraacetic acid, 0.1 mM CaCl$_2$, protease Inhibitor cocktail pellet (Roche).
19. 2X c/nPKC Kinase Buffer: Place 2 μL of Phosphatidyl Serine solution (10 mg/mL in Chloroform, Avanti Polar Lipids) into an Eppendorf tube and dry away all liquid in a vacuum centrifuge for approx 10 min. Add 250 μL of stock PKC kinase buffer without phosphatidyl serine (100 mM Tris-HCl, pH 7.4, 20 mM MgSO$_4$, 200 μM CaCl$_2$, 2 mM dithiothreitol, 2×10^{-7} M Bryostatin1; Calbiochem). Sonicate for 20 sec to break the lipids into microvesicles.
20. BrdU (5-bromo-2′-deoxyuridine) solution: Prepare 10 mg/mL in saline. Aliquot and store at –20°C. Thaw aliquot once and discard unused portion.
21. AOM (azoxymethane, Midwest Research Institute, National Cancer Institute Chemical Carcinogen Reference Standard Repository; Kansas City, MO):

Prepare 1 mg/mL in saline. Aliquot and store at –20°C. Thaw aliquot once, discard unused portion.

22. 4% paraformaldehyde in phosphate buffered saline (toxic, light sensitive, store at 4°C, stable for 2 wk in solution).
23. BCA Protein Assay (Pierce).
24. Fast Green stock: 2 g of Fast Green, 400 mL of methanol, 100 mL of acetic acid, QS to 1 L with water and store at room temperature.
25. Protein A Sepharose beads (Sigma, P-9424).
26. Histone (Sigma): 1.25 µg/µL stock in water.

3. Methods

3.1. Brief Description of the Model System: Transgenic Mice Expressing PKC Isoforms in the Colonic Epithelia

Our laboratory has generated transgenic mice that specifically express different isoforms of PKC in their colonic epithelia. Methods for generating transgenic animals are beyond the scope of this chapter and are described elsewhere *(29)*. Transgene constructs are generated by cloning the full-length human PKC cDNA into a plasmid that places them under the control of the rat liver fatty acid-binding protein promoter (FABP; obtained from Dr. Jeffrey Gordon, Washington University, St. Louis, MO), which has been shown to selectively drive the expression of exogenous genes in the mouse colonic epithelium *(30)*. This plasmid also contains an SV-40 poly A sequence 3′ to the inserted PKC cDNA to ensure proper processing and expression of the message. The entire transgene construct (promoter, cDNA and poly A sequence) is then excised, purified, and microinjected into fertilized mouse embryos.

A critical decision surrounds the choice of embryos for microinjection of the transgene. Classically, this has been performed using F1 hybrid embryos between two inbred strains of mice, for instance, C3H and C57B6. This is caused by the increased viability and ease of microinjecting these cross-strain embryos. However, if F1 hybrid embryos are used, subsequent founders will be of mixed genetic background, requiring a minimum of seven subsequent backcrosses to reestablish the line on a stable inbred strain background. In fact, in order to assure that the genetic background is the same as the strain into which one is backcrossing, at least 10 generations of backcrosses should be performed. This process is costly in time, manpower, and resources, but must be undertaken if one is to perform any long-term carcinogenesis studies as the result of well-documented strain differences in inherent susceptibility to cancer. The presence of strain-specific genetic modifiers complicates the analysis of transgene function.

An alternative to this approach, which eliminates the potential for genetic variability, is to microinject the transgene of interest into inbred embryos of the strain desired. Whereas this approach increases the initial cost of generating the trans-genic founders because of the need to microinject more embryos to ensure that founders are obtained, it more than makes up for this additional cost in time savings by eliminating the need for costly and time-consuming backcrosses and the worry of encountering strain-specific genetic modifiers of phenotype. Most transgenic mouse facilities are capable of successfully microinjecting inbred strain embryos.

Another critical consideration in the generation of transgenic mice is the need to establish and characterize multiple lines of mice containing the same trans-gene. Multiple independent transgenic lines are necessary to eliminate the possi-bility that any phenotype is caused by the site of insertion of the transgene. Obtaining two independent lines expressing the same transgene that exhibit the same or a similar phenotype provides convincing evidence that the pheno-type is caused by the presence of the transgene itself rather than its site of integration. It can also afford the opportunity to analyze the effect of a transgene at several different levels of expression, allowing for a gene-dose effect to be established.

3.2. Animal Husbandry

All transgenic mice are bred and housed in microisolator cages maintained at constant temperature and humidity on a 12 h on/12 h off light cycle in a pathogen-free barrier facility. Mice are provided a standard autoclavable chow (Purina 7012, 5% fat) and autoclaved water ad libitum. Animals in mating cages are fed an increased fat autoclaved chow (Purina 5021, 9% fat). Heterozygous transgenic mice are mated with nontransgenic littermates to propagate the transgenic lines. Pups are weaned at 21 d of age, the tails snipped, genomic DNA isolated, and genotyped as described below.

3.3. Polymerase Chain Reaction (PCR) Genotype Analysis

Genomic tail DNA is screened for the presence of the PKC transgene construct using a PCR detection method. One-hundred nanograms of genomic DNA is combined with transgene-specific forward and reverse primers or beta-globin forward (5′-CCAATCTGCTCACACAGGATAGAGAGGGCAGG-3′) and reverse (5′-CCTTGAGGCTGTCCAAGTATTCAGGCCATCG-3′) and amplified with 2.5 U of Taq DNA polymerase in a total volume of 50 μL in Easy-Start PCR Mix-in-a-Tube PCR tubes (Molecular Bio-Products, Inc) using temperature and cycling conditions optimized for these primer pairs. Amplification of the endogenous beta-globin gene ensures the presence of sufficient DNA to detect an amplified product and controls for experimental conditions common to both reactions.

3.4. AOM Carcinogen Treatment

1. To induce carcinogenesis in mice, female experimental and control animals 5–7 wk of age are injected intraperitoneally with 10 mg/kg body weight of AOM (experimental animals) or an equal volume of saline (control animals) weekly for 4 wk. Injections must be performed in a chemical fume hood and the mice are housed in the fume hood for 48 h after the injections to protect against exposure to expired carcinogen.
2. After 48 h, the mice are transferred to clean cages. The dirty cages are washed with dilute bleach solution to degrade any remaining carcinogen and the bedding is disposed of as a biohazard.
3. Harvests may be performed at any time, but we have found that 12 wk (for ACF and early tumor development) and 40 wk after the final AOM injection to be optimal for pathological observation.

3.5. Dissection and Removal of Mouse Colon Tissue

1. One hour before harvest, the mice are administered 50 mg/kg body weight BrdU by intraperitoneal injection for later measurement of in vivo cellular proliferation.
2. Mice are euthanized by CO_2 asphyxiation (cervical dislocation is also an acceptable means). Resect the entire colon.
3. The colon is inflated and fecal material is removed from the lumen of the colon by flushing with cold phosphate-buffered saline (PBS) using a 10-cc syringe fitted with a gavage needle. Insert the needle into the proximal end of the colon. Slowly and gently inject cold PBS into the colon. In addition to washing out the fecal material, this process inflates the colon and breaks the circular muscles of the colon, facilitating further analysis of the colonic epithelium.
4. At this point, the colon may be used for RNA and protein extraction (*see* **Subheading 3.6.**), fixed and stained for analysis of aberrant crypt foci (*see* **Subheading 3.7.**), or fixed in 4% paraformaldehyde (*see* **Subheading 3.7., step 1** for fixation protocol), embedded and sectioned and used for immunohistochemical analysis (*see* **Subheading 3.7.**).

3.6. Preparation of Colon Tissue for Analysis of PKC mRNA Expression, Kinase Activity, Protein Expression, and Localization

1. The resected and rinsed colon is then measured and dissected by a longitudinal incision and laid flat on a clean, cold glass plate (resting on a bed of ice) that has been pretreated with RNAse Zap to eliminate RNAse activity. Further rinsing of the colonic mucosa with cold PBS may be required at this point.
2. For immunohistochemical analysis of the colon tissue (*see* **Subheading 3.7.**) sections (1 cm in length) are removed from the cecal-colon junction (proximal colon) and the distal end of the colon (distal colon), placed in an embedding cassette and is immersed in either 70% ethanol or 4% paraformaldehyde for tissue fixation.

3. For protein and RNA analysis, the colon (still laying flat on the glass plate) is scraped using a clean, RNAse Zap-treated plastic microscope slide. It is important to apply firm, constant pressure during the scraping procedure. Equal portions of the viscous material are aliquoted into 0.3 mL of modified lysis buffer (MLB, for protein analysis, **Subheadings 2., item 17**) and into 0.1 mL of Trizol (for RNA isolation and analysis, **Subheading 2., item 8**).

3.7. ACF Analysis

1. After the colon has been resected, rinsed and split longitudinally (*see* **Subheading 3.6., step 1**), it is laid flat between two pieces of 3M filter paper and fixed in either 70% Ethanol (4° for 30min) or 4% PFA at 4°C for 4 h. A glass slide is placed over the colon-filter paper sandwich to assure that the colon remains flat during fixation. Fixed colons are then dehydrated by successive ethanol washes as follows:
   a. 50% EtOH rinse, 30 s.
   b. 3 × 20-min 50% EtOH washes.
   c. 3 × 20-min 70% EtOH washes.
   d. Colon sections may now be stored at 4°C in 70% EtOH until staining and ACF analysis.
2. Fixed colons are stained with 0.5% methylene blue in PBS for 5–15 min before mounting on a glass slide for observation at low magnification (×40) on a dissecting light microscope.
3. ACF are identified by the following criteria (*see* **Fig. 1**)
   a. Aberrant crypts are at least two to three times the size of surrounding crypts.
   b. Luminal opening is usually elliptical.
   c. The thickness of the epithelial lining is greater than that separating normal crypts from each other.
4. Colons are scored for total number, position (usually measured in cm from the distal end of the colon), and crypt multiplicity (number of crypts/focus) of the ACF. The observer should be blinded to the treatment status of the animals (e.g. experimental vs control mice).
5. At this point, ACF and adjacent normal tissue can be "plucked" using a dissecting microscope and fine-tipped surgical forceps (*see* **Note 1**). ACF and normal samples may then be placed in MLB for protein isolation or into Trizol reagent for RNA analysis (*see* **Note 2**).

3.8. Immunoblot Analysis of PKCs from Crude Mouse Colonic Epithelial Extracts

1. Scraped colonic epithelium is homogenized in MLB using a Polytron homogenizer. Pellet crude colonic protein extract (in MLB, *see* **Subheading 2., item 17**) for 10 min at high speed in a refrigerated bench-top Microfuge. Remove supernatant, quantitate the protein by the BCA method (Pierce), and store at –20°C.
2. Denature equal amounts of protein in crude extract (extracted in MLB, *see* **Subheading 2., item 17**) in a suitable volume of Laemmli Sample Buffer (LSB) by boiling for 5 min. Samples may be stored at –20°C until use.

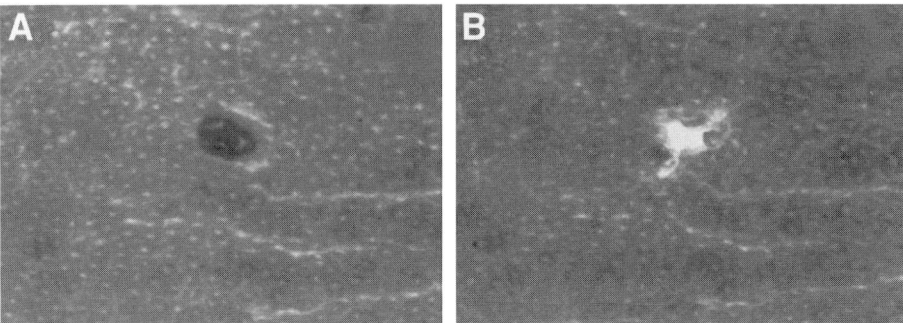

Fig. 1. ACF dissection from an AOM-treated mouse colon. Colons from AOM-injected mice were isolated, fixed flat in 70% ethanol (4°C for at least 30 min), and stained with 1% methylene blue for 5 min. ACF were identified under low magnification (40× by the criteria indicated in **Subheading 3.7., step 3.** This figure illustrates an ACF with a multiplicity of 3 shown before (**A**) and after (**B**) microdissection of the aberrant crypts (*see* **Note 1**). (Figure reprinted by permission from Gokmen-Polar et al., *Cancer Res.* **61,** 1375–1381, 2001).

3. Subject 30 µg of protein per well of samples to sodium dodecyl sulfate poly-acrylamide gel electrophoresis (SDS-PAGE) on a 10% acrylamide gel (see Bio-Rad Protran II System).
4. Transfer proteins to nitrocellulose membrane using the Bio-Rad Transblot System (~200 mAmp for 1.5 h).
5. After transfer, briefly rinse the membrane with water and stain for 5 min with Fast Green reagent. Destain with multiple washes of distilled water until lanes are clearly visible. Examine the blot for proper transfer, bubbles, and appropriately equal loading. Indicate lanes and molecular weight markers on membrane by marking with ink pen.
6. Incubate the membrane in 5% nonfat dry milk in PBS:-0.05% Tween-20 (PBST) for 1 h with rocking at 25°C to inhibit nonspecific binding of antibody to membrane.
7. Briefly rinse away excess blocking solution and incubate the membrane on a rocking platform for 3 h at room temperature with the indicated dilution of primary antibody in PBST (*see* **Table 1**).
8. Rinse membrane three times for 5 min with PBST.
9. Incubate membrane for 1 h with 2° horseradish peroxidase conjugated antibody in 5% milk/PBST on rocking plate at room temperature.
10. Wash membrane with PBST three times for 10 min with shaking. Rinse membrane briefly with PBS to remove detergent.
11. Develop Western blot with Amersham-Pharmacia ECL kit as instructed by the manufacturer.

Table 1
Optimized Primary Antibodies and Dilutions
for PKC Western Blotting From Mouse Colons

Antibody	Company	1° Antibody Dilution	2° Antibody Dilution
PKCβII	Santa Cruz (SC–210)	1 : 1000	1 : 10,000 (rabbit)
PKCβI	Santa Cruz (SC–209)	1 : 1000	1 : 20,000 (rabbit)
PKCα	Fields Lab Generated	1 : 2000	1 : 20,000 (rabbit)
PKCι/λ	BD Transduction Labs	1 : 3000	1 : 10,000 (mouse)
PKCζ	Santa Cruz (SC–726)	1 : 200	1 : 5000 (rabbit)
PKCε	Gibco	1 : 1000	1 : 10,000 (rabbit)
PKCδ	BD Transduction Labs	1 : 1000	1 : 10,000 (rabbit)
PKCη	Santa Cruz (SC–215)	1 : 2000	1 : 10,000 (rabbit)

2° Antibodies purchased from KPL Laboratories and diluted as recommended.

3.9. Immunoprecipitation-Kinase Assay of PKCι/λ Activity in Crude Mouse Colonic Extracts (see Note 3)

1. Quantitate protein concentration (BCA assay, Pierce) in crude extracts extracted with MLB (*see* **Note 4**). Dilute to 2 μg/μL with MLB in a 1.5-mL centrifuge tube.
2. Preclear: Add 50 μL of Protein A Sepharose beads (Sigma) to 0.5 mL of diluted crude extract and incubate at 4°C while tumbling for 1 h. Centrifuge for 1 min at 100g in a bench top microcentrifuge. Move cleared supernatant to a clean tube.
3. Antibody binding: Add 10 μL of PKCι/λ monoclonal antibody (BD Transduction Labs) to cleared extract (supernatant). Incubate at 4°C for 1 h while tumbling.
4. Immunoprecipitation: Add 50 μL of protein A Sepharose beads to the antibody/cleared extract mix. Incubate 10 min at 4°C with tumbling (*see* **Note 5**). Pellet sample at 100g as before, remove, and discard the supernatant.
5. Wash: Resuspend the beads in 1 mL of cold MLB, spin down. Repeat MLB wash step four times. Wash once more with 1 mL of cold 1X Kinase Assay Buffer (1X KAB), remove the supernatant, and go on to **step 6**.
6. Kinase Assay:
   a. Make 2X assay buffer stock (enough for 20 assays).
      a. Start with 250 μL of 2X KAB, add 1 μL of nonisotopically labeled 100 mM ATP (Sigma).
      b. Add 200 μL of 2X KAB/cold ATP mix to 10 μL of γ-^{32}P ATP (3000Ci/mmol, 10 mCi/mL, Amersham-Pharmacia), mix well.
   b. Add 10 μL of 2X assay buffer stock (with radioactive ATP) to immunoprecipitation pellet. Place 10 μL of histone stock solution (1.25 mg/mL in H$_2$O, purchased as a solid from Sigma) in the cap of the Eppendorf tube. Pulse spin all tubes to simultaneously start all reactions.

    c. Incubate for 30 min at room temperature. Stop the reaction by placing 40 µL of LSB into the tube cap and simultaneously spinning all samples.

7. Heat the sample tubes for 5 min in a boiling water bath to denature proteins.

8. Pellet out the Sepharose beads as before and load only the supernatant on the gel. Subject samples to SDS-PAGE on 10% acrylamide gel and transfer to membrane as described in **Subheading 3.8.**

9. Expose membrane to film. Radioactive histone bands will appear at approx 31 kDa. Bands can be quantitated and compared by densitometric analysis or by phosphorimaging analysis of the membrane.

10. Perform Western blot analysis of PKCι/λ levels as described in **Subheading 3.8.** Quantitate immunoprecipitated PKCι/λ protein levels by densitometric analysis and normalize assay results to determine a relative specific activity.

3.10. Reverse Transcription PCR of mRNA Isolated from Mouse Colonic Epithelia

1. Total cellular RNA from normal colonic epithelium (**Subheading 3.6.**), ACF (**Subheading 3.6., step 2**), and colon tumors (*see* **Note 2**) is isolated using Trizol Reagent (Gibco) as instructed by the manufacturer, quantitated and stored at –80°C. The use of RNAse free water, tubes, pipet tips, and other equipment is essential during isolation and storage of the RNA samples.

2. Reverse transcription reaction using Superscript II reverse transcriptase (Invitrogen):

    a. Combine 2 µg of total RNA and 1 µg of oligo-(dT) 12–18mer primer in 10 µL RNAse free water in a 0.5-mL Eppendorf tube. Heat to 65°C for 5 min, then allow samples to cool to 37°C.

    b. Add to denatured RNA/primer mix: 10 mM dithiothreitol, 0.5 mM dNTPs, 10 µL of 5X first-strand buffer (provided with Superscript II reverse transcriptase, Invitrogen) and 200 U Superscript II reverse transcriptase. Dilute to 50 µL final reaction volume with RNAse free water and mix well by gentle pipetting. Samples without reverse transcriptase serve as negative controls.

    c. Incubate at 37°C for 1h. Heat inactivate the sample by heating to 95°C for 5 min and place immediately on ice.

    d. Amplification of the cDNAs was carried out using EasyStart PCR mix-in-a-tube (Molecular Bioproducts), 2.5 U of Taq polymerase, (Promega), and the appropriate primer pairs for each PKC isozyme.

    e. Primers used for PKC isozyme-specific PCR are shown in **Table 2.**

    f. The optimized linear range for each PKC isozyme was determined as 30 cycles for PKCα, 35 cycles for PKCβI and βII, and 25 cycles for β-actin. PCR cycling was as follows: 95°C for 45 s, 60°C for 45 s, and 72°C for 2 min, followed by an incubation of 10 min at 72°C. An initial denaturation step was performed at 95°C for 2 min.

    g. PCR products were separated in 1.8% agarose gel (FMC Bioproducts) and the intensity of ethidium bromide fluorescence was quantitated using an Eagle Eye (Stratagene) densitometer. Quantitation of PKC mRNA expression was

Table 2
Forward and Reverse Primers for Reverse Transcription-PCR Analysis of PKC mRNA from Mouse Colon

Isozyme	Forward Primer	Reverse Primer	Product Length
PKCα	5'TGAATCCTCAGTGGAATGAGT3'	5'GGTTGCTTTCTGTCTTCTGAA3'	325 bp
PKCβI	5'TGTGATGGAGTATGTGAACGGGGG3'	5'TCGAAGTTGGAGGTGTCTCGCTTG3'	640 bp
PKCβII	5'CATCTGGGATGGGGTGACAACC3'	5'CGGTCGAAGTTTCAGCGTTTC3'	420 bp
β actin	5'GTGGGCCGCTCTAGGCACCAA3'	5'CTCTTGATCTCACGCACGATTTC3'	540 bp

normalized to β-actin mRNA levels, which have been shown to be unchanged during colon carcinogenesis *(31)*.

3.11. Immunohistochemical Analysis of PKC Protein Expression in Mouse Colonic Epithelia (see Note 6)

1. Colon sections (1 cm, *see* **Subheading 3.6., step 2**) of normal, tumor, or ACF tissues remain in 4% paraformaldehyde for 4 h at 4°C. They are then dehydrated by successive ethanol washes as follows
   a. 50% EtOH rinse, 30 s.
   b. 3 × 20 min 50% EtOH washes.
   c. 3 × 20 min 70% EtOH washes.
   d. Colon sections may now be stored at 4°C in 70% EtOH until embedding.
2. Tissues are embedded in low melting temperature paraffin and cut into 5-μm sections and placed on poly-L-lysine-coated slides.
3. Sections are deparaffinized and rehydrated as according to standard protocol.
4. At this point the sections are processed for antigen-retrieval by heating to 95°C for 40 min in 1× DAKO Antigen Retrieval Solution (DAKO Inc) allow the sections to cool for 20 min and then wash 3 × 2 min in PBS.
5. Sections are then treated for 10 min with 1% hydrogen peroxide in water to inhibit endogenous peroxidases. Wash with PBS.
6. PKCβII expression is detected using the ABC Staining System and isotype-specific antibody to PKCβII (Santa Cruz Biotechnology). Primary antibody concentrations are variable dependent on the tissue and fixation technique. Generally dilutions of 1 : 100–1 : 1000 are sufficient for accurate detection of PKCs βII and ι.

4. Notes

1. With practice, ACF can be isolated using very fine-tipped surgical forceps (Roboz Inc). Take care to remove the ACF with as little disruption of the surrounding tissue as possible. Examination of the surrounding tissue under the dissection microscope before and after plucking will allow assessment of how precise the removal was and how much normal/ACF tissue was obtained. This technique is illustrated in **Fig. 1**. Multiple plucked ACF may be pooled to acquire enough tissue for protein or RNA analysis.
2. In addition to ACF analysis, AOM-treated mice may be allowed to progress to colonic tumors (40 wk after AOM injections). Tumors may then be dissected away from normal tissue and similar analyses may be performed to look at PKC expression profiles in normal vs ACF vs tumor tissue to follow PKC isoform expression through the tumorigenesis process.
3. This method will work for both atypical PKC isoforms. It can be adapted for use with classical PKC isoforms (alpha, beta, gamma) and for novel isoforms (delta and epsilon). Note that measurement of immunoprecipitation kinase activity levels in classical and novel PKCs will not completely reflect the activity in vivo because these PKCs require calcium (classical only) and/or diacylglycerol

(both classic and novel) for activation. In order to perform this adaptation on PKCs other than ι/λ or ζ substitute 2× c/nPKC kinase buffer (*see* **Subheading 2., item 19**) for 2×KAB in the kinase assay description.

4. It is extremely important to the in vitro activity of PKCι in immunoprecipitation kinase assays that MLB or a similar buffer that does not contain SDS or large amounts of nonionic detergent. It is also important that the kinase assays be performed on the same day as the mice are harvested and extracts are made in order to maintain full activity.

5. **Subheading 3.4., step 4** is also a critical step for maintaining a very low background. Many other immunoprecipitation procedures call for a 1-h incubation. A 10-min incubation is optimal in this assay for maintaining an acceptable signal to noise ratio in the final assay analysis.

6. In addition to immunohistochemical analysis of PKC isozyme expression, the fixed tissues can be stained with hematoxylin and eosin and evaluated for crypt morphometrics *(32)*. The differentiation status of the colonic epithelia can be determined by detection of lectin binding *(32)*. Immunohistochemical detection of BrdU incorporation can be used to detect cell proliferation *(33)* and the percentage of cells undergoing apoptosis can be detected by TdT-mediated dUTP-biotin nick end labeling of fragmented DNA (TUNEL) assay using the apoTACS kit from Trevigen *(32,34)*.

References

1. Guan, R., Zhang, Y., Jiang, J., Baumann, C. A., Black, R. A., Baumann, G., et al. (2001) Phorbol ester- and growth factor-induced growth hormone (GH) receptor proteolysis and GH-binding protein shedding: relationship to GH receptor down-regulation. *Endocrinology* **142,** 1137–1147.
2. Yamanishi, J., Takai, Y., Kaibuchi, K., Sano, K., Castagna, M., and Nishizuka, Y. (1983) Synergistic functions of phorbol ester and calcium in serotonin release from human platelets. *Biochem. Biophys. Res. Commun.* **112,** 778–786.
3. Saxon, M. L., Zhao, X., and Black, J. D. (1994) Activation of protein kinase C isozymes is associated with post-mitotic events in intestinal epithelial cells in situ. *J. Cell Biol.* **126,** 747–763.
4. Gregorio, C. C., Repasky, E. A., Fowler, V. M., and Black, J. D. (1994) Dynamic properties of ankyrin in T lymphocytes: colocalization with spectrin and protein kinase C beta. *J. Cell Biol.* **125,** 345–358.
5. Black, S. C., Fagbemi, S. O., Chi, L., Friedrichs, G. S., and Lucchesi, B. R. (1993) Phorbol ester-induced ventricular fibrillation in the Langendorff-perfused rabbit heart: antagonism by staurosporine and glibenclamide. *J. Mol. Cell Cardiol.* **25,** 1427–1438.
6. Dempsey, E. C., Newton, A. C., Mochly-Rosen, D., Fields, A. P., Reyland, M. E., Insel, P. A., et al. (2000) Protein kinase C isozymes and the regulation of diverse cell responses. *Am. J. Physiol.* **279,** L429–L438.

7. Dorn, G. W., Souroujon, M. C., Liron, T., Chen, C. H., Gray, M. O., Zhou, H. Z., et al. (1999) Sustained in vivo cardiac protection by a rationally designed peptide that causes epsilon protein kinase C translocation. *Proc. Natl. Acad. Sci. USA* **96,** 12,798–12,803.

8. Csukai, M., Chen, C. H., De Matteis, M. A., and Mochly-Rosen, D. (1997) The coatomer protein beta'-COP, a selective binding protein (RACK) for protein kinase Cepsilon. *J. Biol. Chem.* **272,** 29,200–29,206.

9. Johnson, J. A., Gray, M. O., Chen, C. H., and Mochly-Rosen, D. (1996) A protein kinase C translocation inhibitor as an isozyme-selective antagonist of cardiac function. *J. Biol. Chem.* **271,** 24,962–24,966.

10. Sheehy, A. M., Burson, M. A., and Black, S. M. (1998) Nitric oxide exposure inhibits endothelial NOS activity but not gene expression: a role for superoxide. *Am. J. Physiol.* **274,** L833–L841.

11. Craven, P. A. and DeRubertis, F. R. (1992) Alterations in protein kinase C in 1,2-dimethylhydrazine induced colonic carcinogenesis. *Cancer Res.* **52,** 2216–2221.

12. DeRubertis, F. R. and Craven, P. A. (1987) Relationship of bile salt stimulation of colonic epithelial phospholipid turnover and proliferative activity: role of activation of protein kinase C1. *Prev. Med.* **16,** 572–579.

13. La Porta, C. A. and Comolli, R. (1994) Membrane and nuclear protein kinase C activation in the early stages of diethylnitrosamine-induced rat hepatocarcinogenesis. *Carcinogenesis* **15,** 1743–1747.

14. Basu, A., Mohanty, S., and Sun, B. (2001) Differential sensitivity of breast cancer cells to tumor necrosis factor-alpha: involvement of protein kinase C. *Biochem. Biophys. Res. Commun.* **280,** 883–891.

15. Alele, J., Jiang, J., Goldsmith, J. F., Yang, X., Maheshwari, H. G., Black. R. A., et al. (1998) Blockade of growth hormone receptor shedding by a metalloprotease inhibitor. *Endocrinology* **139,** 1927–1935.

16. Hocevar, B. A., Burns, D. J., and Fields, A. P. (1993) Identification of protein kinase C (PKC) phosphorylation sites on human lamin B. Potential role of PKC in nuclear lamina structural dynamics. *J. Biol. Chem.* **268,** 7545–7552.

17. Misra-Press, A., Fields, A. P., Samols, D., and Goldthwait, D. A. (1992) Protein kinase C isoforms in human glioblastoma cells. *Glia* **6,** 188–197.

18. Frey, M. R., Leontieva, O., Watters, D. J., and Black, J. D. (2001) Stimulation of protein kinase C-dependent and -independent signaling pathways by bistratene A in intestinal epithelial cells. *Biochem. Pharmacol.* **61,** 1093–1100.

19. Fields, F. O. and Thorner, J. (1991) Genetic suppression analysis of the function of a protein kinase C (PKC1 gene product) in Saccharomyces cerevisiae cell cycle progression: the SKCd mutations. *Cold Spring Harb. Symp. Quant. Biol.* **56,** 51–60.

20. Hunter, S. E., Seibenhener, M. L., and Wooten, M. W. (1995) Atypical zeta–protein kinase c displays a unique developmental expression pattern in rat brain. *Brain Res. Dev. Brain Res.* **85,** 239–248.

21. Nel, A. E., Schabort, I., Rheeder, A., Bouic, P., and Wooten, M. W. (1987) Inhibition of antibodies to CD3 surface antigen and phytohemagglutinin-mediated T cellular responses by inhibiting Ca2+/phospholipid-dependent protein kinase activity with the aid of 1-(5-isoquinolinylsulfonyl)-2-methylpiperazine dihydrochloride. *J. Immunol.* **139**, 2230–2236.

22. Berra, E., Municio, M. M., Sanz, L., Frutos, S., Diaz-Meco, M. T., and Moscat, J. (1997) Positioning atypical protein kinase C isoforms in the UV-induced apoptotic signaling cascade. *Mol. Cell Biol.* **17**, 4346–4354.

23. Frutos, S., Moscat, J., and Diaz-Meco, M. T. (1999) Cleavage of zetaPKC but not lambda/iotaPKC by caspase-3 during UV-induced apoptosis. *J. Biol. Chem.* **274**, 10,765–10,770.

24. Sanz, L., Sanchez, P., Lallena, M. J., Diaz-Meco, M. T., and Moscat, J. (1999) The interaction of p62 with RIP links the atypical PKCs to NF-kappaB activation. *EMBO J.* **18**, 3044–3053.

25. Cai, H., Smola, U., Wixler, V., Eisenmann-Tappe, I., Diaz-Meco, M. T., Moscat, J., et al. (1997) Role of diacylglycerol-regulated protein kinase C isotypes in growth factor activation of the Raf-1 protein kinase. *Mol. Cell Biol.* **17**, 732–741.

26. Standaert, M. L., Bandyopadhyay, G., Perez, L., Price, D., Galloway, L., Poklepovic, A., et al. (1999) Insulin activates protein kinases C-zeta and C-lambda by an autophosphorylation-dependent mechanism and stimulates their translocation to GLUT4 vesicles and other membrane fractions in rat adipocytes. *J. Biol. Chem.* **274**, 25,308–25,316.

27. Sanz-Navares, E., Fernandez, N., Kazanietz, M. G., and Rotenberg, S. A. (2001) Atypical protein kinase Czeta suppresses migration of mouse melanoma cells. *Cell Growth Differ.* **12**, 517–524.

28. Roy, H. K., Bissonnette, M., Frawley, B. P., Jr., Wali, R. K., Niedziela, S. M., Earnest, D., et al. (1995) Selective preservation of protein kinase C-zeta in the chemoprevention of azoxymethane-induced colonic tumors by piroxicam. *FEBS Lett.* **366**, 143–145.

29. Hogan, B., Beddington, R., Costantini, F., and Lacy, E. (Eds.) (1994) *Manipulating the Mouse Embryo: A Laboratory Manual*, 2nd ed., Cold Spring Harbor Press, Cold Spring Harbor, New York.

30. Simon, T. C., Roth, K. A., and Gordon, J. I. (1993) Use of transgenic mice to map cis-acting elements in the liver fatty acid-binding protein gene (Fabpl) that regulate its cell lineage-specific, differentiation-dependent, and spatial patterns of expression in the gut epithelium and in the liver acinus. *J. Biol. Chem.* **268**, 18,345–18,358.

31. Yoshimi, N., Wang, A., Makita, H., Suzui, M., Mori, H., Okano, Y., et al. (1994) Reduced expression of phospholipase C-delta, a signal-transducing enzyme, in rat colon neoplasms induced by methylazoxymethanol acetate. *Mol. Carcinog.* **11**, 192–196.

32. Murray, N. R., Davidson, L. A., Chapkin, R. S., Gustafson, W. C., Schattenberg, D. G., and Fields, A. P. (1999) Overexpression of protein kinase C betaII induces

colonic hyperproliferation and increased sensitivity to colon carcinogenesis. *J. Cell Biol.* **145,** 699–711.

33. Murray, N. R., Weems, C., Chen, L., Leon, J., Yu, W., Davidson, L. A., et al. (2002) Protein kinase C βII and TGFβRII in δ-3 fatty acid- mediated inhibition of colon carcinogenesis. *J. Cell Biol.* **157,** 915–920.
34. Kerr, J. F., Gobe, G. C., Winterford, C. M., and Harmon, B. V. (1995) Anatomical methods in cell death. *Methods Cell Biol.* **46,** 1–27.

42

Alcohol Addiction

M. Foster Olive and Dorit Ron

1. Introduction

There is increasing evidence for a role of individual protein kinase C (PKC) isozymes in the pathology of various diseases of the brain *(1)*. For example, it was recently demonstrated that a mutation in PKCγ results in Parkinson's disease-like symptoms in rats *(2)*. There is also evidence that alterations in the expression of PKC isoforms may influence alcohol consumption and the behavioral responses to alcohol and other drugs of abuse *(3–7)*. However, the paucity of pharmacological ligands that selectively modulate the activity of individual PKC isozymes has compelled scientists to turn to genetic methods, such as viral gene delivery, antisense oligonucleotide, and targeted gene-deletion techniques to ascertain the function of individual PKC isoforms in vivo.

Alcohol (ethyl alcohol, or ethanol) has been shown to alter the function and activity of numerous types of voltage- and ligand-gated ion channels in the central nervous system. However, it is increasingly apparent that ethanol also affects multiple intracellular signaling pathways. Of these, signaling by the PKC family of enzymes has received considerable attention. In vitro studies have shown that acute alcohol exposure can directly inhibit or enhance PKC activity and alter the subcellular distribution of individual PKC isozymes, whereas chronic exposure to ethanol generally leads to an up-regulation of PKC expression and/or function *(8,9)*.

In this chapter, we discuss several methodologies for determining ethanol consumption patterns and the behavioral effects of ethanol that are suitable for use in PKC isoform "knockout" mice. First, we discuss two separate methods of determining voluntary ethanol intake. We then discuss several behavioral assays for the determination of the acute effects of ethanol on motor behavior.

From: *Methods in Molecular Biology, vol. 233: Protein Kinase C Protocols*
Edited by: A. C. Newton © Humana Press Inc., Totowa, NJ

Next, a liquid diet protocol for producing ethanol dependence is outlined, as well as a procedure for examining withdrawal symptom severity and blood ethanol concentrations. Finally, the conditioned place preference (CPP) paradigm is then introduced as a model for examining the motivational effects of ethanol in mice. Although we have extensively used these protocols for examining alcohol addiction and related behaviors in PKC mutant mice, they can just as easily be applied to mice with targeted gene-deletion of other protein kinase signaling pathways, such as the protein kinase A pathway *(10,11)*. In addition, these protocols can be easily modified to study other drugs of abuse such as psychostimulants and opiates.

Several notes of caution, however, should be taken when interpreting behavioral data in knockout mice. First, because many knockout mice are generated using gene-deletion techniques in embryonic cells, the lack of a particular PKC isoform during development may result in compensatory increases in the levels of other PKC isoforms in adulthood. Protein levels of individual PKC isoforms in the brain and other organs should be routinely analyzed for such reasons. Aditionally, it has been noted that the genetic background of PKCγ-deficient mice is a critical determinant of genotypic differences in alcohol-related behaviors *(12)*. Other background-specific phenotypes of genetically modified mice have also been observed *(13,14)*. Thus, it is important to consider possible contribution of other genes to differences in alcohol-related behaviors found between PKC knockout and wild-type mice. Finally, genotypic differences in behavior may also be extremely sensitive to subtle differences in handling procedures, testing apparatus, lighting conditions, etc. *(15)*. Thus, results obtained from the same knockout mice may differ slightly between individual laboratories. Extreme care should be taken to minimize procedural differences when comparing data with those generated by other investigators.

2. Materials

2.1. Oral Ethanol Self-Administration

2.1.1. Two-Bottle Choice Paradigm

1. 95% v/v ethanol, diluted in tap water to the following concentrations (expressed as v/v): 2% (21 mL of 95% ethanol in 979 mL H_2O), 4% (42 mL of 95% ethanol in 958 mL H_2O), 8% (84 mL of 95% ethanol in 916 mL H_2O), and 10% (105 mL of 95% ethanol in 895 mL H_2O). Solutions are stirred for 30 min using a magnetic stirrer to remove any air bubbles formed by mixing H_2O with ethanol.
2. Nalgene 25-mL conical plastic tubes with graduations marked on the side for reading of fluid levels to the nearest 0.25 mL. One-hole stoppers and 2-inch straight stainless steel drinking tubes are also needed.

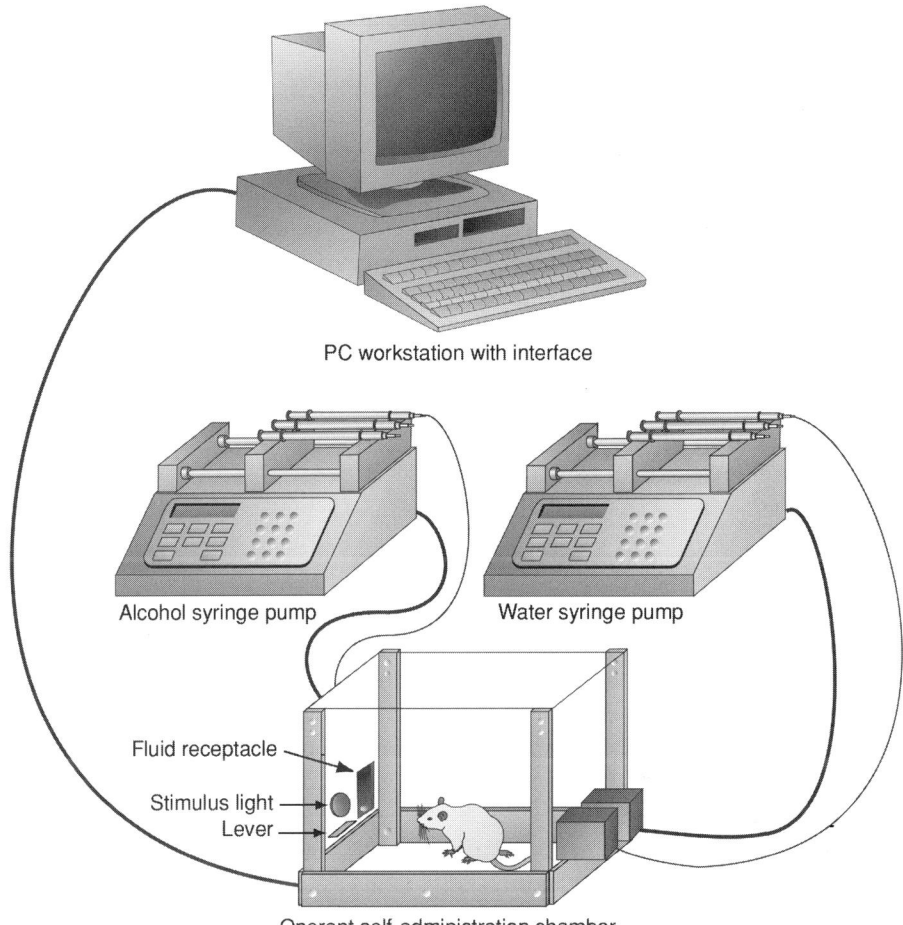

PC workstation with interface

Alcohol syringe pump   Water syringe pump

Fluid receptacle
Stimulus light
Lever

Operant self-administration chamber

Fig. 1. Apparatus for operant ethanol self-administration in mice. A lever press by the mouse activates the syringe pump to deliver the reinforcer (0.01 mL of water or 10% ethanol) into the fluid receptacle. The lever press also illuminates the stimulus light to signal the availability of the reinforcer. A computer records the number and temporal distribution of lever presses over a 16–h overnight self-administration session.

2.1.2. Operant Self-Administration Paradigm

1. Plexiglas operant chambers (Med Associates) measuring $15.9 \times 14 \times 12.7$ cm with stainless steel grid floors (*see* **Fig. 1**). Each chamber is housed in a sound-attenuating cubicle equipped with an exhaust fan to mask external noise and provide ventilation. The left and right wall of each operant chamber are equipped with one ultra-sensitive stainless steel response lever, a stimulus light located

directly above the response lever, and a fluid receptacle attached to a syringe pump. Each chamber also contains a 2.8-watt house light that is always illuminated. The chambers and syringe pumps are controlled using a computer interface and PC software (Med Associates).

2. Ethanol solutions are made as described in **Subheading 2.1.1.** In addition, sucrose solutions (2, 5, and 10% w/v) is made with and without various concentrations of ethanol (*see* **Subheading 3.1.2.**).

3. Solutions are placed in 20-mL plastic syringes to allow fluid measurement to the nearest 0.1 mL (*see* **Note 1**).

2.2. Acute Behavioral Effects

For all acute behavioral studies utilizing intraperitoneal (i.p.) injection of ethanol, 95% ethanol is diluted to 20% v/v in physiological saline (i.e., 21 mL of 95% ethanol in 79 mL of 0.9% NaCl) (*see* **Note 5** for determination of amount needed per injection.)

2.2.1. Locomotor Activity Stimulation

Plexiglas open-field chambers (43×43 cm, Med Associates) located in sound-attenuating cubicles equipped with exhaust fans that masked external noise as well as a 2.8-watt house light. The chambers are equipped with two sets of 16 pulse-modulated infrared photobeams were placed on opposite walls at 2.5 cm centers to record x-y ambulatory movements. Activity chambers are computer interfaced for data sampling at 100-msec resolution.

2.2.2. Motor Coordination

Automated rotarod (commercially available through numerous vendors). The apparatus basically consists of a horizontal rotating cylinder covered with textured rubber, approx 3 cm in diameter. Constant speed rotarods that rotate at approx 5 rpm are available, as are accelerating rotarods that rotate at 4 rpm and gradually increase to a speed of 40 rpm over a 5-min period. Approximately 15 cm below the rotarod is a floor panel that senses the presence of the animal when it falls from the rotarod. Automated timers are also included on many models.

2.2.3. Loss-of-Righting Reflex

Electronic timer with 1-sec resolution.

2.3. Chronic Ethanol Administration

2.3.1. Liquid Diet Consumption

1. Lieber-DeCarli modified high-protein ethanol liquid diet (#710262, Dyets, Inc.). Diet is made fresh daily by adding 67 mL of 95% ethanol to 163.03 g of diet, q.s. to 1 liter with tap water and mixing for 30 s in a blender (final concentration is approximately 6.5% v/v ethanol).

2. Lieber-DeCarli modified high-protein control liquid diet (#710029, Dyets, Inc.). Diet is made fresh daily by adding 252.63 g of diet to 840 mL tap water and mixing for 30 s in a blender.
3. "Liquidiet" 50 mL feeding tubes (Bio-Serv) with 1-mL graduations.

2.3.2. Withdrawal Symptom Severity

Video camera and VCR.

2.4. Blood Ethanol Determination

1. Alcohol diagnostic kit (Sigma #332-C), consisting of 0.38 N trichloroacetic acid, glycine buffer reagent, and nicotinamide adenine dinucleotide-alcohol dehydrogenase (NAD-ADH) reagent
2. Ethanol standards (0.05, 0.10, and 0.30% v/v, Sigma).
3. Heparizined borosilicate glass capillaries (Fisher Scientific) and micropipet bulbs (Alltech Chromatography).
4. Spectrophotometer capable of reading ultraviolet absorbance at 340 nm.

2.5. Motivational Effects of Ethanol (Conditioned Place Preference)

Place conditioning apparatus for mice (Med Associates). The testing apparatus consists of two 16.8- × 12.7-cm compartments with distinct visual and tactile cues (*see* **Fig. 2**). The two compartments are connected by a 7.2- × 12.7-cm center compartment with gray walls and solid plastic flooring. Each chamber is equipped with a 2.8-watt light centered above the compartment, and is housed in a sound-attenuating cubicle with a fan that helps to mask external noise. The center compartment is equipped with two computer-controlled guillotine doors that provided access to one or both of the conditioning compartments.

3. Methods

For all behavioral procedures, animals should be at least 6 to 8 wk of age at the time of testing and housed individually in standard Plexiglas cages with food available ad libitum. Prior to testing, mice should be given at least 1 wk to acclimatize to individual housing conditions and handling. During this period, water should be the only fluid available.

The colony room should be maintained on a 12:12 light–dark cycle with lights on at 0600–0800 h. However, because mice are nocturnal and exhibit peak levels of consummatory and locomotor activity during the dark phase of the light–dark cycle, the investigator may wish to perform experiments during the this period with the colony and testing rooms maintained on a reversed light–dark cycle (i.e., lights off at 0600–0800 h).

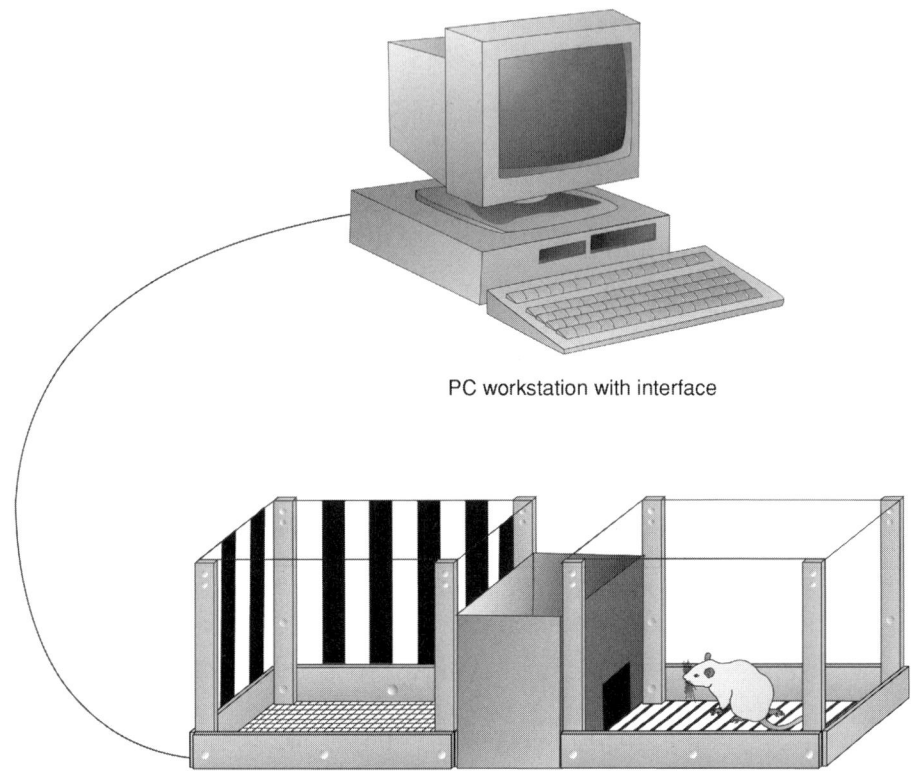

PC workstation with interface

Place conditioning apparatus

Fig. 2. Place preference conditioning apparatus. During conditioning, the animal is injected with saline i.p. and confined to one of the two conditioning chambers for 5 min, and on alternating days is injected with ethanol 2 g/kg i.p. and confined to the other chamber for 5 min. After a total of eight conditioning sessions, the animal is placed in the center compartment and allowed free movement between the two conditioning chambers for 30 min. A computer records time spent and locomotor activity in each chamber. A greater amount of time spent in the chamber in which ethanol injections were received as compared to that in which saline injections were received is considered a conditioned place preference. Note that both visual (wall color pattern) and tactile (floor grid pattern) cues are unique to each chamber.

3.1. Oral Self-Administration

3.1.1. Two-Bottle Choice Paradigm

This paradigm measures voluntary ethanol consumption at increasing ethanol concentrations by giving the animal free choice between water and an ethanol-containing solution in two separate bottles placed on the home cage.

This procedure is generally used for continuous access (i.e., 24 h/d) conditions, but can also be used for limited access (i.e., 2 h/d) procedures to differentiate ethanol consumption from general consummatory behavior.

1. Mice are housed individually in cages equipped with two grommets on one end to allow for the placement of the two solutions.
2. Each day, mice are weighed and placed in individual holding chambers while the fluids are placed in the home cage. Fluid levels are recorded at the beginning and end of each access period.
3. For continuous access procedures, mice are given a choice between concurrently available ethanol (2% v/v) and water 24 h/d, 7 d/wk. After 4 d of exposure to 2% ethanol and water, the ethanol concentration is increased from 2 to 4%. The process is then repeated until mice have had access all concentrations (i.e., 2, 4, 8, and 10%, in that order) for 4 d at each concentration (*see* **Note 2**).
4. For limited-access procedures, mice are deprived of fluids for 22 h and then allowed concurrent access to 10% ethanol and water for 2 h/d for 3 consecutive d. On the third day, body weights were measured prior to the 2-h access period, and fluid levels were recorded at the beginning and end of the 2-h access period (*see* **Note 3**).
5. The position (left or right) of each solution on the cage is alternated daily to control for side preferences.
6. For calculation of ethanol intake in g/kg, *see* **Note 4**.

3.1.2. Operant Self-Administration Paradigm

This procedure measures the ability of an animal to perform an operant task (i.e., pressing a lever) in order to obtain a reinforcer (i.e., a reward, such as water or an ethanol-containing solution). However, owing to its aversive taste, higher concentrations of ethanol are not readily consumed by rodents without training. This training involves the learning of the operant response necessary to obtain the reinforcer (initially a sucrose solution). Subsequently, ethanol is gradually introduced into the sucrose solution, and subsequently the sucrose is "faded" out of the ethanol solution, so that the animals eventually perform the operant task to obtain 10% ethanol as a reinforcer *(16)*. A visual cue (i.e., stimulus light) is also used to signal the availability of the reinforcer (*see* **Fig. 1**).

1. Animals are water-deprived for 24 h before the first training session to increase motivation for fluid-seeking. Animals are then allowed 2 h limited access after each training session.
2. Mice are weighed daily and placed individually into test chambers at approx 1600 h.
3. At the beginning of the first training session, the experimenter manually activates each lever in order to encourage drinking from the fluid receptacle as well as to acclimate the mouse to the sound of the syringe pump. Animals are then allowed to "self-train" to lever-press and drink from the fluid receptacle during overnight

training sessions lasting 16 h. During these training sessions, both levers were active on a concurrent fixed ratio one (CONC FR1 FR1) schedule, where each lever press illuminates the stimulus light for 4 s and activates the syringe pump to deliver 0.01 mL of 10% sucrose or water into the fluid receptacle. The position of each solution is alternated daily to control for side preference. Each animal is trained for 2–3 consecutive d to establish reliable lever-pressing behavior.

4. Animals are then trained to orally self-administer ethanol (10% v/v) using a sucrose substitution procedure *(16)*. Briefly, animals are trained in overnight self-administration sessions where ethanol is gradually introduced into the sucrose solution (i.e., 10% sucrose/2% ethanol for 2–3 d, 10% sucrose/5% ethanol for 2–3 d, and 10% sucrose/10% ethanol for 2–3 d), and subsequently the sucrose concentration is "faded" out (5% sucrose/10% ethanol for 2–3 d, 2% sucrose/10% ethanol for 2–3 d, and ultimately 10% ethanol for the remainder of the experiments). During sucrose fading procedures, both levers are concurrently active, with water presented as the alternative reinforcer.

5. Principal parameters recorded are body weight, total number of lever presses, and mL of ethanol and water consumed (derived from volume of solution in syringe before and after self-administration session). Ethanol consumption and preference are calculated as described in **Subheading 2.1.1.**

6. Once stable baseline responding for 10% ethanol and water has been established, additional treatments such as pharmacological manipulations and deprivation periods can be introduced *(6)*. In addition, increasing the reinforcement ratio (i.e., from 1 to 3–10 lever presses required for each reinforcer delivery) can also be examined as a model of ethanol-seeking behavior.

3.2. Acute Behavioral Effects

The acute behavioral effects of ethanol in rodents range from locomotor stimulation and motor incoordination at low doses to hypothermia and loss-of-righting at higher doses. These parameters can be quantitatively assessed in rodents as a measure of acute sensitivity to the behavioral effects of ethanol.

3.2.1. Locomotor Activity Stimulation

Mice demonstrate a transient increase in locomotor activity after acute administration of low doses of ethanol. However, because mice also demonstrate increased locomotor activity when placed in a novel environment (i.e., a test chamber), a habituation period is generally allowed prior to ethanol administration.

1. Animals are placed in the center of the locomotor activity chamber and allowed to habituate undisturbed for 60 min.

2. After habituation, animals are administered ethanol 2 g/kg i.p. (*see* **Note 5** for dose conversions) and immediately returned to the locomotor testing chambers for an additional 60 min of activity monitoring.

3. Horizontal distance traveled (cm) is recorded in 10-min intervals.
4. Activity chambers are cleaned with 2.5% v/v acetic acid after each experiment to mask pheromones from previous experiments.

3.2.2. Motor Coordination

The rotarod is a commonly used apparatus for determining motor coordination in mice. When placed on the rotarod, animals must continuously walk forward to keep from falling off the rotarod. Because ethanol produces motor incoordination, it reduces the latency to fall. However, significant tolerance to the motor-incoordinating effects of ethanol can develop with repeated exposure, and thus this parameter is often used to assess the development of tolerance to the acute behavioral effects of ethanol.

1. Start rotarod at a constant speed of 5 rpm.
2. Place mouse onto rotarod.
3. Perform three repeated trials (with 30-s rest between each trial) until animal learns to stay on rotarod for 2 min. If necessary, repeat this process over 3–4 d.
4. On test day, inject mouse with saline or 2 g/kg i.p. ethanol (*see* **Note 5** for dose conversions) and place immediately on rotarod. Record latency to fall from rotarod.
5. Retest mice on apparatus every 10 min until they maintain the ability to stay on the rotarod for two consecutive 30-s trials.
6. For tests of tolerance to the motor-incoordinating effects of ethanol, subsequent ethanol injections can be given and the animals retested 60 min later, again determining latency to fall and the time elapsed until the ability to stay on the rotarod for two consecutive 30-s trials is recovered.

3.2.3. Loss-of-Righting Reflex

High doses of ethanol (>3 g/kg i.p.) produce profound sedation, characterized by the inability of animals to right themselves onto all four paws after being placed on their backs. This is known as loss-of-righting reflex (LORR, sometimes inaccurately referred to as "sleep time"), and is an index of behavioral sensitivity to high doses of ethanol.

1. Administer ethanol (4 g/kg i.p.) to mice and start timer.
2. Intermittently place mice on their backs and test for LORR, defined as the inability to complete a righting reflex three times within a 30-s interval (usually 1–3 min).
3. After 1–2 h, mice will start to regain righting reflex, as evidenced by the ability to complete a righting reflex three times within a 30-s interval (*see* **Note 6**). LORR duration is defined as the time interval (in minutes) between loss and return of the righting reflex.

3.3. Chronic Ethanol Administration

Although two-bottle and operant self-administration paradigms can yield high levels of ethanol intake, they generally do not produce physical dependence (as evidenced by physical withdrawal symptoms). Thus, several methods for chronic ethanol administration in rodents have been developed. These include chronic ethanol infusion via gastric intubation, ethanol vapor inhalation, and liquid diet administration. However, gastric intubation is a difficult surgical procedure that is associated with numerous complications and alterations in feeding behavior, and is thus not commonly used.. Chronic exposure of animals to ethanol vapor is an adequate method for producing ethanol dependence, but has been criticized because it produces constant brain and blood ethanol levels that do not fluctuate over a 24-h period as observed in human alcoholism. Thus, we routinely use the Lieber-DeCarli liquid diet *(17,18)* procedure for inducing ethanol dependence in mice. In this method, the sole source of vitamins and nutrients to the animal is an ethanol-containing high protein liquid diet, and in control animals the ethanol content is replaced by an equicaloric amount of maltose dextrin. This method can produce ethanol dependence in mice in as little as 2 wk *(7)*.

3.3.1. Liquid Diet Consumption

1. Remove all food and water sources from animal cages.
2. Give animals 50 mL of freshly prepared control diet for 2 d to acclimate them to the diet.
3. On d 3, in the experimental animal group, replace control diet with ethanol-containing diet (*see* **Note 7**). Control animals should be maintained on control diet (*see* **Note 8**).
4. Weigh animals (*see* **Note 8**), prepare diet, and record fluid levels daily for at least 2 wk.
5. Ideally, blood samples (*see* **Subheading 3.4.**) should be determined at random after at least 1 wk of diet administration to ensure that animals are obtaining significant blood ethanol concentrations.

3.3.2. Withdrawal Symptom Severity

Acute withdrawal from chronic ethanol consumption is characterized behaviorally by increased autonomic activity, arousal, hypothermia, insomnia, tremor, and anxiety-like behavior, and psychologically by agitation, nausea, hallucinations, and hypervigilance. In more severe cases of ethanol withdrawal, increased susceptibility to seizures is observed. This latter symptom is widely used as a behavioral index of ethanol withdrawal severity in animals, and can be quantified using the Handling Induced Convulsions (HIC) scale after chronic ethanol exposure.

1. Maintain animals on ethanol-containing or control liquid diet for at least 2 wk (*see* **Subheading 3.3.1.**).
2. Remove diet from cage.
3. At approx 2 h after diet removal, lift mouse by the tail and videotape it for 10 s while giving the animal a gentle 360° spin.
4. Repeat for each animal at 4, 6, and 8 h after diet removal.
5. Seizure severity should later be assessed on the videotape using the HIC scale (**Note 9**) by at least two investigators blind to both genotype and treatment condition (i.e., control or ethanol diet).

3.4. Blood Ethanol Determination

In any alcohol study in mutant mice, it is important to determine whether any potential genotypic differences in alcohol consumption or alcohol-related behaviors are possibly due to differences in ethanol clearance or metabolism. In addition, it is often necessary to determine if voluntary ethanol consumption in mutant mice results in physiologically significant blood ethanol levels (i.e., >0.10%). Blood ethanol assay kits from commercial vendors (i.e., Sigma) are enzymatic assays that catalyze the conversion of alcohol to its metabolite acetaldehyde by ADH coupled to a trapping agent (NAD). In addition, the reaction is further catalyzed by the use of a glycine buffer reagent (pH 9.0). The absorbance of the reaction product is then quantified in a spectrophotometer. The blood ethanol assay procedures described below have been modified from the manufacturer's directions.

1. At 10 min after a self-administration session or intraperitoneal injection of 3 or 4 g/kg ethanol, clip off approx 5 mm of the tail.
2. Collect approx 20 µL of blood into heparinized borosilicate glass capillary.
3. Expel blood from capillary into 0.5-mL centrifuge tube using micropipet bulb, and store on ice.
4. Repeat for each mouse at subsequent desired time points (i.e., 30, 60, 90, and 180 min postinjection or ingestion). Any scab formed over the tip of the tail can be easily removed to obtain additional blood samples; however, reclipping of the tail should not be necessary.
5. Store blood samples at –20°C until assay.
6. To isolate blood plasma, add 180 µL of trichloroacetic acid per 20 µL of thawed blood sample, vortex, and let stand for 5 min at room temperature.
7. Centrifuge blood samples at 2000g for 5 min.
8. Add 16 mL of glycine buffer to one NAD-ADH vial and invert several times to dissolve.
9. Add 725 µL of NAD-ADH/glycine solution to 1.5-mL centrifuge tubes for each blood sample. Then, add 25 µL of each sample to tube and invert several times to mix.

10. Also, add 725 µL of NAD-ADH/glycine solution to 1.5-mL centrifuge tubes for a blank measurement. Then add 25 µL of distilled water to tube and invert several times to mix.
11. Add 747.5 µL to 1.5-mL centrifuge tubes for a standard ethanol measurement (i.e., 0.05, 0.10, and 0.30%). Then, add 2.5 µL of each standard concentration to tube and invert several times to mix.
12. Allow all tubes to stand at room temperature for 10 min.
13. Pipet 100 µL of contents of blank tube into a quartz cuvette and read absorbance at 340 nm as a blank control.
14. Pipet 100 µL of contents of each ethanol standard into a quartz cuvette and read absorbance at 340 nm (*see* **Note 10**).
15. Pipet 100 µL of contents of each blood sample tube into a quartz cuvette and read absorbance at 340 nm.
16. Rinse cuvette with distilled water between each reading.
17. *see* **Note 11** for calculation of ethanol concentrations from absorbance values.

3.5. Motivational Effects of Ethanol (Conditioned Place Preference)

The CPP paradigm is a well-established model to test the rewarding or reinforcing effects of drugs of abuse. Animals are injected with saline and confined to one of two conditioning compartments, each with unique visual and tactile surfaces (**Fig. 2**). On the next day, the animal is injected with ethanol (or another abused drug) an placed in the other conditioning compartment. This alternating procedure is repeated for approx 2 wk. These repeated pairings of ethanol with one of the environmentally distinct chambers results in the association of the subjective effects of ethanol with a particular environment. On the final (test) day, the animal (without receiving an injection) in placed in the center compartment and allowed to freely move between the two conditioning chambers. A greater amount of time spent in the ethanol/drug-paired chamber is interpreted as a "preference" for that environment and is thought to reflect a secondary association of the chamber with the positive subjective effects of the drug. All substances that are abused by humans produce CPP in rodents. Although the ability of ethanol and some other drugs of abuse to produce CPP in rodents may be dependent on the strain of the animal used (*see* **ref. 19**), we have routinely used the following procedure to produce ethanol CPP in mice of a variety of genetic backgrounds.

1. Mice are weighed daily immediately prior to placement in the CPP apparatus.
2. On the first day of testing (habituation session), mice are placed in the center compartment and given access to both conditioning compartments for 5 min (*see* **Note 12**).
3. Over the next 8 d (conditioning sessions), animals receive alternating i.p. injections of either saline or ethanol (2 g/kg, *see* **Note 5**) and then given access

to only one of the conditioning compartments for 5 min (*see* **Note 13**). Saline- and ethanol-paired environments should be counterbalanced within each group as well as across genotypes (*see* **Note 12**). All injections are given immediately before the beginning of the conditioning sessions.

4. On the final (test) day, animals are placed in the center chamber and given access to both compartments for 30 min, and the amount of time spent in the ethanol- and saline-paired compartments is measured electronically by photobeams placed 1.2 cm apart in the conditioning compartments. Locomotor activity (photobeam crosses) is also measured during the test session.

5. The CPP apparatus should be wiped clean with 2.5% acetic acid after each habituation, conditioning, or test session to mask any olfactory cues left by the previous animal.

4. Notes

1. Other investigators have used lever-activated liquid dippers as the method for fluid delivery from fluid troughs. However, in our experience, this method can lead to inaccuracies in assessing the precise amount of fluid consumed and can lead to evaporation of the ethanol from the solution. The syringe pump methods described in this chapter circumvents these problems.

2. Ethanol has both sweet and bitter taste properties. Thus, if genotypic differences in ethanol consumption and preferences are detected in the two-bottle choice procedures, the investigator may wish to conduct a bitter-sweet taste preference experiment to assess potential differences in taste reactivity and neophobias. This simple protocol involves administration of a sweet substance (i.e., 0.033 and 0.066% saccharin) and a bitter substance (i.e., 0.015 and 0.03 m*M* quinine) vs water in a continuous access two-bottle choice paradigm, with 4 d of access to each concentration (*5*).

3. Acclimation to the fluid deprivation procedures makes it necessary to allow at least 3 d of 2-h fluid access sessions.

4. Data from two-bottle choice procedures are traditionally expressed as ethanol consumed or intake (g/kg/session, calculated as $x_1/100$, where $x_1 = x_2 \times 100 + 0.5$, $x_2 = (1000/\text{body weight}) \times (\text{mL ethanol consumed} \times x_3)$, and x_3 = ethanol concentration $\times 0.0397/5$), and ethanol preference (%, calculated as mL ethanol solution consumed/session divided by total mL fluid consumed/session $\times 100$). For continuous access paradigms, these parameters are calculated daily and averaged over the 4-d access period at each ethanol concentration.

5. Doses of ethanol for intraperitoneal injection are calculated as indicated in the table on the following page.

6. Occasionally mice will regain righting reflex, but when placed on their backs again will not regain the reflex. This makes it necessary to establish the criteria of three righting reflexes within a 30-s period as the recovery of the righting reflex.

7. For mice that display increased sensitivity to ethanol, such as PKCε-deficient mice (*5–7*), it may be necessary to "fade" in the alcohol concentration over a period of 7–8 d (i.e., 0% ethanol on days 1–2, 1.6% on days 3–4, 3.2% on

| | Ethanol Dose | | | |
Body Weight (g)	1 g/kg	2 g/kg	3 g/kg	4 g/kg
20	0.12	0.25	0.38	0.50
22	0.14	0.28	0.42	0.55
24	0.15	0.30	0.45	0.60
26	0.16	0.32	0.48	0.65
28	0.18	0.35	0.52	0.70
30	0.19	0.38	0.57	0.75
32	0.20	0.40	0.60	0.80
34	0.21	0.42	0.62	0.85
36	0.22	0.45	0.68	0.90
38	0.24	0.48	0.72	0.95
40	0.25	0.50	0.75	1.00

Values given are mL of 20% ethanol solution, diluted in saline.

days 5–7, 4.8% on days 8+). In addition, it may be necessary to omit highest concentration of ethanol (6.5%) because mice with increased sensitivity to ethanol will either completely avoid consuming the diet with this concentration of ethanol present and become malnourished. Alternatively, if animals with increased ethanol sensitivity do consume the diet with 6.5% ethanol, it may render the animals unconscious and result in profound weight loss or death *(7)*.

8. It is not uncommon to observe a slight degree (<10%) of weight loss in animals placed on the ethanol-containing liquid diet, which is usually regained by the second week of diet administration. However, if significant weigh loss persists, it may be necessary to prolong the time in which the initial lower concentrations of ethanol are introduced. Alternatively, the experimenter may wish to "pair feed" control animals; that is, animals fed the control diet be administered the average volume of diet as that is consumed daily by the ethanol-fed animals. This ensures that both groups of animals gain the same amount of body weight during the entire diet administration period.

9. The HIC scale on the following page is recommended for assessing withdrawal seizure severity *(7)*. A tonic seizure is defined as a prolonged rigid extension of the extremities, a clonic seizure is defined as repetitive jerky movements of the extremities, and a tonic-clonic seizure is defined as a seizure exhibited both tonic and clonic characteristics.

10. The absorbance of the reaction product is linear up to values of 1.000. Thus, 0.05, 0.10, and 0.30% standards are analyzed only to assure that correct absorbance values are being obtained. If desired, the investigator may wish to construct a standard curve (absorbance vs ethanol standard concentration) and use the resulting equation to calculate the ethanol concentrations from the absorbance values obtained from blood samples.

Score	Description
7	Severe, tonic-clonic convulsions prior to tail lift with rapid onset and long duration; spontaneous or elicited by mild environmental stimuli, such as lifting the cage top
6	Severe, tonic-clonic convulsions when lifted by the tail; rapid onset and long duration, often continuing several seconds after the mouse is released
5	Tonic-clonic convulsion when lifted by the tail; onset delayed by 1–2 s
4	Tonic convulsion when lifted by the tail
3	Convulsion when lifted by the tail after gentle 360° spin
2	No convulsion when lifted by the tail, but tonic convulsion elicited by gentle 360° spin
1.5	Facial grimace when lifted by the tail
1	No convulsion or facial grimace when lifted by the tail, but facial grimace after gentle 360° spin
0	No convulsion or facial grimace

11. Blood alcohol concentration (mg/dL) = $A_{340} \times 223$. Blood alcohol concentration (% v/v) = Blood alcohol concentration (mg/dL) / 1000. Blood alcohol concentration (mmol/L) = Blood alcohol concentration (mg/dL) $\times 0.217$.
12. In general, animals do not display a significant preference for one conditioning chamber over the other prior to conditioning. However, if such a phenomenon is observed in a particular animal, the investigator may choose to use the initial nonpreferred chamber as the one that is repeatedly paired with ethanol.
13. If the investigator wishes to test CPP to other drugs of abuse, such as opiates and psychostimulants, longer conditioning sessions (i.e., 15–20 min) should be used.

References

1. Battaini, F. (2001) Protein kinase C isoforms as therapeutic targets in nervous system disease states. *Pharm. Res.* **44,** 353–361.
2. Craig, N. J., Duran Alonso, M. B., Hawker, K. L., Shiels, P., Glencorse, T. A., Campbell, J. M., et al. (2001) A candidate gene for human neurodegenerative disorders: a rat PKC gamma mutation causes a Parkinsonian syndrome. *Nat. Neurosci.* **4,** 1061–1062.
3. Harris, R. A., McQuilkin, S. J., Paylor, R., Abeliovich, A., Tonegawa, S., and Wehner, J. M. (1995) Mutant mice lacking the gamma isoform of protein kinase C show decreased behavioral actions of ethanol and altered function of gamma-aminobutyrate type A receptors. *Proc. Natl. Acad. Sci. USA* **92,** 3658–3662.
4. Bowers, B. J. and Wehner, J. M. (2001) Ethanol consumption and behavioral impulsivity are increased in protein kinase Cγ null mutant mice. *J. Neurosci.* **21,** RC180:1–5.

5. Hodge, C. W., Mehmert, K. K., Kelley, S. P., McMahon, T., Haywood, A., Olive, M. F., et al. (1999) Supersensitivity to allosteric GABA-A receptor modulators and alcohol in mice lacking PKCε. *Nat. Neurosci.* **2,** 997–1002.

6. Olive, M. F., Mehmert, K. K., Messing, R. O., and Hodge, C. W. (2000) Reduced operant ethanol self-administration and in vivo mesolimbic dopamine responses to ethanol in PKCε deficient mice. *Eur. J. Neurosci.* **12,** 4131–4140.

7. Olive, M. F., Mehmert, K. K., Nannini, M. A., Camarini, R., Messing, R. O., and Hodge, C. W. (2001) Reduced ethanol withdrawal severity and altered withdrawal–induced c–fos expression in various brain regions of mice lacking protein kinase C-epsilon. *Neuroscience* **103,** 171–179.

8. Harris, R. A., Alling, C., Messing, R. O., Pandey, S. C., and Stubbs, C. D. (1996) Protein kinase C: molecular and cellular targets for the action of ethanol. *Alcohol. Clin. Exp. Res.* **20,** 67A–71A.

9. Stubbs, C. D. and Slater, S. J. (1999) Ethanol and protein kinase C. *Alcohol. Clin. Exp. Res.* **23,** 1552–1562.

10. Thiele, T. E., Willis, B., Stadler, J., Reynolds, J. G., Bernstein, I. L., and McKnight, G. S. (2000) High ethanol consumption and low sensitivity to ethanol–induced sedation in protein kinase A–mutant mice. *J. Neurosci.* **20,** RC75.

11. Wand, G., Levine, M., Zweifel, L., Schwindinger, W., and Abel, T. (2001) The cAMP-protein kinase A signal transduction pathway modulates ethanol consumption and sedative effects of ethanol. *J. Neurosci.* **21,** 5297–5303.

12. Bowers, B. J., Owen, E. H., Collins, A. C., Abeliovich, A., Tonegawa, S., and Wehner, J. S. (1999) Decreased ethanol sensitivity and tolerance development in γ-protein kinase C null mutant mice is dependent on genetic background. *Alcohol. Clin. Exp. Res.* **23,** 387–397.

13. Gerlai, R. (1996) Gene–targeting studies of mammalian behavior: is it the mutation or the background genotype? *Trends Neurosci.* **19,** 177–181.

14. Gerlai, R. (2001) Gene targeting: technical confounds and potential solutions in behavioral brain research. *Behav. Brain. Res.* **125,** 13–21.

15. Crabbe, J. C., Wahlsten, D., and Dudek, B. C. (1999) Genetics of mouse behavior: interactions with laboratory environment. *Science* **284,** 1670–1672.

16. Samson, H. H. (1986) Initiation of ethanol reinforcement using a sucrose–substitution procedure in food- and water-sated rats. *Alcohol. Clin. Exp. Res.* **10,** 436–442.

17. Lieber, C. S. and DeCarli, L. M. (1982) The feeding of alcohol in liquid diets: two decades of applications and 1982 update. *Alcohol. Clin. Exp. Res.* **6,** 523–531.

18. Lieber, C. S. and DeCarli, L. M. (1989) Liquid diet technique of ethanol administration: 1989 update. *Alcohol Alcohol.* **24,** 197–211.

19. Cunningham, C. L., Niehus, D. R., Malott, D. H., and Prather, L. K. (1992) Genetic differences in the rewarding and activating effects of morphine and ethanol. *Psychopharmacology* **107,** 385–393.

Index